Lecture Notes in Computer Science 7984

Commenced Publication in 1973
Founding and Former Series Editors:
Gerhard Goos, Juris Hartmanis, and Jan van Leeuwen

T0241014

Alexander Dudin   Koen De Turck (Eds.)

Analytical and Stochastic Modelling Techniques and Applications

20th International Conference, ASMTA 2013
Ghent, Belgium, July 8-10, 2013
Proceedings

 Springer

Volume Editors

Alexander Dudin
Belarusian State University
Department of Applied Mathematics and Computer Science
Nezavisimosti Ave. 4, Minsk 220030, Belarus
E-mail: dudin@bsu.by

Koen De Turck
Ghent University
Department of Telecommunications and Information Processing
Sint-Pietersnieuwstraat 41, 9000 Gent, Belgium
E-mail: kdeturck@telin.ugent.be

ISSN 0302-9743   e-ISSN 1611-3349
ISBN 978-3-642-39407-2   e-ISBN 978-3-642-39408-9
DOI 10.1007/978-3-642-39408-9
Springer Heidelberg Dordrecht London New York

Library of Congress Control Number: 2013941915

CR Subject Classification (1998): C.2, D.2.4, D.2.8, D.4, C.4, H.3, F.1

LNCS Sublibrary: SL 2 – Programming and Software Engineering

Typesetting: Camera-ready by author, data conversion by Scientific Publishing Services, Chennai, India

Printed on acid-free paper

Springer is part of Springer Science+Business Media (www.springer.com)

Preface

It is our pleasure to present the proceedings of the 20th International Conference on Analytical and Stochastic Modelling and Applications (ASMTA 2013) held in Ghent, Belgium, during July 8–10, 2013. ASMTA conferences have become established quality events in the calendar of analytical, numerical and even simulation experts in Europe and well beyond. We were happy to receive interesting contributions from both regular participants and newcomers from such countries as the USA, the UK, Germany, France, Italy, The Netherlands, Belgium, Hungary, Belarus, Latvia, Hungary, Korea and many others.

The quality of this year's program was exceptionally high. The conference committee was extremely selective this year, accepting 30 papers, and rejecting many high-quality contributions. The International Program Committee reviewed the submissions critically and in detail, thereby assisting the Program Chairs in making the final decision as well as in providing the authors with useful comments to improve their papers. We would therefore like to thank every member of the Program Committee for their time and efforts.

We are very grateful for the generous support of the sponsors. The conference was co-sponsored by the IEEE UK-RI Computer Chapter, by ECMS - The European Council for Modelling and Simulation, by the Faculty of Engineering and Architecture of Ghent University, by the SMACS Research Group and by IAP VII/19 DYSCO and IAP VII/23 BESTCOM. Thank you for your contribution to ASMTA 2013.

May 2013

Alexander Dudin
Koen De Turck
Khalid Al-Begain
Herwig Bruneel

Organization

Program Committee

Sergey Andreev	Tampere University of Technology, Finland
Jonatha Anselmi	Basque Center for Applied Mathematics, Spain
Konstantin Avrachenkov	INRIA Sophia-Antipolis, France
Simonetta Balsamo	University Ca' Foscari Venice, Italy
Jeremy Bradley	Imperial College London, UK
Giuliano Casale	Imperial College London, UK
Hind Castel	Institut Telecom, France
Koen De Turck	Ghent University, Belgium
Alexander Dudin	Belarus State University, Belarus
Antonis Economou	University of Athens, Greece
Paulo Fernandes	Pontifical Catholic University of Rio Grande do Sul, Brazil
Dieter Fiems	Ghent University, Belgium
Jean-Michel Fourneau	University of Versailles, France
Rossano Gaeta	University of Turin, Italy
Marco Gribaudo	Politecnico di Milano, Italy
Peter Harrison	Imperial College London, UK
Richard Hayden	Imperial College London, UK
Yezekael Hayel	University of Avignon, France
Andras Horvath	University of Turin, Italy
Helen Karatza	Aristotle University of Thessaloniki, Greece
William Knottenbelt	Imperial College London, UK
Lasse Leskelä	University of Jyväskylä, Finland
Andrea Marin	University Ca' Foscari Venice, Italy
Don Mcnickle	University of Canterbury, UK
Yoni Nazarathy	University of Queensland, Australia
Bo Friis Nielsen	Technical University of Denmark
José Niño-Mora	Universidad Carlos III de Madrid, Spain
Evgeny Osipov	Lulea University of Technology, Sweden
Antonio Pacheco	Instituto Superior Tecnico, Lisbon, Portugal
Tuan Phung-Duc	Tokyo Institute of Technology, Japan
Balakrishna Prabhu	LAAS-CNRS, Toulouse, France
Jacques Resing	Eindhoven University of Technology, The Netherlands
Leonardo Rojas-Nandayapa	University of Queensland, Australia

Table of Contents

The Taylor Series Expansions for Performance Functions of Queues: Sensitivity Analysis

Sofiane Ouazine[1], Karim Abbas[2], and Bernd Heidergott[3]

[1] Department of Mathematics
[2] LAMOS Laboratory,
University of Bejaia, Campus of Targua Ouzemour, 06000 Bejaia, Algeria
[3] Department of Econometrics, Vrije Universiteit Amsterdam, the Netherlands

Abstract. We discuss the application of an efficient numerical algorithm to sensitivity analysis of the GI/M/1 queue. Specifically, we use a numerical approach based on the Taylor series expansion to examine the robustness of the GI/M/1 queue to some specific perturbations in the arrival process: linear and non-linear perturbations. For each kind of perturbation we approximately compute the sensitivity of the main characteristics of the GI/M/1 queue corresponding to the case where the arrival processes are lightly different from that of the nominal queue. Numerical examples are presented to illustrate the accuracy of the proposed approach.

Keywords: Taylor series expansion approach, Sensitivity analysis, GI/M/1 queue, Numerical methods, Performance measures.

1 Introduction

Queueing models play a major role in performance evaluation of computer networks, communication systems and manufacturing systems. Specifically, the GI/M/1 queue is widely used as the model approximates the behaviour of queues that deviate slightly from the statistical assumptions of the standard M/M/1 model. For example, if we hope to analyse a real system such as a queueing model which constitues of a single server queue with infinite capacity and at FCFS discipline, then an analysis of a sample of the inter-arrival times indicate that they are not quite exponentially distributed. The deviations observed in the inter-arrival times may be due to sampling error or they may reflect a true deviation. If it can be shown that the difference is small between the stationary characteristics in the GI/M/1 queue and the queue with the observed deviation, then we say the GI/M/1 queue is insensitive to the deviation, and we can expect the GI/M/1 approximation to the queue to yield reasonable results. Thus, the insensitivity or robustness of the GI/M/1 queue to deviations is an issue of practical importance.

There exists numerous results on the robustness of queueing models in a general framework. The sensitivity of the queue to a perturbation is measured by various metrics of the probability distributions associated with the perturbed

A. Dudin and K. De Turck (Eds.): ASMTA 2013, LNCS 7984, pp. 1–11, 2013.

and nominal queueing processes. These analyses can be found in Kotzurek and Stoyan [13], Whitt [16] and Zolotarev [17], Fricker et al. [9], Núñez-Queija et al. [14], Benaouicha and Aïssani [5] and Abbas and Aïssani [1]. Particularly, different versions of the Taylor series expansion approach have been considered in the literature for analyzing stochastic models. The first work seems to be by Schweitzer in 1968 [15]. For a recent overview, we refer the reader to [12]. More recently, De Turck et al. [7,8] have considered the analysis of a discrete-time queueing systems by a Taylor series expansion approach. Specifically, the use of Taylor series expansions to study the robustness of queues has initiated by Albin [4], where the author has considered several perturbations in the arrival process of the M/M/1 queue. In the same work, the Taylor series expansion is sketched out in a direct manner to estimate the difference between the expected numbers of customers in the M/M/1 and perturbed M/M/1 queueing systems.

Here the proposed framework involves the use of Taylor series expansions to examine the robustness of the GI/M/1 queue to perturbations in the arrival process. Specifically, we analyse numerically the sensitivity of the entries of the stationary distribution vector of the GI/M/1 queue to that perturbations, where we exhibit these entries as polynomial functions of the inter-arrival parameter of the considered queue. A significant contribution of this work is the derivation of the Taylor series coefficients, which are expressed in closed form as functions of the deviation matrix [11]. Therefore, the determination of the mean busy period of a queueing model with a general distribution of the inter-arrival time is very complicated. For this reason, we used this approach to predict accurately a such characteristic of the queue. Especially, the use of the GI/M/1 queue allows for evaluating the potential of our algorithmic approach. The proposed framework has been applied for analyzing of another models with several case studies (see for example [3,12]).

The remainder of this paper is organized as follows. In Section 2, we introduce the necessary notations for analyzing of the considered queueing model, and present closed-form expressions for the sensitivity of the stationary distribution to model parameter as a function of the deviation matrix. In Section 3, we outline the numerical framework to compute the relative absolute error in computing the stationary distribution. Concluding remarks are provided in Section 4.

2   Queueing Model Analysis

Consider the GI/M/1 queueing model, where customers arrive at time points $\{\tau_n; n \geq 0\}$ and get served by a single server. Suppose that the service times are exponentially distributed independent and identically distributed random variables with $S(t) = 1 - e^{-\mu t}; \mu > 0$ and $t \geq 0$. Let $Z_n = \tau_{n+1} - \tau_n, n \geq 1$, be independent and identically distributed random variables with distribution function $A(t)$ with mean a. Finally, the inter-arrival times and service times are mutually independent, and the service discipline is first-come-first-served.

Here we examine the robustness of the GI/M/1 queueing model to a specific perturbation in the arrival process, where we consider two different kinds of perturbations: linear and non-linear. Let $X(t)$ denote the number of customers in the GI/M/1 queueing system at time t (queue length). We choose τ_n as the embedded points and denote $X_n = X_{\tau_n^-}$ to represent the queue length just prior to the nth arrival point, see Gross and Harris [10]. Then $\{X_n; n \in \mathbb{N}\}$ is a Markov chain with transition matrix:

$$
P = \begin{pmatrix}
p_{0,0} & \beta_0 & 0 & 0 & 0 & \cdots \\
p_{1,0} & \beta_1 & \beta_0 & 0 & 0 & \cdots \\
p_{2,0} & \beta_2 & \beta_1 & \beta_0 & 0 & \cdots \\
p_{3,0} & \beta_3 & \beta_2 & \beta_1 & \beta_0 & \cdots \\
\vdots & \vdots & \vdots & \vdots & \vdots & \ddots
\end{pmatrix}, \tag{1}
$$

where β_i denotes the probability of serving i customers during an inter-arrival time given that the server remains busy during this interval (thus there are more than i customers present). To calculate β_i we note that given the duration of the inter-arrival time, say, t, the number of customers served during this interval is Poisson distributed with parameter μt. Hence, we have

$$
\beta_i = \int_{t=0}^{\infty} \frac{(\mu t)^i}{i!} e^{-\mu t} dA(t), \tag{2}
$$

where $A(t)$ is distribution function of the inter-arrival times. Since the transition probabilities from state j should add up to one, it follows that

$$
p_{i,0} = 1 - \sum_{j=0}^{i} \beta_j = \sum_{j=i+1}^{\infty} \beta_j. \tag{3}
$$

Let π denote the stationary distribution of the Markov chain X_n. We define the traffic intensity $\rho =$ (arrival rate)/(service rate) $= 1/a\mu$. It can be shown that the Markov chain X_n is positive recurrent when $\rho < 1$. In this case, π can be computed as the solution of $\pi P = \pi$. This allows for evaluating the potential of our approach. In this paper, we choose a different point of view. We consider π as a mapping of some real-valued parameter θ, in notation π_θ. For example, θ may denote the mean inter-arrival time of the queue. We are interested in obtaining higher-order sensitivity of stationary distribution to parameter θ. The k-order sensitivity of the stationary distribution π_θ to the parameter θ is given in the following Theorem.

Theorem 1. *[11] Let $\theta \in \Theta$ and let $\Theta_0 \subset \Theta$, with $\Theta \subset \mathbb{R}$ a closed interval with θ be an interior point such that the queue is stable on Θ_0. Provided that the entries of P are n-times differentiable with respect to θ, let*

$$K_\theta(n) = \sum_{1 \le m \le n;\, 1 \le l_k \le n;\, l_1 + \cdots + l_m = n} \frac{n!}{l_1! \cdots l_m!} \prod_{k=1}^{m} \left(P_\theta^{(l_k)} D_\theta \right). \tag{4}$$

Then it holds that

$$\pi_\theta^{(n)} = \pi_\theta K_\theta(n), \tag{5}$$

where $P_\theta^{(k)}$ and $\pi_\theta^{(k)}$ are respectively the kth order derivative of P_θ and π_θ with respect to parameter θ.

The matrix D_θ given in the formula (4) is the deviation matrix defined by:

$$D_\theta = \sum_{n=0}^{\infty} (P_\theta^n - \Pi_\theta) = \sum_{n=0}^{\infty} (P_\theta - \Pi_\theta)^n - \Pi_\theta,$$

where Π_θ is a matrix with rows equal to $\pi_\theta^\top$, with $x^\top$ denoting the transposed of vector x.

In the following, we propose a numerical approach to compute the stationary distribution π_θ for some parameter value θ, and we demonstrate how this stationary distribution can be evaluated for the case where the control parameter θ is changed in some interval. In other words, we will compute the function $\pi(\theta + \Delta)$ on some Δ-interval. More specifically, we will compute $\pi(\theta + \Delta)$ by an polynomial in Δ. To achieve this we will use the Taylor series expansion approach established in [11]. Under some mild conditions it holds that $\pi_{\theta + \Delta}$ can be developed into a Taylor series of the following form

$$\pi_{\theta + \Delta} = \sum_{n=0}^{k} \frac{\Delta^n}{n!} \pi_\theta^{(n)}, \tag{6}$$

where $\pi_\theta^{(n)}$ denotes the n-th order derivative of π_θ with respect to θ (see the formula (5)). Under the conditions put forward in Theorem 1 it holds for $k < n$ that

$$\pi_\theta^{(k+1)} = \sum_{m=0}^{k} \binom{k+1}{m} \pi_\theta^{(m)} P_\theta^{(k+1-m)} D_\theta. \tag{7}$$

An explicit representation of first derivatives of π_θ is given by [3]:

$$\pi_\theta^{(1)} = \pi_\theta P_\theta^{(1)} D_\theta$$

and

$$\pi_\theta^{(2)} = \pi_\theta P_\theta^{(2)} D_\theta + 2\pi_\theta (P_\theta^{(1)} D_\theta)^2.$$

Elaborating on the recursive formula for higher order derivatives (7), the second order derivative can be written as

$$\pi_\theta^{(2)} = \pi_\theta P_\theta^{(2)} D_\theta + 2\pi_\theta^{(1)} P_\theta^{(1)} D_\theta.$$

In the same vein, we obtain for the third order derivative

$$\pi_\theta^{(3)} = \pi_\theta P_\theta^{(3)} D_\theta + 3\pi_\theta^{(2)} P_\theta^{(1)} D_\theta + 3\pi_\theta^{(1)} P_\theta^{(2)} D_\theta.$$

For our analysis, we provide two different kinds of perturbation: linear perturbation (Example 1) and non-linear perturbation (Example 2).

Example 1. Consider the $H_2/M/1$ queue with arrival process is hyperexponential with two rates λ_1 and λ_2, and exponential service rate μ. In this case, the density function of the the inter-arrival times is given by:

$$f(t) = \theta\, \lambda_1\, e^{-\lambda_1\, t} + (1-\theta)\, \lambda_2\, e^{-\lambda_2\, t},\; t \geq 0. \tag{8}$$

Then, the probability (2) is given by:

$$\beta_i = \theta\lambda_1 \int_{t=0}^{\infty} \frac{(\mu t)^i}{i!} e^{-(\mu+\lambda_1)t} dt + (1-\theta)\lambda_2 \int_{t=0}^{\infty} \frac{(\mu t)^i}{i!} e^{-(\mu+\lambda_2)t} dt. \tag{9}$$

Example 2. Consider the D/M/1 queue with arrival process is deterministic with rate c and exponential service rate μ. Then, the probability (2) is given by:

$$\beta_i = \frac{(\mu c)^i}{i!} e^{-\mu c}. \tag{10}$$

Remark 1. The Taylor series expansion developed above applies to differentiable Markov kernels. For linear perturbations (Example 1) the transition matrix P_θ of the embedded Markov chain X_n described the state of the $H_2/M/1$ queue, is of form:

$$P_\theta = \theta P_1 + (1-\theta) P_2,\; \theta \in [0,1], \tag{11}$$

where P_1 (resp. P_2) is a transition matrix of the M/M/1 queue embedded at arrival points τ_n with arrival rate λ_1 (resp. λ_2) and with a common service rate μ. In this case, the derivative of the transition matrix P_θ with respect to the parameter of interest θ is given by:

$$P'_\theta = P_1 - P_2. \tag{12}$$

Therefore, the elements of Taylor series expansion are of the form:

$$\pi_\theta^{(n)} = \pi_\theta[(P_1 - P_2)D_\theta]^n. \tag{13}$$

Note that the perturbation considered in Example 2 is significantly more difficult than the linear case of Example 1 as in the case of non-linear perturbation all higher-order derivatives of P_θ maybe different from zero.

3 Numerical Application

In this section, we apply the numerical approach based on the Taylor series expansions introduced above to the GI/M/1 queue, where we consider the two examples of the GI/M/1 queue with perturbed arrival processes. In each case, we estimate numerically the sensitivity of the stationary distribution of the queueing model with respect to the perturbation.

Let $A(\cdot)$ have density mapping $f(\cdot)$. Let $\Theta = (a, b) \subset \mathbb{R}$, for $0 < a < b < \infty$.

(H)   For $i \geq 0$ it holds that β_i is n-times differentiable with respect to θ on Θ.

Under **(H)** it holds that the first n derivatives of P exist. Let $P^{(k)}$ denote the kth order derivative of P with respect to θ, then it holds that

$$P^{(k)}(i,j) = \frac{d^{(k)}}{d\theta^{(k)}} P(i,j), \quad i,j \geq 0, \tag{14}$$

or, more specifically,

$$P^{(k)} = \begin{pmatrix} p_{0,0}(k) & \beta_0(k) & 0 & 0 & 0 & \cdots \\ p_{1,0}(k) & \beta_1(k) & \beta_0(k) & 0 & 0 & \cdots \\ p_{2,0}(k) & \beta_2(k) & \beta_1(k) & \beta_0(k) & 0 & \cdots \\ p_{3,0}(k) & \beta_3(k) & \beta_2(k) & \beta_1(k) & \beta_0(k) & \cdots \\ \vdots & \vdots & \vdots & \vdots & \vdots & \ddots \end{pmatrix}, \tag{15}$$

where

$$\beta_i(k) = \frac{d^k}{d\theta^k} \beta_i, \quad i \geq 0. \tag{16}$$

3.1   Linear Perturbation: The $H_2/M/1$ Queue

Consider the $H_2/M/1$ queue with service rate μ and hyperexponential inter-arrival time with two rates λ_1 and λ_2. First, we present the numerical results obtained by applying our approach to the case of linear perturbation. Therefore, we set $\mu = 2$, $\lambda_1 = 1$ and $\lambda_2 = 2$. For the implementation of our algorithm in MATLAB, we require a finite version of our queueing model. In order to achieve this we set $N = 3$ (N is the truncation level corresponding to the transition matrix P). Figures 1, 2 and 3 depict the relative error on the stationary distributions $\pi_\theta^{(k)}(i)$ for $0 \leq i \leq N$ and $k = 1, 2, 3$, of the $H_2/M/1$ queue versus the perturbation parameter $\Delta \in [-\delta, \delta]$, where $\delta = 0.1$. As expected, the relative error on the stationary distributions decreases as the perturbation parameter θ decreases.

3.2   Non-linear Perturbation: The $D/M/1$ Queue

Here we consider the $D/M/1$ queue with service rate μ and deterministic inter-arrival time $c = \theta$. In the sequel we illustrate our numerical approach applied with a non-linear perturbation. For that, we set $\mu = 2$ and $c = 1$. As the first case (linear perturbation) we need a finite version of our queueing model. Like before we fix the truncation level of the transition matrix N to 3 . Figures 4, 5 and 6 depict the relative error on the stationary distributions $\pi_\theta^{(k)}(i)$ for $0 \leq i \leq N$ and

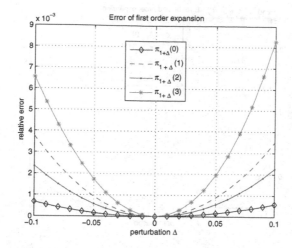

Fig. 1. The relative error in computing $\pi_{1+\Delta}$ by Taylor series of 1st order

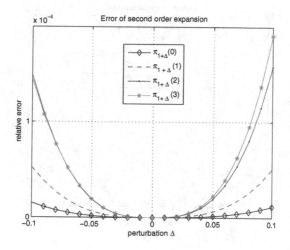

Fig. 2. The relative error in computing $\pi_{1+\Delta}$ by Taylor series of 2nd order

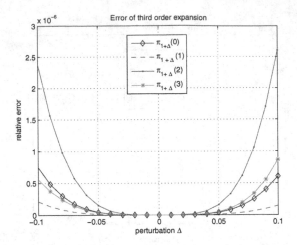

Fig. 3. The relative error in computing $\pi_{1+\Delta}$ by Taylor series of 3rd order

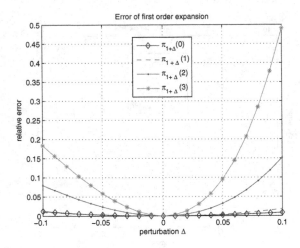

Fig. 4. The relative error in computing $\pi_{\theta+\Delta}$ by Taylor series of 1st order

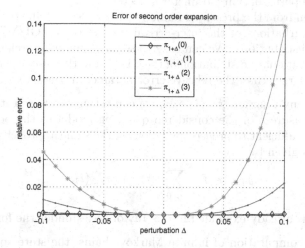

Fig. 5. The relative error in computing $\pi_{\theta+\Delta}$ by Taylor series of 2nd order

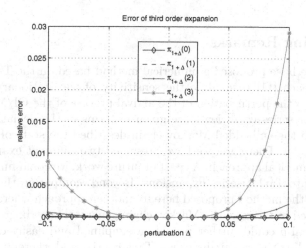

Fig. 6. The relative error in computing $\pi_{\theta+\Delta}$ by Taylor series of 3rd order

$k = 1, 2, 3$, of the D/M/1 queue versus the perturbation parameter $\Delta \in [-\delta, \delta]$, where $\delta = 0.1$. As expected, the relative error on the stationary distributions decreases as the perturbation parameter θ decreases.

As can be seen from the presented numerical examples, we illustrate the effect of different perturbations of the inter-arrival process of the GI/M/1 queue on the stationary distributions. We can easy to see that the made relative error in computing the stationary distribution of the GI/M/1 in the case of the non-linear perturbation is larger than the relative error obtained in the linear case.

Remark 2. For any function $f : \mathbb{N} \to \mathbb{R}$, we can estimate the sensitivity of the performance measure η of the considered queueing model to the perturbation. For this end, its sufficient to proceed by following the same approach used above. This estimate is given by

$$\pi_{\theta+\Delta} \times f = \sum_{i \geq 0} f(i) \times \pi_{\theta+\Delta}(i), \tag{17}$$

where $\pi_{\theta+\Delta}(i)$ is already expanded into a Taylor series under the form (6).

Remark 3. For computation of infinite Markov chains, the state space truncation is frequently required, then this truncation becomes necessary to illustrate the application of the considered approach for analysis of queues with infinite capacity. So, in our numerical computations we note that for N large the stationary distribution of the finite queueing model is almost identical to that of the infinite one.

4   Concluding Remarks

In this paper, we have proposed a numerical method based on the Taylor series expansion approach [11] to analyse the sensitivity of the stationary distributions relatively to the perturbation of the arrival process of the GI/M/1 queue, where the Taylor series coefficients are given in terms of the deviation matrix corresponding to the embedded Markov chain described the state of the nominal queueing model. Therefore, we have presented two different examples that illustrate our numerical approach. As part of future work, we could further investigate multivariate Taylor series expansions. Beyond, comparing the efficiency and accuracy of the method proposed here to another approach where the sensitivity is computed in terms of the Markov chain generator-matrix group inverse [6]. Besides, we also could further provide a simplified and easily computable expression bounding the remainder of the Taylor series and, thereby provide an algorithmic way of deciding which order of the Taylor polynomial is sufficient to achieve a desired precision of the approximation [2,3].

References

1. Abbas, K., Aïssani, D.: Strong Stability of the Embedded Markov Chain in an GI/M/1 Queue with Negative Customers. Applied Mathematical Modelling 34, 2806–2812 (2010)

2. Abbas, K., Heidergott, B.: A Functional Approximation for Queues with Breakdowns (in preparation)
3. Abbas, K., Heidergott, B., Aïssani, D.: A Functional Approximation for the M/G/1/N Queue. Discrete Event Dynamic Systems 23, 93–104 (2013)
4. Albin, S.L.: Analyzing M/M/1 Queues with Perturbations in the Arrival Process. Journal of the Operational Research Society 35, 303–309 (1984)
5. Benaouicha, M., Aïssani, D.: Strong Stability in a G/M/1 Queueing System. Theory of Probability and Mathematical Statistics 71, 22–32 (2004)
6. Campbell, S.L., Meyer Jr., C.D.: Generalized Inverses of Linear Transformations. Dover Publications, Mineola (1991)
7. De Turck, K., De Cuypere, E., Wittevrongel, S., Fiems, D.: Algoritmic Approach to Series Expansions around Transient Markov Chains with Applications to Paired Queuing Systems. In: 6th International Conference on Performance Evaluation Methodologies and Tools (VALUETOOLS 2012), pp. 38–44. IEEE Press, Piscataway (2012)
8. De Turck, K., Fiems, D., Wittevrongel, S., Bruneel, H.: A Taylor Series Expansions Approach to Queues with Train Arrivals. In: 5th International ICST Conference on Performance Evaluation Methodologies and Tools (VALUETOOLS 2011), pp. 447–455. ICST (Institute for Computer Sciences, Social-Informatics and Telecommunications Engineering), Brussels (2011)
9. Fricker, C., Guillemin, F., Robert, P.: Perturbation Analysis of an M/M/1 Queue in a Diffusion Random Environment. Queueing Systems: Theory and Applications 61, 1–35 (2009)
10. Gross, D., Harris, C.: Fundamentals of Queueing Theory. Wiley (1985)
11. Heidergott, B., Hordijk, A.: Taylor Series Expansions for Stationary Markov Chains. Advances in Applied Probability 35, 1046–1070 (2003)
12. Heidergott, B., Hordijk, A., Leder, N.: Series Expansions for Continuous-Time Markov Processes. Operations Research 58, 756–767 (2010)
13. Kotzurek, M., Stoyan, D.: A Quantitative Continuity Theorem for Mean Stationary Waiting Time in GI/G/1. Mathematische Operationsforschung und Statistik 7, 595–599 (1976)
14. Núñez-Queija, R., Altman, E., Avrachenkov, K.: Perturbation Analysis for Denumerable Markov Chains with Application to Queueing Models. Advances in Applied Probability 36, 839–853 (2004)
15. Schweitzer, E.: Perturbation Theory and Finite Markov Chains. Journal of Applied Probability 5, 401–413 (1968)
16. Whitt, W.: Quantitative Continuity Results for the GI/G/1 Queue. Bell Laboratories report (1981)
17. Zolotarev, V.M.: General Problems of the Stability of Mathematical Models. In: 41st International Statistical Institute, New Delhi (1977)

Admission Control to an M/M/1 Queue with Partial Information

Eitan Altman[1] and Tania Jiménez[2]

[1] INRIA, 2004 Route des Lucioles, 06902 Sophia Antipolis, France
[2] Avignon University, LIA EA4128, F-84000, Avignon, France

Abstract. We consider both cooperative as well as non-cooperative admission into an M/M/1 queue. The only information available is a signal that says whether the queue size is smaller than some L or not. We first compute the globally optimal and the Nash equilibrium stationary policy as a function of L. We compare the performance to that of full information on the queue size. We identify the L that optimizes the equilibrium performance.

1 Introduction

This paper is devoted to revisiting the problem of whether an arrival should queue or not in an M/M/1 queue. This is perhaps the first problem to be studied in optimal control of queues, going back to the seminal paper of Pinhas Naor [1]. Naor considered an M/M/1 queue, in which a controller has to decide whether arrivals should enter a queue or not. The objective was to minimize a weighted difference between the average expected waiting time of those that enter, and the acceptance rate of customers. Naor then considers the individual optimal threshold (which can be viewed as a Nash equilibrium in a non-cooperative game between the players) and shows that it is also of a threshold type with a threshold $L' > L$. Thus, under individual optimality, arrivals that join the queue wait longer in average. Finally, he showed that there exists some toll such that if it is imposed on arrivals for joining the queue then the threshold value of the individually optimal policy can be made to agree with the social optimal one. Since this seminal work of Naor there has been a huge amount of research that extend the model. More general interarrival and service times have been considered, more general networks, other objective functions and other queueing disciplines, see e.g. [2–10] and references therein.

In the original work of Naor, the decision maker(s) have full state information when entering the system. However, the fact that a threshold policy is optimal implies that for optimally controlling arrivals we only need partial information - we need a signal to indicate whether the queue exceeds or not the threshold value L. The fact that this much simpler information structure is sufficient for obtaining the same performance as in the full information case motivates us to study the performance of threshold policy and related optimization issues.

We first consider the socially optimal control policy for a given (non-necessarily optimal) threshold value L. When L is chosen non-optimally then the optimal

A. Dudin and K. De Turck (Eds.): ASMTA 2013, LNCS 7984, pp. 12–21, 2013.

policy for the partial information problem does no longer coincide with the policy with full information.

We then study the individual optimization problem with the same partial information: a signal (red) if the queue length exceeds some value L and a green signal otherwise.

For both the social and the individual optimization problems we show that the following structure holds: either whenever the signal is green all arrivals are accepted with probability 1, or whenever the signal is red all arrivals are rejected with probability 1.

We note that by using this signalling approach instead of providing full state information, users cannot choose any threshold policy with parameter different than L. Thus, in the individual optimization case, one could hope that by determining the signalling according to the value L that optimizes the socially optimal problem (in case of full information), one would obtain the socially optimal performance. We show that this is not the case, and determine the value L for which the reaction of the users optimizes the system performance. We compare this to the performance in case of full information.

2   The Model

Assume an M/M/1 queue where the admission rate is $\underline{\lambda}$ for $i \geq L$ and is otherwise $\overline{\lambda}$. Let μ be the service rate and set $\overline{\rho} = \overline{\lambda}/\mu$ and $\underline{\rho} = \underline{\lambda}/\mu$. We shall make the standard stability assumption that $\underline{\rho} < 1$. The balance equations are given

$$\mu\pi(i+1, L) = \lambda\pi(i, L)$$

where $\lambda = \underline{\lambda}$ for $i > L$ and is otherwise given by $\lambda = \overline{\lambda}$. The solution of these equations give

$$\pi(i, L) = \pi(0, L)\overline{\rho}^i$$

for $i \leq L$ and otherwise

$$\pi(i, L) = \pi(L, L)\underline{\rho}^{i-L}. \tag{1}$$

Hence

$$\pi(0, L) = \frac{1}{\sum_{i=0}^{L-1} \overline{\rho}^i + \overline{\rho}^L \sum_{i=0}^{\infty} \underline{\rho}^i}$$

$$= \frac{1}{\frac{1-\overline{\rho}^L}{1-\overline{\rho}} + \frac{\overline{\rho}^L}{1-\underline{\rho}}}$$

Thus

$$\pi(L, L) = \frac{1 - \underline{\rho}}{1 - \left(\frac{(1-\underline{\rho})(1-\overline{\rho}^{-L})}{1-\overline{\rho}}\right)} \tag{2}$$

Assume that an arrival receives the information on whether the size of the queue exceeds $L - 1$ or not. If it does not exceeds we shall say that it receives a "green" signal denoted by G, and otherwise a red one (R). The conditional state probabilities given the signals are denoted by

$$\pi(i, L|R) = (1 - \rho)\rho^{i-L}$$

for $i \geq L$, and is otherwise zero. The conditional tail distribution is

$$P(I > n|R) = \rho^{n+1-L}$$

for $n \geq L$, and is otherwise 1. Thus

$$E(I|R) = (L - 1) + \frac{1}{(1 - \rho)} \tag{3}$$

For a green light we have:

$$\pi(i, L|G) = \frac{1 - \bar{\rho}}{1 - \bar{\rho}^L}\bar{\rho}^i$$

for $0 \leq i < L$ and is otherwise zero. Hence the tail probabilities are

$$P(I > n|G) = \frac{\bar{\rho}^{n+1} - \bar{\rho}^L)}{1 - \bar{\rho}^L}$$

for $n < L$, and is otherwise 0. Hence

$$E(I|G) = \frac{1}{1 - \bar{\rho}^L} \left(\frac{(\bar{\rho}^L - \bar{\rho})}{\bar{\rho} - 1} - (L - 1)\bar{\rho}^L \right) \tag{4}$$

3 The Partially Observed Control Problem

We assume that $\nu < \underline{\lambda}$ is the rate of some uncontrolled Poisson flow. In addition there is an independent Poisson arrival flow of intensity ζ. We restrict to stationary policies, i.e. policies that are only function of the observation. A policy is thus a set of two probabilities: q_s where s is either R or G. q_s is the probability of accepting an arrival when the signal is s. For a given policy, we obtain the framework of the previous section with

$$\underline{\lambda} = \nu + \zeta q_R, \quad \bar{\lambda} = \nu + \zeta q_G.$$

Our goal is to minimize over $\mathbf{q}$

$$J_{\mathbf{q}} = E_{\mathbf{q}}[I] - \gamma T_{acc}(\mathbf{q}) = \sum_{s=G,R} P_{\mathbf{q}}(s)(E_{\mathbf{q}}[I|s] - \gamma T_{acc}(\mathbf{q}))$$

where

$$T_{acc} = \bar{\lambda} * P(G) + \underline{\lambda} * P(R) = \mu[P(R)(\rho - \bar{\rho}) + \bar{\rho}]$$

and $P(R) = P(I \geq L)$ is given by

$$P(R) = \pi(L, L)\frac{1}{1 - \rho} \tag{5}$$

$$E[I] = E[I|R] * P(R) + E[I|G] * P(G) = (E[I|R] - E[I|G]) * P(R) + E[I|G],$$

$$= \left(((L-1) + \frac{1}{\underline{\rho}^L(1-\underline{\rho})}) - (\frac{1}{1-\overline{\rho}^L}\left(\frac{(\overline{\rho}^L - \overline{\rho})}{\overline{\rho}-1} - (L-1)\overline{\rho}^L\right))\right) \times$$

$$\frac{(1-\overline{\rho})\overline{\rho}^L}{(1-\underline{\rho}) + \overline{\rho}^L(\underline{\rho} - \overline{\rho})} + \left(\frac{1}{1-\overline{\rho}^L}\left(\frac{(\overline{\rho}^L - \overline{\rho})}{\overline{\rho}-1} - (L-1)\overline{\rho}^L\right)\right)$$

The expression obtained for $J_{\mathbf{q}}$ is lower semi-continuous in the policy $\mathbf{q} = \{q_s, s = G, R\}$. Hence a minimizing policy $\mathbf{q}^*$ exists.

Lemma 1. *Assume that $\nu > 0$. If $\rho_R \geq 1$ then for any L $E[I]$ is infinite. In particular, if $\underline{\rho} \geq 1$ then for any L and any q, $E[I]$ is infinite.*

Proof. The expected queue length $E[I_t]$ at any time t and for any L is bounded from below by the one obtained by $E[I'_t] - L$ where I'_T is the queue size obtained when replacing q_G with $q_G = q_R$. $E[I'_t]$ corresponds to an M/M/1 queue with a workload $\rho \geq 1$ which is known to have infinite expectation. ∎.

3.1 The Structure of Optimal Policies

Figure 1 shows the values of the two component of the vector ρ corresponding to the optimal policy. We assume that ν and λ are such that $\overline{\rho} = 0.8$ and $\underline{\rho} = 0.3$. We further took $\mu = 1$, $\gamma = 15$, for 4 different values of the threshold L.

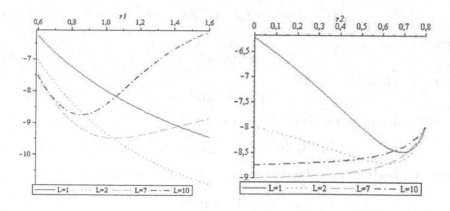

Fig. 1. The performance of different policies for several values of L

We observe the following structure: for any L, the optimal vector ρ satisfies the following property: whenever the minimum cost is achieved at an interior point for one of the components of ρ, then it is achieved on the boundary for the other component. More precisely, the optimal ρ satisfies either $\rho_2 = (r2) = \underline{\rho}$ or $\rho_1 = (r1) = \overline{\rho}$. We shall next prove this structure for the partially observable control problem.

Theorem 1. *Assume that $0 < \nu/\mu < 1$. Then there is a unique optimal station- ary strategy and it has the following property: either $q^*(G) = 1$ or $q^*(R) = 0$.*

Proof. Let $\mathbf{q}$ be optimal. We first show that $\alpha > 0$ where $\alpha := \mu - (\nu + q_R)\zeta)$. Indeed, if it were not the case then we would have $\rho \geq 1$ so by the previous Lemma, the queue length and hence the cost would be infinite. But then $\mathbf{q}$ cannot be optimal since the cost can be made finite by choosing $q_R) = 0$.

Assume that an optimal policy $\mathbf{q}$ does not have the structure stated in the Theorem. This would imply that q_R can be further decreased and q_G increased. In particular, one can perturb $\mathbf{q}$ in that way so that T_{acc} is unchanged. More precisely, note first that T_{acc} is monotone increasing in both $q_R)$ and in q_G. Hence

$$T_{acc}(1, q(R)) \geq T_{acc}(\mathbf{q}) \geq T_{acc}(q_G), 0),$$

Hence, if $T_{acc}(1, 0) < T_{acc}(\mathbf{q})$ then there is some $\mathbf{q_2}$ such that

$$T_{acc}(\mathbf{q_2}) = T_{acc}(\mathbf{q})$$

where either

$$\mathbf{q_2} := (1, q_R^2), \quad \text{or} \quad \mathbf{q_2} := (q_G^2, 0)$$

We have

$$P_{\mathbf{q_2}}(I = 0) = 1 - T_{acc}(\mathbf{q_2}) = 1 - T_{acc}(\mathbf{q})$$

(e.g. from Little's Theorem). From rate balance arguments it follows that

$$P_{\mathbf{q_2}}(I = i) = (1 - T_{acc})(\mathbf{q_2})\overline{\rho}_2^i \qquad \text{for } i \leq L. \tag{6}$$

Hence

$$P_{\mathbf{q_2}}(I \geq i) < P_{\mathbf{q}}(I \leq i) \tag{7}$$

for $i \geq L$. Thus

$$P_{\mathbf{q2}}(R) < P_{\mathbf{q}}(R).$$

By combining this with (1) it follows that

$$P_{\mathbf{q_2}}(I \geq i) = P_{\mathbf{q_2}}(R)\underline{\rho}(\mathbf{q_2})^{i-L} \leq P_{\mathbf{q_2}}(R)\rho(\mathbf{q_2})^{i-L} \leq= P_{\mathbf{q_2}}(I \geq i)$$

Hence (7) holds for all i. Taking the sum over i we thus obtain that

$$E_{\mathbf{q_2}}[I] < E_{\mathbf{q}}[I].$$

Since T_{acc} are the same under $\mathbf{q}$ and $\mathbf{q_2}$, it follows that $J_{\mathbf{q_2}} < J_{\mathbf{q}}$. Hence $\mathbf{q}$ is not optimal, which contradicts the assumption in the beginning of the proof. This establishes the structure of optimal policies. ∎

3.2   Optimizing the Signal

Here we briefly discuss the case of choosing L so as to minimize $J_{\mathbf{q}}$ not only with respect to $\mathbf{q}$ but also with respect to the value L of the threshold.

To that end we first consider the problem of minimizing J over all stationary policies in case that full state information is available. This is a Markov decision process and an optimal policy is known to exists within the pure stationary policies. Moreover, a direct extension of the proof in [1] can be used to show that the structure of the optimal policy is of a threshold type: accept all arrivals as long as the state is below a threshold and reject all controlled arrivals otherwise. Note however that this policy makes use only of the information available also in our cases, i.e. of whether the state exceeds L or not.

We conclude that the problem of optimizing $J_{\mathbf{q}}$ over both L and $\mathbf{q}$ has an optimal pure threshold policy i.e. with $q_R = 0$ and $q_{G=1}$, or in other words $\mathbf{q} = (1,0)$.

The optimal L for our problem can therefore be computed by minimizing $J_{\mathbf{q}}$ over pure threshold policies. Using Figure 2, we compute this optimal L for $\mu = 1$, $\eta = 0.01$, $\lambda = 0.98$ and $\gamma = 1, 5, 10, 15, 20$. and obtain $L = 5$ for $\gamma = 20$.

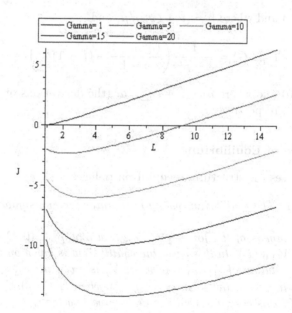

Fig. 2. The optimal performance for several values of L and γ

4   The Game Problem

We again assume that there is some uncontrolled flow ν and a flow of strategic players with intensity ζ. All users receive the signal G or R as before, and we restrict to polices as in the control case.

Assume that an arrival has a reward $\psi > 0$ for being processed in the queue, and a waiting cost of $E[W|s]$ where W is the waiting time. Note that $E[W|s] = E[I|s]/\mu$.

Let $Y(P)$, where $P = P(s)$, $s = R, G$ be the set of best responses of an individual if all the rest use $P(s), s = R, G$, and the system is in the corresponding steady state.

Then q is an equilibrium strategy if and only if $q \in Y(q)$. Note that the cost $J(q, P)$ corresponding to a strategy q of a player, when all others play P satisfies the following in order to be a best response to P: for each s, if $q(s)$ is not pure (is not 0 or 1) then at s, any other probability q' is also a best response.

The cost for a user for entering when the signal is s given that the strategy of other users is $\mathbf{q} = (q_G, q_R)$ is given by

$$V_{\mathbf{q}}(s) = E_{\mathbf{q}}[W|s] - \gamma = E_{\mathbf{q}}[I|s]/\mu - \gamma \tag{8}$$

It is zero if it does not enter. Here, $E_{\mathbf{q}}[I, s]$ are given by

$$E_{\mathbf{q}}(I|R) = (L-1) + \frac{1}{(1-\rho)} \tag{9}$$

where $\mathbf{q} = (1, q_R)$ and where $\rho = (\nu + \zeta q_R)/\mu$, and

$$E_{\mathbf{q}}(I|G) = \frac{1}{1 - \rho^L} \left(\frac{(\rho^L - \rho)}{\rho - 1} - (L-1)\rho^L \right) \tag{10}$$

where $\mathbf{q} = (q_G, 0)$ and where $\rho = (\nu + \zeta q_G)/\mu$. (the derivations of the above are as in (3) and (4), respectively.

4.1   Structure of Equilibrium

The following gives the structure of equilibria policies.

Theorem 2. *(1) The equilibrium policy is to enter for any signal if and only if $V_{(1,1)}(R) \leq 0$*
(2) The equilibrium is of the form $\mathbf{q} = (1, q_R)$ where $q_R \in (0, 1)$ if and only if $V_{(1,1)}(R) > 0 > V_{(1,0)}(R)$. In this case, the equilibrium is given by the $q = (1, q_R)$ where q_R is the solution of $V_{(1,q_R)} = 0$ where V_q is given in (8).
(3) The equilibrium is of the form $\mathbf{q} = (q_G, 0)$ where $q_G \in (0, 1]$ if and only if $V_{(1,0)}(G) \geq 0$. In this case, the equilibrium is given by the $\mathbf{q} = (q_G, 0)$ where q_G is the solution of $V_{(q_G,0)} = 0$ and where $V_{\mathbf{q}}$ is given in (8).

Proof. Follows directly from continuity of the expected queue length with respect to $\mathbf{q}$ and from the fact that $V_{(q_G,q_R)}$ is strictly monotone increasing in both arguments. We establish the continuity in the Appendix using an approach that does not require the exact explicit form of $E[I]$ and thus will be useful when attempting to generalize the results to other models (such as the case of more than a single server).   ∎.

4.2   Numerical Examples

We consider here as an example the parameters $\gamma = 20$, $\mu = 1$, $\lambda = 0.98$ and $\zeta = 0.01$. For all L condition (1) of Theorem 2 does not hold, so $(1,1)$ is not an equilibrium. condition (2) of the Theorem holds for $L \leq 20$. In that case, the equilibrium is given by $(1, q_R)$ where q_R is given in Fig 3. The value at equilibrium is given in Figure 4 for the case of the signal G and is otherwise zero for all $L \leq 20$. For the case of $L > 20$ we have the opposite, i.e. $V_R = 0$. V_G is given by $E[I|G] - \gamma$ where $E[I|G]$ is expressed in (4).

Let L^* denote one plus the largest value L for which $V_{1,0} < 0$. L^* thus separates case (2) and (3) in Theorem 2. Then L^* equals the smallest integer greater than or equal to $\gamma\mu$. In our case it is given by 20 as is seen in Figure 4. For every $L > L^*$ we know that, in fact, $q_R = 0$. Indeed, a red signal in that case would mean that the queue exceeds size $\gamma\mu$ and thus if the individual entered, its expected waiting time would exceed γ. For $L < L^*$ we know that $q_G = 1$ since the expected time of an admitted customer would be smaller than γ. It is then easy to see that for $L = L^*$, the pure threshold policy with parameter L^* is a pure (state dependent) equilibrium in the game with full information.

4.3   Optimizing the Signal

We are interested here in finding the L for which the induced equilibrium gives the best system performance. We plot the system performance J at equilibrium as a function of L in Figure 5.

The optimal L is seen to equal 20 and the corresponding performance measures at equilibrium are $J = -14.13$ and $T_{acc} = 0.83$. We conclude that the policy for which the social cost is minimized has the same performance as the full state information equilibrium policy.

If we take the $L = 5$ which we had computed for optimizing the system performance, and use it in the game setting, we obtain $T_{acc} = 0.95$ and $J^* = -3.49$. which indeed gives a much worse performance than the performance under the $L = 20$.

5   Appendix: Uniform f-Geometric Ergodicity and the Continuity of the Markov Chain

We show continuity of the expected queue length in $\mathbf{q}$ for q_R restricted to some closed interval for which the corresponding value of ρ_R is smaller than 1. (Due to Lemma 1 there exists indeed such an interval such that any policy for which q_R is not in the interval cannot be optimal.

We show that the Markov chain is f-Geometric Ergodic and then use Lemma 5.1 from [11].

Consider the Markov chain embedded at each transition in the queue size. Thus for $I \geq \max(L, 1)$, with probability β the event is a departure and otherwise it is an arrival, where

$$\beta := \frac{\mu}{\mu + \nu + q_R \zeta}.$$

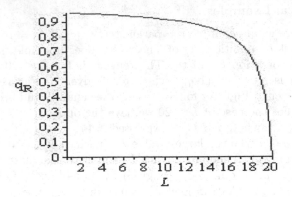

Fig. 3. Equilibrium action q_R as a function of L

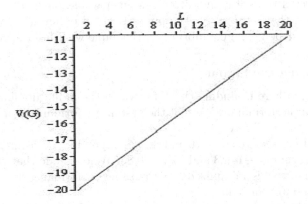

Fig. 4. Equilibrium value V_G for signal G as a function of L. We used case (2) in Theorem 2 and the results are therefore valid only for $L \leq 20$.

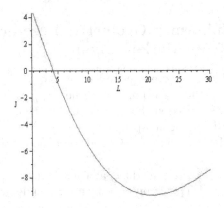

Fig. 5. The social value J at equilibrium as a function of L

Note $\alpha > 0$ implies that $\beta > 1/2$ (α is defined in the proof of Theorem 1).
Define $f(i) = \exp(\gamma i)$, for any $I \geq \max(L, 1)$,

$$E[f(I_{t+1}) - f(I_t)|I_t = i] = \beta \exp[\gamma(i-1)] + (1-\beta)\exp[\gamma(i+1)] - \exp(\gamma i)$$

$$= f(i)\Delta \quad \text{where} \quad \Delta = \beta z^{-1} + (1-\beta)z - 1$$

and where $z := \exp(-\gamma)$. Note that $\Delta = 0$ at

$$z_{1,2} = \frac{1 \pm \sqrt{1 - 4\beta(1-\beta)}}{2(1-\beta)} = \{1, \frac{\beta}{1-\beta}\}$$

Thus $\Delta < 0$ for all γ in the interval $\left(0, \log\left(\frac{\beta}{1-\beta}\right)\right)$ (which is non-empty since
we showed that $1 > \beta > 1/2$). We conclude that for any γ in that interval, f is
a Lyapunov function and the Markov chain is γ-geometrically ergodic uniformly
in $\mathbf{q}$.

References

1. Naor, P.: On the regulation of queueing size by levying tolls. Econometrica 37, 15–24 (1969)
2. Yechiali, U.: On optimal balking rules and toll charges in a $GI|M|1$ queueing process. Operations Research 19, 349–370 (1971)
3. Stidham, S., Weber, R.: Monotonic and insensitive optimal policies for control of queues with undiscounted costs. Operations Research 37, 611–625 (1989)
4. Stidham, S., Rajagopal, S., Kulkarni, V.G.: Optimal flow control of a stochastic fluid-flow system. IEEE Journal on Selected Areas in Communications 13, 1219–1228 (1995)
5. Hsiao, M.T., Lazar, A.A.: Optimal decentralized flow control of Markovian queueing networks with multiple controllers. Performance Evaluation 13, 181–204 (1991)
6. Korilis, Y.A., Lazar, A.: On the existence of equilibria in noncooperative optimal flow control. Journal of the ACM 42(3), 584–613 (1995)
7. Altman, E., Shimkin, N.: Individual equilibrium and learning in processor sharing systems. Operations Research 46, 776–784 (1998)
8. Hordijk, A., Spieksma, F.: Constrained admission control to a queueing system. Advances in Applied Probability 21, 409–431 (1989)
9. Altman, E., Gaujal, B., Hordijk, A.: Admission control in stochastic event graphs. IEEE Automatic Control 45(5), 854–867 (2000)
10. Stidham, S.: Optimal control of admission to a queueing system. IEEE Transactions on Automatic Control 30, 705–713 (1985)
11. Spieksma, F.: Geometrically Ergodic Markov Chains and the Optimal Control of Queues. PhD thesis, Leiden University (1990)

Approximate Transient Analysis of Queuing Networks by Quasi Product Forms

Alessio Angius[1], András Horváth[1], and Verena Wolf[2]

[1] Dipartimento di Informatica, Università di Torino, Italy
[2] Computer Science Department, Saarland University, Saarbrücken, Germany

Abstract. In this paper we deal with transient analysis of networks of queues. These systems most often have enormous state space and the exact computation of their transient behavior is not possible. We propose to apply an approximate technique based on assumptions on the structure of the transient probabilities. In particular, we assume that the transient probabilities of the model can be decomposed into a *quasi product form*. This assumption simplifies the dependency structure of the model and leads to a relatively small set of ordinary differential equations (ODE) that can be used to compute an approximation of the transient probabilities. We provide the derivation of this set of ODEs and illustrate the accuracy of the approach on numerical examples.

1 Introduction

Queuing networks are heavily used in performance evaluation of distributed systems, such as computer networks, telecommunication systems, and manufacturing or logistics infrastructures. Markovian models of such systems can, in principle, be treated by standard transient and steady state analysis techniques developed for Markov chains [22] but this is almost always precluded because of the huge state space. A family of algorithms that can be applied to models with large state space are based on the fact that the steady state probabilities, under particular assumptions, are in product form. The most classical works in this direction include: Jackson networks [16] which are open networks with Poisson arrivals and infinite capacity, exponential servers; Gordon-Newell networks [11] which are closed networks with exponential servers; and BCMP networks [4] which can be either open, closed or mixed and can contain multiple classes of customers. When the probabilities are not in product form and the state space is large, approximate techniques can be used. One category of approximate approaches is decomposition. By this approach, nodes are analyzed in isolation and their output traffic is characterized and forwarded to the subsequent nodes in order to evaluate the whole network. Methods based on decomposition ranges from simple nodes with simple inter-node traffic description [17, 23] to more general nodes with correlated inter-node traffic [14, 15]. Approximate techniques can also be based on hierarchical analysis where a subnetwork is aggregated into a single node. Algorithms that use this approach are presented, for example, in [3, 21] where the concept of flow equivalent server is employed [9].

The vast majority of the above and related methods are developed to compute steady state measures, such as average response time, long-run throughput or long-run system

A. Dudin and K. De Turck (Eds.): ASMTA 2013, LNCS 7984, pp. 22–36, 2013.

utilization. It is for a long time known, however, that steady state measures can be insufficient to describe the behavior of even a single queue. In [24] the GI/M/1 queue is analyzed for what concerns the range of possible fluctuations in the queue-length process for a fixed steady state queue-length distribution. It is shown that queues with equal equilibrium queue-length distribution can have very different second order performance indices, such as, variance of the busy period. For what concerns networks of queues, [20] reveals that equilibrium traffic streams on non-exit arcs in Jackson networks with loops are not Poisson which means that the product form steady state distribution is not sufficient to characterize the traffic in the network. Moreover, modern distributed systems are such complex and dynamic that they often never reach steady state.

For the above mentioned reasons transient analysis of queuing networks is an important topic. Exact computations are possible only for small models or for very particular situations, like networks of infinite server queues [6, 7, 12, 18]. In these networks clients are independent of each other and this leads to the fact that the number of clients at a station follows a Poisson distribution whose mean can be calculated by a set of ODEs. System of ODEs can be used also to develop approximate techniques. Among these we have moment closure techniques [19] which provide approximate moments of the system and fluid approximations that can be used to find the bottlenecks of the network [10]. Methods based on aggregation can also be developed, see, for example, [5]. There are fewer techniques that maintain the original state space of the model and, as a consequence, allow to calculate distributions and not only moments. In [13] an iterative method is suggested to solve the time-dependent Kolmogorov equations of the model but this approach suffers from the state space explosion problem. A memory efficient approach is proposed instead in [25] where the number of ODEs that describe the transient behavior is decreased by assuming a limited dependency structure among the queues of the network.

In this paper we apply an approximate transient analysis technique that maintains the state space of the model. The technique is based on assuming that the transient probabilities can be written in a *quasi product form* (QPF) and it was proposed in [2] to analyze reaction networks. These networks are characterized by the presence of infinite server-like mechanisms combined with switches. Instead in this work, we apply and experiment the approach in the context of queuing networks with finite number of servers.

The QPF assumption leads to a memory efficient description of the transient probabilities and determines a relatively small set of ODEs that provides an approximation of the transient probabilities. A similar approach was adopted in [8] in the context of closed networks to compute passage time distributions. The method that comes closest to ours is the one introduced in [25] and called *partial product form* (PPF) decomposition. Our exposition differs from [25] in the following points. **i)** We apply a more relaxed way of decomposing the transient probabilities into a product form and therefore allow for a more general dependency structure. In particular, the dependency structure considered in [25] is a special case of the dependency structure applied here. **ii)** Our approach is such that the network topology can be used to provide candidates for the way of decomposing the transient probabilities. **iii)** We illustrate the technique on numerical examples and show that it can outperform the one proposed in [25].

The paper is organized as follows. In Section 2 we describe the considered class of queuing networks. Section 3 provides the approximate transient analysis technique based on QPF. In Section 4, characteristics of the proposed technique are discussed. Numerical examples are provided in Section 5. Conclusions are drawn in Section 6.

2   Considered Class of Queueing Networks

We consider an open network of M queues. The maximum number of jobs at queue i is B_i including the job under service with $B_i \in \mathbb{N} \cup \{\infty\}$, i.e., each buffer is either finite or infinite. Clients that arrive to a full buffer are lost. The ith queue of the network receives jobs (clients) entering from outside according to a Poisson-process with intensity λ_i. The routing probabilities are given by r_{ij} with $i, j \in \{1, 2, ..., M\}$ and the probability that a client leaving queue i exits the system is denoted by $r_{i0} = 1 - \sum_{j=1}^{M} r_{ij}$. Service times are exponential and can depend on the length of the local queue. The service intensity of queue i in the presence of j jobs at the station is denoted by $\mu_{i,j}$ (in the following equations we assume that $\mu_{i,0} = 0$).

A given state of the model is an M-dimensional vector $x = |x_1, ..., x_M|$ where x_i denotes the number of jobs at station i. We assume that the arrival intensities, λ_i, and the routing probabilities, r_{ij}, are such that every queue can be reached by the clients. Accordingly, the state space is $\mathcal{S} = \{x = |x_1, ..., x_M| \mid 0 \leq x_i \leq B_i\}$. We use the following notation

$$f_{x,i} = \begin{cases} 1 \text{ if } x_i = B_i \\ 0 \text{ otherwise} \end{cases}$$

to indicate that the ith queue in state x is full and apply also its complement as $\bar{f}_{x,i} = 1 - f_{x,i}$. The number of jobs at queue i at time t is denoted by $Q_i(t)$ and the full system state by $Q(t) = |Q_1(t), ..., Q_M(t)|$. The probability that the system is in state x at time t is denoted by $p(t, x)$, i.e.,

$$p(t, x) = P\{Q(t) = x\} = P\{\wedge_{i \in \{1,2,...,M\}} Q_i(t) = x_i\}$$

As throughout the paper we consider transient analysis, in the following we omit the dependency on time and write Q instead of $Q(t)$ and $p(x)$ instead of $p(t, x)$.

As the network of queues form a continuous time Markov chain, $p(x)$ satisfies the following ODE

$$\frac{dp(x)}{dt} = -p(x) \left(\sum_{i=1}^{M} \bar{f}_{x,i} \lambda_i + \sum_{i=1}^{M} \mu_{i,x_i} \right) + \sum_{i=1}^{M} p(x - h_i) \lambda_i + \tag{1}$$

$$\sum_{i=1}^{M} \sum_{j=0}^{M} p(x + h_i - h_j) \mu_{i,x_i} r_{ij} + \sum_{i=1}^{M} \sum_{j=1}^{M} p(x + h_i) \mu_{i,x_i} \bar{f}_{x,j} r_{ij}$$

where h_0 denotes the vector of zeros, h_i with $1 \leq i \leq M$ the vector with a 1 in the ith place and zeros elsewhere and we assumed that $p(x) = 0$ if $x \notin \mathcal{S}$. In (1) the first term of the right hand side collects the outgoing probabilities of state x; the second the incoming probabilities due to arrivals from the outside; the third the incoming probabilities due to a job leaving station i and joining station j; and the fourth is similar to the third but when station j is full and the job gets lost.

3 Quasi Product Form Approximation

In [1] we applied an approximate transient analysis method for stochastic reaction networks arising in systems biology. This method is based on the assumption that the transient probabilities are in product form, i.e., we can write

$$p(x) = P\{Q = x\} = \prod_{i=1}^{M} P\{Q_i = x_i\}. \tag{2}$$

This assumption leads to a set of ODEs that can be used to compute an approximation of the transient probabilities. The number of ODEs in this set is much lower than the number of states of the system. Roughly speaking, the number of ODEs grows linearly with the number of queues while the number of states grows in an exponential manner. This means that the space complexity of the resulting algorithm is much lower than that of computing the transient probabilities by the classical and widely used uniformisation approach (see, for example, [22]).

The product form assumption given in (2) leads to exact results if the model corresponds to a network of M/M/∞ queues. In [1] we showed that the approximation is satisfactory if the model resembles a network of M/M/∞ queues but can give imprecise results in other cases. In this paper we apply a more relaxed assumption that leads to a good approximation for a wider range of models. In particular, we will assume that there exist sets of queues whose conditional probabilities depend only on a set of other queues and not on all the rest of the queues. For example, if we assume that the conditional probabilities of the number of clients in queue 1 and 2 depend only on the number of clients in queue 3, 4 and 5 then we can write

$$P\{Q_1 = x_1, Q_2 = x_2 | Q_3 = x_3, Q_4 = x_4, ..., Q_M = x_M\} =$$
$$P\{Q_1 = x_1, Q_2 = x_2 | Q_3 = x_3, Q_4 = x_4, Q_5 = x_5\}.$$

A set of assumptions like the one above allows us to decompose the probability $P\{\wedge_{i \in \{1,2,...,M\}} Q_i = x_i\}$ into a product. As this product is not in the classical product form given in (2), we will refer to it as *quasi product form* and in the following we provide its formal description.

The QPF decomposition of the transient probabilities is conveniently described by a directed acyclic graph (DAG), denoted by $\mathcal{F}$. The set of nodes is denoted by $\mathcal{V}$ and a node $v \in \mathcal{V}$ represents a subset of the queues. The index set of the queues represented by node v is denoted by $I(v)$. The set $\mathcal{V}$ must be such that it provides a partitioning of the set of queues, i.e., $\bigcup_{v \in \mathcal{V}} I(v) = \{1, 2, ..., M\}$ and $\forall v_1, v_2 \in \mathcal{V}, v_1 \neq v_2 :$ $I(v_1) \cap I(v_2) = \emptyset$. The edge set of the DAG, denoted by $\mathcal{E}$, provides the assumed dependency structure of the transient probabilities. Specifically, if $e = (u, v) \in \mathcal{E}$ (denoted as $u \longrightarrow v$) then the conditional probability of the queues in v *depends* on those queues that are present in u. The set of queues present in the predecessors of v will be denoted by $P(v)$, i.e., $P(v) = \bigcup_{u:u \longrightarrow v} I(u)$. The conditional probability of the queues in $I(v)$ is independent of those queues that are not present in $P(v)$, i.e.,

$$P\{\wedge_{i \in I(v)}(Q_i = x_i) | \wedge_{j \in \{1,2,...,M\}/I(v)} (Q_j = x_j)\} =$$
$$P\{\wedge_{i \in I(v)}(Q_i = x_i) | \wedge_{j \in P(v)} (Q_j = x_j)\}$$

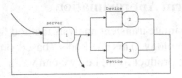

Fig. 1. Open central server network

By considering every node of the tree, the probability of a given state of the system, $|x_1, ..., x_M|$, can be written as

$$P\{\wedge_{i\in\{1,2,...,M\}}(Q_i = x_i)\} = \prod_{v\in\mathcal{V}} P\{\wedge_{i\in I(v)}(Q_i = x_i)|\ \wedge_{j\in P(v)}(Q_j = x_j)\} =$$

$$\prod_{v\in\mathcal{V}} \frac{P\{\wedge_{i\in D(v)}(Q_i = x_i)\}}{P\{\wedge_{j\in P(v)}(Q_j = x_j)\}} \tag{3}$$

where we applied the notation $D(v) = I(v) \bigcup P(v)$. In the following we give three examples for the DAG $\mathcal{F}$.

Example 1. The assumption of complete product form would be expressed by the DAG with M nodes, $v_1, ..., v_M$, such that $I(v_i) = \{i\}$, and an empty set of arcs, $\mathcal{E} = \emptyset$. With this DAG the probabilities are in the form given in (2).

Example 2. The decomposition called *partial product form* in [25] would be expressed by a DAG in which the nodes provide a partitioning of the queues and the set of arcs, $\mathcal{E}$, is empty.

Example 3. Let us consider the simple network depicted in Fig. 1 which will also serve as a numerical example in Section 5. The topology suggests to use a DAG with three nodes, v_1, v_2 and v_3, corresponding to the three stations, respectively, and two arcs as $\mathcal{E} = \{(v_1, v_2), (v_1, v_3)\}$. With this DAG the transient probabilities are approximated in the form

$$P\{\wedge_{j\in\{1,2,3\}}(Q_j = x_j)\} = \tag{4}$$
$$P\{Q_1 = x_1\}P\{Q_2 = x_2|Q_1 = x_1\}P\{Q_3 = x_3|Q_1 = x_1\} =$$
$$P\{Q_1 = x_1\}\frac{P\{Q_1 = x_1, Q_2 = x_2\}}{P\{Q_1 = x_1\}}\frac{P\{Q_1 = x_1, Q_3 = x_3\}}{P\{Q_1 = x_1\}}$$

In order to compute the transient probabilities based on the QPF assumption expressed by the DAG $\mathcal{F}$, we need the quantities appearing in (3). Since $P(v) \subseteq D(v)$, the quantities in the denominator can be computed simply by appropriate summation of the quantities in the numerator. The quantities in the numerator can instead be computed by the ODEs provided by the following theorem.

Theorem 1. *If the transient probabilities satisfy the QPF decomposition expressed by the DAG $\mathcal{F}$, then the following ODE holds for all nodes $v \in \mathcal{V}$ and every possible values of $x_i, i \in D(v)$:*

$$\frac{dP\{\wedge_{i\in D(v)}(Q_i=x_i)\}}{dt} = \sum_{\substack{|y_1,\dots,y_M|\,:\\ \forall k\in D(v),\, y_k=x_k}} \Bigg(\tag{5}$$

$$- \prod_{v\in\mathcal{V}} \frac{P\{\wedge_{l\in D(v)}(Q_l=y_l)\}}{P\{\wedge_{m\in P(v)}(Q_m=y_m)\}} \left(\sum_{i=1}^{M} \bar{f}_{y,i}\lambda_i + \sum_{i=1}^{M}\mu_{i,y_i}\right) +$$

$$\sum_{i=1}^{M} \prod_{v\in\mathcal{V}} \frac{P\{\wedge_{l\in D(v)}(Q_l=y_l-h_{il})\}}{P\{\wedge_{m\in P(v)}(Q_m=y_m-h_{im})\}}\lambda_i +$$

$$\sum_{i=1}^{M}\sum_{j=0}^{M} \prod_{v\in\mathcal{V}} \frac{P\{\wedge_{l\in D(v)}(Q_l=y_l+h_{il}-h_{jl})\}}{P\{\wedge_{m\in P(v)}(Q_m=y_m+h_{im}-h_{jm})\}}\mu_{i,y_i} r_{ij} +$$

$$\sum_{i=1}^{M}\sum_{j=1}^{M} \prod_{v\in\mathcal{V}} \frac{P\{\wedge_{l\in D(v)}(Q_l=y_l+h_{il})\}}{P\{\wedge_{m\in P(v)}(Q_m=y_m+h_{im})\}}\mu_{i,y_i} f_{y,j} r_{ij} \Bigg)$$

where h_{ij} denotes the jth entry of h_i.

Proof. It is easy to see that we have

$$\frac{dP\{\wedge_{i\in D(v)}(Q_i=x_i)\}}{dt} = \frac{d}{dt} \sum_{\substack{|y_1,\dots,y_M|\,:\\ \forall k\in D(v),\, y_k=x_k}} P\{\wedge_{i\in\{1,\dots,M\}}(Q_i=y_i)\} \tag{6}$$

where the order of the derivative and the summation can be exchanged. By applying (1) and the QPF assumption given in (3) the theorem follows.

Given a DAG, the set of equations provided by Theorem 1 can be redundant. This happens when we have two nodes, v and u, such that $D(v) \subset D(u)$ because in this case the probabilities $P\{\wedge_{i\in D(v)}(Q_i=x_i)\}$ can be computed by appropriate summation of the quantities $P\{\wedge_{i\in D(u)}(Q_i=x_i)\}$.

In the following example we apply Theorem 1 to the network shown in Fig. 1 with the DAG described in Example 3.

Example 4. Since we have $D(v_1) \subset D(v_2)$ the equations associated with node v_1 can be discarded. For what concerns v_2, assuming that the queues of the network are with infinite buffer, we have for all $x_1 \geq 0, x_2 \geq 0$

$$\frac{dP\{Q_1=x_1, Q_2=x_2\}}{dt} = \sum_{x_3\geq 0} \Bigg($$

$$- P\{|Q_1,Q_2,Q_3| = |x_1,x_2,x_3|\} \left(\lambda_1 + \sum_{i=1}^{3}\mu_{i,x_i}\right) +$$

$$P\{|Q_1,Q_2,Q_3| = |x_1-1,x_2,x_3|\}\lambda_1 +$$

$$P\{|Q_1,Q_2,Q_3| = |x_1+1,x_2,x_3|\}\mu_{1,x_1+1}r_{10} +$$

$$P\{|Q_1,Q_2,Q_3| = |x_1+1,x_2-1,x_3|\}\mu_{1,x_1+1}r_{12} +$$

$$P\{|Q_1,Q_2,Q_3| = |x_1+1,x_2,x_3-1|\}\mu_{1,x_1+1}r_{13} +$$

$$P\{|Q_1,Q_2,Q_3| = |x_1-1,x_2+1,x_3|\}\mu_{2,x_2+1} +$$

$$P\{|Q_1,Q_2,Q_3| = |x_1-1,x_2,x_3+1|\}\mu_{3,x_3+1} \Bigg)$$

which, by applying the assumed QPF decomposition given in (4) and using

$$P\{|Q_1,Q_2| = |x_1,x_2|\} = \sum_{x_3 \geq 0} P\{|Q_1,Q_2,Q_3| = |x_1,x_2,x_3|\}$$

can be written as

$$\frac{dP\{Q_1 = x_1, Q_2 = x_2\}}{dt} = \tag{7}$$

$$-P\{|Q_1,Q_2| = |x_1,x_2|\}\left(\lambda_1 + \sum_{i=1}^{2}\mu_{i,x_i}\right)-$$

$$P\{|Q_1,Q_2| = |x_1,x_2|\}\sum_{x_3 \geq 0}\frac{P\{Q_1 = x_1, Q_3 = x_3\}}{P\{Q_1 = x_1\}}\mu_{3,x_3}+$$

$$P\{|Q_1,Q_2| = |x_1 - 1, x_2|\}\lambda_1+$$

$$P\{|Q_1,Q_2| = |x_1 + 1, x_2|\}\mu_{1,x_1+1}r_{10}+$$

$$P\{|Q_1,Q_2| = |x_1 + 1, x_2 - 1|\}\mu_{1,x_1+1}r_{12}+$$

$$P\{|Q_1,Q_2| = |x_1 + 1, x_2|\}\mu_{1,x_1+1}r_{13}+$$

$$P\{|Q_1,Q_2| = |x_1 - 1, x_2 + 1|\}\mu_{2,x_2+1}+$$

$$P\{|Q_1,Q_2| = |x_1 - 1, x_2|\} \times$$

$$\sum_{x_3 \geq 0}\frac{P\{Q_1 = x_1 - 1, Q_3 = x_3 + 1\}}{P\{Q_1 = x_1 - 1\}}\mu_{3,x_3+1}$$

The above ODE provides a clear interpretation of the impact of applying the QPF decomposition. The events whose intensity does not depend on the number of jobs at the third queue (i.e., arrival from outside, service of clients at station 1 and 2) are considered in an exact manner. The event of serving a job at station 3, whose intensity certainly depends on the number of clients at station 3, is seen from the point of view of the 1st and the 2nd queue as it was independent of the number of clients at station 2. Indeed, the summations in the 2nd and the 8th term of the right hand side of (7) correspond to the intensity of the event of serving a client at station 3 provided that the number of clients at station 1 is x_1 and independent of the number of clients at station 2.

In order to have a complete set of ODEs for the network in Fig. 1 with the DAG described in Example 3, Theorem 1 must be applied to node v_3 as well. The resulting equations, which provide the ODEs for the quantities $P\{Q_1 = x_1, Q_3 = x_3\}$ with $x_1 \geq 0, x_3 \geq 0$, are symmetric to the those presented for v_2 in (7).

4   Characteristics of the Approximation

4.1   Number of Equations

Assuming that the highest considered buffer level for queue i is $\tilde{B}_i$, the number of equations in the original set of ODEs given in (1) is $\prod_{i=1}^{M}(\tilde{B}_i + 1)$. The number of equations describing the QPF in (5) can be determined as follows. Let Y denote that subset of the nodes of the DAG that provides a complete and not redundant set of ODEs, i.e., $Y = \{v \mid v \in \mathcal{V}, \nexists u \in \mathcal{V}$ such that $D(v) \subset D(u)\}$. Having defined Y, the necessary number of ODEs for the QPF is given by $\sum_{v \in Y} \prod_{i \in D(v)} (\tilde{B}_i + 1)$. This means that the original M-dimensional state space is reduced to $\max_{v \in Y} |D(v)|$ dimensions.

4.2 Choice of DAG

There are two simple factors that can be used to construct the DAG that describes the QPF decomposition. The first is the topology together with the routing probabilities. As considering only this first factor can lead to large system of ODEs, a second factor can also be considered, namely, the dimensionality reduction we want to obtain.

More sophisticated arguments to construct the DAG can be based on the theoretical results presented in the papers dealing with networks of infinite servers [6, 7, 12, 18] that enjoy product form in transient. According to these results, a crucial aspect for the presence of transient product form is that traffic flows are inhomogeneous Poisson processes. This suggests that if the output of a queue is close to an inhomogeneous Poisson process then the dependency between this queue and those that receive its outgoing jobs can be neglected. A necessary condition to have Poisson output in transient is that jobs are not queuing at the station. This condition can be fulfilled by a station with finite number of servers only if its load is low. This implies that if some dependencies of connected queues must be neglected in order to keep low the number of ODEs, then it is convenient to neglect dependencies on the queues with lower loads.

In Section 5 we show examples for which we considered only the topology to guide the choice of the DAG and others for which more factors were taken into account.

4.3 Coherence of the Set of ODEs

By looking at (5), one can check that the ODEs provided by Theorem 1 maintain unity of the total probability, i.e., for every node $v \in \mathcal{V}$ summing $P\{\wedge_{i \in D(v)}(Q_i = x_i)\}$ for every possible values of $x_i, i \in D(v)$, gives one.

If there exists a queue whose index, i, is present in both $D(u)$ and $D(v)$ with $u, v \in \mathcal{V}, u \neq v$, then the marginal probabilities of queue i, i.e., $P\{Q_i = x_i\}$, can be derived using the quantities associated with u, i.e., $P\{\wedge_{j \in D(u)}(Q_j = y_j)\}$, or using the quantities associated with v, i.e., $P\{\wedge_{j \in D(v)}(Q_j = y_j)\}$. By considering the derivative $dP\{Q_i = x_i\}/dt$ which can be computed based on both u and v by summation of their associated ODEs given in (5), it is easy to show that the different ways of calculating $P\{Q_i = x_i\}$ lead to the same result. The above reasoning can be generalized to any marginal distribution of the model.

4.4 Limiting Behavior

In principle, the steady state determined by the QPF assumption can be calculated by setting the left hand side of (5) to zero and solving the resulting set of equations. In practice, this is not feasible because the set of equations is not linear and the number of equations can be large. Exact steady state is determined by the equations given in (1) by setting the left hand side to zero. One can observe that the right hand side of (5) contains summations of the right hand side of (1) for given set of states. This implies that the limiting behavior of the QPF approximation is such that it satisfies sums of those equations that determine the exact steady state. Moreover, if the exact steady state is uniquely determined by these sums of equations then the limiting behavior of the approximation is exact.

5 Numerical Examples

In this section we apply the QPF approximation to three models with various settings of the parameters. The first example has a small state space and thus allows us to compare the results obtained by the approximation to the numerical solution of the underlying CTMC. We use this small example to show that the QPF decomposition can outperform the PPF decomposition [25]. In case of the last two models with large state spaces we compare the QPF approximation to results obtained using Monte Carlo simulation.

The algorithm based on the QPF assumption has been implemented in JAVA using the odeToJava package[1] for the solution of the system of ODEs. In particular, we applied the "explicit Runge-Kutta triple" solver of the package. The accuracy of this method is determined by two parameters, called relative and absolute tolerance, and we set these values to 10^{-10} and 10^{-12}, respectively. The reported run times refer to these settings and by choosing less restrictive values the computation times can be lowered significantly, by about one order of magnitude. All the experiments have been performed on an Intel Centrino Dual Core with 4Gb of RAM.

5.1 Open Central Server Network

As first, we consider a network representing a server connected to two devices (Fig. 1). Jobs arrive to the server with a constant rate λ_1 and compete for the resources of the system. After each service at the first station, the job can leave the system with probability $r_{1,0}$ or use one of the two devices with probabilities $r_{1,2}$ and $r_{1,3}$, respectively. Jobs leaving the two devices go back to the server. We tested the approximation with the parameters $r_{1,2} = 0.2$, $r_{1,3} = 0.3$, $\lambda_1 = 0.6$, $\mu_1 = 1$, $\mu_2 = 5$, $\mu_3 = 1$. We assume that the server is triplicated, i.e., three jobs can be under service at a time at the server. Note that with single server policy the system would not be stable due to the presence of the loop. The number of jobs at a queue is at most 15, further arrivals are lost.

We analyzed the model by using its original CTMC, by all possible partial product form decompositions, and by the QPF decomposition given in Example 3. With this QPF decomposition the set of required marginal distributions are the probabilities $P\{Q_1 = x_1, Q_2 = x_2\}$ and $P\{Q_1 = x_1, Q_3 = x_3\}$. We considered two situations: starting with empty queues and starting with five jobs at the server. In Fig. 2 and 3 we depicted the mean and the variance of the number of jobs at the server as function of time. Starting with empty queues, both the QPF and PPF approximations lead to good results with the exceptions of the third PPF decomposition that fails to provide an accurate estimate of the variance. Starting with five jobs at the server, the approximations are worse and only the QPF decomposition captures the mean correctly and approximate the variance well.

---

[1] Available at http://www.netlib.org/ode/ and developed by M. Patterson and R. J. Spiteri.

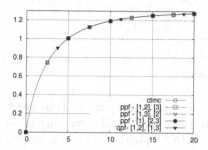
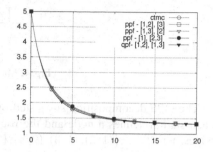

Fig. 2. Open server network: mean number of jobs at the server vs. time; starting with empty queues (left), with five jobs at the server (right)

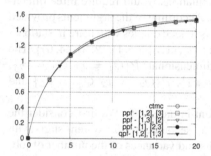
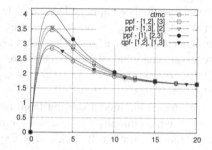

Fig. 3. Open server network: variance of number of jobs at the server vs. time; starting with empty queues (left), with five jobs at the server (right)

The original CTMC is composed of $16^3 = 4096$ states. The PPF approximations lead to $16^2 + 16 = 272$ ODEs while the QPF uses $2 \times 16^2 = 512$ ODEs. The number of equation of the QPF decomposition is about two times more but the dimensionality of the the approaches are the same as both consider the dependencies of at most two stations. The calculations with QPF decomposition required about 1 second of CPU time.

5.2 Multi-path Network

As second example, we consider a more complex pipeline where each request can be satisfied by following four different production paths. The network is depicted in Fig. 4. Requests arrive with a fixed rate $\lambda_1 = 0.6$ to the first station that constitutes a pre-processing step. After that, the possible paths for a request are stations 2-4, 2-5, 3-4 and 3-5 followed by a post-processing phase that takes place at station 6. The parameters that we consider are such that the routing probabilities are symmetric with $r_{1,2} = r_{2,4} = r_{3,5} = 0.5$ but the processing intensities are asymmetric with $\mu_2 = \mu_4 = 0.4$ and $\mu_3 = \mu_5 = 2$, i.e., the branch composed by stations 2 and 4 is slow while the one containing stations 3 and 5 is fast. The pre-processing and the post-processing stations are serving the jobs with intensity $\mu_1 = \mu_6 = 1$. The maximum number of clients for each queue is 50.

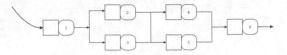

Fig. 4. Multi-path network

The QPF decomposition suggested by the topology is given by the DAG in which each station defines a node and the routing of the network coincides with the set of edges, i.e., $\mathcal{E} = \{(1,2),(1,3),(2,4),(2,5),(3,4),(3,5),(4,6),(5,6)\}$. It is easy to see that the corresponding set of necessary marginal distributions are composed of the probabilities $P\{Q_1 = x_1, Q_2 = x_2\}$, $P\{Q_1 = x_1, Q_3 = x_3\}$, $P\{Q_2 = x_2, Q_3 = x_3, Q_4 = x_4\}$, $P\{Q_2 = x_2, Q_3 = x_3, Q_5 = x_5\}$, $P\{Q_4 = x_4, Q_6 = x_6\}$ and $P\{Q_5 = x_5, Q_6 = x_6\}$. Since this way the calculations would require three dimensional marginals we apply a DAG with less edges. According to the arguments provided in Sec. 4.2, we neglect the dependencies on the less loaded queues which are in station 3 and 5. This way the edges of the DAG are $\mathcal{E} = \{(1,2),(1,3),(2,4),(2,5),(4,6)\}$. The necessary marginals with this DAG are all two dimensional: $P\{Q_1 = x_1, Q_2 = x_2\}$, $P\{Q_1 = x_1, Q_3 = x_3\}$, $P\{Q_2 = x_2, Q_4 = x_4\}$, $P\{Q_2 = x_2, Q_5 = x_5\}$ and $P\{Q_4 = x_4, Q_6 = x_6\}$.

We tested the model starting from two different initial states. We first consider the case in which initially all queues are empty. In the second case the system starts with 10 requests in stations 2, 3, 4, and 5. Expectations and variances of the number of jobs at stations 4, 5 and 6 for the two cases are depicted in Fig.s 5 and 6. By comparing the two figures, one can observe to what extent starting from the second initial state penalizes station 4. After about 10 time units the average number of clients at station 4 is about 14 while the same average never exceeds 3 starting from an empty system. The longer run effect can be seen instead looking at the variances: for station 4 this quantity is increasing up to about 100 time units and for station 5 as well it reaches much higher values than in case of starting from an empty network. All these behaviours are captured well by the QPF approximation. In Fig. 7 we depicted the probabilities of having no clients at station 4 and 6. On these curves as well one can observe the effect of the choice of the initial state.

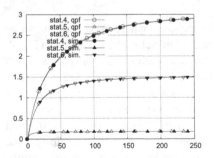
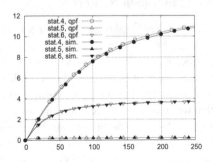

Fig. 5. Multi-path: expectations (left) and variances (right) of the stations 4, 5 and 6 vs. time starting from state $\{0,0,0,0,0,0\}$

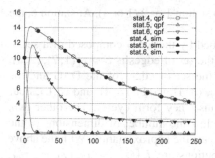
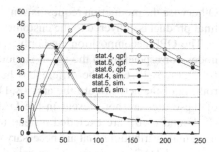

Fig. 6. Multi-path: expectations (left) and variances (right) of the stations 4, 5, and 6 vs. time starting from state $\{0, 10, 10, 10, 10, 0\}$

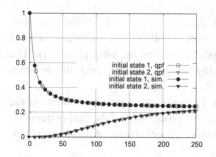
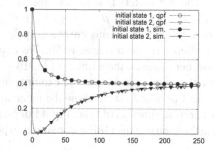

Fig. 7. Multi-path: Probability of empty queue in station 4 and 6 vs. time starting from the two different initial states

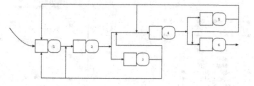

Fig. 8. On-demand production system

The network is composed of $51^6 = 1.76 \times 10^{10}$ states. The number of ODEs for the QPF approximation is $5 \times 51^2 = 13005$. The presented results were calculated in about one minute.

5.3 On-Demand Production System

As last example, we propose a network representing a production pipeline. The network is depicted in Fig. 8. Customer requests arrive at station 1 where they are processed and sent to the production stations. In the optimal case, each customer request requires only two production steps that are performed at station 2 and 4. The product undergoes also a quality check at station 2 and 4. When this check at station 2 (4) is not successful the product is sent to station 3 (5) where it is monitored. If the problem is a false positive or fixable, then the product is reintegrated at the next station of the production line.

Otherwise the production request is sent back to station 1 or 2 (1 or 4). At station 6, the final product is delivered to the customer.

The QPF decomposition we apply is described by the DAG composed of a set of nodes corresponding to the set of stations and the set of edges is $\mathcal{E} = \{(1,2), (2,3), (2,4), (4,5), (4,6)\}$. The differences between the set of edges of the DAG and the routing of the network are the arcs $(3,1)$, $(3,2)$, $(3,4)$, $(5,1)$, $(5,4)$, and $(5,6)$. Four of these arcs are not present in the DAG because they form cycles. The other two, $(3,4)$ and $(5,6)$, are excluded instead to keep low the number of dimensions of the marginal distributions that are necessary to compute the approximation. With the above described DAG the necessary marginal distributions are $P\{Q_1 = x_1, Q_2 = x_2\}$, $P\{Q_2 = x_2, Q_3 = x_3\}$, $P\{Q_2 = x_2, Q_4 = x_4\}$, $P\{Q_4 = x_4, Q_5 = x_5\}$ and $P\{Q_4 = x_4, Q_6 = x_6\}$.

The parameters that we use are $\lambda_1 = 0.4$, $r_{2,3} = r_{4,5} = 0.5$, $r_{3,1} = r_{5,1} = 0.2$, $r_{3,2} = r_{5,4} = 0.7$, $\mu_1 = \mu_2 = \mu_4 = \mu_6 = 1$, and $\mu_3 = \mu_5 = 1.5$. Note that with these parameters clients enter the cycles with high probability. The maximum number of clients for each queue is 50. We assume that the queues are empty initially. The number of states, the number of ODEs representing the QPF and the computation times are the same as in case of the example in Sec. 5.2.

In Fig. 9 and 10 we show the mean and the variance of the number of clients at the stations as function of time. Fig. 11 depicts instead the probability of the empty queue. It can be seen that the QPF approximation provides quite a precise view of these quantities.

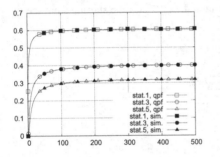
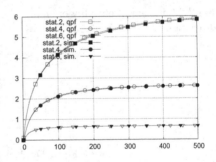

Fig. 9. On-demand production system: expectation of the number of jobs at the stations vs. time

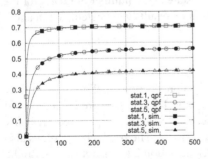
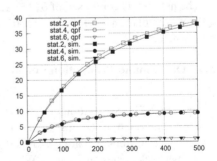

Fig. 10. On-demand production system: variance of the number of jobs at the stations vs. time

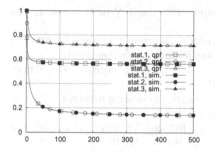 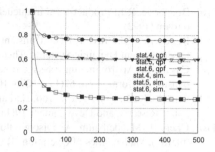

Fig. 11. On-demand production system: probability of empty queue at the stations vs. time

6 Conclusions

In this paper we applied an approximate transient analysis technique for networks of queues. The technique is based on the assumption that the transient probabilities can be expressed approximately in QPF. This assumption leads to a memory efficient algorithm to analyze networks even with huge state space. We discussed the properties of the algorithm and provided several numerical examples to illustrate its accuracy.

In the future we plan to develop solvers that are specific to the ODEs arising from the QPF approximation. In this paper we intentionally concentrated on a relatively simple class of networks. The technique can be applied to networks in which arrival streams and service processes are correlated in a Markovian manner. Moreover, time dependent arrival rates and service processes can also be incorporated into the ODEs that describe the approximation. We plan to implement the resulting method and apply it to a wide range of cases.

Acknowledgments. This work has been supported in part by project grant Nr. 10-15-1432/HICI from the King Abdulaziz University of Kingdom of Saudi Arabia and by project grant AMALFI (Advanced Methodologies for the AnaLysis and management of the Future Internet) financed by the Intesa Sanpaolo banking group.

References

[1] Angius, A., Horváth, A.: Product form approximation of transient probabilities in stochastic reaction networks. Electronic Notes on Theoretical Computer Science 277, 3–14 (2011)

[2] Angius, A., Horváth, A., Wolf, V.: Quasi product form approximation for markov models of reaction networks. In: Priami, C., Petre, I., de Vink, E. (eds.) Transactions on Computational Systems Biology XIV. LNCS, vol. 7625, pp. 26–52. Springer, Heidelberg (2012)

[3] Anselmi, J., Casale, G., Cremonesi, P.: Approximate solution of multiclass queuing networks with region constraints. In: Proc. of 15th Int. Symp. on Modeling, Analysis, and Simulation of Computer and Telecommunication Systems (MASCOTS 2007), pp. 225–230 (2007)

[4] Baskett, F., Chandy, K.M., Muntz, R.R., Palacios, G.: Open, closed, and mixed networks of queues with different classes of customers. Journal of the ACM 22(2), 248–260 (1975)

[5] Bazan, P., German, R.: Approximate transient analysis of large stochastic models with WinPEPSY-QNS. Computer Networks 53, 1289–1301 (2009)

[6] Boucherie, R.J.: Product-form in queueing networks. PhD thesis, Vrije Universiteit, Amsterdam (1992)

[7] Boucherie, R.J., Taylor, P.G.: Transient product form distributions in queueing networks. Discrete Event Dynamic Systems: Theory and Applications 3, 375–396 (1993)

[8] Casale, G.: Approximating passage time distributions in queueing models by Bayesian expansion. Perform. Eval. 67(11), 1076–1091 (2010)

[9] Chandy, K.M., Herzog, U., Woo, L.: Parametric analysis of queueing networks. IBM Journal of Research and Development 19(1), 36–42 (1975)

[10] Chen, H., Mandelbaum, A.: Discrete flow networks: Bottleneck analysis and fluid approximations. Mathematics of Operations Research 16(2), 408–446 (1991)

[11] Gordon, W.J., Newell, G.F.: Cyclic queueing networks with exponential servers. Operations Research 15(2), 254–265 (1967)

[12] Harrison, J.M., Lemoine, A.J.: A note on networks of infinite-server queues. J. Appl. Probab. 18(2), 561–567 (1981)

[13] Harrison, P.G.: Transient behaviour of queueing networks. Journal of Applied Probability 18(2), 482–490 (1981)

[14] Horváth, A., Horváth, G., Telek, M.: A traffic based decomposition of two-class queueing networks with priority service. Computer Networks 53, 1235–1248 (2009)

[15] Horváth, A., Horváth, G., Telek, M.: A joint moments based analysis of networks of MAP/MAP/1 queues. Performance Evaluation 67, 759–778 (2010)

[16] Jackson, J.R.: Jobshop-like queueing systems. Management Science 10(1), 131–142 (1963)

[17] Kuehn, P.: Approximate analysis of general queuing networks by decomposition. IEEE Transactions on Communications 27(1), 113–126 (1979)

[18] Massey, W.A., Whitt, W.: Networks of infinite-server queues with nonstationary Poisson input. Queueing Systems 13, 183–250 (1993)

[19] Matis, T.I., Feldman, R.M.: Transient analysis of state-dependent queueing networks via cumulant functions. Journal of Applied Probability 38(4), 841–859 (2001)

[20] Melamed, B.: Characterizations of Poisson traffic streams in jackson queueing networks. Advances in Applied Probability 11(2), 422–438 (1979)

[21] Sauer, C.H.: Approximate solution of queueing networks with simultaneous resource possession. IBM Journal of Research and Development 25(6), 894–903 (1981)

[22] Stewart, W.J.: Introduction to the Numerical Solution of Markov Chains. Princeton University Press (1995)

[23] Whitt, W.: The queueing network analyzer. Bell System Technical Journal 62(9), 2779–2815 (1983)

[24] Whitt, W.: Untold horrors of the waiting room. What the equilibrium distribution will never tell about the queue-length process. Management Science 29(4), 395–408 (1983)

[25] Whitt, W.: Decomposition approximations for time-dependent Markovian queueing networks. Operations Research Letters 24, 97–103 (1999)

Average Delay Estimation in Discrete-Time Systems with Periodically Varying Parameters

Evgeny Bakin[1], Sergey Andreev[2], Anna Solovyeva[1], and Andrey Turlikov[1]

[1] State University of Aerospace Instrumentation (SUAI), St. Petersburg, Russia
{jenyb,solovyeva,turlikov}@vu.spb.ru
[2] Tampere University of Technology (TUT), Tampere, Finland
sergey.andreev@tut.fi

Abstract. In this paper, we consider communication systems with periodically varying parameters, such as success probability when a user requests service. These systems may be described by a Markov chain of a cyclic structure. We concentrate on investigating the average delay value in such systems. In the general case, a direct solution is given with its complexity being linear of the number of Markov chain states. For the case when the parameters of a system are varying slowly, a fast algorithm of the average delay calculation is proposed. Some applications of our results to contemporary communication systems are demonstrated.

Keywords: Communication systems, Markov chains, average delay, varying parameters.

1 Introduction

According to many predictions, the proportion of traffic transmitted over broadband communication networks is expected to grow considerably in the very near future [1]. Consequently, the currently deployed wireless technologies are very likely to face serious overloads resulting in a dramatic degradation in the levels of quality of experience for their users. Therefore, it is increasingly important to understand what we can do in order to to better adjust to the varying user demand. One decisive step in this direction is the recent work in [2], where the authors explicitly address the changes in the mobile user behavior to study the dynamic load of a wireless cellular network. In particular, [2] argues that user behavior is tightly coupled with the external environment.

This important discussion on how the increasing demand from mobile users should be accounted for to improve wireless resource allocation has been started as long as in [3]. There, while studying some dynamic population models, it has been established that the load on a wireless network is often *periodic*. This is rooted in the way the user population interacts with various attractors. More recently, [4] deepened our knowledge on periodically varying load of contemporary cellular wireless networks. Motivated by these works, we believe that by specifically accounting for the periodic nature of communication environment it

A. Dudin and K. De Turck (Eds.): ASMTA 2013, LNCS 7984, pp. 37–51, 2013.

is possible to reduce the complexity associated with performance evaluation of modern and future wireless technologies.

In this paper, we investigate a model of a communication system where some parameters may vary periodically. As mentioned previously, one important example is a cellular wireless network under periodic user load. Another application area is periodic channel variation, which is not limited to cellular networks, but is also applicable for mobile ad hoc networks, and wireless sensor networks. Accounting for channel variation may be important to further improve channel-aware resource allocation strategies [5].

Interestingly, some real-world communication channels inherently exhibit periodic behavior, such as the one in [6], which may be explicitly used to design specific access control procedures [7]. In sensor networks, wireless channel quality variation may also be periodic due to the specific properties of the environment [8]. This is particularly true in several specialized applications, such as given by [9], where periodic variation in the signal strength is due to rotating radio transmitters. Finally, cognitive radio networks also contribute to the set of potential applications for our model where the occupancy of a channel is sometimes found to be completely periodic [10].

Not limited to periodic load and channel variations, our approach is also suitable in the cases when the channel access protocol itself takes advantage of the periodic structure [11], [12]. Therefore, a variety of distributed election algorithms can also fall into the scope of our model due to periodic events in the environment [13]. In conclusion, we believe that our delay estimation scheme for the systems with periodically varying parameters will be useful in a wide range of important practical applications and across a rich set of wireless communication technologies.

The remainder of this paper is organized as follows. In Section 2, we outline the considered system and the corresponding Markov chain based model. In Section 3, a direct approach to the average delay calculation is detailed. Section 4 introduces an improved method of the average delay calculation suitable for communication systems with slowly varying parameters. The proposed approach results in significantly lower computational complexity. In Section 5, the stability of our solution is assessed. Sections 6 and 7 demonstrate some areas where our approach may be applied and present the respective numerical results. Finally, conclusions are given in Section 8.

2   Model Description

We consider a discrete-time communication system where users request some resource before initiating their data transmission. System time is divided into equal slots. In a particular slot, a user may send its request into the system. Depending on the current system state, this request may either be accepted or rejected. If the request is rejected (e.g., when the system is overloaded), the user retransmits its request in the next slot. In practical systems, the probability of successful request transmission is typically varying with time (for instance, due to daily activity of the system users). In what follows, we assume this probability to vary periodically with the fixed period of N slots.

Therefore, such system may be described by a Markov chain consisting of $N+1$ states. In particular, N of these states correspond to different slots in a period. The last $(N+1)$-th state is an absorbing state (trap) which corresponds to the moment of request acceptance by the system. Each of N non-absorbing states is characterized by two transitions: (i) request acceptance and (ii) request rejection. The probabilities of these transitions in the i-th state are denoted as a_i and r_i respectively $(a_i + r_i = 1, i = \overline{0, N-1})$. The considered Markov chain is illustrated in Figure 1.

For simplicity of further derivations, the indexes of a_i and r_i are considered to be $i \mod N$ by default, i.e. $a_{i+kN} = a_i$ and $r_{i+kN} = r_i$ for $k \in Z$.

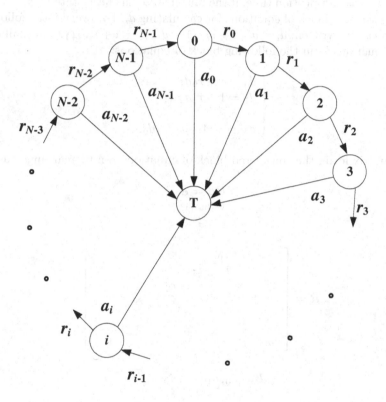

Fig. 1. Structure of the considered Markov chain

The absorption time in the described Markov chain corresponds to the request transmission delay. Generally, it depends on the initial system state at the moment of request generation by the user. In the next section, we detail a direct approach to the average delay evaluation in the considered system.

3   Direct Solution

Evaluation of the average absorption time of a Markov chain is a known problem considered by e.g. [14]. Accordingly, this evaluation may be achieved with the following equation:

$$\bar{d} = \sum_{i=0}^{N-1} s_i d_i. \tag{1}$$

Here, s_i denotes the probability that a request arrived in the i-th slot $\left(\sum_{i=0}^{N-1} s_i = 1 \right)$, d_i is the average absorption time, if the initial state has been state i.

Let us set up a block of equations for calculating d_i. For every i, the following equation can be written: $d_i = a_i \cdot 1 + r_i \cdot (1 + d_{i+1}) = 1 + r_i d_{i+1}$. Thus, all of N similar equations form the following block of equations:

$$\begin{cases} d_0 = 1 + r_0 d_1 \\ d_1 = 1 + r_1 d_2 \\ \dots \\ d_{N-1} = 1 + r_{N-1} d_0. \end{cases} \tag{2}$$

In the matrix form, the considered block of equations can be rearranged as:

$$\boldsymbol{R}\boldsymbol{d} = \boldsymbol{e}. \tag{3}$$

Here:

$$\boldsymbol{R} = \begin{bmatrix} 1 & -r_0 & & & & \\ & 1 & -r_1 & & & \\ & & 1 & -r_2 & & \\ & & & \ddots & & \\ & & & & 1 & -r_{N-2} \\ -r_{N-1} & & & & & 1 \end{bmatrix} \tag{4}$$

$$\boldsymbol{d} = [d_0, d_1, \dots, d_{N-1}]^T \tag{5}$$

$$\boldsymbol{e} = [1, 1, \dots, 1]^T \tag{6}$$

Hence, the sought solution is $\boldsymbol{d} = \boldsymbol{R}^{-1}\boldsymbol{e}$. The specific structure of $\boldsymbol{R}$ allows finding the solution to (2) with the complexity of $O(N)$, i.e. linear of N. The corresponding recursive algorithm is given in Appendix A.

Unfortunately, in the case when N is large, the direct solution of block (2) may still be prohibitive even if the proposed recursive algorithm is used. However, as we demonstrate below for the case when the rate of system state variation is relatively low, the solution can be significantly simplified.

4   Particular Solution for Slowly Varying Parameters

If parameter variation rate is relatively low, the period in the system can be further split into several sub-periods during which the success probability can be regarded as constant. Note that durations of sub-periods may generally be different. Let us denote the number of sub-periods as K. In i-th sub-period, the number of slots, the probabilities of request acceptance and rejection are denoted as N_i, p_i, and q_i respectively ($i = \overline{0, K-1}$, $\sum_{i=0}^{K-1} N_i = N$). Figure 2 illustrates the considered case.

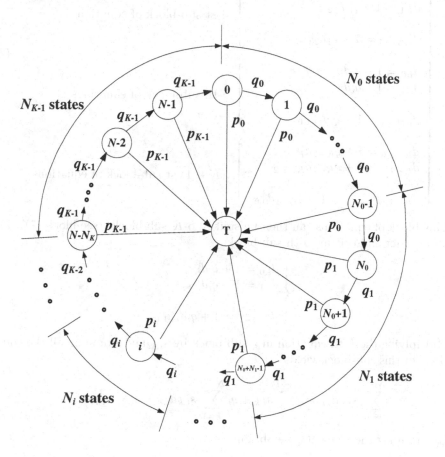

Fig. 2. Markov chain for the system with slowly varying parameters

In order to describe our simplified solution, let us also introduce the following extra notation: $d_{i,j}$ is the average absorption time, if the initial state is state j in the i-th sub-period, $s_{i,j}$ is the probability that a request arrived in j-th slot of the i-th sub-period ($i = \overline{0, K-1}$, $j = \overline{0, N_i - 1}$).

Using this notation and accounting for the fact that the reject probability is constant during a sub-period, the block of equations (2) can be rearranged as:

$$
\begin{cases}
\left.\begin{array}{l}
d_{0,0} = 1 + q_0 d_{0,1} \\
d_{0,1} = 1 + q_0 d_{0,2} \\
\ldots \\
d_{0,N_0-1} = 1 + q_0 d_{1,0}
\end{array}\right] \text{0-th sub-block of equations} \\[4pt]
\left.\begin{array}{l}
d_{1,0} = 1 + q_1 d_{1,1} \\
d_{1,1} = 1 + q_1 d_{1,2} \\
\ldots \\
d_{1,N_1-1} = 1 + q_1 d_{2,0}
\end{array}\right] \text{1-st sub-block of equations} \\[4pt]
\ldots \\
\left.\begin{array}{l}
d_{i,0} = 1 + q_i d_{i,1} \\
d_{i,1} = 1 + q_i d_{i,2} \\
\ldots \\
d_{i,N_i-1} = 1 + q_i d_{i+1,0}
\end{array}\right] \text{i-th sub-block of equations} \\[4pt]
\ldots \\
\left.\begin{array}{l}
d_{K-1,0} = 1 + q_{K-1} d_{K-1,1} \\
d_{K-1,1} = 1 + q_{K-1} d_{K-1,2} \\
\ldots \\
d_{K-1,N_{K-1}-1} = 1 + q_{K-1} d_{0,0}
\end{array}\right] \text{$(K-1)$-st sub-block of equations}
\end{cases}
\tag{7}
$$

This block of equations can then be split into K sub-blocks (see block (7)). Let us consider an arbitrary i-th sub-block:

$$
\begin{cases}
d_{i,0} = 1 + q_i d_{i,1} \\
d_{i,1} = 1 + q_i d_{i,2} \\
\ldots \\
d_{i,N_i-1} = 1 + q_i d_{i+1,0}
\end{cases}
$$

Multiplying the k-th equation in a sub-block by $s_{i,k}$ and summing all the equations in this sub-block yields:

$$
\sum_{k=0}^{N_i-1} s_{i,k} d_{i,k} = \sum_{k=0}^{N_i-1} s_{i,k} + q_i \sum_{k=1}^{N_i-1} s_{i,k} d_{i,k} + s_{i,N_i-1} q_i d_{i+1,0}.
$$

By grouping the elements, we obtain:

$$
(1 - q_i) \sum_{k=0}^{N_i-1} s_{i,k} d_{i,k} = \sum_{k=0}^{N_i-1} s_{i,k} - s_{i,0} q_i d_{i,0} + s_{i,N_i-1} q_i d_{i+1,0}.
$$

Considering that $1 - q_i = p_i$ and dividing both sides of the above equation by p_i, we derive:

$$
\sum_{k=0}^{N_i-1} s_{i,k} d_{i,k} = \frac{1}{p_i} \sum_{k=0}^{N_i-1} s_{i,k} - \frac{1}{p_i} s_{i,0} q_i d_{i,0} + \frac{1}{p_i} s_{i,N_i-1} q_i d_{i+1,0}.
$$

Summing over all K sub-blocks results in:

$$\sum_{i=0}^{K-1}\sum_{k=0}^{N_i-1} s_{i,k}d_{i,k} = \sum_{i=0}^{K-1}\left(\frac{1}{p_i}\sum_{k=1}^{N_i-1} s_{i,k}\right) - \sum_{i=0}^{K-1}\left(\frac{1}{p_i}s_{i,0}q_id_{i,0}\right) + \sum_{i=0}^{K-1}\left(\frac{1}{p_i}s_{i,N_i-1}q_id_{i+1,0}\right).$$

It can be easily seen that the expression on the left side of the equation is equal to $\bar{d}$ (see (1)):

$$\bar{d} = \sum_{i=0}^{K-1}\left(\frac{1}{p_i}\sum_{k=1}^{N_i-1} s_{i,k}\right) - \sum_{i=0}^{K-1}\left(\frac{1}{p_i}s_{i,0}q_id_{i,0}\right) + \sum_{i=0}^{K-1}\left(\frac{1}{p_i}s_{i,N_i-1}q_id_{i+1,0}\right). \quad (8)$$

In the particular case when $s_{i,k}$ is constant during a sub-period and equals to S_i ($i = \overline{0,K-1}$, $k = \overline{0,N_i-1}$), we obtain:

$$\bar{d} = \sum_{i=0}^{K-1}\frac{N_iS_i}{p_i} - \sum_{i=0}^{K-1}\left(\frac{1}{p_i}S_iq_id_{i,0}\right) + \sum_{i=0}^{K-1}\left(\frac{1}{p_i}S_iq_id_{i+1,0}\right).$$

Grouping the elements on the right side around the terms $d_{i,0}$ gives us:

$$\bar{d} = \sum_{i=0}^{K-1}\frac{N_iS_i}{p_i} + \sum_{i=0}^{K-1}d_{i,0}\left(\frac{S_{i-1}q_{i-1}}{p_{i-1}} - \frac{S_iq_i}{p_i}\right). \quad (9)$$

In another particular case, when the initial state is uniformly distributed over the period ($S_0 = S_1 = \cdots = S_{K-1} = 1/N$), we have:

$$\bar{d} = \frac{1}{N}\sum_{i=0}^{K-1}\frac{N_i}{p_i} + \frac{1}{N}\sum_{i=0}^{K-1}d_{i,0}\left(\frac{1}{p_{i-1}} - \frac{1}{p_i}\right). \quad (10)$$

Equation (10) indicates that $\bar{d}$ can be compactly expressed with only $d_{i,0}$ ($i = \overline{0,K-1}$). Let us set a block of equations for finding $d_{i,0}$. It can be easily shown that $d_{i,0} = 1\cdot p_i + 2\cdot q_ip_i + 3\cdot q_i^2p_i + \cdots + N_i\cdot q_i^{N_i-1}p_i + (N_i + d_{i+1,0})\cdot q_i^{N_i}$. Grouping the elements, we establish that $d_{i,0} = \left[\frac{1-q_i^{N_i}}{p_i} - N_iq_i^{N_i}\right] + (N_i + d_{i+1,0})\cdot q_i^{N_i}$.

Finally, $d_{i,0} = \frac{1-q_i^{N_i}}{p_i} + q_i^{N_i}d_{i+1,0}$. The block of equations can thus be rewritten as follows:

$$\begin{cases} d_{0,0} = \frac{1-q_0^{N_0}}{p_0} + q_0^{N_0}d_{1,0} \\[2mm] d_{1,0} = \frac{1-q_1^{N_1}}{p_1} + q_1^{N_1}d_{2,0} \\[2mm] \cdots \\[2mm] d_{i,0} = \frac{1-q_i^{N_i}}{p_i} + q_i^{N_i}d_{i+1,0} \\[2mm] \cdots \\[2mm] d_{K-1,0} = \frac{1-q_{K-1}^{N_{K-1}}}{p_{K-1}} + q_{K-1}^{N_{K-1}}d_{0,0} \end{cases} \quad (11)$$

Note that the block of equations (11) has the same structure as the block (2). Hence, using the same solving algorithm allows estimating $\bar{d}$ with the complexity $O(K)$, which can be significantly less than $O(N)$.

5   Solution Adequacy Analysis

The dimension of the block (2) can be rather large. Consequently, this may lead to numerical instability of our solution in the case when the probabilities r_i in the block (2) are known with errors. Hence, instead of the correct matrix $\boldsymbol{R}$, another matrix $\hat{\boldsymbol{R}}$ is accounted for. This may cause incorrectness of the established solution $\hat{\boldsymbol{d}}$ instead of the appropriate $\boldsymbol{d}$. Let us denote the errors in the matrix as $\Delta\boldsymbol{R} = \hat{\boldsymbol{R}} - \boldsymbol{R}$ and the errors in the solution as $\Delta\boldsymbol{d} = \hat{\boldsymbol{d}} - \boldsymbol{d}$.

It is known that $\frac{\|\Delta\boldsymbol{d}\|}{\|\boldsymbol{d}\|} \leq c_R \frac{\|\Delta\boldsymbol{R}\|}{\|\boldsymbol{R}\|}$. Here, c_R denotes the condition number of the matrix $\boldsymbol{R}$, where $\|\cdot\|$ denotes the norm of the matrix or vector (in this work, the infinity norm is used) [15]. As one can see, the relative error in the block solution is c_R times higher than the relative error in the block coefficients. Let us now estimate the adequacy of the solution by means of analyzing the condition number of the matrix $\boldsymbol{R}$.

By definition, the condition number $c_R = \|\boldsymbol{R}\| \, \|\boldsymbol{R}^{-1}\|$. First, let us calculate the inverse matrix $\boldsymbol{R}^{-1}$.

Theorem 1. The elements of the inverse matrix may be obtained by the following expression:

$$\left(\boldsymbol{R}^{-1}\right)_{i,j} = \frac{1}{1 - \prod_{k=0}^{N-1} r_k} \prod_{k=0}^{j-i-1} r_{i+k}. \tag{12}$$

We remind that the index of r_i is considered to be $i \mod N$, i.e. $r_{i+kN} = r_i$ for $k \in Z$. The general structure of the matrix $\boldsymbol{R}^{-1}$ is as follows:

$$\boldsymbol{R}^{-1} = \frac{1}{1 - \prod_{k=0}^{N-1} r_k} \times \tag{13}$$

$$\times \begin{bmatrix}
1 & r_0 & r_0 r_1 & r_0 r_1 r_2 & \cdots & \prod_{i=0}^{N-2} r_i \\[2ex]
r_1 r_2 \ldots r_{N-1} & 1 & r_1 & r_1 r_2 & \cdots & \prod_{i=1}^{N-2} r_i \\[2ex]
r_2 r_3 \ldots r_{N-1} & r_2 r_3 \ldots r_{N-1} r_0 & 1 & r_2 & \cdots & \prod_{i=2}^{N-2} r_i \\[2ex]
r_3 r_4 \ldots r_{N-1} & r_3 r_4 \ldots r_{N-1} r_0 & \prod_{i=3}^{N-1} r_i \cdot r_0 r_1 & 1 & \cdots & \prod_{i=3}^{N-2} r_i \\[2ex]
\vdots & \vdots & \vdots & \vdots & \ddots & \vdots \\[2ex]
\prod_{i=k-1}^{N-1} r_i & \prod_{i=k-1}^{N-1} r_i \cdot r_0 & \prod_{i=k-1}^{N-1} r_i \cdot r_0 r_1 & \prod_{i=k-1}^{N-1} r_i \cdot r_0 r_1 r_2 & \cdots & \prod_{i=k-1}^{N-2} r_i \\[2ex]
\vdots & \vdots & \vdots & \vdots & \ddots & \vdots \\[2ex]
r_{N-1} & r_{N-1} r_0 & r_{N-1} r_0 r_1 & r_{N-1} r_0 r_1 r_2 & \cdots & 1
\end{bmatrix}$$

The proof of Theorem 1 is given in Appendix B.

Let us denote the maximum rejection probability as $r_{max} = \max(r_0, r_1, \ldots, r_{N-1})$ and the average geometric value of r as $r_{geo} = \sqrt[N]{r_0 r_1 \ldots r_{N-1}}$. Then, the following theorem can be formulated.

Theorem 2. The condition number c_R of the matrix $\boldsymbol{R}$ satisfies the following inequality:

$$c_R \leq \frac{1 + r_{max}}{1 - r_{max}} \frac{1 - r_{max}^N}{1 - r_{geo}^N}. \tag{14}$$

Proof. By definition, the infinity norm of any matrix $\boldsymbol{A}$ is the maximum sum of the absolute values of elements in a row, i.e. $\|\boldsymbol{A}\| = \max_{i=\overline{1,N}} \sum_{j=1}^{N} |A_{i,j}|$. Then, $\|\boldsymbol{R}\| = 1 + r_{max}$ (see (4)).

Analyzing the matrix (13), we notice that for $\forall i$,

$$\sum_{j=1}^{N} \left(\boldsymbol{R}^{-1}\right)_{i,j} \leq \frac{1}{1 - \prod_{k=0}^{N-1} r_k} \left(1 + r_{max} + r_{max}^2 + \ldots + r_{max}^{N-1}\right). \tag{15}$$

Therefore, $\|\boldsymbol{R}^{-1}\| \leq \frac{1}{1 - \prod_{k=0}^{N-1} r_k} \frac{1 - r_{max}^N}{1 - r_{max}}$.

Concluding, $c_R = \|\boldsymbol{R}\| \|\boldsymbol{R}^{-1}\| \leq \frac{1 + r_{max}}{1 - r_{max}} \frac{1 - r_{max}^N}{1 - \prod_{k=0}^{N-1} r_k}$. Remembering that $\prod_{k=0}^{N-1} r_k = r_{geo}^N$, we obtain the inequality (14), which completes the proof.

The proposed estimation of c_R allows accounting for the effect of errors caused by either (i) partial a priori uncertainty or (ii) rounding of the values of r_i according to the model described in Section 2.

6 Application of the Proposed Solution

6.1 Communication System with Delivery Acknowledgments

We continue by presenting several potential applications of our solution. First, we consider communication systems where every data packet transmission is followed by a respective acknowledgment. If the packet is received successfully, then a positive acknowledgment (ACK) is forwarded to the sender and the following data packet may be attempted. Otherwise, a negative acknowledgement (NAK) is generated and the packet transmission has to be repeated (see Figure 3 and details in [16]).

In the considered system, the probability of successful packet delivery depends on the quality of the underlying communication channel expressed as its bit-error rate (BER). We may identify several practical systems where the value of BER varies periodically, including:

1. Microwave (radiorelay) communication lines where the channel pathloss depends on daytime [17], [18].

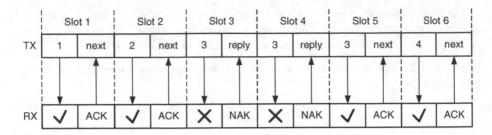

Fig. 3. Communication system with delivery acknowledgments

2. Communication systems with a scanning antenna pattern. Here, BER depends on the current transmitter antenna gain in the direction of receiver [19].

The above analysis as per (9) and (10) can be applied to calculate the average delay of data packet delivery. In particular, the inequality (14) allows accounting for inaccuracies in the estimation of BER.

6.2   Slotted Decentralized Systems

Second, we address slotted distributed networks as another example application. There are many decentralized algorithms based on time-division multiple access to arbitrate medium access across a number of channel users. Below we concentrate on one of them, named Node Activation Multiple Access (NAMA) [11], [20].

Assume that the communication system comprises u_{max} users. Every once in a while each of the users connects to the network and then leaves it. When the user connects, it is assigned a unique identification number (ID). We denote the number of users connecting to the network in the i-th slot as u_i. In every slot, each user generates a random value $prio_k$, named its priority, using the following rule:

$$prio_k = \text{Hash}(k|i)|k, \tag{16}$$

where i is the slot number and k is the user ID.

Here, *Hash* is a function that generates uniformly distributed pseudo-random numbers with a given seed. The notation "|" refers to the concatenation of two arguments. Note that the *Hash*-function may return identical values for more than one input argument. However, the concatenation with a unique user ID ensures that all the priorities are different. The user with the highest priority is assumed to win the competition over the current slot and obtains an opportunity to transmit its data packet there. An example of NAMA operation is demonstrated in Figure 4.

In the figure, the communication system is interpreted as a graph. The users are denoted with letters (a to g). The priorities of users in the current slot are shown next to the users. In this example, we consider four consecutive time slots. The winner of the competition over the current slot (a user which transmits its data in this slot) is highlighted in gray.

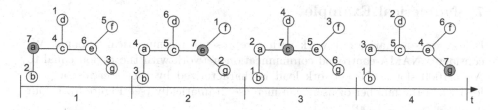

Fig. 4. Algorithm NAMA example operation

According to the algorithm NAMA, the probability of winning the competition depends only on the number of connected users and equals to $p_i = \frac{1}{u_i}$. In real-world systems, the number of users typically varies in time due to their daily activity. Therefore, the proposed approach to delay analysis can be followed again.

We denote the average number of connected users in the system as $\bar{u}$. Hence, the equation (10) transforms into the following:

$$\bar{d} = \bar{u} + \frac{1}{N} \sum_{i=0}^{K-1} d_{i,0} \Delta u_i. \tag{17}$$

Here, $\Delta u_i = u_{i-1} - u_i$ is a difference in the number of users between the adjacent slots. The given equation expresses the average delay for the hypothetical user which chooses slot for data transmission uniformly over the period.

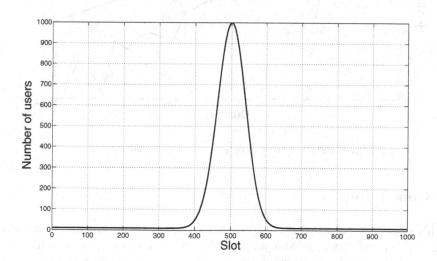

Fig. 5. Variation in the number of connected users

7   Numerical Example

Below, we summarize our work with a representative numerical example. We consider a NAMA-controlled communication network with the period equal to $N = 1000$ slots. The network load is characterized by the intervals of peak load when the number of users can increase dramatically (see Figure 5). In our example, $u_{max} = 1000$.

In these conditions, we calculate the average delay according to (17). Further, we introduce an error in the collected data by quantizing the request rejection probability with a given quantum size. The curves corresponding to the true relative error and the relative error estimation with the use of (14) are given in Figure 6 depending on the size of the quantum. As one can see from the example the proposed bound is quite tight and can be used for preliminary estimation of negative effect connected with possible inaccuracy in system parameters definition. Also it allows choosing an appropriate quantum size depending on preassigned level of relative error.

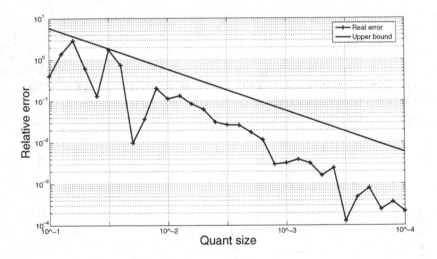

Fig. 6. Relative error dependency on the quantum size

8   Conclusion

In this paper, we proposed two approaches to estimate the average delay associated with request transmission and suitable for a wide class of communication systems with periodically varying parameters. We specifically focused on systems with slowly varying parameters. We also estimated the accuracy of our solution in the situations when the initial system parameters are known with some error. A comprehensive numerical example confirms the precision of the proposed estimation.

References

1. Cisco Visual Networking Index: Global Mobile Data Traffic Forecast Update 2011-2016 (February 2012)
2. Dudin, A., Osipov, E., Dudin, S., Schelén, O.: Socio-behavioral scheduling of time-frequency resources for modern mobile operators. In: Dudin, A., Klimenok, V., Tsarenkov, G., Dudin, S. (eds.) BWWQT 2013. CCIS, vol. 356, pp. 69–82. Springer, Heidelberg (2013)
3. Bermudez, V., Rodriguez, D., Molina, C., Basu, K.: Adaptability theory modeling of time variant subscriber distribution in cellular systems. In: Proc. of the IEEE Vehicular Technology Conference, vol. 3, pp. 1779–1784 (1999)
4. Liu, F., Xiang, W., Zhang, Y., Zheng, K., Zhao, H.: A novel QoE-based carrier scheduling scheme in LTE-Advanced networks with multi-service. In: Proc. of the IEEE Vehicular Technology Conference (2012)
5. Liu, Q., Wang, X., Giannakis, G.: A cross-layer scheduling algorithm with QoS support in wireless networks. IEEE Transactions on Vehicular Technology 55(3), 839–847 (2006)
6. Corripio, F., Arrabal, J., del Rio, L., Munoz, J.: Analysis of the cyclic short-term variation of indoor power line channels. IEEE Journal on Selected Areas in Communications 24(7), 1327–1338 (2006)
7. Yonge, L., Katar, S., Kostoff, S., Earnshaw, W., Blanchard, B., Afkhamie, H., Mashburn, H.: Channel adaptation synchronized to periodically varying channel (2005)
8. Lal, D., Manjeshwar, A., Herrmann, F., Uysal-Biyikoglu, E., Keshavarzian, A.: Measurement and characterization of link quality metrics in energy constrained wireless sensor networks. In: Proc. of the IEEE Global Telecommunications Conference, vol. 1, pp. 446–452 (2003)
9. Wang, K.-C., Jacob, J., Tang, L., Huang, Y., Gu, F.: Error pattern analysis for data transmission of wireless sensors on rotating industrial structures. In: Proc. of the Sensors and Smart Structures Technologies for Civil, Mechanical, and Aerospace Systems, vol. 6932 (2008)
10. Harrold, T., Cepeda, R., Beach, M.: Long-term measurements of spectrum occupancy characteristics. In: Proc. of the IEEE Symposium on New Frontiers in Dynamic Spectrum Access Networks, pp. 83–89 (2011)
11. Bao, L., Garcia-Luna-Aceves, J.: A new approach to channel access scheduling for Ad Hoc networks. In: Proc. of the 7th Annual International Conference on Mobile Computing and Networking, pp. 210–221 (2001)
12. Bao, L., Garcia-Luna-Aceves, J.: Distributed dynamic channel access scheduling for ad hoc networks. Journal of Parallel and Distributed Computing 63(1), 3–14 (2003)
13. Rajendran, V., Obraczka, K., Garcia-Luna-Aceves, J.: Energy-efficient collision-free medium access control for wireless sensor networks. Wireless Networks 12(1), 63–78 (2006)
14. Feller, W.: An Introduction to Probability Theory and Its Applications, 3rd edn., vol. 1. John Wiley and Sons, London (1968)
15. Kaw, A., Kalu, E.: Numerical Methods with Applications: Abridged. Lulu.com (2011)
16. Sklar, B.: Digital Communications: Fundamentals and Applications, 2nd edn. Prentice Hall (2001)

17. Sirkova, I.: Path loss calculation for a surface duct statistics. In: Proc. of the International Conference on Environmental Science and Technology, pp. 53–56 (2009)
18. Kerans, A., Woods, G., Lensson, E., French, G.: Propagation at 10.6 GHz over a long path in the tropical evaporation duct. In: Proc. of the Workshop on the Applications of Radio Science (2002)
19. Skolnik, M.: Introduction to Radar Systems, 3rd edn. McGraw-Hill Science (2002)
20. Park, S., Sy, D.: Dynamic control slot scheduling algorithms for TDMA based mobile Ad Hoc networks. In: Proc. of the IEEE Military Communications Conference (2008)

Appendix A: Recursive Algorithm to Solve the Block of Equations

In the block of equations (2), by substituting the second equation into the first one, we obtain $d_0 = 1 + r_0 (1 + r_1 d_2)$. The substitution of the third equation into this equation gives us $d_0 = 1 + r_0 (1 + r_1 (1 + r_2 d_3))$, etc. Finally, $d_0 = 1 + r_0 (1 + r_1 (1 + ...r_{N-2} (1 + r_{N-1} d_0)))$. Hence,

$$d_0 = \frac{1 + r_0 (1 + r_1 (1 + ...r_{N-3} (1 + r_{N-2})))}{1 - \prod_{i=0}^{N-1} r_i}.$$

Therefore, $N - 1$ backward substitutions let us establish all other roots of the block. For the proposed solution, $3N - 3$ multiplications are required ($2N - 2$ for d_0 calculation and $N - 1$ for backward substitutions). Thus, the computational complexity of the proposed algorithm is $O(N)$, i.e. linear of N.

Appendix B: Proof of Theorem 1

For finding $\boldsymbol{R}^{-1}$, the following known equality can be used:

$$\boldsymbol{R}^{-1} = \sum_{i=0}^{\infty} (I_N - \boldsymbol{R})^i.$$

Here, I_N denotes $[N \times N]$ identity matrix. Let us denote $\tilde{\boldsymbol{R}} = I_N - \boldsymbol{R}$.

$$\tilde{\boldsymbol{R}} = \begin{bmatrix} & r_0 & & & & \\ & & r_1 & & & \\ & & & r_2 & & \\ & & & & \ddots & \\ & & & & & r_{N-2} \\ r_{N-1} & & & & & \end{bmatrix}$$

Using this notation,

$$\boldsymbol{R}^{-1} = \sum_{i=0}^{\infty} \tilde{\boldsymbol{R}}^i.$$

The expression on the right side of the above equation can be decomposed as:

$$\sum_{i=0}^{\infty} \tilde{R}^i = \sum_{i=0}^{N-1} \tilde{R}^i \sum_{i=0}^{\infty} \left(\tilde{R}^N\right)^i.$$

For further processing of the above equation, the following definition and lemma are introduced.

Definition. The t-shift diagonal matrix with the initial vector x is a matrix that can be generated from the diagonal matrix $diag(x)$ by a cyclic shift of its columns on t positions. Let us denote such matrix as $diag(x, t)$. For example, using this notation, $\tilde{R} = diag(r, 1)$, where $r = [r_0, r_1, \cdots, r_{N-1}]$.

Lemma. The multiplication of $[M \times M]$ t-shift diagonal matrix $diag(x, t)$ on $[M \times M]$ 1-shift diagonal matrix $diag(y, 1)$ gives $(t + 1)$-shift diagonal matrix $diag(z, t+1)$, where $z = [x_0 y_t, x_1 y_{t+1}, \cdots, x_{M-t-1} y_{M-1}, x_{M-t} y_0, \cdots, x_{M-1} y_{t-1}]$ is an element-wise multiplication of vector x with cyclically shifted vector y on t positions.

Corollary 1. The matrix $\hat{R}^N = \prod_{i=0}^{N-1} r_i \cdot I_N$.

Corollary 2. For $k < N$, the matrix $\tilde{R}^k = diag(g, k)$, where the i-th element of the vector g is equal to $\prod_{j=0}^{k-1} r_{i+j}$. We remind that the index of r_i is considered to be $i \mod N$, i.e. $r_{i+kN} = r_i$ for $k \in Z$.

Using Corollary 1, we obtain:

$$\sum_{i=0}^{\infty} \left(\tilde{R}^N\right)^i = \left[1 + \prod_{i=0}^{N-1} r_i + \left(\prod_{i=0}^{N-1} r_i\right)^2 + \left(\prod_{i=0}^{N-1} r_i\right)^3 + \ldots\right] I_N = \frac{1}{1 - \prod_{i=0}^{N-1} r_i} I_N.$$

Using Corollary 2, we can see that $\sum_{i=0}^{N-1} \tilde{R}^i$ forms a square $[N \times N]$ matrix which element (i, j) is equal to $\prod_{k=0}^{j-i-1} r_{i+k}$.

Combining the two above expressions completes the proof.

Modelling Retrial-Upon-Conflict Systems with Product-Form Stochastic Petri Nets

Simonetta Balsamo, Gian-Luca Dei Rossi, and Andrea Marin

Università Ca' Foscari di Venezia
Dipartimento di Scienze Ambientali, Informatica e Statistica
via Torino 155, Venezia
{balsamo,deirossi,marin}@dais.unive.it

Abstract. In this paper we consider a particular class of stochastic Petri nets that admits a product-form stationary distribution under general conditions (independent exponentially distributed firing time). We show that the structure of these stochastic Petri nets is appropriate for modelling systems in which several components compete for a resource and conflicts may arise during the contention that require to newly perform the operation after a recovery time. Examples of these systems are wireless stations competing for a channel or processes operating on the same database concurrently. We derive the performance indices as functions of the model parameters, i.e., firing rates, probabilities of conflicts, number of competing components.

1 Introduction

Petri nets (PNs) and their timed extension Stochastic Petri nets (SPNs) [18,17] are widely used to model concurrent systems in which fork and join synchronisations can occur. Informally, Petri nets are bipartite graphs consisting of places and transitions. Arcs connect places with transitions or vice versa and they are associated with a natural number that represent the weight. When all the arcs have weight 1 we say that the Petri nets is ordinary. A marking associates a natural number with each place and represents the state of the net. The transitions determine the dynamic behaviour of the net according to the semantic of firing which is formally defined in Section 2. The problem of reachability, i.e., deciding if given the initial marking another marking is reachable, is known to belong to the class of EXPSPACE.

In a SPN each transition fires after an exponentially distributed firing time which is independent of the firing times of all the other transitions. This implies that the underlying stochastic process is a Continuous Time Markov Chain (CTMC) and the chain's state space corresponds to the state space of the corresponding PN. Once the CTMC is derived, the performance indices can be computed using standard algorithms. However, it can be shown that the cardinality of the model's state space can grow more than exponentially with the structure of the model, i.e., the number of places and transitions and hence the whole generation is time and space

A. Dudin and K. De Turck (Eds.): ASMTA 2013, LNCS 7984, pp. 52–66, 2013.

consuming. Even worse, the numerical algorithms for deriving the steady-state performance indices become numerically unstable and prohibitive in terms of computation time. In order to overcome this problem, known as *state space explosion*, a range of techniques has been proposed, from state space reduction through aggregation (lumping) to approximation techniques.

Product-form theory takes another approach: applying a *divide et impera* paradigm, SPNs can be efficiently solved through the analysis of their components (i.e., the places) in isolation. Product-form SPNs were first introduced in [12], and then generalised in [3,11,5]. More recently, new theoretical developments have been proposed connected to system biology [14]. We base our paper on the methodology and results developed in [1,15], in which a general approach to design and recognise product-form stochastic Petri nets, based on the Reversed Compound Agent Theorem (RCAT) [8] and its extensions [9,10], is described. The analysis of the class of SPNs proposed in [11,5] relies on a so called *rank theorem* that imposes a strict condition of the transition rates which is quite hard to interpret from the modelling point of view. In [1,15] the authors show that that class of product-form SPNs (extended to open nets) can be analysed in terms of decomposition into simpler structures and hence the rate condition can be interpreted locally rather than on the whole SPN.

The aim of this paper is to analyse a class of SPNs which is useful to model systems in which concurrent activities can lead to conflicts, requiring a recovery phase before a new execution of the same activities is retried. Instances of this kind of systems are quite frequent in the real world, for example in computer networks, databases and operating systems. We provide a formal model for these systems in terms of SPNs and show that they belong to the class studied in [1,15]. Moreover, we prove two interesting properties for such a class of SPNs: first, their product-form does not require any condition on the transition rates and, second, the joint state space is the Cartesian product of the states that are reachable by each of the model's places. The former property enhances the applicability of the proposed model, while the latter allows us to derive the normalised stationary distributing in a straightforward way for open models. The model that we proposed can be combined with other quasi-reversible components maintaining the product-form property of the joint steady-state distribution (see, e.g., [2,13,7]).

Structure of the paper. The article is structured as follows: first, in Section 2, in order to keep the paper self-contained, we give some basic background notion on Stochastic Petri Nets and on product-form; in Section 3 we analyse the structure and behaviour of a specific class of SPNs with feedback and we show that, for those models, the steady state solution is in product form. In Section 4 we give two applicative examples of the previously described class of SPNs, and we compute performance indices for these models. Finally, Section 5 recapitulates the results of the article.

2   Theoretical Background

In this section we give some basic notions about *stochastic Petri nets* and *building blocks*, which will be used thorough the paper.

2.1 Stochastic Petri Nets

A stochastic Petri net [18,17] is a tuple, SPN $= (\mathcal{P}, \mathcal{T}, \chi(\cdot), \mathbf{I}(\cdot), \mathbf{O}(\cdot), \mathbf{m}_0)$ where:

- $\mathcal{P} = \{P_1, \ldots, P_N\}$ is a set of N places,
- $\mathcal{T} = \{T_1, \ldots, T_M\}$ is a set of M transitions,
- $\chi : \mathcal{T} \to \mathbb{R}^+$ is a positive valued function that associates a firing rate with every transition; we usually write χ_i as an abbreviation for $\chi(T_i)$,
- $\mathbf{I} : \mathcal{T} \to \mathbb{N}^N$ associates an input vector with every transition,
- $\mathbf{O} : \mathcal{T} \to \mathbb{N}^N$ associates an output vector with every transition.

A *marking* of the model is a vector $\mathbf{m} \in \mathbb{N}^N$ that represents the numbers of tokens m_i in each place P_i, $i = 1, \ldots, N$, and thus it identifies the current state of the model. The initial marking is called $\mathbf{m}_0$. A transition T_i is *enabled* by $\mathbf{m}$ if $\mathbf{m} - \mathbf{I}(t_i)$ has non-negative components. An enabled transition T_i *fires* after an exponentially distributed random time with rate χ_i. In this case, the new state $\mathbf{m}'$ is $\mathbf{m} - \mathbf{I}(T_i) + \mathbf{O}(T_i)$. If the input and output vector domains are $\{0,1\}^N$, i.e. tokens move one by one, the net is called *ordinary*.

The graphical representation of SPNs uses circles for places and bars for transitions. If the j-th component of $\mathbf{I}(T_i)$ (respectively $\mathbf{O}(T_i)$) is $k > 0$, we draw an arc from P_j (respectively T_i) to T_i (respectively P_j) and we label it with k (for ordinary nets we omit the labels).

The reachability set $RS(\mathbf{m}_0)$ is the set of all the possible states of the net, given the initial marking $\mathbf{m}_0$. In general, the problem of determining the reachability set of a *SPN* is NP-hard and has an exponential space requirement [6]. The nodes in the *reachability graph* are the states of the reachability set and there is an arc from every node $\mathbf{m}'$ to $\mathbf{m}''$ for which there exists a transition T such that $\mathbf{m}'' = \mathbf{m}' - \mathbf{I}(T) + \mathbf{O}(T)$. The incidence matrix $\mathbf{A}$ of a SPN is an $M \times N$ matrix, row i of which is defined as $\mathbf{O}(T_i) - \mathbf{I}(T_i)$.

The reachability graph can be either finite or infinite and from it, the continuous time Markov chain (CTMC) underlying the SPN model can be derived simply (either lazily or in a parameterised way if the state space is infinite) [17]. Henceforth we consider models whose underlying CTMCs are ergodic and so admit a unique, equilibrium, state probability distribution. Calculating this can be a difficult computational task because of the state space explosion problem, which causes even a structurally small net to have a reachability set with high cardinality. In such cases, solution of the global balance equations rapidly becomes numerically intractable.

We call the fundamental structure that we use to analyse SPNs in product-form a *building block* (BB). We now formally define a BB and give an expression for its product-form solution, together with sufficient conditions for it to exist.

2.2 Building Blocks

According to [1], a BB consists of a set of places $P_1, \ldots P_N$, a set $\mathcal{T}_I$ of input transitions whose input vectors are null (i.e. $\mathbf{0} = (0, \ldots, 0)$), and a set $\mathcal{T}_O$ of output transitions whose output vectors are null. All the arcs have weight 1.

In the BBs for each input transition T_y there must exist an output transition T_y' whose input vector is equal to the output vector of T_y.

Definition 1 (Building block (BB)). *Given an ordinary (connected) SPN S with set of transitions $\mathcal{T}$ and set of N places $\mathcal{P}$, then S is a building block if it satisfies the following conditions:*

1. *For all $T \in \mathcal{T}$ then either $\mathbf{O}(T) = \mathbf{0}$ or $\mathbf{I}(T) = \mathbf{0}$. In the former case we say that $T \in \mathcal{T}_O$ is an output transition while in the latter we say that $T \in \mathcal{T}_I$ is an input transition. Note that $\mathcal{T} = \mathcal{T}_I \cup \mathcal{T}_O$ and $\mathcal{T}_I \cap \mathcal{T}_O = \emptyset$, where $\mathcal{T}_I$ is the set of input transitions and $\mathcal{T}_O$ is the set of output transitions.*
2. *For each $T \in \mathcal{T}_I$, there exists $T' \in \mathcal{T}_O$ such that $\mathbf{O}(T) = \mathbf{I}(T')$ and vice versa.*
3. *Two places $P_i, P_j \in \mathcal{P}, 1 \leq i, j \leq N$, are connected, written $P_i \sim P_j$, if there exists a transition $T \in \mathcal{T}$ such that the components i and j of $\mathbf{I}(T)$ or of $\mathbf{O}(T)$ are non-zero. For all places $P_i, P_j \in \mathcal{P}$ in a BB, $P_i \sim^* P_j$, where $\sim^*$ is the transitive closure of $\sim$.*

Note that in an isolated BB, if two or more input (output) transitions have the same output (input) vector, we can fuse them in one transition whose rate is the sum of the rates of the original transitions. Therefore, without loss of generality, we assume that all the input (output) transitions have different output (input) vectors. Finally, to simplify the notation, we use T_y (T_y') to denote an input (output) transition, where y is the set of place-indices of the non-zero components in the output (input) vector of T_y (T_y'). We now recall Theorem 1 that gives sufficient condition for the product-form of a BB.

Theorem 1 (Theorem 2 of [1]). *Consider a BB S with N places. Let $\rho_y = \lambda_y/\mu_y$, where λ_y, μ_y are the firing rates for $T_y, T_y' \in \mathcal{T}, |y| \geq 1$, respectively. If the following system of equations has a unique solution ρ_i, $(1 \leq i \leq N)$:*

$$\begin{cases} \rho_y = \prod_{i \in y} \rho_i & \forall y : T_y, T_y' \in \mathcal{T} \wedge |y| > 1 \\ \rho_i = \frac{\lambda_i}{\mu_i} & \forall i : T_i, T_i' \in \mathcal{T}, 1 \leq i \leq N \end{cases} \tag{1}$$

then the net's balance equations – and hence stationary probabilities when they exist – have product-form solution:

$$\pi(m_1, \ldots, m_N) \propto \prod_{i=1}^{N} \rho_i^{m_i}. \tag{2}$$

Another interesting result is proved in [15] where the throughput (namely, the reversed rate) of the output transition of a product-form BBs are derived:

Lemma 1. *In a product-form BB that satisfies the conditions of Theorem 1, the throughput (reversed rate) of every output transition labelled T_y' is λ_y, i.e., the rate of the corresponding input transition.*

3   The Conflict Model

Consider a SPN consisting of a set of interconnecting building blocks: a *main building block* (MBB) with l places (from here onward MBB(L)) and a certain number of *conflict building blocks* (CBB).

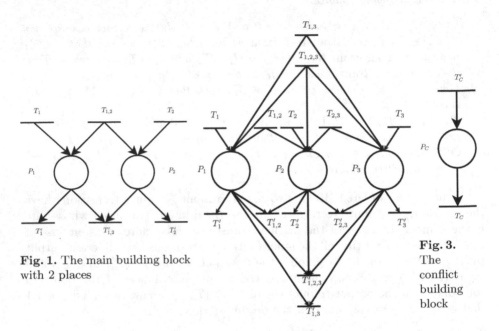

Fig. 1. The main building block with 2 places

Fig. 3. The conflict building block

Fig. 2. The main building block with 3 places

As previously stated, the building block MBB(L) has a set L of l places $L = \{P_1, \ldots, P_l\}$. For each place P_i, there are both an incoming transition T_i with rate λ_{P_i} and an outgoing transition T_i' with rate μ_{P_i}. Examples of the main building block with 2 or 3 places, respectively, are given in Figures 1 and 2. For each subset C of two or more places, $C \subseteq L, |C| \geq 2$, we define an incoming transition T_C with rate λ_C and an outgoing transition T_C' with rate μ_C. For each of those pairs of transitions, there is a conflict building block defined as follows: an arc connects transition T_C' to the place P_C, which is in turn connected to transition T_C. Notice that the firing semantics of transitions T_C, with $|C| \geq 2$, can be single server or infinite servers [16]. A conflict building block in isolation is shown in Figure 3, while a complete model for $l = 2$ is depicted in Figure 4.

If $|C| = k$, there are $\binom{l}{k}$ of such CBBs, and thus the total number of CBBs is $\sum_{k=2}^{l} \binom{l}{k} = 2^l - l - 1$. The total number of places in the model is then $|\mathcal{P}| = 2^l - 1$, and the total number of transitions is $|\mathcal{T}| = 2|\mathcal{P}| = 2^{l+1} - 2$.

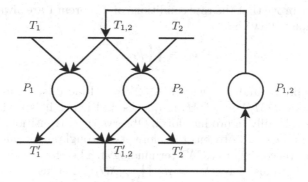

Fig. 4. A complete model for $l = 2$

Example 1 (Conflict model for 2 classes). Let us consider the SPN model depicted by Figure 4. Customers arrive at P_1 and P_2 according to independent homogeneous Poisson processes modelled by T_1 and T_2, respectively. Successful customers' services are modelled by transitions T_1' and T_2'. Transition $T_{1,2}'$ models the conflict events, and is enabled only when both P_1 and P_2 are non-empty. After a conflict event a token is removed from each of the places P_1 and P_2 and one is put into $P_{1,2}$ to represent the recovery phase. Transition $T_{1,2}$ models the recovery time. This model is in product-form because the arrivals modelled by transitions T_1 and T_2 can be replaced by the output process of other product-form models (such as BCMP queues [2], G-queues [7] or MSCCC [13]) and, conversely, the firing of transitions T_1' and T_2' can be used to model the arrival process at other product-form models.

The following proposition shows that conflict model is in product-form according to Theorem 1.

Proposition 1 (Product-form of the conflict model). *The conflict model consists of building blocks satisfying the structural conditions of Theorem 1. Moreover, in stability, it yields without any rate-constraint the following product-form solution:*

$$\pi(\mathbf{m}) = \prod_{C \in 2^L \setminus \emptyset} g_C(m_C)$$

where m_C is the component of the joint state associated with place P_C and

$$g_C(m_C) =$$

$$\begin{cases} (1 - \frac{\lambda_P}{\mu_P})(\frac{\lambda_P}{\mu_P})^{m_P} & \text{if } C = \{P\} \\ (1 - \frac{\mu_C}{\lambda_C} \prod_{P \in C} \frac{\lambda_P}{\mu_P})(\frac{\mu_C}{\lambda_C} \prod_{P \in C} \frac{\lambda_P}{\mu_P})^{m_C} & \text{if } |C| \geq 2 \text{ and } T_C \text{ is single server} \\ (\frac{\mu_C}{\lambda_C} \prod_{P \in C} \frac{\lambda_P}{\mu_P})^{m_C} \exp(-\frac{\mu_C}{\lambda_C} \prod_{P \in C} \frac{\lambda_P}{\mu_P}) \frac{1}{m_C!} & \text{if } |C| \geq 2 \text{ and } T_C \text{ is } \infty \text{ servers} \end{cases}$$

Proof. First, we prove that the rate conditions of Theorem 1 are always satisfied. Basically, we must show that:

$$\rho_C = \prod_{i \in C} \rho_i,$$

for any choice of λ_C and μ_C, with $C \in 2^L \setminus \emptyset$. Indeed, the building blocks P_C behave like a $M/M/1$ or $M/M/\infty$ queues and hence it is well-known that they are unconditionally in product-form. Observe that, transition T_C fires only when P_C has at least one token, therefore its throughput is *not* λ_C but its reversed rate $\overline{\lambda}_C$ (according to RCAT terminology). Therefore, in order to satisfy Conditions (1) we must have $\overline{\lambda}_C = \mu_C \prod_{i \in C} \lambda_i / \mu_i$. Let us now consider the rate equation corresponding to the reversed rate (throughput) of transition T'_C. Using Lemma 1, we immediately derive that its throughput, i.e., the arrival rate perceived by P_C, is $\overline{\lambda}_C$. Observe that, analogously to what happens in closed queueing networks, the flows of tokens due to the firing of T_C and T'_C with $C \geq 2$ give the identity $\lambda_C = \overline{\lambda}_C$ and hence any non-trivial solution can be taken and, here we take: $\overline{\lambda}_C = \prod_{i \in C}(\lambda_i / \mu_i)\mu_C$ [1]. However, differently form closed queueing networks, the conflict model's state space is such that each of its places can have from 0 to ∞ tokens, and the joint state space is the Cartesian product of the state spaces of each of its components. Therefore, the normalisation can be carried out by normalising the stationary distribution of each place considered in isolation.

For the stability conditions, the following proposition holds:

Proposition 2. *The conflict model is stable if the following conditions hold:*

$$\forall i \in \{1, \ldots, l\} \quad \lambda_i < \mu_i, \tag{3}$$

for the places of the main building block, while for the places of conflict building blocks P_C whose corresponding T_C is single server, we have that

$$\forall C \subseteq L \quad \overline{\mu}_C = \mu_C \prod_{P_i \in C} \rho_{P_i} < \lambda_C, \tag{4}$$

where $\overline{\mu}_C$ identifies the throughput (reversed rate) of transition T'_C.

Proof. The proof relies on deriving the necessary and sufficient conditions required for normalising the stationary probabilities of Proposition 1.

Notice that Condition (4) can be rewritten as

$$\forall C \subseteq L \quad \mu_C \frac{\prod_{P_i \in C} \lambda_{P_i}}{\prod_{P_i \in C} \mu_{P_i}} < \lambda_C \tag{5}$$

If all the transitions T_C have a single server semantics, we can compute the average number of customers in the system as

$$E[N] = \sum_{i=1}^{l} \frac{\rho_{P_i}}{1 - \rho_{P_i}} + \sum_{C \subseteq L, |C| \geq 2} |C| \frac{\rho_C}{1 - \rho_C} \tag{6}$$

Now we consider the case where all the transitions T_C, with $|C| \geq 2$ have an *infinite server* semantics, i.e. they can be seen as delay stations [16]. In this case the average number of customers for each place P_C becomes ρ_C, and thus the average number of customer for the whole system is

$$E[N] = \sum_{i=1}^{l} \frac{\rho_{P_i}}{1 - \rho_{P_i}} + \sum_{C \subseteq L, |C| \geq 2} |C| \rho_C \qquad (7)$$

In both cases, the average response time of the system is given by Little's law, i.e., under their respective stability conditions:

$$E[R] = \frac{E[N]}{\sum_{i=1}^{l} \lambda_i} \qquad (8)$$

4 Applications

4.1 A Computer Network with Collisions

We want to model a computer network in which there is a set L of l transmitting stations $L = \{s_1, \ldots, s_l\}$. Packets become ready to be sent from each station s_i according to an homogeneous Poisson distribution with parameter λ_i. The global rate at which packets arrive to the system is thus $\lambda = \sum_{i=1}^{l} \lambda_i$. The time that a packet takes to ben transmitted from each station s_i is exponentially distributed with parameter μ_i^*. The channel, by itself, is capable of transmitting with a global rate of $M = \sum_{i=1}^{l} \mu_i^*$. During the transmission of a packet, a collision, i.e., the simultaneous transmission of more than one packet using the same physical medium, can occur, thus making the transmission ineffective. A collision can occur between any combination of k stations, $2 \leq k \leq L$, with probability $p_k(L)$. After a collision between the transmissions of a subset of stations $C \subseteq L, |C| \geq 2$, an exponentially-distributed recovery time, with parameter μ_C is performed. After that time, a new transmission is retried.

We can abstract the previously described system as a *conflict model* (Section 3), in which, since there is no queueing for recovery phases, the conflict building blocks have an infinite-server firing semantics.

If we assume that $\mu_{s_i} = \mu_1$, $\lambda_{s_i} = \lambda_1$, $\forall s_i \in L$ and that $\lambda_C = \lambda_{|C|}$ and $\mu_C = \mu_{|C|}$, $\forall C \subseteq L, |C| \geq 2$, i.e., all the transitions representing collisions between the same number of stations have the same rate, we can simplify Equation (6) as

$$E[N] = l \frac{\rho_1}{1 - \rho_1} + \sum_{k=2}^{l} \binom{l}{k} k \rho_k \qquad (9)$$

We now have to make further assumptions on the behaviour of the system in order to parametrise the model and thus compute the numerical value of the interested performance indices.

Let $q \in [0,1]$ be the probability that a certain station is transmitting on the channel. Here we assume that q can be computed as:

$$q = \frac{\lambda_1}{M}, \tag{10}$$

such that q^2 represents the probability of a collision between two nodes. We define $p_C(L)$, i.e., the probability of having a collision between a specific set C of k stations in a system with L stations, as

$$p_C(L) = q^k(1-q)^{L-k} \tag{11}$$

and, since, for any given station, there are $\binom{L-1}{k-1}$ possible collision sets of k station, the probability $p_k(L)$ of having a collision among k, having chosen one of the station, is

$$p_k(L) = \binom{L-1}{k-1} q^k(1-q)^{L-k} \tag{12}$$

Thus, the probability $\bar{p}$ of not having any collision for a chosen station is

$$\bar{p} = 1 - \sum_{k=2}^{L} \binom{L-1}{k-1} q^k(1-q)^{L-k} \tag{13}$$

We can now compute the various service rates μ_C, $C \subset L, |C| = k \geq 2$ as

$$\mu_k = \mu^* q^k(1-q)^{L-k} \tag{14}$$

and for μ_1

$$\mu_1 = \mu^* \left(1 - \sum_{k=2}^{L} \binom{L-1}{k-1} q^k(1-q)^{L-k} \right) \tag{15}$$

A Numerical Example. Consider an Ethernet-like network having a shared channel with a bandwidth of 100Mbps. We assume that the time to send a frame is exponentially distributed, with an expected value that is the time to send 800 bytes at full speed, and that, on average, the backoff time is the same as the one required for transmitting 512 bits. We can model the network as described in Section 4.1, having a set of parameters given in Table 1,

Fixed a value for L, i.e., the number of transmitting stations, we set the other parameters according to Equations (10), (14) and (15), while $\mu_1^* = ML^{-1}$

Table 1. Parameter values for the example of Section 4.1

Parameter Name	Value
M	$10^8/(800 \cdot 8) \approx 15625$
λ_C	$10^8/512 \approx 195313$

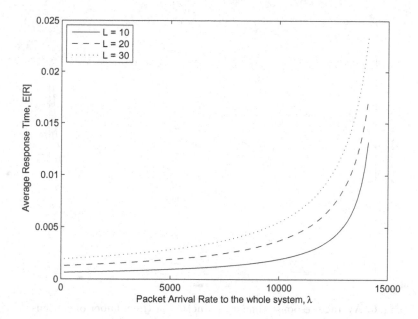

Fig. 5. Average response time as a function of packet arrival rate, with different number of stations

In Figure 5 we see the behaviour of the average response time, i.e., the average time a packet has to spend to be successfully transmitted, as the value of λ increases, with three different numbers of stations L. We can see that, the response time increases both as the arrival rate increases, due to queueing, and as the number of stations increases, due to collisions.

In Figure 6 we can see how, given a fixed arrival rate at each station, $\lambda_1 = 300$, the average response time increases with the number of transmitting stations, due to the increased number of collisions.

4.2   A Transactional Database System

Consider a database system in which there is a set L of l processors $L = \{s_1, \ldots, s_l\}$. The transaction requests to be processed arrive to the processor s_i according to an homogeneous Poisson distribution with parameter λ_i. The time that a transaction takes to be processed by s_i is exponentially distributed with parameter μ_i^*. During the execution of a transaction, a conflict, i.e., a concurrent access to the same data in which at least one of the access is a write operation, can occur, thus requiring a strategy to resolve the conflict. A collision can occur between any combination of k stations, $2 \leq k \leq L$, with probability $p_k(L)$. In order to implement this strategy, after a conflict occurs between the

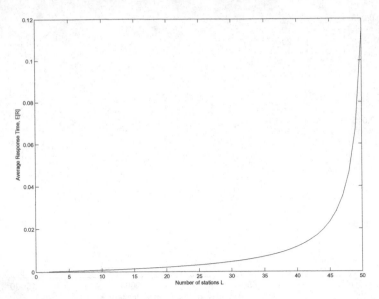

Fig. 6. Average response time as a function of the number of stations

transactions running on a subset of processors $C \subseteq L$, $|C| \geq 2$, an exponentially-distributed recovery time, with parameter μ_C is performed. If the same set of processors is engaged in the resolution of another conflict, the resolution requests are enqueued. After the conflict is recovered, the transaction is retried.

As for the example of Section 4.1, we can model the described system as a conflict model. However, in this case, since recovery requests are enqueued, the conflict building blocks have the ordinary firing semantics of SPNs.

As for the previous example, we can assume that $\mu_{s_i} = \mu_1$, $\lambda_{s_i} = \lambda_1$, $\forall s_i \in L$ and that $\lambda_C = \lambda_{|C|}$ and $\mu_C = \mu_{|C|}$, $\forall C \subseteq L, |C| \geq 2$, i.e., transitions of conflict building blocks representing recoveries between the same number of processors have the same rate. Thus we can simplify Equation (6) as

$$E[N] = \sum_{k=1}^{l} \binom{L}{k} k \frac{\rho_k}{1 - \rho_k} \tag{16}$$

Moreover, we observe that the same arguments made for the collision probabilities could be made for transaction conflicts, assuming that the collision probabilities, here, mean the probability of having a conflicting transaction between k processors. Thus, Equations 14 and 15 still hold. However, here the choice of a suitable parameter q is free.

A Numerical Example. Consider a database system like the one described in Section 4.2, in which each processor is capable of serving, on average, 100

Table 2. Parameter values for the example of Section 4.2

Parameter Name	Value
μ_1	100
λ_C	10
q	10^{-2}

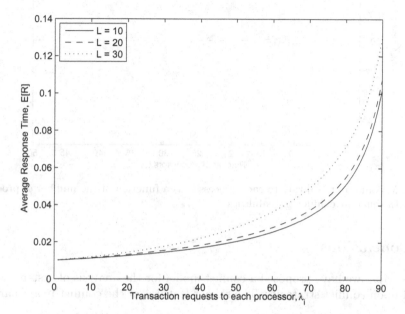

Fig. 7. Average response time as a function of the arrival rate of transactions, with different number of processors

transactions per second and the system can recover 10 transactions per second. The parameters are shown in Table 2, where q is chosen arbitrarily.

Figure 7 shows the average response time as a functions of the arrival rate of transactions to each processor.

Moreover, in systems where q is constant, from Equations (3) and (15) we directly obtain the maximum admissible arrival rate to each processor λ_1 as a function of q, μ and L. Figure 8 shows a numerical example for $\mu_1 = 1$.

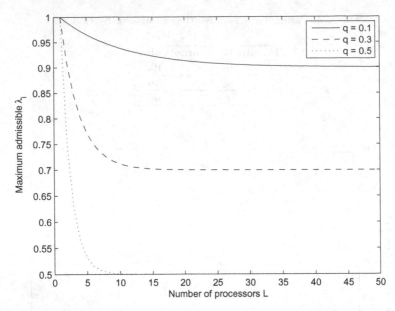

Fig. 8. Maximum arrival rate to each processor as a function of the number of processors, with different conflict probabilities

5   Conclusions

In this paper we have proposed a conflict model for the analysis of systems with retrial-upon-conflict strategies. The salient features of the conflict models are:

- It is formally defined in terms of stochastic Petri nets.
- The stationary distribution can be computed efficiently because the conflict model is product-form. Moreover, it can be composed with other models that are in product-form by RCAT or that have been proved to be quasi-reversible, including the BCMP stations [2], the G-queues [7] and other more sophisticated queueing stations [13].
- We have shown that, differently from other SPNs, the conflict model is unconditionally in product-form. In many other cases, when dealing with product-form SPNs, rate conditions arise (see [11,5,14,1,15]).

We have shown the application of the conflict model to study some realistic systems and have given some numerical results which show examples on how those models behave under some conditions.

Future Works. We plan to extend our work in order to consider also closed networks with conflict models, i.e., systems where the total number of customers remain constant. In these cases the problem of the computation of the normalising constant must be addressed. The well-known algorithms for the direct computation of the normalising constant (e.g., the convolution [4,5]) or for the computation of the average performance indices [19,20] should be considered. Moreover, the problem of parametrising the model according to the behaviour of the system, especially due to the combinatorial nature of these model, should be studied in depth.

References

1. Balsamo, S., Harrison, P.G., Marin, A.: Methodological construction of product-form stochastic Petri nets for performance evaluation. Journal of Systems and Software 85(7), 1520–1539 (2012)
2. Baskett, F., Chandy, K.M., Muntz, R.R., Palacios, F.G.: Open, closed, and mixed networks of queues with different classes of customers. J. ACM 22(2), 248–260 (1975)
3. Boucherie, R.J.: A characterisation of independence for competing Markov chains with applications to stochastic Petri nets. IEEE Trans. on Software Eng. 20(7), 536–544 (1994)
4. Buzen, J.P.: Computational algorithms for closed queueing networks with exponential servers. Commun. ACM 16(9), 527–531 (1973)
5. Coleman, J.L., Henderson, W., Taylor, P.G.: Product form equilibrium distributions and a convolution algorithm for Stochastic Petri nets. Perf. Eval. 26(3), 159–180 (1996)
6. Esparza, J., Nielsen, M.: Decidability Issues for Petri Nets - a Survey. Bulletin of the European Association for Theoretical Computer Science 52, 245–262 (1994)
7. Gelenbe, E.: Product form networks with negative and positive customers. J. of Appl. Prob. 28(3), 656–663 (1991)
8. Harrison, P.G.: Turning back time in Markovian process algebra. Theoretical Computer Science 290(3), 1947–1986 (2003)
9. Harrison, P.G.: Reversed processes, product forms and a non-product form. Linear Algebra and Its Applications 386, 359–381 (2004)
10. Harrison, P.G., Lee, T.T.: Separable equilibrium state probabilities via time reversal in Markovian process algebra. Theoretical Computer Science 346(1), 161–182 (2005)
11. Henderson, W., Lucic, D., Taylor, P.G.: A net level performance analysis of Stochastic Petri Nets. J. Austral. Math. Soc. Ser. B 31, 176–187 (1989)
12. Lazar, A.A., Robertazzi, T.G.: Markovian Petri Net Protocols with Product Form Solution. Perf. Eval. 12(1), 67–77 (1991)
13. Le Boudec, J.Y.: A BCMP extension to multiserver stations with concurrent classes of customers. In: SIGMETRICS 1986/PERFORMANCE 1986: Proc. of the 1986 ACM SIGMETRICS Int. Conf. on Computer Performance Modelling, Measurement and Evaluation, pp. 78–91. ACM Press, New York (1986)
14. Mairesse, J., Nguyen, H.-T.: Deficiency Zero Petri Nets and Product Form. In: Franceschinis, G., Wolf, K. (eds.) PETRI NETS 2009. LNCS, vol. 5606, pp. 103–122. Springer, Heidelberg (2009)
15. Marin, A., Balsamo, S., Harrison, P.G.: Analysis of stochastic Petri nets with signals. Perform. Eval. 69(11), 551–572 (2012)
16. Marsan, M.A., Balbo, G., Conte, G., Donatelli, S., Franceschinis, G.: Modelling with generalized stochastic Petri nets. Wiley (1995)
17. Molloy, M.K.: Performance analysis using stochastic Petri nets. IEEE Trans. on Comput. 31(9), 913–917 (1982)

18. Murata, T.: Petri nets: Properties, analysis and applications. Proc. of the IEEE 77(4), 541–580 (1989)
19. Resiser, M., Lavenberg, S.S.: Mean Value Analysis of closed multichain queueing network. J. ACM 27(2), 313–320 (1980)
20. Sereno, M., Balbo, G.: Mean Value Analysis of stochastic Petri nets. Perform. Eval. 29(1), 35–62 (1997)

Discriminatory Processor Sharing from Optimization Point of View*

Jozsef Biro[2], Tamás Bérczes[1], Attila Kőrösi[3], Zalan Heszberger[3],
and János Sztrik[1]

[1] University of Debrecen, Faculty of Informatics
[2] Inter-University Centre for Telecommunications and Informatics
Kassai u. 26., Debrecen, Hungary
[3] MTA-BME Information Systems Research Group
Budapest University of Technology and Economics
Magyar tudosok krt. 2., Budapest, Hungary
biro@tmit.bme.hu

Abstract. Discriminatory Processor Sharing models play important
role in analysing bandwidth allocation schemes in packet based communication systems. Users in such systems usually have access rate
limitations which also influence their bandwidth shares. This paper is
concerned with DPS models which incorporate these access rate limitations in a bandwidth economical manner.

In this paper the interlock between access rate limited Discriminatory Processor Sharing (DPS) models and some constrained optimization problems is investigated. It is shown, that incorporating the access
rate limit into the DPS model is equivalent to extending the underlying constrained optimization by constraints on the access rates. It also
means that the available bandwidth share calculation methods for the
access rate limited DPS are also non-conventional solution methods for
the extended constrained optimization problem.

We also foreshadow that these results might be important steps towards obtaining efficient pricing and resource allocation mechanism when
users are selfish and subject to gaming behavior when competing for
communication resources.

1 Introduction

Processor Sharing (PS) models have been long studied in the queueing systems
literature. One of the first reports on processor sharing have been performed
in [11] motivated mainly by the modeling of time-shared computer systems. In
this seminal work and also in [14] not only the egalitarian sharing (equal service
rates are allocated for customers), but the (multi-class) discriminatory processor
sharing has also been analyzed. In DPS the division of the processing capacity

---

* The publication was supported by the TAMOP-4.2.2.C-11/1/KONV-2012-0001
project. The project has been supported by the European Union, co-financed by
the European Social Fund.

A. Dudin and K. De Turck (Eds.): ASMTA 2013, LNCS 7984, pp. 67–80, 2013.

C can be controlled by a set of weights $(g_1, g_2, \ldots, g_K)=:\mathbf{g}$ in the following way. If there are $n_1, n_2, \ldots, n_K=:\mathbf{n}$ customers from the classes in the queuing system, they are served simultaneously, and the (instantaneous) service rate of a class$-i$ customer is $c_i = \frac{g_i}{\sum_{j=1}^{K} g_j n_j} C$. Fayolle et al. in [5] completed the early efforts in solving discriminatory processor sharing models. They gave the correct characterizations for the M/G/1-PS system with respect to the steady-state average response times (by integro-differential equations), and also showed that in the special case of exponential distribution of the service time requirements, the steady-state average response times can be obtained by solving a system of linear equations. In [16] Rege and Sengupta showed how to obtain the moments of the queue length distributions as the solutions to linear equations in case of exponential service time requirements, and they also presented a heavy-traffic limit theorem for the joint queue length distribution. These results were extended to phase type distributions by van Kessel et al. [9]. A further remarkable milestone in DPS analysis is [1] in which the authors showed that the mean queue lengths of all classes are finite under reasonable stability conditions, regardless of the higher moments of the service requirements. It was also shown that the conditional response times of the different classes are stochastically ordered according to the DPS weights. In [3] Cheung, van den Berg and Boucherie for the egalitarian PS model obtained an exact and analytically tractable decomposition. This decomposition was applied to discriminatory PS to obtain an efficient and analytically tractable approximation of the queue length distribution and mean sojourn times.

In the discriminatory processor sharing models above, a common characteristics is that the users from the same class get equal share of the capacity allocated to that class, and the service capacity allocations of classes are proportional to their weights and to their number of users.

The other valuable direction of further developments of processor sharing model is to introduce capacity limits by which the customers can receive service from the server. This is mainly motivated by involving access rate limitations of users (e.g. in DSL-type access systems) in to the modeling framework. In [13] Lindberger introduced and analyzed the so-called M/G/R-PS model, which is a single-class processor sharing model with access rate limit b on the users ($R := C/b$ is the "number of servers" in this system). Several improvements of this model were studied for dimensioning purposes of IP access networks, e.g. in [17] and [4] still remaining at the single-class models.

Introducing access rate limitations in multi-class discriminatory processor sharing inherently raises the question of bandwidth re-distribution. This means that if customers in a class can not fully utilize their prospective service capacity share (bandwidth share) due to their access rate limit, how this unused (left) bandwidth is re-distributed among the other classes. In one of the extreme cases, there is no re-distribution at all meaning that the possible remaining unused bandwidth due to rate limits is wasted. One can also interpret this as the server capacity may not be fully utilized, even in those cases when there is "enough" customers in the system. This approach is followed for example in the papers [12], [2]. In the other extreme case of multi-class discriminatory processor sharing, all the unused bandwidth by access rate limited customers are fully utilized by the other (non-limited) customers. In this scenario even the bandwidth shares can not be easily determined for given **n** and **g**. Some of the classes will share certain amount of capacity proportional to their weights and number of users, whilst users in other classes are saturated at their access rate limits. Efficient algorithms have been presented for computing the capacity shares in [15] and [10].

In the original DPS model, the capacity shares of users for given **n** and **g** are solutions of a constrained optimization problem (see next chapter for details). Furthermore, because this solution is a proportional bandwidth allocation (among classes), this has a bridging role between the utilitarian social welfare maximization and individual payoff maximization, when classes of users are competing for the bandwidth shares.

2 Discriminatory Processor Sharing with Access Rate Limitations

Let n_i denote the number of class-i ($i = 1 \ldots K$) flows (users, jobs) in the C capacity processor sharing system. In the original discriminatory processor sharing model [5] the server share its capacity among the classes in a discriminatory fashion by weights g_i, that is, a class-i flow bandwidth share from the total capacity C is

$$c_i = \frac{g_i}{\sum_{j=1}^{K} g_j n_j} C \ . \tag{1}$$

This weight-proportional share of bandwidth implies two important properties: For every pair of classes i, j the ratio of the service rates allocated to class-i and class-j users is equal to the ratio of the class weights, that is,

$$\frac{c_i}{c_j} = \frac{g_i}{g_j} \ , \ \forall i, j \in 1, \ldots, K \ . \tag{2}$$

The total amount of capacity in a non-empty system used by the users of classes is C, that is,

$$\sum_{i=1}^{K} n_i c_i = C \ . \tag{3}$$

These properties can also be considered as requirements for a type of bandwidth share, which are uniquely fulfilled by (1).

It can also be shown (e.g. using Lagrangian multiplier method) that the solution of the following constrained optimization problem

$$\max_{\mathbf{x}} \sum_{i=1}^{K} n_i g_i \log x_i \ , \quad \text{s.t.} \ \sum_{i=1}^{K} n_i x_i = C \ , \quad x_i \geq 0 \tag{Opt}$$

is exactly that one in (1). Here it is worth noting that solving this optimization problem might be tedious at first glance, in fact, its solution is a closed-form proportional allocation-like formula in (1).

There can be several ways to introduce access rate limitations in the DPS model. One reasonable and simple enough approach is the following. Compute the bandwidth shares of class-i users according to (1) and cut at the access rate limits b_i, i.e.

$$c_i = \min \left(\frac{g_i}{\sum_{j=1}^{K} g_j n_j} C, b_i \right) . \tag{4}$$

which is in full accordance with the first extreme case (there is no bandwidth re-distribution at all) mentioned in the introductory section. Such kind of models were analyzed in [12], [2] and in network context in [6]. From computational point of view the benefit of this case lies in the very simple calculation of service rates of classes, and as a consequence, the set of uncompressed (limited by their access rates) and the compressed (non-limited by their access rates, i.e. they cannot reach their upper limit of their access rates b_i, hence the name "compressed") traffic classes directly follows from the service rate computation. The drawback of this simple access rate involvement is the possible waste of resources, that is it can occur that $\sum_{i=1}^{K} n_i c_i < C$ even if $\sum_{i=1}^{K} n_i b_i > C$. One can also interpret, that (3) as a requirement can not always be fulfilled by the bandwidth share governed by (4). [1]

Intuitively, a very straightforward way of introducing service rate limits would be to include further constraints $c_i \in [0, b_i]$ into the underlying optimization task, that is

$$\max_{\mathbf{x}} \sum_{i=1}^{K} n_i g_i \log x_i , \quad \text{s.t.} \quad \sum_{i=1}^{K} n_i x_i = C \text{ and } x_i \in [0, b_i] , \quad i = (1, \ldots, K) .$$

(OptBounds)

Involving the access rate constraints into the optimization one may expect bet-ter utilization of the server capacity, when some users can not obtain its original proportional bandwidth share due to their access rate limits. In other words, the constraints of the optimization problem ensures that such an allocation repre-sented by the optimal solution never waste server capacity if $\sum_{i=1}^{K} n_i b_i > C$, i.e. $\sum_{i=1}^{K} n_i c_i = C$ will hold.

Unfortunately, directly solving this extended optimization task can not be completed by Lagrange-multipliers, in this way it is not helpful in obtaining nice bandwidth share formula of the bandwidth-economical access-rate limited DPS. However, standard numerical methods could be applicable.

3   Determining the Bandwidth Shares

Instead of solving directly this optimization problem, first we recall two methods (presented in [15] and [10]) which can be used to determine the bandwidth shares of users in the bandwidth-economical access rate-limited DPS model.

---

[1] Henceforward we assume that $\sum_{i=1}^{K} n_i b_i > C$, that is, there is always at least one class, say class-j, for which $c_j < b_j$.

The first method is quite intuitive and provides an easily computable numerical method to determine the bandwidth shares. This is based on modifying the equation (2) into the following one:

$$c_j = \min\left(b_i, \frac{g_j}{g_i}c_i\right) , \ \forall i \in \mathcal{Z} \tag{5}$$

and keeping the capacity constraint as

$$\sum_{i=1}^{K} n_i c_i = C . \tag{6}$$

(Where $\mathcal{Z}$ denotes the set of compressed class indexes, that is, $\mathcal{Z} = \{i : c_i < b_i, i = 1, \ldots, K\}$). An important observation is that if $\sum_{i=1}^{n} n_i b_i > C$ then there is always at least one compressed class (e.g. the class which has the smallest g_i/b_i ratio), let its index be K. Summing up both sides by j with $i = K$ we get (using (6))

$$C = \sum_{j=1}^{K} \min\left(b_K, \frac{g_j}{g_K}c_K\right) \tag{7}$$

from which c_K can be numerically determined (note that the right hand side is monotone increasing by c_K) and using (5) all the other c_i's can be determined. For more detailed analysis, see [15]. For the time being we don't know whether this is a solution of (OptBounds), we can only state that the equations (5) and (6) are fulfilled.

In the second, more deductive method (presented in [10]) , let us assume first that

$$\frac{g_1}{b_1} \geq \frac{g_2}{b_2} \geq \ldots \geq \frac{g_K}{b_K} . \tag{8}$$

Further, assume that $\left\{\sum_{i=1}^{K} n_i b_i > C\right\}$ holds, i.e. there is at least one class whose flows are compressed. In what follows, a method is given to determine the set of compressed flows $\mathcal{Z}$. For this, first consider the following inequality with respect to class-1:

$$\frac{g_1}{\sum_{j=1}^{K} n_j g_j}C \geq b_1$$

which could be rearranged as

$$\frac{g_1}{b_1} \geq \frac{\sum_{j=1}^{K} n_j g_j}{C} .$$

If this inequality holds, it means that class-1 is surely uncompressed, in other words, class-1 can not utilize its bandwidth share completely, there is excess bandwidth to be re-distributed among other classes.[2] It means that $C - n_1 b_1$ capacity is shared among the classes $\{2, \ldots, K\}$. Since

---

[2] Also observe that if the inequality above would not hold, class-1 could exceptionally be attributed as surely compressed, because due to (8) all the other classes would also be compressed.

$$\frac{\sum_{j=1}^{K} n_j g_j - n_1 g_1}{C - n_1 b_1}$$

is constant, by following the order given in (8), class 2 has the highest chance to be uncompressed in the set $\{2, \ldots, K\}$. Thus, the following inequality should be checked in sequence:

$$\frac{g_2}{b_2} \geq \frac{\sum_{j=1}^{K} n_j g_j - n_1 g_1}{C - n_1 b_1} .$$

If it holds, class-2 is also uncompressed, and the similar inequalities should be checked for class-3.

It can be seen that for determining the compressed classes the following inequalities are to be checked, in increasing order of indexes

$$\frac{g_i}{b_i} \geq \frac{\sum_{j=i}^{K} n_j g_j}{C - \sum_{k=1}^{i-1} n_k b_k} , \quad i = 1, 2, \ldots$$

Suppose that i^* is the last index for which the inequality above holds. On one hand, it follows that every class-i, $i \leq i^*$ is uncompressed, on the other hand, for every class-i, $i > i^*$ the inequalities

$$\frac{g_i}{b_i} < \frac{\sum_{j=i^*+1}^{K} n_j g_j}{C - \sum_{k=1}^{i^*} n_k b_k} , \quad i = i^* + 1, \ldots, K$$

hold, that is, the set of compressed flows is $\mathcal{Z} = \{i^* + 1, \ldots, K\}$ and their bandwidth shares are

$$c_i = \frac{g_i}{\sum_{j=i^*+1}^{K} n_j g_j} (C - \sum_{k=1}^{i^*} n_k b_k) , \quad i \in \mathcal{Z} . \tag{9}$$

It can also be shown in a straightforward manner that the c_i in (9) and $c_j = b_j$, $\forall j \in \mathcal{U}$ uniquely satisfy the requirements in (5) and (6).

4   The Main Result

Now, we are ready to state and prove the main result of the paper.

Theorem 1. *The solution of the optimization problem (OptBounds) $\underline{c} = (c_1, \ldots, c_K)$ is the following:*

1. *If $\sum_{k=1}^{K} n_k b_k \leq C$, then $c_k = b_k$ for $1 \leq k \leq K$.*
2. *If $b_k \geq \frac{g_k}{\sum_{i=1}^{K} g_i n_i} C$ for any $1 \leq k \leq K$, then $c_k = \frac{g_k}{\sum_{i=1}^{K} g_i n_i} C$ for $1 \leq k \leq K$.*
3. *If there exists an i^*, $1 \leq i^* \leq K - 1$ such that*

$$\sum_{k=1}^{i^*} n_k b_k + \sum_{k=i^*+1}^{K} n_k g_k \frac{b_{i^*}}{g_{i^*}} \leq C \text{ and } \sum_{k=1}^{i^*+1} n_k b_k + \sum_{k=i^*+2}^{K} n_k g_k \frac{b_{i^*+1}}{g_{i^*+1}} > C,$$

then

$$c_k = b_k, \ \text{if } k \leq i^* \ \text{and } c_k = \frac{g_k}{\sum_{i=i^*+1}^{K} g_i n_i} \left(C - \sum_{j=1}^{i^*} n_j b_j \right), \ \text{if } \ i^* < k.$$

Proof. We apply induction for the number of classes. If $K = 1$ the statement of the theorem is true. Now, assume that for $K - 1$ classes the statement holds for arbitrary parameter setup. We prove the statement for K classes. By scaling g_i for each $1 \leq i \leq K$ with the same positive multiplicative factor does not affect the system. Hence without loss of generality we assume that the weights are chosen such that

$$\sum_{k=1}^{K} g_k n_k = 1. \tag{10}$$

We separate three cases.

In Case 1 we assume $b_1 \geq g_1 C$. This corresponds to the setup when all classes are compressed.

In Case 2 we assume that $b_1 < g_1 C$ and there exists $x_1^* \in [0, b_1]$ for which

$$c_2^{\mathcal{S} \setminus \{1\}, C - n_1 x_1} < b_2,$$

where $c_i^{\mathcal{S} \setminus \{1\}, C - n_1 x_1}$, $i \in \mathcal{S} \setminus \{1\}$ denotes the optimal solution of (OptBounds) with classes $\mathcal{S} \setminus \{1\}$ and capacity $C - n_1 x_1$. This corresponds to the setup when except the first class all classes are compressed.

In Case 3 we assume that $b_1 < g_1 C$ and for any $x_1 \in [0, b_1]$ we have

$$c_2^{\mathcal{S} \setminus \{1\}, \ C - n_1 x_1} = b_2.$$

This corresponds to the setup when the first and the second class are compressed. Here, we have two subcases

a) $b_2 \leq g_2 C$.
b) $b_2 > g_2 C$.

Case 1. Assume $b_1 \geq g_1 C$ holds. In this case the optimal solution of (Opt) is equal to the optimal solution of (OptBounds). Indeed, the optimal solution of (Opt) is $c_i = g_i C \ \forall i$. Since $g_1 C \leq b_1$ and (8) it follows that $\forall i \ g_i C \leq b_i$. Thus, $\forall i \ c_i = g_i C$ is optimal solution of (OptBounds) as well. We remark that we have not used induction in this case.

Case 2. Assume that $b_1 < g_1 C$ and there exists $x_1^* \in [0, b_1]$ for which

$$c_2^{\mathcal{S} \setminus \{1\}, C - n_1 x_1^*} < b_2.$$

We will show that under these conditions the optimal solution of the optimization problem is

$$c_1 = b_1, \ \text{and } c_i = \frac{g_i}{\sum_{j=2}^{K} n_j g_j} (C - n_1 b_1), \ i \geq 2, \tag{11}$$

where $c_i < b_i$ for $i \geq 2$.

First, we prove the following.

Lemma 1. *For any $1 \leq i \leq K$ we have $c_i \in [0, b_i]$.*

Proof. We clearly have $c_1 = b_1 \in [0, b_1]$. In the system with classes $S = \{2, 3, \ldots, K\}$ and with capacity $C - n_1 x_1^*$ the optimal bandwidth for class 2 is smaller than b_2 since the assumption of Case 2. So class 2 is compressed. Using the induction one gets that each class is compressed, so

$$c_i^{S \setminus \{1\}, C - n_1 x_1^*} = \frac{g_i}{\sum_{j=2}^{K} n_j g_j} (C - n_1 x_1^*), \ 2 \leq i \leq K.$$

By the assumption $\frac{g_2}{\sum_{j=2}^{K} n_j g_j} (C - n_1 x_1^*) < b_2$ and by (8) we have

$$\frac{C - n_1 b_1}{\sum_{j=2}^{K} n_i g_i} \leq \frac{C - n_1 x_1 *}{\sum_{j=2}^{K} n_j g_j} < \frac{b_2}{g_2} \leq \frac{b_i}{g_i}, \ i \geq 2$$

$$\frac{g_i}{\sum_{j=2}^{K} n_j g_j} (C - n_1 b_1) < b_i, \ i \geq 2.$$

In the followings, we show that the optimal solution of (OptBounds) is given in (11). One can easily check that the following inequality holds

$$\max_{\substack{x_1 \in [0, b_1] \\ x_i \in [0, b_i], i \geq 2}} \sum_{i=1}^{K} n_i g_i \log x_i \leq \max_{\substack{x_1 \in [0, b_1] \\ x_i \geq 0, i \geq 2}} \sum_{i=1}^{K} n_i g_i \log x_i. \tag{12}$$

For $0 < x < C/n_1$ let

$$M(x) = \max_{\substack{x_1 = x \\ x_i \geq 0, i \geq 2}} \sum_{i=1}^{K} n_i g_i \log x_i.$$

Since the solution of problem (Opt) is known we have

$$M(x) = n_1 g_1 \log x + \sum_{i=2}^{K} n_i g_i \log \left(\frac{g_i}{\sum_{j=2}^{K} n_j g_j} (C - n_1 x) \right).$$

We will prove the following

Lemma 2. *$M(x)$ is increasing if and only if $0 < x < g_1 C$.*

Lemma 2 implies that on $[0, b_1]$ takes its maximum at $x = b_1$ since $b_1 < g_1 C$. Hence,

$$\max_{\substack{x_1 \in [0, b_1] \\ x_i \geq 0, i \geq 2}} \sum_{i=1}^{K} n_i g_i \log x_i \leq \max_{\substack{x_1 = b_1 \\ x_i \geq 0, i \geq 2}} \sum_{i=1}^{K} n_i g_i \log x_i.$$

From (12) we have

$$\max_{\substack{x_1 \in [0, b_1] \\ x_i \in [0, b_i], i \geq 2}} \sum_{i=1}^{K} n_i g_i \log x_i \leq \max_{\substack{x_1 = b_1 \\ x_i \geq 0, i \geq 2}} \sum_{i=1}^{K} n_i g_i \log x_i. \tag{13}$$

The optimal solution of the right hand side of (13) is c_i, $1 \leq i \leq K$ given in (11). c_i, $1 \leq i \leq K$ is a possible solution of the left side of (13) by Lemma 1, and because of the inequality (13) it is the optimal solution.

Proof (Proof of Lemma 2.). First, we show that $M(x)$ is increasing if $0 < x < g_1 C$. Take the derivative of $M(x)$:

$$M'(x) = n_1 g_1 \frac{1}{x} + \sum_{i=2}^{K} n_i g_i \frac{-n_1}{C - n_1 x} = n_1 g_1 \frac{1}{x} - n_1 \frac{\sum_{i=2}^{K} n_i g_i}{C - n_1 x}.$$

Find the interval on which $M'(x)$ is positive:

$$n_1 g_1 \frac{1}{x} - n_1 \frac{\sum_{i=2}^{K} n_i g_i}{C - n_1 x} > 0$$

$$g_1 \frac{1}{x} > \frac{\sum_{i=2}^{K} n_i g_i}{C - n_1 x}$$

$$g_1 C - n_1 g_1 x > x \sum_{i=2}^{K} n_i g_i$$

$$g_1 C > x \sum_{i=1}^{K} n_i g_i$$

$$g_1 C > x.$$

Case 3. Assume that $b_1 < g_1 C$ and for any $x_1 \in [0, b_1]$ we have $c_2^{S \setminus \{1\}, C - n_1 x_1} = b_2$ holds. In this case for any $x_1 \in [0, b_1]$ class 2 is uncompressed. Take away class 2, that is, consider the reduced system with classes $\{1, 3, \ldots, K\}$ and capacity $C - n_2 b_2$. Now, by induction the theorem holds for the reduced system. *We have to prove that in the reduced system class 1 is uncompressed.* If this holds, then Theorem 1 holds for the original system with classes $\{1, \ldots, K\}$ and capacity C. Indeed, if class 1 and 2 are uncompressed, then the badwidth shares of class 1 and class 2 are b_1 and b_2, respectively. We can apply the induction for the reduced system with classes $\{2, 3, \ldots, K\}$ and capacity $C - n_1 b_1$, that is, either case 1 or case 3 of Theorem 1 holds for the reduced system in the form:

1) If $\sum_{k=2}^{K} n_k b_k \leq C - n_1 b_1$, then $c_k = b_k$ for $2 \leq k \leq K$. For the original system we can rewrite the statement using simple rearrangement and using the fact that class 1 is uncompressed with bandwidth share b_1:
   If $\sum_{k=1}^{K} n_k b_k \leq C$, then $c_k = b_k$ for $1 \leq k \leq K$.

2) For the reduced system: If there exists an i^*, $2 \leq i^* \leq K - 1$ such that

$$\sum_{k=2}^{i^*} n_k b_k + \sum_{k=i^*+1}^{K} n_k g_k \frac{b_{i^*}}{g_{i^*}} \leq C - n_1 b_1$$

and

$$\sum_{k=2}^{i^*+1} n_k b_k + \sum_{k=i^*+2}^{K} n_k g_k \frac{b_{i^*+1}}{g_{i^*+1}} > C - n_1 b_1,$$

then

$$c_k = b_k, \text{ if } 2 \le k \le i^* \text{ and } c_k = \frac{g_k}{\sum_{i=i^*+1}^{K} g_i n_i} \left(C - \sum_{j=1}^{i^*} n_j b_j \right), \text{ if } i^* < k.$$

Using simple rearrangement and using the fact that class 1 is uncompressed with bandwidth share b_1, we can rewrite the statement and get the form that is written in the statement of Theorem 1.

In th rest of the proof we show that class 1 is uncompressed.

Case 3a. Assume further that $b_2 \le g_2 C$ holds. Assume that class 1 is compressed, that is,

$$\frac{g_1}{1 - g_2 n_2}(C - n_2 b_2) < b_1.$$

Taking into consideration condition that we have assumed $b_1 < g_1 C$, we have

$$\frac{g_1}{1 - g_2 n_2}(C - n_2 b_2) < g_1 C.$$

Rearranging the terms, we get

$$g_1(C - n_2 b_2) < g_1 C(1 - g_2 n_2)$$

$$g_1 C - g_1 n_2 b_2 < g_1 C - g_1 C g_2 n_2$$

$$b_2 > C g_2$$

which contradicts to assumption $b_2 \le g_2 C$.

Case 3b. Now, assume that $b_2 > g_2 C$ holds. The assumption, for any $x_1 \in [0, b_1]$ we have $c_2^{S \setminus \{1\}, \ C - n_1 x_1} = b_2$ in the reduced system with classes $\{2, \ldots, K\}$, means that for any $x_1 \in [0, b_1]$ we have

$$\frac{g_2}{\sum_{k=2}^{K} g_k n_k}(C - n_1 x_1) = \frac{g_2}{1 - n_1 g_1}(C - n_1 x_1) \ge b_2. \tag{14}$$

Especially for $x_1 = b_1$ we have

$$\frac{g_2}{1 - n_1 g_1}(C - n_1 b_1) \ge b_2$$

$$\frac{g_2}{1 - n_1 g_1}(C - n_1 b_1) - g_2 C \ge b_2 - g_2 C$$

$$g_2 C \left(\frac{1}{1 - n_1 g_1} - 1 \right) - \frac{g_2}{1 - n_1 g_1} n_1 b_1 \ge b_2 - g_2 C$$

$$\frac{n_1 g_2}{1 - n_1 g_1}(g_1 C - b_1) \ge b_2 - g_2 C. \tag{15}$$

Now, assume class 1 is compressed in the system with classes $\{1, 3, \ldots, K\}$ and capacity $C - n_2 b_2$, that is,

$$\frac{g_1}{1 - g_2 n_2}(C - n_2 b_2) < b_1$$

$$\frac{g_1}{1 - g_2 n_2}(n_2 b_2 - C) > -b_1$$

$$\frac{g_1}{1 - g_2 n_2}(n_2 b_2 - C) + g_1 C > g_1 C - b_1$$

$$\frac{n_2 g_1}{1 - g_2 n_2}(b_2 - g_2 C) > g_1 C - b_1. \tag{16}$$

From (15) and (16):

$$g_1 C - b_1 < \frac{n_2 g_1}{1 - n_2 g_2}(b_2 - g_2 C) \leq \frac{n_2 g_1}{1 - n_2 g_2}\frac{n_1 g_2}{1 - n_1 g_1}(g_1 C - b_1)$$

$$1 < \frac{n_2 g_1}{1 - n_2 g_2}\frac{n_1 g_2}{1 - n_1 g_1}$$

$$(1 - n_1 g_1)(1 - n_2 g_2) < n_1 n_2 g_1 g_2$$

$$1 < n_1 g_1 + n_2 g_2 \leq \sum_{i=1}^{K} n_i g_i = 1$$

which is a contradiction.

5 Discussion

For simplifying the following presentation let us introduce $w_i = n_i g_i$, the bandwidth share of the class $C_i = n_i c_i$.

The bandwidth allocation in the original DPS can also be considered as a mechanism, in which players (the classes of users in our setup) are competing to the resources C, they give payments (also called bids) $w_i (> 0)$ to a central entity (often referred to as resource manager), and this manager chooses an allocation d_i, $i = 1, \ldots, K$ by charging each class (within the class the charge always distributed evenly among the users) by the same price μ (assuming there is no price discrimination between the players). Because we already assumed that the manager tries to allocate the whole capacity of the resources, we have

$$\sum_{j=1}^{K} \frac{w_j}{\mu} = C \tag{17}$$

from which we get for the price

$$\mu = \frac{\sum_{j=1}^{K} w_j}{C}. \tag{18}$$

Applying this price for the payment w_i the allocation is

$$C_i = \frac{w_i}{\sum_{j=1}^{N} w_j} C. \tag{19}$$

Note that we also assume here that the players are *price takers*, that is they accept the price and the resulted allocations.

This proportional allocation above is also the solution of the following optimization problem.

$$\max_{\mathbf{x}} \sum_{i=1}^{N} w_i \log x_i , \quad \text{s.t.} \quad \sum_{i=1}^{K} x_i = C , \quad x_i \geq 0 \tag{20}$$

which is a simple rewrite of (Opt) according to our new quantities w_i and C_i.

The proportional allocation and the price taker characteristic have a much more fundamental property, which can be explained through the notion of *competitive equilibrium* between users and the resource manager [8].

In competitive equilibrium players (classes in our case) $(i = 1, \ldots, K)$ act to maximize their payoff functions $P(w_i, \mu)$ for a given price μ over w_i, which payoff consists of the utility $U_i(\frac{w_i}{\mu})$ by obtaining $\frac{w_i}{\mu}$ resources minus the payment w_i for that amount of resource to the manager, that is $P(w_i, \mu) := U_i(\frac{w_i}{\mu}) - w_i$. A pair $(\mathbf{w}, \mu)$, $\mathbf{w} \geq 0, \mu > 0$ is said to be a competitive equilibrium between the users and the network, if users maximize their payoff functions and the network determines the price as

$$\mu = \frac{\sum_{i=1}^{N} w_i}{C} . \tag{21}$$

It can also be shown that if $((\mathbf{w}, \mu))$ forms a competitive equilibrium, then the proportional allocation $C_i = \frac{w_i}{\sum_i w_i} C$ is a solution of the following *utilitarian social welfare*[3] maximization

$$\max_{\mathbf{x}} \sum_{i=1}^{K} U_i(x_i) , \quad \text{s.t.} \quad \sum_{i=1}^{K} x_i = C , \quad x_i \geq 0 \tag{OptSocial}$$

An important observation here is that the proportional allocation (19) spans a bridge between the individual user payoff maximization and the social welfare maximization, that is the two maximization coincide when proportional allocation applied. This remarkable property of proportional allocation also makes possible to investigate the case when players are *price anticipating*, that is the players take into account that the price is set according to (21) and they maximize their payoff

$$U_i \left(\frac{w_i}{\sum_{k=1}^{N} w_k} C \right) - w_i \tag{22}$$

with respect to their payments w_i [7].

As we can have seen through Theorem 1, the allocation in case of access rate limited shares (which is partially proportional) is the solution of (OptBounds), which is a quite natural extension of (Opt) by the constraints $x_i \leq b_i$. Based

---

[3] With a wide class of possible utility functions, for details see e.g. [8]

on this, we conjecture that the (appropriately modified) competitive equilibrium and the social welfare can be coupled with this allocation in case of limited access to the resource, that is

Conjecture:

The competitive equilibrium $(\mathbf{w}^*, \mu^*)$ exists, in which w_i^* maximizes the payoff function

$$P_i(w_i, \mu^*) = U_i\left(\max(\frac{w_i}{\mu^*}, n_i b_i)\right) - w_i \qquad (23)$$

where

$$\mu^* = \frac{\sum_{j=i^*+1}^{K} w_j}{C - \sum_{k=1}^{i^*} n_k b_k}, \qquad (24)$$

and the previously calculated allocation (9)

$$C_k = n_k b_k, \text{ if } k \leq i^* \ (k \in \mathcal{U}) \qquad (25)$$

and

$$C_k = \frac{w_k}{\sum_{i=i^*+1}^{K} w_i}\left(C - \sum_{j=1}^{i^*} n_j b_j\right) = \frac{w_k}{\mu^*}, \text{ if } \quad i^* < k \ (k \in \mathcal{Z}) . \qquad (26)$$

is a solution of the following utilitarian social welfare optimization extended by the access limits $x_i \leq n_i b_i$.

$$\max_{\mathbf{x}} \sum_{i=1}^{K} U_i(x_i) , \quad \text{s.t.} \quad \sum_{i=1}^{K} x_i = C , \quad x_i \in [0, n_i b_i] \qquad \text{(OptSocialBounds)}$$

6   Conclusion

In this paper we have proven that the bandwidth shares of the bandwidth-economical access rate limited discriminatory processor sharing is the solution of an extended constrained optimization task. The significance of this result lies in the fact that the DPS-related proportional bandwidth allocation play central role in analysing users in strategic environment when they are selfish and competing for the bandwidth. Our result could be used in analysing allocation games in which the players also have limits on their access to the network resources.

Acknowledgment. The research of the second author was also realized in the frames of TAMOP 4.2.4.A/2-11-1-2012-0001 National Excellence Program Elaborating and operating an inland student and researcher personal support system The project was subsidized by the European Union and co-financed by the European Social Fund.

The work of the third and fourth author was partially supported by the European Union and the European Social Fund through project FuturICT.hu (grant no.: TAMOP-4.2.2.C-11/1/KONV-2012-0013).

References

1. Avrachenkov, K., Ayesta, U., Brown, P., Nez-Queija, R.: Discriminatory processor sharing revisited. In: Proc. IEEE Infocom 2005, Miami, FL, pp. 784–795 (2005)
2. Ayesta, U., Mandjes, M.: Bandwidth-sharing networks under a diffusion scaling. Annals Operation Research 170(1), 41–58 (2009)
3. Cheung, S., van den Berg, H., Boucherie, R.: Decomposing the queue length distribution of processor-sharing models into queue lengths of permanent customer queues. Performance Evaluation 62(1-4), 100–116 (2005)
4. Fan, Z.: Dimensioning Bandwidth for Elastic Traffic. In: Gregori, E., Conti, M., Campbell, A.T., Omidyar, G., Zukerman, M. (eds.) NETWORKING 2002. LNCS, vol. 2345, pp. 826–837. Springer, Heidelberg (2002)
5. Fayolle, G., Mitrani, I., Iasnogorodski, R.: Sharing a processor among many job classes. J. ACM 27(3), 519–532 (1980)
6. Frolkova, M., Reed, J., Zwart, B.: Fixed-point approximations of bandwidth sharing networks with rate constraints. Performance Evaluation Review 39(3), 47–49 (2011)
7. Johari, R.: The Price of Anarchy and the Design of Scalable Resource Allocation Mechanisms. In: Algorithmic Game Theory, pp. 543–568. Cambridge Univ. Press (2007)
8. Kelly, F.: Charging and rate control for elastic traffic. European Transactions on Telecommunications 8, 33–37 (1997)
9. van Kessel, G., Nunez-Queija, R., Borst, S.: Asymptotic regimes and approximations for discriminatory processor sharing. ACM SIGMETRICS Performance Evaluation Review 32(2), 44–46 (2004)
10. Kőrösi, A., Székely, B., Vámos, P., Bíró, J.: Characterization of peak-rate limited DPS with Pareto-efficient bandwidth sharing. Annales Mathematicae et Informaticae 39 (2012)
11. Kleinrock, L.: Time-shared systems: A theoretical treatment. J. of ACM 14(2), 242–261 (1967)
12. Lakshmikantha, A., Srikant, R., Beck, C.: Differential equation models of flow-size based priorities in internet routers. International Journal of Systems, Control and Communications 2(1), 170–196 (2010)
13. Lindberger, K.: Balancing quality of service, pricing and utilisation in multiservice networks with stream and elastic traffic. In: ITC-16: International Teletraffic Congress, pp. 1127–1136 (1999)
14. O'Donovan, T.M.: Direct solutions of M/G/1 processor-sharing models. Operations Research 22, 1232–1235 (1974)
15. Pályi, P.L., Kőrösi, A., Székely, B., Bíró, J., Rácz, S.: Characterization of peak-rate limited bandwidth efficient DPS. Acta Polytechnica Hungarica 9 (2012)
16. Rege, K., Sengupta, B.: Queue length distribution for the discriminatory processor-sharing queue. Operation Research 44, 653–657 (1996)
17. Riedl, A., Bauschert, T., Frings, J.: A framework for multi-service IP network planning. In: International Telecommunication Network Strategy and Planning Symposium (Networks), pp. 183–190 (2002)

A Central Limit Theorem
for Markov-Modulated Infinite-Server Queues

Joke Blom[1], Koen De Turck[2,*], and Michel Mandjes[3,1,**]

[1] CWI, P.O. Box 94079, 1090 GB Amsterdam, The Netherlands
[2] TELIN, Ghent University, St.-Pietersnieuwstraat 41, B9000 Gent, Belgium
[3] Korteweg-de Vries Institute for Mathematics, University of Amsterdam,
Science Park 904, 1098 XH Amsterdam, The Netherlands
joke.blom@cwi.nl, kdeturck@telin.ugent.be, M.R.H.Mandjes@uva.nl

Abstract. This paper studies an infinite-server queue in a Markov environment, that is, an infinite-server queue with arrival rates and service times depending on the state of a Markovian background process. Scaling the arrival rates λ_i by a factor N and the rate q_{ij} of the background process by a factor N^α, with $\alpha \in \mathbb{R}^+$, we establish a central limit theorem as N tends to ∞. We find different scaling regimes, based on the value of α. Remarkably, for $\alpha < 1$, we find a central limit theorem with a non-square-root scaling but rather with $N^{\alpha/2}$; in the expression for the variance deviation matrices appear.

Keywords: Infinite-server queues, Markov modulation, central limit theorem, deviation matrices.

1 Introduction

Infinite-server queues have found widespread use in various application domains, often as an approximation for models with many servers. In these systems jobs arrive, are served in parallel, and leave when their service is completed; the jobs do not interfere with each other. The infinite-server queue was originally developed (over a century ago) to model the dynamics of the number of calls in progress in a communication network. More recently, however, applications in various other domains have been explored, such as road traffic [11] and biology [9,10].

In the standard infinite-server model, jobs arrive according to a Poisson process with rate λ, where their service times form a sequence of independent and identically distributed (i.i.d.) random variables (distributed as a random variable

---

* Work done while K. de Turck was visiting Korteweg-de Vries Institute for Mathematics, University of Amsterdam, the Netherlands, with greatly appreciated financial support from *Fonds Wetenschappelijk Onderzoek / Research Foundation – Flanders*.
** M. Mandjes is also with Eurandom, Eindhoven University of Technology, Eindhoven, the Netherlands, and IBIS, Faculty of Economics and Business, University of Amsterdam, Amsterdam, the Netherlands.

A. Dudin and K. De Turck (Eds.): ASMTA 2013, LNCS 7984, pp. 81–95, 2013.

B with finite first moment), independent of the call arrival process. A key result states that in this $M/G/\infty$ queue the stationary number of jobs in the system obeys a Poisson distribution with mean $\lambda\,\mathbb{E}B$; there is insensitivity, in that the stationary distribution depends on B only through its mean.

In many practical situations, however, the assumptions are not realistic: there is no constant arrival rate, and the jobs do not stem from a single distribution. A model that allows the input process to exhibit some sort of 'burstiness' is the *Markov-modulated* infinite-server queue. In this system, a finite-state irreducible continuous-time Markov process (often referred to as the *background process*) modulates the input process: if the background process is in state i, the arrival process is a Poisson process with rate, say, λ_i, while the service times are distributed as a random variable, say, B_i (while the obvious independence conditions are imposed). Often the B_is are assumed exponential with mean μ_i^{-1}.

The Markov-modulated infinite-server queue has attracted (relatively limited) attention over the past decades. The main focus in the literature so far has been on characterizing (through the derivation of moments, or even the full probability generating function) of the steady-state number of jobs in the system; see [3,5,7,8] and references therein. Interestingly, under an appropriate time-scaling [1,6] in which the transitions of the background process occur at a faster rate than the Poisson arrivals, we retrieve the Poisson distribution for the steady-state number of jobs in the system. Recently, transient results have been obtained as well, under specific scalings of the arrival rates and transition times of the modulating Markov chain [1,2].

Contribution. The present paper considers one of the scalings studied in [1,2]: the arrival rates λ_i are scaled by a factor N and the rate q_{ij} of the background process by a factor N^α. However, where in [1,2] the situation of $\alpha > 1$ was considered, we now allow α to be any positive number, which results in significantly richer limit behavior, as we will demonstrate. We focus the number of jobs in the scaled system at time t, denoted by $M^{(N)}(t)$, aiming at deriving a central limit theorem (CLT) for $M^{(N)}(t)$ as well as for its stationary counterpart $M^{(N)}$. Interestingly, we find different scaling regimes, based on the value of α: for $\alpha > 1$ the variance of $M^{(N)}(t)$ scales essentially linearly in N, while for $\alpha < 1$ it behaves as $N^{2-\alpha}$.

The approach is as follows. We first derive differential equations for the probability generating functions (pgfs) of $M^{(N)}(t)$ and $M^{(N)}$. Then we establish weak laws of large numbers for both random quantities, so that we know how to centre them. Finally, the resulting centered random variables are scaled, so as to obtain a CLT. The proofs rely on (non-trivial) manipulations of the differential equations that describe the pgfs; interestingly *deviation matrices* [4] play a crucial role here.

Organization. The organization of the rest of this paper is as follows. In Section 2, we explain the model in detail and introduce some notation. In Section 3, systems of differential equations are derived that describe the stationary and transient distribution of the number of jobs in the system. Then in Sections 4–5, we state and prove the main results of this paper, viz. the CLTs mentioned above, for the stationary and transient distribution, respectively. The final section of the paper, Section 6, contains a discussion and concluding remarks.

2 Model Description, Preliminaries, and Motivation

As described in the introduction, this paper studies an infinite-server queue with Markov-modulated Poisson arrivals and general service times. In full detail, the model is described as follows.

Consider an irreducible continuous-time Markov process $(J(t))_{t \in \mathbb{R}}$ on a finite state space $\{1, \ldots, d\}$, with $d \in \mathbb{N}$. Its rate matrix is given by $Q := (q_{ij})_{i,j=1}^{d}$; the q_{ij} are nonnegative if $i \neq j$, and $q_{ii} = -\sum_{j \neq i} q_{ij}$. Let π_i the stationary probability that the background process is in state i, for $i = 1, \ldots, d$. The time spent in state i (often referred to as the *transition time*) has an exponential distribution with mean $1/q_i$, where $q_i := -q_{ii}$.

While the process $(J(t))_{t \in \mathbb{R}}$, often referred to as the *background process* or *modulating process*, is in state i, jobs arrive according to a Poisson process with rate $\lambda_i \geq 0$. The service times are assumed to be exponentially distributed with rate μ_i, but, importantly this statement can be interpreted in two ways:

▷ In the first variant the service rate results is determined the background state upon its arrival;
▷ In the second variant all jobs present at a certain time instant t are subject to a hazard rate determined by the state of background chain at time t, regardless of when they arrived.

We focus in this paper on the latter variant, but analogous claims can be established for the former variant.

For notational convenience, we introduce the diagonal matrices Λ and M, where $[\Lambda]_{ii} = \lambda_i$ and $[M]_{ii} = \mu_i$. We denote the invariant distribution corresponding to the transition matrix Q by the (row-)vector $\boldsymbol{\pi}$; we follow the convention that we write vectors in bold fonts. In the sequel we frequently use the 'time-average arrival rate' $\lambda_\infty := \sum_{i=1}^{d} \pi_i \lambda_i = \boldsymbol{\pi}\boldsymbol{\lambda}$ and 'time average departure rate' $\mu_\infty := \sum_{i=1}^{d} \pi_i \mu_i = \boldsymbol{\pi}\boldsymbol{\mu}$.

In this paper, we consider a scaling in which the arrival rates and the background process are sped up at a possibly distinct rate. The arrival rates are scaled linearly, that is, as $\lambda_i \mapsto N\lambda_i$, whereas the background chain is scaled as $q_{ij} \mapsto N^\alpha q_{ij}$, for some non-negative α. We call the resulting background process $(J^{(N)}(s))_{s \in \mathbb{R}}$, to stress the dependence on N. Under this scaling the infinite-server system exhibits non-standard behavior, where it turns out to matter whether α is smaller than, equal to or larger than 1. Letting the system start off empty at time 0, we consider the number of jobs present at time t, denoted by $M^{(N)}(t)$; we write $M^{(N)}$ for its stationary counterpart. Our main result is a 'non-standard CLT': with $\varrho(t) := (\lambda_\infty/\mu_\infty) \cdot (1 - e^{-\mu_\infty t})$,

$$\frac{M^{(N)}(t) - N\varrho(t)}{N^\beta}$$

converges in distribution to a zero-mean Normal distribution with a certain variance, say, $\sigma^2(t)$. Here, importantly, for $\alpha > 1$ we have that the scaling parameter β equals the usual $\frac{1}{2}$, while for $\alpha \leq 1$ it has the uncommon value $1 - \frac{\alpha}{2}$. A similar dichotomy holds for the stationary counterpart $M^{(N)}$.

In Fig. 1 we see typical sample paths of the number of jobs in the system. In the left panel the background process evolves on a substantially slower time scale than the arrival process (α close to 0), so that the number of jobs converges to a local equilibrium during each transition time. In the right panel the background process is faster than the arrival process (α substantially larger than 1), so that the process essentially behaves as a (non-modulated) M/M/∞ system.

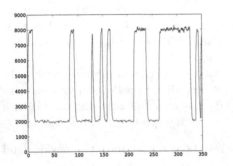 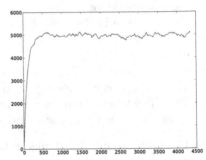

Fig. 1. Evolution of number of jobs in the system. Left: background process is *slow* relative to arrival process; right: background process is *fast* relative to arrival process.

In addition, we prove that for $\alpha > 1$ the variance $\sigma^2(t)$ equals $\varrho(t)$. The intuition here is that in this regime the background process jumps essentially faster than the arrival process, so that the arrival stream is nearly Poisson with parameter λ_∞. The resulting system is therefore, as $N \to \infty$, close to an M/M/∞, in which the transient distribution is Poissonian, thus explaining the fact that both the normalized mean and the normalized variance equal $\varrho(t)$. If $\alpha < 1$ the background process is essentially slower than the arrival process. Here the computations are substantially more complex: $\sigma^2(t)$ turns out to be a linear combination of the entries of the so-called deviation matrix D of the transition rate matrix Q; see for a number of fundamental properties of deviation matrices [4] and references therein.

We illustrate the above dichotomy by determining the asymptotic variance $\sigma^2 := \lim_{t \to \infty} \sigma^2(t)$ through an elementary computation, which reveals some of the key steps of the (considerably more elaborate) derivations later in this paper. We focus on $\mathbb{V}\mathrm{ar}M^{(N)}$. To this end, we recall the formula in [8] for the n-th factorial moment:

$$\mathbb{E}\left[M^{(N)}(M^{(N)} - 1) \cdot \ldots \cdot (M^{(N)} - n + 1)\right] = n!N^n \boldsymbol{\pi}\Lambda X_1 \Lambda X_2 \Lambda \cdot \ldots \cdot X_{n-1}\Lambda X_n \mathbf{1},$$

where $X_n := (nM - N^\alpha Q)^{-1}$. To keep this introductory derivation as focused as possible, we consider the special case that the service rates in each of the states are identical, i.e., $M = \mu I$ for some $\mu > 0$.

As a first elementary computation, we compute $\mathbb{E}[M^{(N)}]$. According to the above formula, this mean equals $N\pi\Lambda X_1 \mathbf{1}$. Realize that, for any $n \in \mathbb{N}$, by virtue of $Q^i \mathbf{1} = \mathbf{0}$ for $i \in \{1, 2, \ldots\}$,

$$X_n \mathbf{1} = \frac{1}{n\mu}\left(I - N^\alpha Q \frac{1}{n\mu}\right)^{-1}\mathbf{1} = \sum_{i=0}^{\infty}\left(N^\alpha Q \frac{1}{n\mu}\right)^i \mathbf{1} = \frac{1}{n\mu}\mathbf{1}.$$

It now follows that $\mathbb{E}[M^{(N)}] = N\varrho$, with $\varrho := \lambda_\infty/\mu$.

Now concentrate on the variance. The relation in the previous display yields that

$$\mathbb{E}\left[M^{(N)}(M^{(N)} - 1)\right] = \frac{1}{\mu}N^2\pi\Lambda X_1 \Lambda \mathbf{1}.$$

To evaluate this expression, we recall some concepts pertaining to the deviation matrix of Markov processes; see e.g. [4]. In particular, we let $\Pi := \mathbf{1}\pi$ denote the *ergodic matrix*. We also define the *fundamental matrix* $F := (\Pi - Q)^{-1}$ and the *deviation matrix* $D := F - \Pi$. We will frequently use the identity $QF = FQ = \Pi - I$, as well as the fact that $\Pi D = D\Pi = 0$ (here 0 is to be read as an all-zeros $d \times d$ matrix).

Lemma 1. *We have that (i) $X_n\Pi = (n\mu)^{-1}\Pi$ and (ii) $X_n = (n\mu)^{-1}\Pi + N^{-\alpha}D + O(N^{-2\alpha})$.*

Proof. First note that, $n\mu X_n - N^\alpha X_n Q = I$. By multiplying both sides by Π, claim (i) follows immediately. Also, noting that $QD = \Pi - I$, we find

$$X_n = X_n\Pi + N^{-\alpha}(I - n\mu X_n)D = \frac{1}{n\mu}\Pi + N^{-\alpha}(I - n\mu X_n)D.$$

Iterating this recursion, we obtain

$$X_n = \frac{1}{n\mu}\Pi + N^{-\alpha}(I - n\mu X_n\Pi)D + O(N^{-2\alpha}),$$

which yields claim (ii) because of $\Pi D = 0$.   □

With this lemma in place, computing the variance is fairly straightforward. Note that

$$\mathbb{E}\left[M^{(N)}(M^{(N)} - 1)\right] = \frac{1}{\mu}N^2\pi\Lambda X_1\Lambda\mathbf{1} = N^2\varrho^2 + N^{2-\alpha}\frac{1}{\mu}\pi\Lambda D\Lambda\mathbf{1} + O(N^{2-2\alpha}),$$

which leads to

$$\mathbb{V}\mathrm{ar}[M^{(N)}] = \mathbb{E}\left[M^{(N)}(M^{(N)} - 1)\right] - \mathbb{E}\left[M^{(N)}\right]^2 + \mathbb{E}\left[M^{(N)}\right]$$
$$= N^{2-\alpha}\sigma_m^2 + N\varrho + O(N^{2-2\alpha}),$$

where

$$\sigma_m^2 := \frac{1}{\mu}\boldsymbol{\pi}\Lambda D\Lambda\mathbf{1} = \frac{1}{\mu}\sum_{i=1}^{d}\sum_{j=1}^{d}\pi_i\lambda_i\lambda_j[D]_{ij}.$$

From this expression, we observe that for $\alpha > 1$ the variance behaves as $N\varrho$ (so that we have 'Poisson-like' behavior), while for $\alpha < 1$ we obtain a variance that features the deviation matrix D. The objective of this paper is now to verify whether this observation for the variance (for the special case of identical service rates μ_i) translates into fully-fledged CLT s, both under stationarity and in the transient case. The above computation suggests that we have the ordinary $\sqrt{N}$ scaling if $\alpha > 1$, while it is anticipated that we have to scale by $N^{1-\alpha/2}$ for $\alpha < 1$.

We use a fairly classical approach to proving these CLT s: we show that under the appropriate scaling, the moment generating function converges to that of the Normal distribution. The general outline of the proofs in this paper is as follows. We use following three vector-valued generating functions throughout the paper: $\boldsymbol{p}$ denotes the unscaled probability generating function (pgf); $\tilde{\boldsymbol{p}}$ denotes the corresponding moment generating function (mgf), scaled and centered appropriately for the central limit theorem at hand; and $\bar{\boldsymbol{p}}$ denotes the mgf under the law of large numbers scaling. For the transient cases, these generating functions involve an extra argument t to incorporate time. All three generating functions are vector-valued (of dimension $1 \times d$ as we consider distributions jointly with the background process $J^{(N)}(\cdot)$). Lastly, ϕ denotes the (scalar) mgf under the scaling. Then the approach consists of the following steps:

▷ We derive a differential equation of the pgf $\boldsymbol{p}$.
▷ We establish the weak law of large numbers under the scaling by making use of the mgf $\bar{\boldsymbol{p}}$, so as to establish the mean behavior.
▷ We scale and center the pgf so as to obtain a differential equation in terms of the mgf $\tilde{\boldsymbol{p}}$.
▷ By discarding asymptotically vanishing terms, we show that the differential equation has in the limit the unique solution, $\phi(\theta) = \exp(\theta^2\sigma^2)$, for some σ^2; this corresponds to a zero-mean Normal distribution with variance σ^2. Due to Lévy's continuity theorem, pointwise convergence of characteristic functions implies convergence in distribution. In this way we derive the CLT.

3 Stationary and Transient Distribution

In this section we derive systems of differential equations for the pgf of the number of jobs in the system, both for the stationary and time-dependent behavior. There is a direct relation with the results on stationary factorial moments, as presented in [8]; our analysis characterizes the pgf, and in addition also includes the transient case. For ease we consider the unscaled model (that is, $N = 1$); the differential equations can be translated easily into those for the N-scaled process introduced in Section 2.

We consider the process $(J^{(1)}(t), M^{(1)}(t))_{t\in\mathbb{R}}$ which is a Markov process on the state space $\{1, \ldots, d\} \times \mathbb{N}$. It has the infinite generator matrix

$$
\begin{pmatrix}
Q - \Lambda & \Lambda & & & \\
M & Q - M - \Lambda & \Lambda & & \\
& 2M & Q - 2M - \Lambda & \Lambda & \\
& & 3M & Q - 3M - \Lambda & \Lambda \\
& & & & \ddots & \ddots & \ddots
\end{pmatrix}.
$$

We set out to find the invariant distribution $(\boldsymbol{p}_k)_{k=0}^{\infty}$, where $\boldsymbol{p}_k$ is a vector of length d whose entries are defined by $[\boldsymbol{p}_k]_j := \mathbb{P}(M^{(1)} = k, J^{(1)} = j)$. The (vector-)pgf $\boldsymbol{p}(z)$ is then given by

$$
\boldsymbol{p}(z) := \sum_{k=0}^{\infty} \boldsymbol{p}_k z^k.
$$

Proposition 1. *The pgf $\boldsymbol{p}(z)$ satisfies the following differential equation:*

$$
\boldsymbol{p}(z)Q = (z-1)[\boldsymbol{p}'(z)M - \boldsymbol{p}(z)\Lambda].
$$

Proof. We immediately have that

$$
\boldsymbol{p}_{k-1}\Lambda + \boldsymbol{p}_k(Q - \Lambda - kM) + (k+1)\boldsymbol{p}_{k+1}M = 0, \tag{1}
$$

for all $k \in \mathbb{N}$, if we conveniently set $\boldsymbol{p}_{-1} = 0$. From the standard relations

$$
\sum_{k=0}^{\infty}(k+1)\boldsymbol{p}_{k+1}z^k = \boldsymbol{p}'(z), \text{ and } \sum_{k=0}^{\infty}k\boldsymbol{p}_k z^k = z\boldsymbol{p}'(z),
$$

we obtain by multiplying both sides of (1) by z^k and summing over $k \in \mathbb{N}$, after some elementary manipulations,

$$
z\boldsymbol{p}(z)\Lambda + \boldsymbol{p}(z)(Q - \Lambda) - z\boldsymbol{p}'(z)M + \boldsymbol{p}'(z)M = 0.
$$

The claim follows directly. $\qquad\square$

Substituting $z = 1$ gives us: $\boldsymbol{p}(1)Q = 0$, so that $\boldsymbol{p}(1) = \boldsymbol{\pi}$, as desired: the stationary distribution of the background chain in the entire Markov chain must be the same as the stationary distribution of the background chain in isolation. We can find the factorial moments of the queue content by repeated differentiations and subsequently substituting $z = 1$. Our results agree with those in [8], in particular the formula for the factorial moments as mentioned in Section 2.

We now present the analogous differential equation for the transient case.

Proposition 2. *The generating function $\boldsymbol{p}(t, z)$ satisfies the following differential equation:*

$$
\frac{\partial \boldsymbol{p}}{\partial t} = \boldsymbol{p}Q + (z-1)\left(\boldsymbol{p}\Lambda - \frac{\partial \boldsymbol{p}}{\partial z}M\right).
$$

Proof. By virtue of the Chapman-Kolgomorov equation, we have that

$$\frac{\partial \boldsymbol{p}_k}{\partial t} = \boldsymbol{p}_{k-1}(t)\Lambda + \boldsymbol{p}_k(t)(Q - \Lambda - kM) + (k+1)\boldsymbol{p}_{k+1}(t)M.$$

for all $k \in \mathbb{N}$, if we put $\boldsymbol{p}_{-1}(t) = 0$ for all $t \geq 0$. From this point on, we can follow the lines of the proof of Prop. 1. □

4   Limit Results for Stationary Distribution

Under the scaling $\Lambda \mapsto N\Lambda$ and $Q \mapsto N^\alpha Q$, we have the following modified differential equation:

$$\boldsymbol{p}(z)Q = N^{-\alpha}(z-1)[\boldsymbol{p}'(z)M - N\boldsymbol{p}(z)\Lambda]; \tag{2}$$

realize that $\boldsymbol{p}(\cdot)$ depends on N. We rewrite the equation in the previous display so as to introduce the fundamental matrix F:

$$\boldsymbol{p}(z) = \boldsymbol{p}(z)\Pi + N^{-\alpha}(z-1)[N\boldsymbol{p}(z)\Lambda - \boldsymbol{p}'(z)M]F. \tag{3}$$

We first establish the mean number of jobs in the system in stationarity. More specifically, we prove the following claim.

Lemma 2. $N^{-1}M^{(N)}$ *converges in probability to* $\varrho = \lambda_\infty/\mu_\infty$ *as* $N \to \infty$.

Proof. We introduce the scaled moment generating function $\bar{\boldsymbol{p}}(\theta) := \boldsymbol{p}(z(\theta))$, with $z \equiv z(\theta) = \exp(\theta N^{-1})$. Evidently,

$$\frac{\mathrm{d}\bar{\boldsymbol{p}}}{\mathrm{d}\theta} = \frac{\mathrm{d}\boldsymbol{p}}{\mathrm{d}z}\frac{\mathrm{d}z}{\mathrm{d}\theta} = N^{-1}\frac{\mathrm{d}\boldsymbol{p}}{\mathrm{d}z}.$$

Substituting these expressions in Eqn. (3), and noting that $z = 1 + \theta N^{-1} + O(N^{-2})$, we obtain

$$\bar{\boldsymbol{p}} = \bar{\boldsymbol{p}}\Pi + N^{-\alpha}\theta\left[\bar{\boldsymbol{p}}\Lambda - \bar{\boldsymbol{p}}'M\right]F + o(N^{-\alpha}).$$

Note that $\bar{\boldsymbol{p}} = \bar{\boldsymbol{p}}\Pi + O(N^{-\alpha})$ and $\bar{\boldsymbol{p}}' = \bar{\boldsymbol{p}}'\Pi + O(N^{-\alpha})$, so that by right-multiplying the previous display by $\mathbf{1}$,

$$0 = N^{-\alpha}\theta\left[\bar{\boldsymbol{p}}\mathbf{1}\lambda_\infty - \bar{\boldsymbol{p}}'\mathbf{1}\mu_\infty\right] + o(N^{-\alpha}),$$

realizing that $F\mathbf{1} = \mathbf{1}$. We thus find a differential equation in $\phi(\theta) := \bar{\boldsymbol{p}}(\theta)\mathbf{1}$, with the solution

$$\phi(\theta) = K\exp\left(\frac{\lambda_\infty}{\mu_\infty}\theta\right);$$

here K is an integration constant, which equals 1 as $\phi(\theta)$ is a mgf. We have thus found the mgf of the constant ϱ. By Lévy's continuity theorem, we have convergence in distribution of $N^{-1}M^{(N)}$ to ϱ, but convergence in probability to a constant implies convergence in distribution to the same constant. This completes the proof. □

Now we know that $N^{-1}M^{(N)}$ can be centered by subtracting ϱ. In the next result we identify the right scaling such that the centered random variable obeys a CLT.

Theorem 1. *The random variable*

$$N^{\beta/2}\left(\frac{M^{(N)}}{N} - \varrho\right)$$

converges to a Normal distribution with zero mean and variance σ^2 as $N \to \infty$; here the scaling parameter β equals $\min\{\alpha, 1\}$, and $\sigma^2 := \sigma_m^2 1_{\{\alpha \leq 1\}} + \varrho 1_{\{\alpha \geq 1\}}$, with $\sigma_m^2 := \mu_\infty^{-1}\boldsymbol{\pi}(\Lambda - \varrho M)D(\Lambda - \varrho M)\mathbf{1}$.

Proof. The proof strategy is as follows. We will establish that the moment generating function $\phi(\theta)$ of the scaled sequence of rv's satisfies in the limit $N \to \infty$) the differential equation $\phi' = \theta\sigma^2\phi$. This differential equation has a unique mgf among its solutions, viz. the mgf of the zero-mean Normal distribution with variance σ^2. By Lévy's continuity theorem this implies convergence in distribution to this random variable.

To this end, we first introduce the centered and scaled mgf $\tilde{\boldsymbol{p}}(\theta)$. We would like to perform a change of variables in Eqn. (3) so as to obtain a differential equation in $\tilde{\boldsymbol{p}}(\theta)$. Note that

$$\tilde{\boldsymbol{p}}(\theta) = \exp(-\varrho\theta N^{\beta/2})\boldsymbol{p}(\exp(\theta N^{-1+\beta/2})),$$

which can be written as

$$\boldsymbol{p}(z) = \exp(\varrho\theta N^{\beta/2})\tilde{\boldsymbol{p}}(\theta),$$

where $z \equiv z(\theta) = \exp(\theta N^{-1+\beta/2})$. Consider the second order Taylor expansions of z and z^{-1}:

$$z = 1 + \theta N^{-1+\beta/2} + \frac{1}{2}\theta^2 N^{-2+\beta} + O(N^{-3+3\beta/2});$$

$$z^{-1} = 1 - \theta N^{-1+\beta/2} + \frac{1}{2}\theta^2 N^{-2+\beta} + O(N^{-3+3\beta/2}).$$

Note that

$$\frac{\mathrm{d}\boldsymbol{p}}{\mathrm{d}z}\frac{\mathrm{d}z}{\mathrm{d}\theta} = \exp(\varrho\theta N^{\beta/2})\left(\varrho N^{\beta/2}\tilde{\boldsymbol{p}} + \frac{\mathrm{d}\tilde{\boldsymbol{p}}}{\mathrm{d}\theta}\right),$$

where

$$\frac{\mathrm{d}z}{\mathrm{d}\theta} = N^{-1+\beta/2}\exp(\theta N^{1-\beta/2}) = N^{-1+\beta/2}z.$$

Combining the above, we conclude that

$$\frac{\mathrm{d}\boldsymbol{p}}{\mathrm{d}z} = z^{-1}\exp(\varrho\theta N^{\beta/2})\left(N^{1-\beta}\varrho\tilde{\boldsymbol{p}} + N^{1-\beta/2}\frac{\mathrm{d}\tilde{\boldsymbol{p}}}{\mathrm{d}\theta}\right).$$

Now perform the change of variables and substitute the expressions for $p(z)$ and $p'(z)$ into Eqn. (3):

$$\tilde{p} = \tilde{p}\Pi + N^{1-\alpha}(z-1)\tilde{p}\Lambda F - N^{1-\alpha}\left(1 - \frac{1}{z}\right)\lambda_\infty \tilde{p}F$$

$$- N^{1-\alpha-\beta/2}\left(1 - \frac{1}{z}\right)\tilde{p}'MF. \tag{4}$$

The next step is to introduce the Taylor expansions for z and z^{-1}. We assume that $\beta \leq 1$ and $\beta \leq \alpha$, as is consistent with the proof statement. By deleting every term that has a provably smaller order than $N^{-\alpha}$, we obtain

$$\tilde{p} = \tilde{p}\Pi + \theta N^{\beta/2-\alpha}\tilde{p}(\Lambda - \varrho M)F + \frac{\theta^2 N^{\beta-1-\alpha}}{2}\tilde{p}(\Lambda + \varrho M)F - \theta N^{-\alpha}\tilde{p}'MF, \tag{5}$$

with an error term that is $o(N^{-\alpha})$. It takes some careful but elementary steps to verify that this is indeed justified; the restrictions imposed on β are intensively used here. Indeed, the third-order Taylor term for the second and third term of the right-hand side of Eqn. (4) has order $N^{1-\alpha-3+3\beta/2}$, which is indeed smaller than $N^{-\alpha}$. The fourth term has as a second order Taylor term with degree $N^{-1-\alpha+\beta/2}$, which is also smaller than $N^{-\alpha}$.

We now transform the differential equation in $\tilde{p}$ into one in $\phi(\theta) = \tilde{p}(\theta)\mathbf{1}$. Note that due to the restrictions on β, we have that $\tilde{p} = \tilde{p}\Pi + o(1) = \phi\pi + o(1)$, as well as $\tilde{p}' = \tilde{p}'\Pi + o(1) = \phi'\pi + o(1)$. Hence if we right-multiply Eqn. (5) by $\mathbf{1}$ and divide by $\theta N^{-\alpha}$,

$$0 = N^{\beta/2}\tilde{p}(\Lambda - \varrho M)\mathbf{1} + \frac{\theta N^{\beta-1}}{2}\tilde{p}(\Lambda + \varrho M)\mathbf{1} - \phi'\pi M\mathbf{1} + o(1),$$

where we use $F\mathbf{1} = \mathbf{1}$. We further manipulate this expression so that occurrences of $\tilde{p}$ are replaced by $\phi(\theta)$, by substituting Eqn. (5). We obtain

$$0 = N^{\beta/2}\phi\pi(\Lambda - \varrho M)\mathbf{1} + \theta N^{\beta-\alpha}\phi\pi(\Lambda - \varrho M)F(\Lambda - \varrho M)\mathbf{1}$$

$$+ \frac{\theta N^{\beta-1}}{2}\phi\pi(\Lambda + \varrho M)\mathbf{1} - \phi'\pi M\mathbf{1} + o(1).$$

Now realize that $\pi(\Lambda - \varrho M)\mathbf{1} = \lambda_\infty - \varrho\mu_\infty = 0$, and likewise

$$F(\Lambda - \varrho M)\mathbf{1} = (D + \Pi)(\Lambda - \varrho M)\mathbf{1} = D(\Lambda - \varrho M)\mathbf{1}.$$

Using the definitions of σ_m^2, we obtain the differential equation

$$\phi' = \theta\left(N^{\beta-\alpha}\sigma_m^2 + N^{\beta-1}\varrho\right)\phi + o(1).$$

First, note that if we choose β smaller than both α and 1, we do not obtain a CLT, but rather that the random variable under study converges in distribution to the constant 0. Hence, we take $\beta = \min(\alpha, 1)$, in which case the largest term dominates, with both terms contributing if $\alpha = 1$. We have thus established the claim.   □

From Thm. 1 we conclude that the variance σ^2 equals ϱ for $\alpha > 1$, in agreement with the intuition presented earlier: the system essentially behaves as a normal $M/M/\infty$ system, with mean and variance roughly equalling $N\varrho$. If $\alpha < 1$ the timescale of the background process is relatively slow, so that the variance of $M^{(N)}$ is more than linear. In addition we note that if $\alpha = 1$ *both* terms appear in σ^2: we then have that $\sigma^2 = \sigma_m^2 + \varrho$.

In case $\mu_i = \mu$ for all $i \in \{1, \dots, d\}$, we find, using $D\mathbf{1} = 0$ and $\pi D = 0$,

$$\pi(\Lambda - \varrho M)D(\Lambda - \varrho M)\mathbf{1} = \pi \Lambda D \Lambda \mathbf{1},$$

so that $\sigma_m^2 = \mu^{-1}\pi \Lambda D \Lambda \mathbf{1}$, agreeing with the findings presented in Section 2.

5 Limit Results for Transient Distribution

The derivation of the transient CLT is a bit more complicated, but follows essentially the same lines as the stationary CLT. Therefore, we provide an appropriately abridged derivation in this section. We assume that at time 0, the system starts off empty, while the background chain is in equilibrium. Due to the scaling, the latter assumption does not change the results, and is quite straightforwardly accounted for. Under the scaling $\Lambda \mapsto N\Lambda$ and $Q \mapsto N^\alpha Q$, we have the following differential equation:

$$\frac{\partial \boldsymbol{p}}{\partial t} = N^{-\alpha}\boldsymbol{p}Q + (z-1)\left(N\boldsymbol{p}\Lambda - \frac{\partial \boldsymbol{p}}{\partial z}M\right). \tag{6}$$

As before, this equation can be written in terms of the fundamental matrix F:

$$\boldsymbol{p} = \boldsymbol{p}\Pi + N^{-\alpha}(z-1)\left(N\boldsymbol{p}(z)\Lambda - \frac{\partial \boldsymbol{p}}{\partial z}M\right)F - N^{-\alpha}\frac{\partial \boldsymbol{p}}{\partial t}F. \tag{7}$$

Notice that this differential equation is basically the same as in the previous section, except for the last term (containing the partial derivative with respect to t). To make the proofs compact, we primarily concentrate on this new term. We start by analyzing the mean number in the system at time t; recall that $\varrho := \lambda_\infty/\mu_\infty$.

Lemma 3. $N^{-1}M^{(N)}(t)$ *converges in probability to* $\varrho(t) = \varrho \cdot (1 - e^{-\mu_\infty t})$.

Proof. We follow the lines of the proof of Lemma 2 closely, and introduce the transient scaled moment generating function $\bar{\boldsymbol{p}}(t, \theta)$:

$$\bar{\boldsymbol{p}}(t, \theta) = \boldsymbol{p}(t, \exp(\theta N^{-1})).$$

The expressions for z and the derivative with respect to θ do not change, except for the fact that the ordinary derivative becomes a partial derivative. The derivative with respect to t is simply $\partial \bar{p}/\partial t = \partial p/\partial t$. We obtain

$$\bar{\boldsymbol{p}} = \bar{\boldsymbol{p}}\Pi + N^{-\alpha}\left(\theta\bar{\boldsymbol{p}}\Lambda F - \theta\frac{\partial \bar{\boldsymbol{p}}}{\partial \theta}M - \frac{\partial \bar{\boldsymbol{p}}}{\partial t}\right)F + o(N^{-\alpha}).$$

We again have that $\bar{p} = \bar{p}\Pi + o(1)$, so that right-multiplying the previous display by $\mathbf{1}$ yields

$$0 = N^{-\alpha}\left(\bar{p}\mathbf{1}\theta\lambda_\infty - \mu_\infty\theta\frac{\partial\bar{p}}{\partial\theta}\mathbf{1} - \frac{\partial\bar{p}}{\partial t}\mathbf{1}\right) + o(N^{-\alpha}).$$

We thus obtain a partial differential equation in $\bar{p}(t,\theta)\mathbf{1}$. It is straightforward to check that $\bar{p}(t,\theta)\mathbf{1} = \exp(\theta\varrho\cdot(1 - \exp(-\mu_\infty)))$ satisfies the equation as well as the boundary conditions $\bar{p}(t,0)\mathbf{1} = 1$ and $\bar{p}(0,\theta) = 1$.   □

Now that we have derived the weak law of large numbers, we proceed with stating and proving the corresponding CLT.

Theorem 2. *The random variable*

$$N^{\beta/2}\left(\frac{M^{(N)}(t)}{N} - \varrho(t)\right)$$

converges to a Normal distribution with zero mean and variance $\sigma^2(t)$ as $N \to \infty$; here the scaling parameter β equals $\min\{\alpha, 1\}$, and $\sigma^2(t) := \sigma_m^2(t)\mathbf{1}_{\{\alpha\leq1\}} + \varrho(t)\mathbf{1}_{\{\alpha\geq1\}}$, with

$$\sigma_m^2(t) := 2e^{-2\mu_\infty t}\int_0^t e^{2\mu_\infty s}\pi(\Lambda - \varrho(s)M)D(\Lambda - \varrho(s)M)\mathbf{1}\,\mathrm{d}s.$$

Proof. The structure of the proof follows that of the stationary case, but we finally obtain a *partial* (rather than an ordinary) differential equation. We again concentrate on the new term $-N^{-\alpha}\cdot\partial p/\partial t\cdot F$. We introduce the mgf $\tilde{p}(t,\theta)$, scaled and centered as in the claim. Note that

$$\tilde{p}(t,\theta) = \exp(-\varrho(t)\theta N^{\beta/2})p(t,\exp(\theta N^{-1+\beta/2})),$$

The expressions for z and the derivative with respect to θ are the same as in the stationary case, except for the type of derivative (ordinary versus partial), while ϱ should obviously be replaced by $\varrho(t)$. Also,

$$\frac{\partial p}{\partial t} = \exp(\varrho(t)\theta N^{\beta/2})\left(\varrho'(t)\theta N^{\beta/2}\tilde{p} + \frac{\partial\tilde{p}}{\partial t}\right).$$

As there is no 'Tayloring' required in this term, we can fastforward to the equivalent of (5), which has two extra terms:

$$\tilde{p} = \tilde{p}\Pi + \theta N^{\beta/2-\alpha}\tilde{p}(\Lambda - \varrho M - \varrho'(t))F + \frac{\theta^2 N^{\beta-1-\alpha}}{2}\tilde{p}(\Lambda + \varrho M)F$$

$$- \theta N^{-\alpha}\frac{\partial\tilde{p}}{\partial\theta}MF - N^{-\alpha}\frac{\partial\tilde{p}}{\partial t}F + o(N^{-\alpha}). \tag{8}$$

We now transform the differential equation in $\tilde{p}$ into one in $\phi(t,\theta) = \tilde{p}(t,\theta)\mathbf{1}$. We have that

$$\tilde{p} = \phi\pi + o(1); \qquad \frac{\partial\tilde{p}}{\partial\theta} = \frac{\partial\phi}{\partial\theta}\pi + o(1); \qquad \frac{\partial\tilde{p}}{\partial t} = \frac{\partial\phi}{\partial t}\pi + o(1).$$

Hence right-multiplying Eqn. (8) by $\mathbf{1}$ and dividing by $N^{-\alpha}$, leads to

$$0 = \theta N^{\beta/2} \tilde{\boldsymbol{p}} \left(\Lambda - \varrho(t)M - \varrho'(t)I \right) \mathbf{1} + \frac{\theta^2 N^{\beta-1}}{2} \tilde{\boldsymbol{p}} (\Lambda + \varrho(t)M) \mathbf{1} - \theta \mu_\infty \frac{\partial \phi}{\partial \theta} - \frac{\partial \phi}{\partial t},$$

up to an $o(1)$ error term. We further manipulate this expression so that occurrences of $\tilde{\boldsymbol{p}}$ are replaced by $\phi(\theta)$, by substituting Eqn. (8), to obtain, with $G(t) := \Lambda - \varrho(t)M - \varrho'(t)I$ and $H(t) := \Lambda + \varrho(t)M$,

$$0 = \theta N^{\beta/2} \phi \boldsymbol{\pi}\, G(t)\, \mathbf{1} + \theta^2 N^{\beta-\alpha} \phi \boldsymbol{\pi}\, G(t)\, F\, G(t) \mathbf{1}$$
$$+ \frac{\theta^2 N^{\beta-1}}{2} \phi \boldsymbol{\pi}\, H(t)\, \mathbf{1} - \theta \mu_\infty \frac{\partial \phi}{\partial \theta} - \frac{\partial \phi}{\partial t} + o(1).$$

It can be verified that $\boldsymbol{\pi}\, G(t)\, \mathbf{1} = 0$ and

$$\boldsymbol{\pi}\, G(t)\, F\, G(t) \mathbf{1} = \boldsymbol{\pi} (\Lambda - \varrho(t)M) D (\Lambda - \varrho(t)M) \mathbf{1};$$
$$\frac{1}{2} \boldsymbol{\pi} H(t) \mathbf{1} = \lambda_\infty \left(1 - \frac{e^{-\mu_\infty t}}{2} \right),$$

so that we arrive at the partial differential equation

$$\frac{\partial \phi}{\partial t} + \theta \mu_\infty \frac{\partial \phi}{\partial \theta} = \theta^2 g(t) \phi + o(1),$$

with

$$g(t) := N^{\beta-\alpha} \boldsymbol{\pi} (\Lambda - \varrho(t)M) D (\Lambda - \varrho(t)M) \mathbf{1} + N^{\beta-1} \lambda_\infty \left(1 - \frac{e^{-\mu_\infty t}}{2} \right).$$

We propose the *ansatz*

$$\phi(t, \theta) = \exp \left(\frac{1}{2} \theta^2 e^{-2\mu_\infty t} f(t) \right),$$

for some unknown function $f(t)$; recognize the characteristic function associated with the Normal distribution. This leads to the following ordinary differential equation for $f(t)$:

$$f'(t) = 2 e^{2\mu_\infty t} g(t),$$

which is obviously solved by integrating the right-hand side. From this we immediately find the expression for the variance $\sigma^2(t)$ of the Normal distribution as given in the statement of the theorem, where it is mentioned that the interval of the integral must be $[0, t]$ due to $\sigma^2(0) = 0$; realize that, as before, whether $\alpha < 1$ or $\alpha > 1$ determines which terms end up in $\sigma^2(t)$ (and for $\alpha = 1$ both).

We can now explicitly compute this integral. After some elementary manipulations we find that

$$\sigma^2(t) = e^{-2\mu_\infty t} f(t) = N^{\beta-\alpha} \left(\sigma_m^2 + v e^{-\mu_\infty t} + w t e^{-2\mu_\infty t} \right) + N^{\beta-1} \varrho(t),$$

where v is given by $2\mu_\infty^{-1} \varrho \left(\boldsymbol{\pi} M D (\Lambda - \rho M) \mathbf{1} + \boldsymbol{\pi} (\Lambda - \rho M) D M \mathbf{1} \right)$, whereas w denotes $2\varrho^2 \boldsymbol{\pi} M D M \mathbf{1}$. As $t \to \infty$, we have that $\sigma_m^2(t) \to \sigma_m^2$, as expected. $\qquad \square$

6 Discussion and Conclusion

In this paper we derived central limit theorems (CLT s) for infinite-server queues with Markov-modulated input. In our approach the modulating Markov chain is sped up by a factor N^α (for some non-negative α), while the arrival process is sped up by N. Interestingly, there is a *phase transition* in the sense that the scaling to be used in the CLT depends on the value of α: rather than the standard normalization by $\sqrt{n}$, it turned out that the centered process should be divided by $N^{1-\beta/2}$, with β equal to $\min\{\alpha, 1\}$. We have proved this by first establishing systems of differential equations for the (transient and stationary) distribution of the number of jobs in the system, and then studying their behavior under the scaling described above.

In the model studied in this paper, the duration of the jobs present at a certain time t are subject to a hazard rate determined by the state of background chain at t. In an other variant, a job's service time is determined by the state of the background process seen upon arrival. To study this alternative mechanism, in principle the same approach can be followed: first setting up a system of differential equations, and then evaluating these under the scaling that we imposed.

In addition, similar techniques as the ones we have developed in this paper can be used to reveal a CLT for the *multivariate* distribution of the number of jobs present at different time instants, cf. the analysis for $\alpha > 1$ in [1]. We anticipate weak convergence to an Ornstein-Uhlenbeck process with appropriate parameters, but establishing such a claim will require different techniques.

References

1. Blom, J., Kella, O., Mandjes, M., Thorsdottir, H.: Markov-modulated infinite server queues with general service times (2012); Submitted
2. Blom, J., Mandjes, M., Thorsdottir, H.: Time-scaling limits for Markov-modulated infinite-server queues. Stochastic Models 29, 112–127 (2013)
3. D'Auria, B.: M/M/∞ queues in semi-Markovian random environment. Queueing Systems 58, 221–237 (2008)
4. Coolen-Schrijner, P., van Doorn, E.: The deviation matrix of a continuous-time Markov chain. Probability in the Engineering and Informational Sciences 16, 351–366 (2002)
5. Fralix, B., Adan, I.: An infinite-server queue influenced by a semi-Markovian environment. Queueing Systems 61, 65–84 (2009)
6. Hellings, T., Mandjes, M., Blom, J.: Semi-Markov-modulated infinite-server queues: approximations by time-scaling. Stochastic Models 28, 452–477 (2012)
7. Keilson, J., Servi, L.: The matrix M/M/∞ system: retrial models and Markov modulated sources. Advances in Applied Probability 25, 453–471 (1993)
8. O'Cinneide, C., Purdue, P.: The M/M/∞ queue in a random environment. Journal of Applied Probability 23, 175–184 (1986)

9. Schwabe, A., Dobrzyński, M., Rybakova, K., Verschure, P., Bruggeman, F.: Origins of stochastic intracellular processes and consequences for cell-to-cell variability and cellular survival strategies. Methods in Enzymology 500, 597–625 (2011)
10. Schwabe, A., Dobrzyński, M., Bruggeman, F.: Transcription stochasticity of complex gene regulation models. Biophysical Journal 103, 1152–1161 (2012)
11. van Woensel, T., Vandaele, N.: Modeling traffic flows with queueing models: a review. Asia-Pacific Journal of Operational Research 24, 235–261 (2007)

Transformation of Acyclic Phase Type Distributions for Correlation Fitting

Peter Buchholz, Iryna Felko, and Jan Kriege

Department of Computer Science, TU Dortmund
{peter.buchholz,iryna.felko,jan.kriege}@udo.edu

Abstract. In this paper similarity transformations for *Acyclic Phase Type Distributions* (APHs) are considered, and representations maximizing the first joint moment that can be reached when the distribution is expanded into a *Markovian Arrival Process* (MAP) are investigated. For the acyclic case the optimal representation corresponds to a hyperexponential representation, which is optimal among all possible representations that can be reached by similarity transformations. The parameterization aspect for the possible transformation of APHs into a hyperexponential form is revealed, together with corresponding transformation rules. For the case when APHs cannot be transformed into a hyperexponential representation a heuristic optimization method is presented to obtain good representations, while transformation methods to increase the first joint moment by adding additional phases are derived.

1 Introduction

In stochastic modeling, the appropriate representation of dynamic processes behavior is of major importance to built realistic models. Phase type distributions (PHDs) and Markovian arrival processes (MAPs) have developed to versatile modeling tools which can be applied to capture a wide range of different stochastic behaviors and can be used in queuing network models [18], renewal theory [15], reliability, and healthcare modeling [9]. PHDs and MAPs allow for applying efficient matrix analytic methods, can approximate any continuous distribution and can be used to model empirical data containing a wide range of stochastic behaviors.

To capture real behavior the parameters of a PHD or MAP have to be estimated according to some observations measured in a real system [13]. Parameter estimation for PHDs and MAPs is a nontrivial task resulting in a nonlinear optimization problem which becomes even more complex since the matrix representation of a PHD or MAP is redundant [20]. To reduce the complexity fitting algorithms for PHDs are often tailored to specific subclasses and canonical representations such that the number of free parameters in the initial distribution vector $\mathbf{p}$ and the transition rates matrix $\mathbf{D}_0$ is minimized [4,22,10]. For the same reason MAP fitting is often performed in two steps [4,11]. In a first step a PHD $(\mathbf{p}, \mathbf{D}_0)$ is fitted and in a second step a matrix $\mathbf{D}_1$ is constructed to model the correlation such that the resulting MAP $(\mathbf{D}_0, \mathbf{D}_1)$ has the stationary distribution $\mathbf{p}$. Obviously, the entries in $(\mathbf{p}, \mathbf{D}_0)$ have a large influence when fitting $\mathbf{D}_1$, since the entries put constraints on the possible values of entries of $\mathbf{D}_1$, thereby limiting the possible range of autocorrelation one can achieve. Since the representation of a PH distribution is

A. Dudin and K. De Turck (Eds.): ASMTA 2013, LNCS 7984, pp. 96–111, 2013.
© Springer-Verlag Berlin Heidelberg 2013

non-unique [20] one tries to find an equivalent representation of $(\mathbf{p}, \mathbf{D}_0)$ by transforming the distribution that allows for the largest flexibility before fitting $\mathbf{D}_1$. However, only little is known about which representation of a PHD is favorable when used in a two-step MAP fitting approach. Obviously, the canonical form of a PHD that is often used for PH fitting is not suitable since it has only a single exit state and does not allow for any flexibility when fitting $\mathbf{D}_1$. The same holds for representations that only have a single entry state. Consequently, existing transformations [3,4,11] aim at increasing the number of entry and exit states. However, in an empirical study it has been shown recently [16] that in general these transformations do not find the optimal representation that allows for the maximal flexibility. In this paper we present a similarity transformation approach in the field of equivalent representations for PHDs and we try to shed some light on which representation of a PHD is favorable for a subsequent MAP fitting step. In particular, we show which representation of a PHD allows for the expansion into a MAP with maximal first order joint moment and present transformations to obtain this representation.

The paper is organized as follows: In Sect. 2 we give a short introduction to PHDs and MAPs. In Sect. 3 we describe the transformation algorithm for the case with two phases in detail. Sect. 4 contains descriptions for the general case. In Sect. 5 a technique for state space expansion with the aim of finding a representation allowing larger flexibility is presented. In Sect. 6 we present some examples to demonstrate the effectiveness of the approach. The paper ends with the conclusions.

2   Background

We first introduce the basic notation and define PHDs and MAPs, then we present the basic results for similarity transformations.

2.1   Basic Definitions for PH Distributions and MAPs

We consider acyclic phase type distributions (APHs) [8] which are defined by a vector matrix pair $(\mathbf{p}, \mathbf{D}_0)$ where $\mathbf{p}$ is a n dimensional row vector containing the initial probability distribution (i.e., $\mathbf{p} \geq \mathbf{0}$, $\mathbf{p}\mathbb{I} = 1$ where $\mathbb{I} = (1, \ldots, 1)^T$ of length n). $\mathbf{D}_0$ is an upper triangular matrix of order n with negative diagonal elements, non-negative elements above the diagonal and row sums ≤ 0. We denote the diagonal entries by $-\lambda_i < 0$ and the non-diagonal entries by $\lambda_{i,j} \geq 0$. Since the matrix is upper triangular, we have $\lambda_{i,j} = 0$ for $i > j$. Furthermore, we denote by $\mathbf{\Lambda}$ a column vector of length n that contains in position i the value λ_i. We use the terms state and phase interchangeably for APHs.

$(\mathbf{p}, \mathbf{D}_0)$ defines an absorbing CTMC and the time to absorption is phase type distributed. The density function, distribution function and moments are given by

$$f(x) = \mathbf{p}e^{\mathbf{D}_0 x}\mathbf{d}_1, \qquad F(x) = 1 - \mathbf{p}e^{\mathbf{D}_0 x}\mathbb{I} \qquad \text{and} \qquad \mu_k = k!\mathbf{p}\mathbf{M}^k\mathbb{I},$$

respectively, where $\mathbf{d}_1 = -\mathbf{D}_0\mathbb{I}$ is the vector of exit rates, $x \geq 0$, $\mathbf{M} = (-\mathbf{D}_0)^{-1}$ is the moment matrix and $\mathbf{m}^k = \mathbf{M}^k\mathbb{I}$ is the vector of the conditional kth moments. For $\mathbf{m}^1$ we use the notation $\mathbf{m}$.

APHs can be expanded to Markovian Arrival Processes (MAPs) [17] which are defined by a pair of two matrices $(\mathbf{D}_0, \mathbf{D}_1)$ such that $\mathbf{D}_0 + \mathbf{D}_1$ is a generator of an irreducible CTMC. A MAP $(\mathbf{D}_0, \mathbf{D}_1)$ is an expansion of an APH $(\mathbf{p}, \mathbf{D}'_0)$ if $\mathbf{D}_0 = \mathbf{D}'_0$ and the embedded distributions at event times are identical which means $\mathbf{p} = \mathbf{pMD}_1$ holds. The expansion of APHs to MAPs is used in so called two phase fitting approaches [5,4,11] where in a first step an APH is fitted and afterwards correlation is inserted by finding an appropriate matrix $\mathbf{D}_1$. This approach is more efficient than the combined fitting of both matrices but has the disadvantage that the generated APH might not be flexible enough.

The joint density and the joint moments of a MAP are given by

$$f(x_1, \ldots, x_m) = \mathbf{p} \left(\prod_{i=1}^{m} e^{\mathbf{D}_0 x_i} \mathbf{D}_1 \right) \mathbb{I} \quad \text{and} \quad \mu_{k,l} = k! l! \, \mathbf{pM}^k \mathbf{MD}_1 \mathbf{M}^l \mathbb{I}$$

for $x_i \geq 0$. The lag k autocorrelation coefficients and (as special case) the lag 1 autocorrelation are computed as

$$\rho_k = \frac{\mu_1^{-2} \mathbf{pM} (\mathbf{MD}_1)^k \mathbf{M} \mathbb{I} - 1}{2\mu_1^{-2} \mathbf{pM}^2 \mathbb{I} - 1} \quad \text{and} \quad \rho_1 = \frac{\mu_1^{-2} \mu_{1,1} - 1}{\mu_1^{-2} \mu_2 - 1}. \tag{1}$$

For a given APH ρ_1 of an expanded MAP is directly proportional to the first joint moment $\mu_{1,1}$ since the other quantities in (1) are determined by the distribution. Observe that if $(\mathbf{p}, \mathbf{D}_0)$ is an APH and $(\mathbf{D}_0, \mathbf{D}_1)$ is a MAP with first joint moment $\mu_{1,1}$ that is expanded from the APH, then $(\mathbf{D}_0, \alpha \mathbf{D}_1 + (1 - \alpha) \mathbf{d}_1 \mathbf{p})$ for $\alpha \in [0, 1]$ is a MAP with first joint moment $\alpha \mu_{1,1} + (1 - \alpha) \mu_1 \mu_1$ which implies that for every first joint moment between $\mu_{1,1}$ and $\mu_1 \mu_1$ a MAP can be expanded from the APH. Consequently, it is important to know the MAP with the maximal or minimal first joint moment that can be expanded from an APH. The maximum case is usually more important since positive correlation is more common in practice than negative correlation. Therefore we consider in the following MAPs with a maximal first joint moment that can be expanded from APHs. However, approaches to derive MAPs with a minimal first joint moment are similar.

2.2 Similarity Transformation Results

As already mentioned the representation of a PHD is not unique and different representations can be transformed into one another. In principle one can use every non-singular matrix $\mathbf{C}$ with unit row sum and generate a new representation $(\mathbf{pC}, \mathbf{C}^{-1} \mathbf{D}_0 \mathbf{C})$ [20]. This representation usually is not an APH and it is often not even a Markov representation. A restricted set of similarity transformations is used in [8] to transform an APH in canonical form. Several canonical forms have been defined by Cumani [8] which have only a single exit state (i.e, $\mathbf{d}_1(i) > 0$ for exactly one i) or single entry state (i.e., $\mathbf{p}(i) = 1$ for exactly one i). Both representations are not suited for the expansion of the APH to a MAP since a single entry or exit state implies that the MAP's correlation matrix $\mathbf{D}_1$ is completely defined by the APH and the MAP has no correlation. However, the transformation rules of Cumani can be used in a more general way which has been done in [4,11] to transform an APH in canonical representation into a representation that is more appropriate for expansion into a MAP. The used methods are heuristics which apply a single transformation step that depends on a free parameter for all rows of the matrix. Experiments show that the transformation often does not work appropriately, in particular for APHs with a larger number of states. This has recently been investigated systematically in [16] where it could be shown that in many cases the transformation is not satisfactory and it is unclear how to set the free parameter. However, the results in [16] also show that an APH which can be expanded to MAPs with a wide range of joint moments $\mu_{1,1}$, in general also implies that the resulting MAPs are flexible in reaching other correlation quantities like higher order joint moments or lag k autocorrelations for $k > 1$. Thus, the optimization of the representation of APHs is important for a good approximation of general traffic processes including correlation.

We first consider the basic transformation step to modify APHs which is based on the ideas presented in [8]. The approach uses the representation of an exponential distribution by a Coxian distribution with 2 states. Consider an exponential distribution with rate λ_1, then this distribution can be represented by the following APH with 2 states.

$$\mathbf{p} = (1 - \lambda_1/\lambda_2, \lambda_1/\lambda_2), \quad \mathbf{D}_0 = \begin{pmatrix} -\lambda_1 & \lambda_1 \\ 0 & -\lambda_2 \end{pmatrix} \tag{2}$$

with $\lambda_2 \geq \lambda_1$. The resulting representation has two entry and one exit state. A variant of the transformation with one entry and two exit states has been proposed in [12]. Based on (2), the following transformations can be defined which modify the representation but not the distribution and still result in an APH. The following equations apply the transformation to two states $i < j$. We assume that for $i < j$ $\lambda_i \leq \lambda_j$ which can be always achieved by elementary transformations following from [8] that are implemented in the BuTools [21]. $\mathbf{p}^\delta$ and $\mathbf{D}_0^\delta$ are the vector and matrix after the transformation with parameter δ has been applied. Proofs for the validity of transformation steps can be found in [3].

$$\mathbf{p}^\delta(k) = \begin{cases} \mathbf{p}(i) + \delta & \text{for } k = i \\ \mathbf{p}(j) - \delta & \text{for } k = j \\ \mathbf{p}(k) & \text{otherwise} \end{cases} \qquad \lambda_{k,l} = \begin{cases} \lambda_{i,j}\frac{\mathbf{p}(i)}{\mathbf{p}(i)+\delta} - \frac{(\lambda_j - \lambda_i)\delta}{\mathbf{p}(i)+\delta} & \text{for } k = i \text{ and } l = j \\ \lambda_{i,l}\frac{\mathbf{p}(i)}{\mathbf{p}(i)+\delta} & \text{for } k = i \text{ and } l > i \wedge l < j \\ \lambda_{i,l}\frac{\mathbf{p}(i)}{\mathbf{p}(i)+\delta} + \lambda_{j,l}\frac{\delta}{\mathbf{p}(i)+\delta} & \text{for } k = i \text{ and } l > j \\ \lambda_{k,i}\frac{\mathbf{p}(i)+\delta}{\mathbf{p}(i)} & \text{for } k < i \text{ and } l = i \\ \lambda_{k,j} - \lambda_{k,i}\frac{\delta}{\mathbf{p}(i)} & \text{for } k < i \text{ and } l = j \\ \lambda_{k,l} & \text{otherwise} \end{cases}$$

$$\tag{3}$$

The diagonal entries λ_k are not modified by the transformation. For the exit vector $\mathbf{d}_1$, $\mathbf{d}_1^\delta(k) = \mathbf{d}_1(k)$ for $k \neq i$ and

$$\mathbf{d}_1^\delta(i) = \frac{\mathbf{p}(i)\mathbf{d}_1(i) + \delta\mathbf{d}_1(j)}{\mathbf{p}(i) + \delta} = \mathbf{d}_1(i) + \delta\frac{\mathbf{d}_1(j) - \mathbf{d}_1(i)}{\mathbf{p}(i) + \delta}. \tag{4}$$

To compute a valid APH, parameter δ has to be chosen from the following interval to assure that the rates and probabilities remain non-negative.

$$\left[\max\left(-\mathbf{p}(i), \min_{l > j, \lambda_{j,l} > 0}\left(-\frac{\mathbf{p}(i)\lambda_{i,l}}{\lambda_{j,l}} \right), -\frac{\mathbf{p}(i)\mathbf{d}_1(i)}{\mathbf{d}_1(j)} \right), \min\left(\mathbf{p}(j), \min_{k < i, \lambda_{k,i} > 0}\left(\frac{\mathbf{p}(i)\lambda_{k,j}}{\lambda_{k,i}} \right), \frac{\mathbf{p}(i)\lambda_{i,j}}{\lambda_j - \lambda_i} \right) \right]$$

$$\tag{5}$$

If $\mathbf{d}_1(j) = 0$, then the last term for the lower bound becomes $-\infty$ and is not used. If $\lambda_j = \lambda_i$, then the last term in the upper bound evaluates to ∞. Furthermore, if δ is set to $-\mathbf{p}(i)$, then the rates of the new APH become infinite.

3   The Two Phase Case

In this section we consider APHs with two states and compute a representation that allows the expansion to a MAP with a maximal first joint moment $\mu_{1,1}$. The APH is given by

$$\mathbf{p} = (\pi, 1 - \pi), \ \mathbf{D}_0 = \begin{pmatrix} -\lambda_1 & \lambda_{1,2} \\ 0 & -\lambda_2 \end{pmatrix}.$$

We assume $\lambda_2 \geq \lambda_1$. If this is not the case, we use the representation

$$\mathbf{p} = \left(\frac{(1-\pi)\lambda_1 + \pi\lambda_{1,2}}{\lambda_1}, 1 - \frac{(1-\pi)\lambda_1 + \pi\lambda_{1,2}}{\lambda_1} \right), \ \mathbf{D}_0 = \begin{pmatrix} -\lambda_2 & \frac{\pi\lambda_2\lambda_{1,2}}{(1-\pi)\lambda_1 + \pi\lambda_{1,2}} \\ 0 & -\lambda_1 \end{pmatrix}$$

Equivalence of both representations can be proved by a comparison of their elementary series [3,8]. Furthermore, the time scale can be modified such that λ_1 becomes 1 and we obtain a representation

$$\mathbf{p} = (\pi, 1 - \pi), \ \mathbf{D}_0 = \begin{pmatrix} -1 & \lambda_{1,2} \\ 0 & -\lambda_2 \end{pmatrix} \text{ where } 0 \leq \pi, \lambda_{1,2} \leq 1 \text{ and } \lambda_2 \geq 1. \tag{6}$$

The transformations (3) simplify in the 2 state case (6) to

$$\mathbf{p}^\delta = \left(\pi^\delta, 1 - \pi^\delta \right) = (\pi + \delta, 1 - \pi - \delta), \ \lambda_{1,2}^\delta = \frac{\pi\lambda_{1,2}}{\pi + \delta} - \frac{(\lambda_2 - 1)\delta}{\pi + \delta} \tag{7}$$

where $\delta \in \left[\max \left(-\pi, -\frac{\pi(1 - \lambda_{1,2})}{\lambda_2} \right), \min \left(1 - \pi, \frac{\pi\lambda_{1,2}}{\lambda_2 - 1} \right) \right]$. We denote the boundaries of the interval as δ^- and δ^+, respectively. APH(δ) is the APH ($\mathbf{p}^\delta, \mathbf{D}_0^\delta$) with the parameters in (7) computed for δ and diagonal entries $-1, -\lambda_2$ in matrix $\mathbf{D}_0^\delta$.

Let $\mu_{1,1}^*(\delta)$ be the maximal joint moment $\mu_{1,1}$ that can be reached with a MAP that is expanded from APH(δ). This raises two questions, namely, what is the best δ, i.e., $\delta^* = \arg\max_{\delta \in [\delta^-, \delta^+]} \left(\mu_{1,1}^*(\delta) \right)$, and what are the parameters of the MAP that results in $\mu_{1,1}^*(\delta)$ for a given δ?

Before we answer the questions, a slightly different representation for $\mu_{1,1}$ will be introduced. In vector matrix representation, the first joint moment is given by

$$\mu_{1,1} = \mathbf{p}\mathbf{M}\mathbf{M}\mathbf{D}_1\mathbf{M}\mathbb{1}$$

where $\mathbf{M} = (-\mathbf{D}_0)^{-1}$. Let $a_{i,j}^\delta$ be the probability that the APH starts in state i and j is the last state before the event occurs (i.e., before absorption if $\mathbf{D}_0$ is interpreted as generator matrix of an absorbing CTMC). We have

$$a_{1,1}^\delta = \pi^\delta(1 - \lambda_{1,2}^\delta), \ a_{1,2}^\delta = \pi^\delta\lambda_{1,2}^\delta, \ a_{2,1}^\delta = 0, \text{ and } a_{2,2}^\delta = 1 - \pi^\delta.$$

Let furthermore $v_{i,j}$ be the mean duration if the process starts in i and j is the last state before the event occurs, then

$$v_{1,1} = 1, v_{1,2} = 1 + (\lambda_2)^{-1}, \text{ and } v_{2,2} = (\lambda_2)^{-1}.$$

$v_{2,1}$ is not available since an APH can not start in state 2 and visit state 1 before an event occurs. To define a MAP we use the probabilities b_i which describe the probability that the MAP starts in state 1 after i was the last state before absorption. Matrix $\mathbf{D}_1$ is then given by

$$\mathbf{D}_1 = \begin{pmatrix} 1 - \lambda_{1,2}^\delta & 0 \\ 0 & \lambda_2 \end{pmatrix} \begin{pmatrix} b_1 & 1 - b_1 \\ b_2 & 1 - b_2 \end{pmatrix}$$

To observe $\mathbf{pMD}_1 = \mathbf{p}$, the following relation has to hold

$$\pi^\delta = a_{1,1}^\delta b_1 + (a_{1,2}^\delta + a_{2,2}^\delta) b_2 = \pi^\delta (1 - \lambda_{1,2}^\delta) b_1 + (\pi^\delta \lambda_{1,2}^\delta + 1 - \pi^\delta) b_2 \qquad (8)$$

The equation shows that only one value b_i can be set, the other one is fixed to observe the initial probability. If we fix b_2, it can be chosen from

$$\left[\frac{\pi^\delta \lambda_{1,2}^\delta}{1 - \pi^\delta (1 - \lambda_{1,2}^\delta)}, \min\left(1, \frac{\pi^\delta}{1 - \pi^\delta (1 - \lambda_{1,2}^\delta)} \right) \right]$$

to assure that $b_1 \in [0, 1]$. With these notations we can write down an equation for $\mu_{1,1}(\delta)$

$$\mu_{1,1}(\delta) = b_1 a_{1,1}^\delta v_{1,1} (a_{1,1}^\delta v_{1,1} + a_{1,2}^\delta v_{1,2})/\pi^\delta + (1 - b_1) a_{1,1}^\delta v_{1,1} v_{2,2} +$$
$$b_2 (a_{1,2}^\delta v_{1,2} + a_{2,2}^\delta v_{2,2})(a_{1,1}^\delta v_{1,1} + a_{1,2}^\delta v_{1,2})/\pi^\delta +$$
$$(1 - b_2)(a_{1,2}^\delta v_{1,2} + a_{2,2}^\delta v_{2,2}) v_{2,2}$$

The equations results from considering all possible sequences of states with the corresponding probabilities and mean durations in the MAP.

After some elementary but still lengthy substitutions and transformations we obtain the following representation for $\mu_{1,1}(\delta)$ in terms of δ and b_2.

$$\mu_{1,1}(\delta) = \frac{y_1 + b_2\left(y_{21} + \delta y_{22} + \frac{1}{\pi + \delta} y_{23} + \frac{\delta}{\pi + \delta} y_{24} + \frac{\delta^2}{\pi + \delta} y_{25}\right)}{\lambda_2^2}$$

where $y_1 = 1 + \pi(\lambda_2^2 + (\lambda_2 + 1)\lambda_{1,2} - 1)$
$y_{21} = \pi((\lambda_2 - 1)^2 + 2\lambda_{1,2}(\lambda_2 - 1)) - (\lambda_2 - 1)^2$
$y_{22} = -(\lambda_2 - 1)^2 \qquad y_{23} = \pi \lambda_{1,2}(1 + \pi \lambda_{1,2} - \lambda_2)$
$y_{24} = (\lambda_2 - 1 - 2\pi \lambda_{1,2})(\lambda_2 - 1) \qquad y_{25} = (\lambda_2 - 1)^2$

Observe that the coefficients y_{xy} do not contain δ or b_i and are defined for the original APH without modifications due to δ. Define

$$y_2(\delta) = y_{21} + \delta y_{22} + \frac{1}{\pi + \delta} y_{23} + \frac{\delta}{\pi + \delta} y_{24} + \frac{\delta^2}{\pi + \delta} y_{25} \qquad (9)$$

The representation implies that for $y_2(\delta) < 0$ b_2 should be as small as possible to maximize $\mu_{1,1}(\delta)$ (i.e., find $\mu_{1,1}^*(\delta)$) and for $y_2(\delta) > 0$ b_2 should be as large as possible for $\mu_{1,1}^*(\delta)$. $y_2(\delta) = 0$ implies that the APH has no flexibility in adopting joint moments.

Now consider the case $\delta = 0$ such that $y_2(0) = y_{21} + y_{23}/\pi$. $y_2(0) = 0$ for $\pi = \frac{\lambda_2 - 1}{\lambda_2 + \lambda_{1,2} - 1}$, $y_2(0) < 0$ for $\pi < \frac{\lambda_2 - 1}{\lambda_2 + \lambda_{1,2} - 1}$ and $y_2(0) > 0$ for $\pi > \frac{\lambda_2 - 1}{\lambda_2 + \lambda_{1,2} - 1}$. Interestingly, the relations remain if we substitute π by π^δ and $\lambda_{1,2}$ by $\lambda_{1,2}^\delta$ since then

$$\pi + \delta = \frac{(\lambda_2 - 1)(\pi + \delta)}{\lambda_2(\pi + \delta) + \pi\lambda_{1,2} - (\lambda_2 - 1)\delta - (\pi + \delta)} = \frac{(\pi + \delta)(\lambda_2 - 1)}{\pi(\lambda_2 + \lambda_{1,2} - 1)}.$$

The latter term equals $\pi + \delta$ if π observes the above equality. This implies that an APH belongs to one of three classes depending on the relation between π and $\frac{\lambda_2 - 1}{\lambda_2 + \lambda_{1,2} - 1}$. The case with equality is not interesting since it allows no flexibility. However, it may be handled by expanding the state space as shown in Section 5. Now we consider the remaining two cases in some detail.

We begin with **Case 1** $\pi < \frac{\lambda_2 - 1}{\lambda_2 + \lambda_{1,2} - 1}$. To obtain $\mu_{1,1}^*(0)$ b_2 is set to the minimum, i.e., $b_2 = \frac{\pi\lambda_{1,2}}{1 - \pi(1 - \lambda_{1,2})}$ and $b_1 = 1$ which follows from substituting b_2 into (8). Denote by b_2^δ the value of b_2 for varying δ which equals

$$b_2^\delta = \frac{(\pi\lambda_{1,2} - (\lambda_2 - 1)\delta)}{1 - \pi + \pi\lambda_{1,2} - \lambda_2\delta} \geq 0.$$

The derivative with respect to δ equals

$$\frac{db_2^\delta}{d\delta} = \frac{\left(\pi - \frac{\lambda_2 - 1}{\lambda_2 + \lambda_{1,2} - 1}\right)(\lambda_2 - 1 + \lambda_{1,2})}{(\pi\lambda_{1,2} - \pi + 1 - \lambda_2\delta)^2} < 0$$

The denominator is always positive and the numerator is negative for *Case 1* as we will show. The first term of the numerator is obviously negative by the assumption for *Case 1*. For the second term $\lambda_2 + \lambda_{1,2} - 1 \geq 0$ holds since $\lambda_2 \geq 1$ is assumed. The derivative of $y_2(\delta)$ equals

$$\frac{dy_2(\delta)}{d\delta} = -(\lambda_2 - 1)^2 + \frac{-\pi\lambda_{1,2}(1 + \pi\lambda_{1,2} - \lambda_2) + \pi(\lambda_2 - 1 - 2\pi\lambda_{1,2})(\lambda_2 - 1) + (2\delta\pi + \delta^2)(\lambda_2 - 1)^2}{(\pi + \delta)^2}$$
$$= \frac{\pi((\lambda_2 + \lambda_{1,2} - 1)((1 - \pi)(\lambda_2 - 1) - \pi\lambda_{1,2}))}{(\pi + \delta)^2} \qquad > 0 \tag{10}$$

Since the denominator and the first part of the numerator are positive, we only have to consider the last term, namely $(1 - \pi)(\lambda_2 - 1) - \pi\lambda_{1,2}$. Since *Case 1* implies $\lambda_{1,2} < \frac{(\lambda_2 - 1)(1 - \pi)}{\pi}$, we have $(1 - \pi)(\lambda_2 - 1) - \pi\lambda_{1,2} > (1 - \pi)(\lambda_2 - 1) - (1 - \pi)(\lambda_2 - 1) = 0$. Then

$$\frac{d\mu_{1,1}(\delta)}{d\delta} = \frac{db_2^\delta}{\delta}\frac{y_2(\delta)}{\lambda_2^2} + \frac{b_2^\delta}{\lambda_2^2}\frac{dy_2(\delta)}{\delta} > 0$$

since the first product consists of two negative values and the second of two positive values such that the sum is positive and δ should be chosen as the maximum value. We set $\delta = \frac{\lambda_{1,2}\pi}{\lambda_2 - 1}$ which implies $1 - \pi > \delta$ since $1 - \pi > \frac{\lambda_{1,2}\pi}{\lambda_2 - 1} \Leftrightarrow \frac{(1 - \pi)(\lambda_2 - 1)}{\pi} > \lambda_{1,2}$ which holds by assumption in *Case 1*. Furthermore,

$$\lambda_{1,2}^\delta = \frac{\pi\lambda_{1,2} - (\lambda_2 - 1)\frac{\lambda_{1,2}\pi}{\lambda_2 - 1}}{\pi + \frac{\lambda_{1,2}\pi}{\lambda_2 - 1}} = 0. \tag{11}$$

This implies that the distribution where the maximum value for $\mu_{1,1}(\delta)$ can be reached is a hyperexponential distribution.

In **Case 2** we have $\pi > \frac{\lambda_2-1}{\lambda_2+\lambda_{1,2}-1}$. To obtain $\mu_{1,1}^*(0)$ b_2 is set to the maximum, i.e., $b_2 = \min\left(1, \frac{\pi}{1-\pi(1-\lambda_{1,2})}\right)$. Two cases have to be distinguished depending whether b_2^δ is bounded by 1 or by the second term. If $b_2^\delta = 1$, then the derivative of b_2^δ with respect to δ is 0. This implies that

$$\frac{d\mu_{1,1}(\delta)}{d\delta} = b_2^\delta \frac{dy_2(\delta)}{\delta} < 0$$

since $b_2^\delta > 0$ and $\lambda_{1,2} > \frac{(\lambda_2-1)(1-\pi)}{\pi}$ holds such that the $>$ in (11) changes to a $<$.

In the second case

$$b_2^\delta = \frac{\pi+\delta}{1-\pi+\pi\lambda_{1,2}-\lambda_2\delta} \quad \text{and} \quad \frac{db_2^\delta}{d\delta} = \frac{1+\pi(\lambda_2+\lambda_{1,2}-1)}{(1-\pi+\lambda_{1,2}\pi-\lambda_2\delta)^2}.$$

$y_2(\delta)$ and its derivative are as given in (9) and (10), respectively. Thus, we obtain after some simplifications

$$\frac{d\mu_{1,1}(\delta)}{d\delta} = \frac{db_2^\delta}{\delta}\frac{y_2(\delta)}{\lambda_2^2} + \frac{b_2^\delta}{\lambda_2^2}\frac{dy_2(\delta)}{\delta} = \frac{\pi\lambda_2\left((\pi-1)(\lambda_2^2-2\lambda_2+1)+\pi\lambda_{1,2}(2\lambda_2-2+\lambda_{1,2})+\lambda_{1,2}(1-\lambda_2)\right)}{\lambda_2^2\left(1-\pi+\pi\lambda_{1,2}-\delta\lambda_2\right)^2}$$

The value of the derivative is positive, as we will show now. First, the denominator and the first term in the numerator are positive such that we have to show that the second term of the numerator (i.e., the term in the brackets) is positive. Observe that π is multiplied in this term with non-negative factors because $\lambda_2 > 1$ such that we can substitute π by the lower bound to obtain a lower bound for the whole term.

$$(\pi-1)\left(\lambda_2^2-2\lambda_2+1\right)+\pi\lambda_{1,2}(2\lambda_2-2+\lambda_{1,2})+\lambda_{1,2}(1-\lambda_2) \quad >$$
$$\frac{-\lambda_{1,2}}{\lambda_2+\lambda_{1,2}-1}\left(\lambda_2^2-2\lambda_2+1\right)+\frac{(\lambda_2-1)\lambda_{1,2}}{\lambda_2+\lambda_{1,2}-1}(2\lambda_2-2+\lambda_{1,2})+\lambda_{1,2}(1-\lambda_2) = \frac{\lambda_2\lambda_{1,2}}{\lambda_2+\lambda_{1,2}-1} \geq 0$$

Thus, the behavior is as follows: As long as $\pi^\delta/(1-\pi^\delta(1-\lambda_{1,2}^\delta)) < 1$, the derivative with respect to δ is positive such that δ has to be increased to increase $\mu_{1,1}(\delta)$. If $\pi^\delta/(1-\pi^\delta(1-\lambda_{1,2}^\delta)) > 1$, the derivative is negative and δ should decreased to increase $\mu_{1,1}(\delta)$. At $\pi^\delta/(1-\pi^\delta(1-\lambda_{1,2}^\delta)) = 1$, the derivative from the left is positive and from the right is negative such that this point is optimal which implies $\delta = (\lambda_{1,2}\pi+1-2\pi)/(1+\lambda_2)$.

Example: We consider three simple examples. The first belongs to *Case 1* with $\delta^* = 0.3$ and is given by the following $\mathbf{p}$ and $\mathbf{D}_0$ and the resulting $\mathbf{p}^{\delta^*}$ and $\mathbf{D}_0^{\delta^*}$

$$\mathbf{p} = (0.6, 0.4), \ \mathbf{D}_0 = \begin{pmatrix} -1.0 & 0.5 \\ 0 & -2.0 \end{pmatrix}, \qquad \mathbf{p}^{\delta^*} = (0.9, 0.1), \ \mathbf{D}_0^{\delta^*} = \begin{pmatrix} -1.0 & 0 \\ 0 & -2.0 \end{pmatrix}.$$

The second example belongs to *Case 2* with $\delta^* = 0.112$ and the original and transformed matrices are as follows:

$$\mathbf{p} = (0.6, 0.4), \ \mathbf{D}_0 = \begin{pmatrix} -1.0 & 0.8 \\ 0 & -1.5 \end{pmatrix}, \qquad \mathbf{p}^{\delta^*} = (0.712, 0.288), \ \mathbf{D}_0^{\delta^*} = \begin{pmatrix} -1.0 & 0.5955 \\ 0 & -2.0 \end{pmatrix}.$$

The third example is given by

$$\mathbf{p} = \left(\frac{2}{3}, \frac{1}{3}\right), \quad \mathbf{D}_0 = \begin{pmatrix} -1.0 & 0.25 \\ 0 & -1.5 \end{pmatrix}.$$

In this example $\pi = \frac{\lambda_2 - 1}{\lambda_2 + \lambda_{2,1} - 1}$ such that $y_2(\delta) = 0$ and the APH has no flexibility according to the expansion to a MAP. The last example, in fact, describes an exponential distribution.

4   The General Case

It is very hard to extend the detailed analysis we presented in the previous section to APHs with an arbitrary number of states. We were only able to prove one of the two cases from the previous section for n states. For the second case, we present below a heuristic optimization algorithm.

Theorem 1. *If an APH can be transformed into an hyperexponential representation with* $\mathbf{p}(i) > 0$ *using similarity transformations (3)-(4), then this representation results in the maximal value* $\mu_{1,1}^*$.

Proof. Note, that for some APH with representation $(\tilde{\mathbf{p}}, \tilde{\mathbf{D}}_0, \tilde{\mathbf{d}}_1)$ which results from the hyperexponential distribution $(\mathbf{p}, \mathbf{D}_0, \mathbf{d}_1)$ using the similarity transformation defined in (3)-(4) these transformations can be collected in a non-singular transformation matrix $\mathbf{C}$ with $\mathbf{C}\mathbb{I} = \mathbb{I}$ such that $\tilde{\mathbf{p}} = \mathbf{p}\mathbf{C}$, $\tilde{\mathbf{D}}_0 = \mathbf{C}^{-1}\mathbf{D}_0\mathbf{C}$ and $\tilde{\mathbf{d}}_1 = \mathbf{C}^{-1}\mathbf{d}_1$. The key idea of the proof is to formulate a linear program that maximizes the first joint moment of $(\tilde{\mathbf{p}}, \tilde{\mathbf{D}}_0, \tilde{\mathbf{d}}_1)$ and from which it becomes visible that the maximal value equals $\mu_{1,1}^*$. The complete proof can be found in the online companion [2].

The proof of the theorem indicates that the hyperexponential representation is not only optimal for the acyclic case, it is optimal among all possible representations that can be reached by similarity transformations. For a given APH the hyperexponential distribution can be generated by a repeated use of the transformation steps (3)-(4) where parameter δ is chosen such that $\lambda_{i,j}$ becomes zero. If a transformation step is not possible because the interval for δ is too small, the distribution cannot be transformed into a hyperexponential form. Formally, the theorem shows the optimality of a hyperexponential representation with $\mathbf{p}(i) > 0$ for all i. However, since $\mu_{1,1}$ is continuous in the transformations (3)-(4), the limiting cases with $\mathbf{p}(i) = 0$ are optimal too and allow the generation of a representation with less states because a hyperexponential phase with initial probability 0 is redundant.

For the case that the APH cannot be transformed into a hyperexponential form, we were not yet able to extend the 2-state case. Our first and natural idea to apply the results for 2 states also in the general case fails in most cases. The transformation often stucks in local maxima that may be far from the global maximum.

Therefore, we used, similar to [16], a heuristic approach to find an optimal APH representation. In our case, simulated annealing [7] is applied for optimization by choosing random indices i, j with $i < j$ and $\lambda_{i,j}$, choose randomly a δ from the interval $[\delta^-, \delta^+]$, perform the transformation, compute the new value for $\mu_{1,1}$ which is kept if it is larger than the previous value or, if it smaller, it is kept with some probability according to the simulated annealing approach. With this approach we usually obtain good representations which, however, needs not be optimal.

Examples: Consider the following APH in canonical form and the corresponding hyperexponential representation that results by applying the transformation steps:

$$\mathbf{p} = \left(\frac{3}{32}, \frac{15}{32}, \frac{7}{16}\right), \ \mathbf{D}_0 = \begin{pmatrix} -0.5 & 0.5 & 0 \\ 0 & -1 & 1 \\ 0 & 0 & -4 \end{pmatrix} \quad \mathbf{p}' = \left(\frac{3}{14}, \frac{1}{2}, \frac{4}{14}\right), \ \mathbf{D}_0' = \begin{pmatrix} -0.5 & 0 & 0 \\ 0 & -1 & 0 \\ 0 & 0 & -4 \end{pmatrix}.$$
(12)

The transformed hyperexponential representation has a maximal lag 1 autocorrelation coefficient of 0.21429.

As a second example we consider the following APH $(\mathbf{p}, \mathbf{D}_0)$ in canonical form which has no hyperexponential representation. If we apply the transformations towards a hyperexponential representation we obtain the representation $(\mathbf{p}', \mathbf{D}_0')$:

$$\mathbf{p} = \left(\frac{2}{5}, \frac{19}{40}, \frac{1}{8}\right), \ \mathbf{D}_0 = \begin{pmatrix} -1 & 1 & 0 \\ 0 & -2 & 2 \\ 0 & 0 & -4 \end{pmatrix} \quad \mathbf{p}' = (0.925, 0.075, 0), \ \mathbf{D}_0' = \begin{pmatrix} -0.5 & 0 & 0.45946 \\ 0 & -2 & 1.4375 \\ 0 & 0 & -4 \end{pmatrix}.$$

The maximal lag 1 autocorrelation which can be achieved with a MAP generated from this representation is 0.0024. With the simulated annealing approach the following representation is generated.

$$\mathbf{p} = (0.783847, 0.182691, 0.033463), \ \mathbf{D}_0 = \begin{pmatrix} -1.00000 & 0.04171 & 0.95732 \\ 0.00000 & -2.00000 & 0.00000 \\ 0.00000 & 0.00000 & -4.00000 \end{pmatrix}. \quad (13)$$

This representation can be expanded to a MAP with lag 1 autocorrelation 0.0751 which is still small but more than with the APH resulting from the previous transformation.

5   State Space Expansion

If the autocorrelation which can be achieved with a given APH is not sufficient, then it is possible to enlarge the state space to find another representation of the APH with a larger flexibility. The extension of the state space is done by cloning single states. The approach uses an equivalence relation between APHs which has been defined in a more general context in [6]. Let $(\mathbf{p}, \mathbf{D}_0)$ and $(\mathbf{p}', \mathbf{D}_0')$ be two APHs of order n and $n + 1$, respectively. $\mathbf{V}$ is an $(n + 1) \times n$ matrix with unit row sums (i.e., $\mathbf{V}\mathbb{I} = \mathbb{I}$). If $\mathbf{p}'\mathbf{V} = \mathbf{p}$ and $\mathbf{D}_0'\mathbf{V} = \mathbf{V}\mathbf{D}_0$, then both APHs are equivalent, i.e., are different representations of the same distribution [6].

We use one specific matrix $\mathbf{V}$ to realize the expanded APH. Let $\mathbf{V} = \begin{pmatrix} \mathbf{I} \\ \mathbf{0} \ 1 \end{pmatrix}$ be the $(n + 1) \times n$ transformation matrix describing the equivalence. $(\mathbf{p}, \mathbf{D}_0)$ is the original n-dimensional APH. We define a $n + 1$ dimensional APH $(\mathbf{p}', \mathbf{D}_0')$ with

$$\mathbf{p}'(i) = \begin{cases} \mathbf{p}(i) \text{ if } i < n \\ 0 \quad \text{ if } i = n \\ \mathbf{p}(n) \text{ if } i = n+1 \end{cases} \quad \text{and } \mathbf{D}'_0(i,j) = \begin{cases} \mathbf{D}_0(i,j) & \text{if } i < n \text{ and } j < n \\ \mathbf{a}(i)\mathbf{D}_0(i,n) & \text{if } i < n \text{ and } j = n \\ (1-\mathbf{a}(i))\mathbf{D}_0(i,n) & \text{if } i < n \text{ and } j = n+1 \\ \mathbf{D}_0(n,n) & \text{if } i,j \in \{n,n+1\} \text{ and } i = j \\ 0 & \text{if } i = n \text{ and } j = n+1 \end{cases}$$

(14)

for some vector $\mathbf{a}$ of length n with elements out of $[0,1]$. It is easy to show that the required relation between the two APHs holds. Although the above transformation works for all vectors $\mathbf{a}$ with elements from $[0,1]$ we assume in the sequel that $\mathbf{a} = \mathbb{I}$.

Let $(\mathbf{D}_0, \mathbf{D}_1)$ be a MAP expanded from $(\mathbf{p}, \mathbf{D}_0)$ with a maximal first joined moments $\mu^*_{1,1}$. We define a MAP $(\mathbf{D}'_0, \mathbf{D}'_1)$ with matrix $\mathbf{D}'_0$ as in (14) with vector $\mathbf{a} = \mathbb{I}$ and

$$\mathbf{D}'_1(i,j) = \begin{cases} \mathbf{D}_1(i,j) + \dfrac{\mathbf{D}_1(n,j)}{\sum_{k=1}^{n-1}\mathbf{D}_1(n,k)}\mathbf{D}_1(i,n) & \text{if } i < n \text{ and } j < n \\ 0 & \text{if } i < n \text{ and } j \geq n \\ \dfrac{\sum_{k=1}^{n}\mathbf{D}_1(n,k)}{\sum_{l=1}^{n-1}\mathbf{D}_1(n,l)}\mathbf{D}_1(n,j) & \text{if } i = n \text{ and } j < n \\ 0 & \text{if } i = j = n \\ 0 & \text{if } i \leq n \text{ and } j = n+1 \\ \lambda_n & \text{if } i = j = n+1 \end{cases}$$

Observe that if the denominator in the first case becomes zero, then also the numerator is zero and the second term of the sum becomes zero. We assume that $\mathbf{D}_1(n,j) > 0$ for at least one $j < n$. If this is not the case, then the transformation cannot be applied and one should use the alternative expansion described below.

In the online companion [2] we present the complete proofs which show that $(\mathbf{D}'_0, \mathbf{D}'_1)$ describes a valid MAP and that $\mu^*_{1,1} \leq \mu'_{1,1}$ holds.

The above transformation results in a representation with a first joint moment which is not smaller than the first joint moment of the original APH. The transformation cannot be applied if $\mathbf{D}_1(n,j) = 0$ for all $j < n$ or $\mathbf{D}_1(i,n) = 0$ for all $i < n$. This is for example the case for hyperexponential APHs. In this case we introduce a second expansion approach that adds two states by first representing one of the exponential phases by a two state representation (2) and afterwards cloning the second state. This results in the following representation of an exponential distribution with three states.

$$\mathbf{p}' = \left(\frac{\lambda_2 - \lambda_1}{\lambda_2}, 0, \frac{\lambda_1}{\lambda_2}\right) \text{ and } \mathbf{D}'_0 = \begin{pmatrix} -\lambda_1 & \lambda_1 & 0 \\ 0 & -\lambda_2 & 0 \\ 0 & 0 & -\lambda_2 \end{pmatrix}$$

(15)

By simple calculations it can be shown that the maximum first joined moment is reached by a MAP with

$$\mathbf{D}'_1 = \begin{pmatrix} 0 & 0 & 0 \\ \lambda_2 & 0 & 0 \\ 0 & 0 & \lambda_2 \end{pmatrix}$$

for $\lambda_2 = 2\lambda_1$. In this case, the lag 1 autocorrelation coefficient equals 0.25. The above construction can, of course, be extended, by representing the last exponential phase by a three state representation with rate $\lambda_3 = 2\lambda_2$ such that the overall distribution has 5 states and a lag 1 autocorrelation of 0.28125. This representation of correlated exponentially distributed random variables using a MAP is an alternative to the approach proposed in [1] which uses bivariate APHs.

Representation (15) can be used to substitute an arbitrary state in a APH. If the last state is substituted then the ordering of states with increasing rates is kept. If another state is substituted, it might be necessary to compute for the resulting APH the canonical representation, because the rates are no longer increasing, and start the transformation process afterwards again. This is easy for hyperexponential distributions but requires the use of the simulated annealing based optimization in general. A state which is substituted by the 3 state representation should have a larger initial probability to assure an effect on the flexibility of the representation.

Example: We apply the transformations to the APHs (12) and (13). Example (12) has a hyperexponential representation with 3 states and can be extended to a MAP with lag 1 autocorrelation coefficient $\rho_1 = 0.21429$. Due to the hyperexponential structure the first transformation cannot be applied. If we substitute the third state for the second transformation, we obtain a representation with 5 states which can be expanded to a MAP with $\rho_1 = 0.21684$ which is only a minor expansion. If the first state is substituted, the resulting APH can be expanded to MAP with $\rho_1 = 0.33641$. However, in this case the representation resulting from the expansion has to be first transformed to canonical form and then the simulated annealing based optimization has to be applied to find the right representation.

Example (13) cannot be transformed to an hyperexponential representation and MAP which can be expanded from the representation has $\rho_1 = 0.0751$. The first transformation results in an APH with 4 states which can be expanded to a MAP with $\rho_1 = 0.0889$. The second transformation cannot be applied to the last state since the initial probability of this state is 0. Expansion of the first state results in an APH with 5 states which can be transformed to a representation which is expanded to a MAP with $\rho_1 = 0.18079$. Again the representation resulting from the transformation first has to be transformed to the canonical form and then the simulated annealing based optimization is applied.

6 Examples

To demonstrate the effect of different PH transformations on the parameter estimation of MAPs we fitted various PHDs to real-world traffic data, transformed the resulting distributions and used it as input for a MAP fitting approach. In particular, the following experiment setup was used: We chose two different traces with interarrival times of network packets, i.e. the well-known benchmark trace *LBL-TCP-3* [19] from the Internet Traffic Archive (http://ita.ee.lbl.gov/) and the newer trace (*TUDo*) that was recorded at the computer science department at TU Dortmund [14]. Acyclic PHDs were fitted to the two traces using the approach from [4] that estimates the parameters of the distribution according to the empirical moments from the trace. The approach results in a PHD in series canonical form, such that no correlation can be modeled with this representation. Consequently, we applied different transformations to the representation and used the resulting distributions as input for a two-step MAP fitting approach that constructs matrix $\mathbf{D}_1$ according to the empirical autocorrelation coefficients of the trace file [14]. As a measure to assess the fitting quality we present plots of the autocorrelation coefficients and the likelihood values of the resulting MAPs. The transformations applied to the PHD are the ones described in Sect. 4, i.e. simulated annealing (denoted by Rand in the plots) and the transformation into a hyperexponential distribution (denoted by HExp). If necessary, we expanded the state space as described in Sect. 5 (denoted by an additional + in the plots). For comparison we also used the transformation from [4] (denoted by QEST) that chooses two phases i and j in each iteration and sets $\delta = 0.9 \cdot \delta^*$ where δ^* is the upper bound of the interval in Eq. 5. It was shown in [16] that this transformation usually does not find the optimal representation.

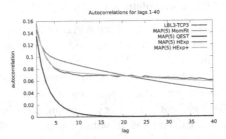

Fig. 1. Autocorrelation coefficients for the Trace *LBL-TCP-3* and PHDs of order 5

Fig. 1 shows the results for a PHD of order 5 fitted to the trace *LBL-TCP-3*. The plot contains the autocorrelation for the trace and 4 MAPs that resulted from expanding the transformed PH representation into a MAP. The MAPs are sorted in increasing order according to the maximal lag 1 autocorrelation that could be modeled with the corresponding PHD. For the original untransformed representation (labeled MomFit) of course no autocorrelation could be modeled. The transformation from [4] (QEST) and transformation into a hyperexponential distribution (HExp) both resulted in a representation that allows for modeling a lag 1 correlation that is (slightly) larger than the one of the trace, albeit HExp resulted in the larger of the two correlation values. Consequently, the hyperexponential representation allowed for a better estimation of the correlation from the trace. Since the transformation into hyperexponential representation worked for this distribution we omitted the transformation using simulated annealing and instead expanded the state space by adding additional phases, which further increased the maximal lag 1 correlation and allowed for the best estimation of the autocorrelation coefficients.

Fig. 2 shows the autocorrelation coefficients and the likelihood values for the trace *LBL-TCP-3* and models of order 6. The transformation from [4] and the transformation into the hyperexponential distribution failed to provide a representation that is adequate for MAP fitting. Since the PHD could not be transformed into a hyperexponential representation we applied the transformation using simulated annealing that resulted in the representation with the largest maximal lag 1 autocorrelation that was suited best for

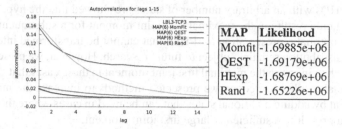

Fig. 2. Autocorrelation coefficients and likelihood values for the Trace *LBL-TCP-3* and PHDs of order 6

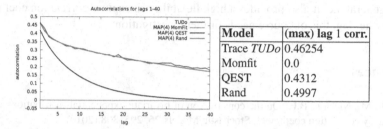

Fig. 3. Autocorrelation coefficients and maximal lag 1 correlation for the Trace *TUDo* and PHDs of order 4

MAP fitting. Fig. 2 also shows the likelihood values which confirm the observations from the plot of the autocorrelation coefficients.

For the last example shown in Fig. 3 the trace *TUDo* was used. Since the original distribution could not be transformed into hyperexponential representation we omitted its curve in the plot. Fig. 3 shows the maximal lag 1 correlation that could be achieved with the representations resulting from the different transformation approaches. As one can see the maximal correlation resulting from the simulated annealing approach is a little larger than the correlation from the trace, while the maximal correlation from the transformation from [4] is slightly lower. However, this has a large impact when fitting the autocorrelation coefficients as shown in the plot.

The experiments show that PH transformations can help to find an adequate representation of a PHD for a subsequent MAP fitting step. Moreover, the results suggest that a larger value for the maximal lag 1 correlation also allows for a better fitting of higher lags, i.e. the representations appear to be more flexible with increasing maximal correlation value.

7 Conclusions

In this article we considered different transformations and representations for acyclic PHDs to compute a representation that allows the expansion to a MAP with a maximal first joint moment. For PHDs of order 2 we distinguished two cases depending on the values of the initial probabilities and transition rates and presented optimal representations for both cases.

For acyclic PHDs with an arbitrary number of states we showed that the hyperexponential representation results in the maximal first joint moment for a subsequent MAP fitting step. The optimal representation for PHDs that cannot be transformed into a hyperexponential distribution remains subject to future research. However, we presented a heuristic approach to increase the maximal first joint moment in these cases that showed good experimental results. Moreover, we presented methods to increase the maximal first joint moment by adding additional states that can be used in cases where the transformations do not result in a sufficiently large first joint moment.

Our considerations focused on the first joint moment which is related to the lag 1 autocorrelation coefficient. Although our experiments suggest that a representation that allows for expansion into a MAP with a maximal lag 1 correlation coefficient is favorable in general, i.e. it also provides a high flexibility to capture correlation coefficients with a higher lag, this of course needs further investigation.

References

1. Bladt, M., Nielson, B.F.: On the construction of bivariate exponential distributions with an arbitrary correlation coefficient. Stochastic Models 26, 295–308 (2010)
2. Buchholz, P., Felko, I., Kriege, J.: Online Companion to the Paper Transformation of Acyclic Phase Type Distributions for Correlation Fitting (2013),
   http://www4.cs.tu-dortmund.de/download/kriege/publications/
   asmta2013_online_companion.pdf
3. Buchholz, P., Kriege, J.: Equivalence Transformations for Acyclic Phase Type Distributions. Technical Report 827, Dep. of Informatics, TU Dortmund (2009)
4. Buchholz, P., Kriege, J.: A heuristic approach for fitting MAPs to moments and joint moments. In: QEST, pp. 53–62 (2009)
5. Buchholz, P., Panchenko, A.: A two-step EM algorithm for MAP fitting. In: Aykanat, C., Dayar, T., Körpeoğlu, İ. (eds.) ISCIS 2004. LNCS, vol. 3280, pp. 217–227. Springer, Heidelberg (2004)
6. Buchholz, P., Telek, M.: Stochastic Petri nets with matrix exponentially distributed firing times. Performance Evaluation 67(12), 1373–1385 (2010)
7. Corana, A., Marchesi, M., Martini, C., Ridella, S.: Minimizing multimodal functions of continuous variables with the "simulated annealing" algorithm. ACM Trans. Math. Softw. 13(3), 262–280 (1987)
8. Cumani, A.: On the canonical representation of homogeneous Markov processes modeling failure-time distributions. Micorelectronics and Reliability 22(3), 583–602 (1982)
9. Fackrell, M.: Modelling healthcare systems with phase-type distributions. Health Care Management Science 12(1), 11–26 (2009)
10. Horváth, A., Telek, M.: PhFit: A general phase-type fitting tool. In: Field, T., Harrison, P.G., Bradley, J., Harder, U. (eds.) TOOLS 2002. LNCS, vol. 2324, pp. 82–91. Springer, Heidelberg (2002)
11. Horváth, G., Telek, M., Buchholz, P.: A MAP fitting approach with independent approximation of the inter-arrival time distribution and the lag-correlation. In: QEST (2005)
12. Iversen, V.B., Nielsen, B.F.: Some properties of Coxian distributions with applications. In: Modeling Techniques and Tools for Performance Analysis, pp. 61–66. Elsevier (1986)
13. Klemm, A., Lindemann, C., Lohmann, M.: Modeling IP traffic using the batch Markovian arrival process. Perform. Eval. 54(2), 149–173 (2003)

14. Kriege, J., Buchholz, P.: An Empirical Comparison of MAP Fitting Algorithms. In: Müller-Clostermann, B., Echtle, K., Rathgeb, E.P. (eds.) MMB & DFT 2010. LNCS, vol. 5987, pp. 259–273. Springer, Heidelberg (2010)
15. Lipsky, L.: Queueing Theory: A Linear Algebraic Approach. Springer (2010)
16. Mészáros, A., Telek, M.: A two-phase MAP fitting method with APH interarrival time distribution. In: Winter Simulation Conference. ACM (2012)
17. Neuts, M.F.: A versatile Markovian point process. Journ. of Applied Probability 16 (1979)
18. Neuts, M.F.: Matrix-geometric solutions in stochastic models. Johns Hopkins University Press (1981)
19. Paxson, V., Floyd, S.: Wide-area traffic: The failure of Poisson modeling. IEEE/ACM Transactions on Networking 3, 226–244 (1995)
20. Telek, M., Horváth, G.: A minimal representation of Markov arrival processes and a moments matching method. Perform. Eval. 64(9-12), 1153–1168 (2007)
21. Telek, M., et al.: Butools - program packages, webspn.hit.bme.hu/~butools/
22. Thümmler, A., Buchholz, P., Telek, M.: A novel approach for phase-type fitting with the EM algorithm. IEEE Trans. Dep. Sec. Comput. 3(3), 245–258 (2006)

Studying Mobile Internet Technologies with Agent Based Mean-Field Models

Marco Gribaudo[1], Daniele Manini[2], and Carla Chiasserini[3]

[1] Dip. di Elettronica e Informazione, Politecnico di Milano,
via Ponzio 34/5, 20133 Milano, Italy
gribaudo@elet.polimi.it
[2] Dip. di Informatica, Università di Torino,
Corso Svizzera 185, 10149 Torino, Italy
manini@di.unito.it
[3] Dip. di Elettronica e Telecomunicazioni, Politecnico di Torino,
Corso Duca degli Abruzzi 24, 10129 Torino, Italy
chiasserini@polito.it

Abstract. We analyze next generation cellular networks, offering connectivity to mobile users through LTE as well as WiFi. We develop a framework based on the Markovian agent formalism, which can model several aspects of the system, including the dynamics of user traffic and the allocation of the network radio resources. In particular, through a mean-field solution, we show the ability of our framework to capture the system behavior in flash-crowd scenarios, i.e., when a burst of traffic requests takes place in some parts of the network service area.

1 Introduction

One of the most evident and urgent challenges in the field of communication networks today is coping with the exponential growth of the wireless data traffic: the average smartphone is expected to generate 2.6 GB of traffic per month by 2016, with a global mobile data traffic that is slated to increase 18–fold by that time [4]. To accommodate such high data-traffic loads, new technologies, such as Long-Term Evolution (LTE), have been introduced to increase the capacity of cellular networks.

The fast uptake of mobile data services, however, indicates that these solutions are not sufficient to meet the intense user demand in many high-density settings all over the world. Thus, a new trend, usually referred to as *mobile data offloading*, has emerged. That is, while the cellular infrastructure will continue to provide essential wide-area coverage and support for high-mobility users, it will be complemented with WiFi hotspots, toward which data traffic should be offloaded whenever possible [9,13].

Such a scenario calls for a new access network architecture, composed of base station units (BSs) that may host several radio interfaces, hence provide Internet connectivity to mobile users through different communication technologies (e.g., LTE and WiFi) [5,2]. Beside meeting the users demand, this network paradigm

A. Dudin and K. De Turck (Eds.): ASMTA 2013, LNCS 7984, pp. 112–126, 2013.

will imply low real-estate and deployment costs, as well as low upgrading and security costs, thus saving the operators millions of dollars in captial and operational expensitures (CAPEX/OPEX).

It is important to stress that wireless communication technologies, like LTE and WiFi, are already mature, and that commercial products implementing the aforementionwd paradigm are available on the market, e.g., [2,1]. What is missing, however, is the definition of algorithms to make the above access network work efficiently. In particular, the functionalities of the system should be optimized so as to support the huge amount of data that wireless users are expected to consume/generate, while meeting the requirements in terms of quality of service, energy consumption and cost.

In this paper, we focus on the above aspect and develop a framework for the analysis of different policies regulating the Internet connectivity of mobile users. Specifically, we consider that the network service area is covered by a number of BSs, each of them hosting both an LTE and a WiFi radio interface. A user can connect to the Internet through either technologies, provided that enough radio resources are available to serve the user. Typically, widely popular communication devices, such as smartphones, implement a simple connectivity policy, which is known as "WiFi first": a user always connects to a WiFi hotspot when available. The rational behind this choice is that WiFi connectivity is much less costly than the cellular one: through WiFi the per-byte cost of data transfers can be reduced by 70% per one estimate [3]. However, WiFi may offer a much slower data transfer than LTE, especially when several users are accessing the same hotspot, or the radio propagation conditions are not favourable because of the user mobility or the presence of obstacles [12,15,5]. This clearly motivates the need for a framework that allows to evaluate various connection policies, beside "WiFi first".

In the following, we make a first step toward the solution of this issues. Specifically, we propose an analytical model and show how mean-field analysis can be successfully applied to the study of the system dynamics. We remark that a similar study carried out through simulation would imply very long computation times, as each communication node in the network (either user or BS) would have to be equipped with two radio interfaces, each of which would require different models representing the signal propagation as well as the node protocol stack.

2   Network Scenario and Mobile Internet Technologies

We consider a urban area (typically characterized by high user density) and covered by an LTE cellular network. The network system is composed of several BSs, each of them covering an area that we will be referred as *coverage area*. Colocated with the LTE interface, there is a WiFi radio (IEEE 802.11a/g/n), so as to implement a hostspot whose coverage coincides with that provided by the LTE technology.

We only consider data transfers, such as content dowloading or video streaming, as voice calls cannot be supported through LTE. Also, we focus on downlink data transfers (from the BS to the users) since traffic is typically highly asymmetric, with a large amount of data flowing from the Internet towards the users. Clearly, users that are under the coverage of more than one BS, can access any of them, although with different quality of service. In particular, it is fair to assume that the link quality increases as the distance between a user and a BS decreases, and that the better the link quality, the higher the transmission rate that the link end points can use. Furthermore, we consider that the users will not significantly move while receiving a data transfer from the Internet.

For clarity, before presenting the model we developed, we summarize the main characteristics of the two technologies that are at the basis of our analysis.

WiFi. The WiFi technology is specified by the IEEE 802.11 standard. It allows for the implementation of a wireless local network (WLAN) or hotspot, composed of an access point (AP) and user communication devices. Users that are connected to a hotspot can download/upload traffic from/to the Internet, or exchange traffic between each other; all traffic however is handled by the AP. In order to transmit data over the wireless channel, users and AP employ a totally distributed scheme, namely, a carrier sense multiple access (CSMA) technique enhanced with a collision avoidance mechnism. In a nutshell, whenever a network node wishes to transfer data, it senses the channel. If idle, it will transmit; otherwise, it defers its transmission by a random time. Clearly, the higher the traffic load within the WLAN, the higher the data latency and the probability that two or more transmissions start at the same time and, thus, fail. It is easy to show that, on the long run, such a channel access scheme provides equal access opportunities to all nodes (both users and AP).

Furthermore, it is important to note that at the physical layer the nodes may use different transmission speeds (i.e., data rates), depending on the propagation conditions between the sender and the intended receiver: the higher the packet error probability that a node experiences, the lower the rate it uses to transmit its data. As an example, in the "a" version of the 802.11 standard 8 values of data rate are possible, ranging from 6 up to 54 Mb/s. It follows that the transmission time of the data packets will depend on both the packet size and the rate used by the sender to transmit toward the intended recepient. This aspect may have severe consequences on the throughput experienced by the users connected to the same hotspot. Indeed, as noted in [14], "slower" senders occupy the channel longer preventing others from transmitting. Since the WiFi technology provides equal channel access opportunities to all nodes, as the number of slow senders increases, the fast ones will be able to transmit less often thus experiencing a low throughput even if they can employ a high data rate.

Finally, we remark that different frequency channels can be used by the WiFi technology. Again, with reference to the 802.11a standard, 8 disjoint channels around 2.4 GHz are available, and each hotspot can select one among these frequency bands. Typically, nearby hotspots employ different channels so as to avoid co-channel interference.

LTE. The LTE cellular technology is marketed as 4G and is already available in several regions, such as Australia, North America, and Scandinavian countries. Thanks to the use of multiple antennas at the transmitter and receiver, it implements multiple-input-multiple-output (MIMO) trasnmission techniques, i.e., the simultaneous transmissions of a number of flows as high as the number of antennas available at the communication end points. The transmission speed is therefore significantly higher than in the previous cellular technologies, and, in principle, it can reach up to 300 Mb/s in downlink (from the BS to the users) and 50 Mb/s in uplink (from user to BS). Current implementations, however, allow for much lower data rates, leading to a throughput of up to 30 Mb/s in downlink and of about 14 Mb/s in uplink.

As in most of the previous cellular technologies, a frequency division duplex (FDD) technique is used for data transmission: two separate frequency channels, say of 1.4 MHz each, are used for uplink and downlink traffic, respectively. Also, within an area covered by a BS, the access to the uplink channel is controlled by the BS itself, which schedules the user transmissions. Focusing on the downlink direction, an orthogonal frequency division multiple access (OFDMA) technique is employed, jointly with a time division scheme. Indeed, the downlink frequency channel is divided into several narrow-band subchannels. Disjoint subsets of such channels are then used by the BS to simultaneously transmit towards different users. Each subchannel is 15-MHz wide and each subset is composed of 12 subchannels. Time is divided into frames that are 10 ms long; each of them is further divided into 10 subframes. In the following, we consider the usage of a subset of 12 subchannels for a 1-ms duration (i.e., a subframe) as the granularity used by the BS to allocate radio resources to a data flow, i.e., the so-called Physical Radio Block (PRB). We remark that the BS may allocate one or more PRBs to transmit at the same time towards the same user. Also, as in the case of the WiFi APs, LTE interfaces at neighboring BSs should use different frequency channels so as to avoid interference.

3 Modelling LTE and WiFi Technologies

As outlined above, the system under study is composed of two main types of communication nodes, interacting with each other: users and BSs. The spatial distribution of such nodes plays a critical role, as a user can access a BS only if its position is within the coverage area of that BS. These peculiariries of the system prevent the use of standard state-space modeling techniques, such as Queueing Networks, Stochastic Petri Net, or Process Algebras, since all of them would suffer from the problem of state-space explosion.

In this work, we therefore resort to a technique based on the Markovian Agent formalism [10], and exploits the Mean Field Analysis [6,7]. In particular, our approach leverages the methodology presented in [8] to compute the model solution.

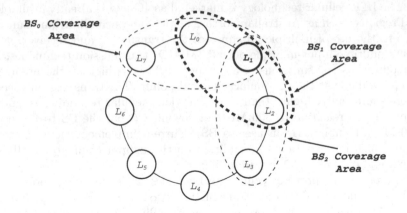

Fig. 1. An example of the ring topology under study

3.1 Network Topology and Model

To simplify the presentation, we consider a network with ring topology, composed of N geographical areas, hereinafter called locations and denoted by $L_0, \ldots, L_{N-1}$. Given the topology, each location L_i has two neighboring locations, L_{i-1} and L_{i+1}, where the generic pedex i has to be intended as $i \bmod N$. At the center of each location L_i, there is a BS, denoted by BS_i $(i = 0, \ldots, N-1)$, whose coverage area extends over three locations, namely, L_i, L_{i-1} and L_{i+1}. It follows that users located within L_i can access BS_i, or equivalently, BS_{i-1} and BS_{i+1}. Fig. 1 shows an example of the ring topology used in our model, for $N = 8$. In particular, circles represent the locations $L_0, \ldots, L_7$, while each dotted ellipse corresponds to the coverage area of a BS.

We call *agent* a portion of the model that describes the behavior of the communication nodes within a given location L_i. Specifically, we talk about user or BS agent depending on whether an agent refers to the behavhior of the users or of the BS. We then define as *access group i* (also denoted by AG_i) the the set of users that access the network using BS_i.

All users agents have exactly the same structure and include the same states, as shown in the left side of Fig. 2. Following the definition given in [8], the agents in the different locations correspond to different agent classes, all deriving from the same metaclass. As depicted in the figure, each agent is composed of 7 states. The first six account for the particular technology that a user is using to access the Internet (namely, LTE or WiFi), while the last state is used to count the user connection attempts that fail due to the unavailability of radio resources. The location from which the technology is accessed is denoted by the following labels: *local* for the BS in the location the user agent refers to, and *cw* and *ccw* for the BS in, respectively, the clockwise and the counter-clockwise neighboring location. In particular, we consider BS_{i+1} and BS_{i-1} being, respectively, the clockwise and counter-clockwise neighbors of BS_i. Finally, the label *Loss* denotes the state counting the service requests that have not been accomodated. The interaction

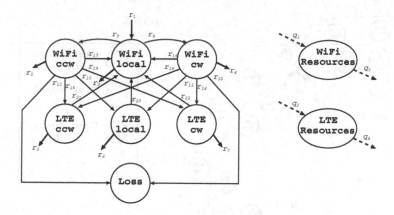

Fig. 2. Mean-field models of a user agent (left) and of a BS agent (right)

among the users is determined by the transition rates, which are functions of the number of users in the each state of the local and of the neighboring locations. Since the transition rates are the key element of the model, they will be described in depth in Section 3.2.

Similarly to the user agent, we define another metaclass to describe the BS status. Again, each BS agent corresponds to a class deriving from the same metaclass. Thus, all BS agents have the same struture: they include two states, each counting the number of allocated radio resources, one for WiFi and the other for LTE (as illustrated in the right side of Fig. 2). Note that by radio resources we mean frequency channels in WiFi and PRBs in LTE. The incoming/outcoming arcs account for the dynamic allocation of BS radio resources.

Fig. 3 further clarifies the relationship among user agents, locations, and access groups. Recall that each access group AG_i ($i = 0, 1, 2$ in the figure) refers to the set of users accessing BS_i, while each agent includes the possible states taken by users located within L_i. For instance, AG_1 includes states that belong to three different agents (each denoted by a different color), as BS_1 can be accessed by users that are in L_0, L_1 and L_2. Thus, AG_1 contains those states of the three user agents that correspond to the usage of radio resource handled by BS_1 (e.g., for the agent referring to L_0: WiFi cw and LTE cw).

3.2   Functional Rates

The interaction among the agents is determined by the transition rates; in this section we present the functions that allow to compute their values.

With regard to the BS agents, the control of the resource allocation at each BS is described through the rates q_1, q_2, q_3 and q_4. In the following, we will consider a static resource allocation, hence we set the above rates to the same constant value.

The evolution of a user agent through the model states is determined by the values assigned to the following rates (see also Fig. 2):

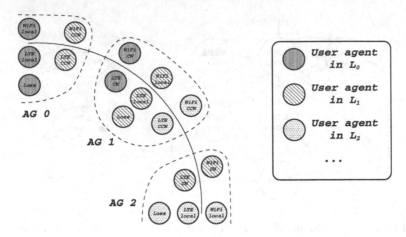

Fig. 3. Relation among user agents, locations, and access groups; different user agents are denoted by different colors

- Arrival of users requests (r_1);
- Service of requests (r_2, r_3 and r_4 for WiFi; r_5, r_6 and r_7 for LTE);
- WiFi switching from the local BS to a neighboring one (r_8, r_9);
- WiFi switching to LTE (r_{10}, r_{11} from neighboring BSs to the local one; r_{14}, r_{15}, r_{18}, r_{19} for neighbouring BSs);
- Loss of requests (r_{12}, r_{13});
- Upgrading: WiFi switching from a neighboring BS to the local one (r_{16}, r_{17}), and technology switching from LTE to WiFi (r_{20}, r_{21}, r_{22}).

Below, we introduce the main concepts and, in few cases, also the equations that we used to derive the rates. The computation details are reported in Table 1, where the following notations are exploited:

$I_1(condition) = 1$ if $condition$ is true, 0 otherwise

$I_2(a, b) = 1$ if $a > b$, 0.5 if $a = b$, 0 if $a < b$[1]

and where $n_{\tau-j}^{[i]}$ denotes the number of users in $\tau - j$ state of the user agent referring to location L_i, with $j \in \{local, cw, ccw\}$ and $\tau \in \{WiFi, LTE\}$.

Next, we describe all the possible events that trigger a transition between two states of the user agent, and the corresponding rates. Note that for what concern technology switching criteria, we model a specific system behavior based on arbitrary rules which can be easily refined to study more complex dynamics.

---

[1] To improve the smoothness of the solution, in the experiments we used for $I_2(\cdot)$ the *Sigmod* function defined as follows:

$$I_2(a, b) = \frac{1}{1 + e^{\alpha(b-a)}}$$

with α very large (we used $\alpha = 100$).

Arrivals. We denote by $\lambda_i(t)$ the number of requests per second generated by users in location L_i; note that such parameter has both a spatial and a time dependency. For the generic user agent referring to location L_i, we set $r_1 = \lambda_i(t)$. Clearly, the request rates are not affected by the system status. While deriving our results, we will generate a burst of requests in location subsets, in order to observe the network response.

Services. The WiFi service rates, to be assigned to r_2, r_3 and r_4, are derived through the following formula:

$$\sigma_{WiFi}^{[i]} = \frac{\mu_{WiFi} \times n_{WiFi-resources}^{[i]}}{n_{WiFi-local}^{[i]} + \alpha(n_{WiFi-cw}^{[i-1]} + n_{WiFi-ccw}^{[i+1]})} \tag{1}$$

where:

- μ_{WiFi} is the WiFi throughput (connection speed) when there is only one user accessing the interface;
- $n_{WiFi-resources}^{[i]}$ is the number of WiFi resources available at BS_i;
- $n_{WiFi-local}^{[i]}$ is the number of users in L_i accessing the WiFi interface at BS_i;
- $n_{WiFi-cw}^{[i-1]}$ is the number of users in location L_{i-1} that access the WiFi interface at BS_i, i.e., at their clockwise (cw) BS;
- $n_{WiFi-ccw}^{[i+1]}$ is the number of users in location L_{i+1} that access the WiFi interface at BS_i, i.e., at their counter-clockwise (ccw) BS;
- $\alpha \geq 1$ is a factor taking into account the throughput reduction due to far-away users accessing the WiFi interface at BS_i; such far-away users are those in locations L_{i-1} and L_{i+1}.

In particular, we have: $r_2 = \sigma_{WiFi}^{[i-1]}$, $r_3 = \sigma_{WiFi}^{[i]}$ and $r_4 = \sigma_{WiFi}^{[i+1]}$. We remark that the factor α is necessary in order to account for the anomaly effect typical of the WiFi technology, as mentioned in Section 2.

Let us call $\sigma_{LTE}^{[i]}(j)$ the LTE service rate assigned to a user in L_j accessing BS_i. For $i = 0, \ldots, n-1$, we have $r_5 = \sigma_{LTE}^{[i-1]}(i)$, $r_6 = \sigma_{LTE}^{[i]}(i)$ and $r_7 = \sigma_{LTE}^{[i+1]}(i)$. The LTE service rate, can be defined as:

$$\sigma_{LTE}^{[i]}(j) = \frac{\mu_{LTE}^{i,j} \times n_{LTE-resources}^{[i]}}{n_{LTE-local}^{[i]} + n_{LTE-cw}^{[i-1]} + n_{LTE-ccw}^{[i+1]}} \tag{2}$$

where:

- $\mu_{LTE}^{i,j}$ is the LTE throughput (connection speed) corresponding to one radio resource assigned from BS_i to a user accessing from L_j. Note that in LTE a radio resource is assigned exclusively to one user, however, depending on the radio propagation conditions, the user can employ a higher or a lower transmission rate. This motivates the dependency of the parameter on the user location (L_j), as the farther away the user is, the lower its connection speed;

- $n^{[i]}_{LTE-resources}$ is the number of LTE channels at the local BS (BS$_i$);
- $n^{[i]}_{LTE-local}$ is the number of users in L_i accessing the LTE interface at BS$_i$;
- $n^{[i-1]}_{LTE-cw}$ is the number of users in location L_{i-1} that access the LTE interface at BS$_i$, i.e., at their clockwise (cw) BS;
- $n^{[i+1]}_{LTE-ccw}$ is the number of users in location L_{i+1} that access the LTE interface at BS$_i$, i.e., at their counter-clockwise (ccw) BS.

WiFi Switching to Neighbouring BSs. The switching from local to a neighbouring WiFi interface is forced if the local service rate r_3 drops below the minimum threshold (μ_{min}). The neighbouring BS with the lowest load is selected (r_8, r_9), if the two locations have the same load the destination is randomly selected between them. The latency due to the interface switching is denoted by $1/\mu_{BS-Sw}$.

Switching from Neighboring WiFi to Local LTE. When the WiFi service rate in neighboring BSs drops below a minimum threshold (μ_{min}), the users attempt to migrate to the LTE interface of the local BS (transitions with associated rates r_{10} and r_{11}). Indeed, although connectivity through LTE is more costly, we expect that users are willing to pay a higher price provided that they can obtain a sufficient connection speed. The transition success depends on the actual load of the local LTE interface; in particular, the following condition must hold:

$$n^{[i]}_{LTE-local} + n^{[i-1]}_{LTE-cw} + n^{[i+1]}_{LTE-ccw} < n^{[i]}_{LTE-resources} \times B \times \gamma_{min} \qquad (3)$$

where $n^{[i]}_{LTE-local}$, $n^{[i-1]}_{LTE-cw}$, $n^{[i+1]}_{LTE-ccw}$ and $n^{[i]}_{LTE-resources}$ are defined as above, while B is number of 1-ms time intervals (subframes) for each LTE frequency subset (see Section 2), and γ_{min} is the utilization threshold of LTE resources at a BS. The user migration takes place (with a latency due to the to technology switch, $1/\mu_{\tau-Sw}$) if the utilization level of the LTE resources at the target BS is under γ_{min}. This avoids overloading the LTE interface and, thus, the need to move traffic flows back again to neighboring WiFi interfaces after some time.

Switching from WiFi to LTE at Neighboring BSs. The switch from neighboring WiFi to local LTE can fail if condition (3) is false. In this case, users may try to move towards an LTE interface at a neighboring BS. The switch is successful if either one of the neighboring BSs exhibits a utilization level of their LTE resources lower than γ_{min}. If both BSs can accomodate the user, the rate function selects the BS with the lower LTE load. In case of a tie, the destination BS is randomly selected. At last, we recall that this algorithm is used for deriving r_{14}, r_{15}, r_{18} and r_{19}.

Loss of Requests. A user data request fails when it cannot be accomodated at any BSs (using either LTE or WiFi), i.e., there are no resources available.

Specifically, when the WiFi service rate is below the μ_{min} threshold, the function for rates r_{12} and r_{13} verify the condition (3) for the LTE interface at the local as well as at the neighboring BSs. If the condition is always false, the request cannot be accomodated and the number of request failures is incremented.

Upgrade. The upgrading to the local WiFi interface can take place from that at neighbouring BSs as well as from the LTE interface at either local or neighbouring BSs. Specifically, when rate r_3 is beyond a given threshold (μ_{max}), the users can migrate to the local WiFi interface. Note that such a transition indeed represents an upgrade since using the (non-overloaded) WiFi inteface at the local BS implies high speed connectivity at low cost. The time required by the transition depends on the type of switch that is performed, i.e., the technology does not change but the BS does ($1/\mu_{BS-Sw}$ for r_{16} and r_{17}) or the technology (and possibly the BS) changes ($1/\mu_{T-Sw}$ for r_{20}, r_{21} and r_{22}).

4   Analysis and Numerical Results

The model proposed in Section 3 is solved using the techniques described in [8]. In particular, the model of the agent presented in Fig. 2 is used to determine, for each location L_i, two matrices and a vector: the transition matrix $\mathbf{C}^{[i]}$, the death matrix $\mathbf{D}^{[i]}$ and the birth vector $\mathbf{b}^{[i]}$. The resulting matrices and vectors are as follows:

$$
\mathbf{C}^{[i]} = \begin{vmatrix} -() & r_{17} & 0 & r_{15} & r_{10} & r_{19} & r_{13} \\ r_9 & -() & r_8 & 0 & 0 & 0 & 0 \\ 0 & r_{16} & -() & r_{18} & r_{11} & r_{14} & r_{12} \\ 0 & r_{21} & 0 & -() & 0 & 0 & 0 \\ 0 & r_{20} & 0 & 0 & -() & 0 & 0 \\ 0 & r_{22} & 0 & 0 & & -() & 0 \\ 0 & 0 & 0 & 0 & 0 & 0 & 0 \end{vmatrix} \quad \mathbf{D}^{[i]} = diag \begin{vmatrix} r_4 \\ r_3 \\ r_2 \\ r_5 \\ r_6 \\ r_7 \\ 0 \end{vmatrix} \quad \mathbf{b}^{[i]} = \begin{vmatrix} r_1 \\ 0 \\ 0 \\ 0 \\ 0 \\ 0 \\ 0 \end{vmatrix}^T \quad (4)
$$

Matrix $\mathbf{C}^{[i]}$ must be an infinitesimal generator: all its rows must sum up to zero. To simplify the presentation, we have used the notation "$-()$" to identify the sum of the other elements in the row, changed of sign. The row represents, respectively, the following states of the agent: $WiFi-ccw$, $WiFi-local$, $WiFi-cw$, $LTE-ccw$, $LTE-local$, $LTE-cw$, $Loss$. Note that, since the $Loss$ state is absorbing, the corresponding row is zero. The number of users in a given state is collected in a row vector $\mathbf{n}^{[i]} = |n^{[i]}_{WiFi-ccw}, \ldots, n^{[i]}_{Loss}|$, and the evolution of the system is computed by solving the following equations[2], one for each location L_i ($i = 0, \ldots, n-1$):

$$
\frac{d\mathbf{n}^{[i]}}{dt} = \mathbf{n}^{[i]} \left(\mathbf{C}^{[i]} - \mathbf{D}^{[i]} \right) + \mathbf{b}^{[i]}. \tag{5}
$$

---

[2] Note that all vectors and matrices depend on time. However, we have omitted the time dependency in order to simplify the notation.

Table 1. Rate functions

Rates	Functions
r_1	$\lambda_i(t)$
r_2	$\sigma_{WiFi}^{[i-1]}$
r_3	$\sigma_{WiFi}^{[i]}$
r_4	$\sigma_{WiFi}^{[i+1]}$
r_5	$\sigma_{LTE}^{[i-1]}(i)$
r_6	$\sigma_{LTE}^{[i]}(i)$
r_7	$\sigma_{LTE}^{[i+1]}(i)$
r_8	$\mu_{\tau-Sw} \times I_1(\sigma_{WiFi}^{[i]} < \mu_{min}) \times$ $I_2(\sigma_{WiFi-local}^{[i+1]}, sigma_{WiFi-local}^{[i-1]})$
r_9	$\mu_{\tau-Sw} \times I_1(\sigma_{WiFi}^{[i]} < \mu_{min}) \times$ $I_2(\sigma_{WiFi-local}^{[i-1]}, sigma_{WiFi-local}^{[i+1]})$
r_{10}	$\mu_{\tau-Sw} \times I_1(\sigma_{WiFi}^{[i-1]} < \mu_{min}) \times$ $I_1(n_{LTE-local}^{[i]} + n_{LTE-cw}^{[i-1]} + n_{LTE-ccw}^{[i+1]} < n_{LTE-resources}^{[i]} \times B \times \gamma_{min})$
r_{11}	$\mu_{\tau-Sw} \times I_1(\sigma_{WiFi}^{[i+1]} < \mu_{min}) \times$ $I_1(n_{LTE-local}^{[i]} + n_{LTE-cw}^{[i-1]} + n_{LTE-ccw}^{[i+1]} < n_{LTE-resources}^{[i]} \times B \times \gamma_{min})$
r_{12}	$\mu_{fail} \times I_1(\sigma_{WiFi}^{[i+1]} < \mu_{min}) \times$ $I_1(n_{LTE-local}^{[i]} + n_{LTE-cw}^{[i-1]} + n_{LTE-ccw}^{[i+1]} \geq n_{LTE-resources}^{[i]} \times B \times \gamma_{min}) \times$ $I_1(n_{LTE-local}^{[i+1]} + n_{LTE-cw}^{[i]} + n_{LTE-ccw}^{[i+2]} \geq n_{LTE-resources}^{[i+1]} \times B \times \gamma_{min}) \times$ $I_1(n_{LTE-local}^{[i-1]} + n_{LTE-cw}^{[i]} + n_{LTE-ccw}^{[i-2]} \geq n_{LTE-resources}^{[i-1]} \times B \times \gamma_{min}) \times$
r_{13}	$\mu_{fail} \times I_1(\sigma_{WiFi}^{[i-1]} < \mu_{min}) \times$ $I_1(n_{LTE-local}^{[i]} + n_{LTE-cw}^{[i-1]} + n_{LTE-ccw}^{[i+1]} \geq n_{LTE-resources}^{[i]} \times B \times \gamma_{min}) \times$ $I_1(n_{LTE-local}^{[i+1]} + n_{LTE-cw}^{[i]} + n_{LTE-ccw}^{[i+2]} \geq n_{LTE-resources}^{[i+1]} \times B \times \gamma_{min}) \times$ $I_1(n_{LTE-local}^{[i-1]} + n_{LTE-cw}^{[i]} + n_{LTE-ccw}^{[i-2]} \geq n_{LTE-resources}^{[i-1]} \times B \times \gamma_{min}) \times$
r_{14}	$\mu_{\tau-Sw} \times I_1(\sigma_{WiFi}^{[i+1]} < \mu_{min}) \times$ $I_1(n_{LTE-local}^{[i]} + n_{LTE-cw}^{[i-1]} + n_{LTE-ccw}^{[i+1]} \geq n_{LTE-resources}^{[i]} \times B \times \gamma_{min}) \times$ $I_2(n_{LTE-local}^{[i+1]} + n_{LTE-cw}^{[i]} + n_{LTE-ccw}^{[i+2]}, n_{LTE-local}^{[i-1]} + n_{LTE-ccw}^{[i]} + n_{LTE-cw}^{[i-2]}) \times$ $I_1(n_{LTE-local}^{[i+1]} + n_{LTE-cw}^{[i]} + n_{LTE-ccw}^{[i+2]} < n_{LTE-resources}^{[i+1]} \times B \times \gamma_{min})$
r_{15}	$\mu_{\tau-Sw} \times I_1(\sigma_{WiFi}^{[i-1]} < \mu_{min}) \times$ $I_1(n_{LTE-local}^{[i]} + n_{LTE-cw}^{[i-1]} + n_{LTE-ccw}^{[i+1]} \geq n_{LTE-resources}^{[i]} \times B \times \gamma_{min}) \times$ $I_2(n_{LTE-local}^{[i-1]} + n_{LTE-ccw}^{[i]} + n_{LTE-cw}^{[i-2]}, n_{LTE-local}^{[i+1]} + n_{LTE-cw}^{[i]} + n_{LTE-ccw}^{[i+2]}) \times$ $I_1(n_{LTE-local}^{[i-1]} + n_{LTE-ccw}^{[i]} + n_{LTE-cw}^{[i-2]} < n_{LTE-resources}^{[i-1]} \times B \times \gamma_{min})$
r_{16}, r_{17}	$\mu_{BS-Sw} \times I_1(\sigma_{WiFi}^{[i]} > \mu_{max})$
r_{18}	$\mu_{\tau-Sw} \times I_1(\sigma_{WiFi}^{[i+1]} < \mu_{min}) \times$ $I_1(n_{LTE-local}^{[i]} + n_{LTE-cw}^{[i-1]} + n_{LTE-ccw}^{[i+1]} \geq n_{LTE-resources}^{[i]} \times B \times \gamma_{min}) \times$ $I_2(n_{LTE-local}^{[i-1]} + n_{LTE-ccw}^{[i]} + n_{LTE-cw}^{[i-2]}, n_{LTE-local}^{[i+1]} + n_{LTE-cw}^{[i]} + n_{LTE-ccw}^{[i+2]}) \times$ $I_1(n_{LTE-local}^{[i-1]} + n_{LTE-ccw}^{[i]} + n_{LTE-cw}^{[i-2]} < n_{LTE-resources}^{[i-1]} \times B \times \gamma_{min})$
r_{19}	$\mu_{\tau-Sw} \times I_1(\sigma_{WiFi}^{[i-1]} < \mu_{min}) \times$ $I_1(n_{LTE-local}^{[i]} + n_{LTE-cw}^{[i-1]} + n_{LTE-ccw}^{[i+1]} \geq n_{LTE-resources}^{[i]} \times B \times \gamma_{min}) \times$ $I_2(n_{LTE-local}^{[i+1]} + n_{LTE-cw}^{[i]} + n_{LTE-ccw}^{[i+2]}, n_{LTE-local}^{[i-1]} + n_{LTE-ccw}^{[i]} + n_{LTE-cw}^{[i-2]}) \times$ $I_1(n_{LTE-local}^{[i+1]} + n_{LTE-cw}^{[i]} + n_{LTE-ccw}^{[i+2]} < n_{LTE-resources}^{[i+1]} \times B \times \gamma_{min})$
r_{20}, r_{21}, r_{22}	$\mu_{\tau-Sw} \times I_1(\sigma_{WiFi}^{[i]} > \mu_{max})$

Equation (5) can be solved using a suitable numerical algorithm. In our work, we have used the *Runge-Kutta with adaptive step-size control* discretization method [16].

Note that some rates include indicator functions that can cause problems to the mean field technique. In current experiments we found that the introduction of sigmoid functions instead of indicators is enough to obtain accurate results. However more general approaches based on higher moment approximations such as [11] can provide better results at the expense of greater number of equations.

4.1   Results: Scenario with Request Burst in One BS

Here, we show the ability of the proposed framework to capture the system dynamics. We consider the topology in Fig. 1 and represent a flash crowd traffic scenario in location L_1. To this end, we set $\lambda_i(t) = 2$ requests/s, $\forall t \in [0, 100]$, for $i \neq 1$, and $\lambda_1(t) = 12$ when $t \in [0, 60)$ and $\lambda_1(t) = 2$ when $t \in [60, 100]$; this corresponds to a request burst affecting L_1 for 60% of the observation time. Recall that the radio resource allocation at the BSs is static, and that all BSs have the same number of WiFi and LTE resources available.

Fig. 4 highlights the impact of the above flash crowd traffic scenario on the load level of the different BSs and on the service provided by the network system. Specifically, the plots depict, for every location, the temporal evolution of the number of user agents in each state. Fig. 4(a) refers to L_1 (i.e., the location affected by the request burst), Fig. 4(b) refers to the neighboring locations L_2 and L_0, Fig. 4(c) represents the behavior of the users in L_3 and L_7, and finally Fig. 4(d) reflects the situation of the users in all the remaining locations. Note that, due to the ring shape of the topology and the considered traffic scenario, the users in locations L_0 and L_2 exhibit the same behavior (except for the fact that cw and ccw states are inverted); the same observation holds for the pairs (L_3, L_7) and (L_4, L_6), as well as for L_5, L_4 and L_6. Thus, for simplicity, in the following we will refer only to L_1, L_2, L_3 and L_4.

To ease the discussion of the results, in the plots we added vertical lines to indicate the following main events that take place in the system.

Event A: Because of the high traffic load, local WiFi resources at BS_1 saturate (red line in Fig. 4(a)). As a consequence, the local service rate drops below μ_{min} and the users migrate toward the neighboring WiFi intefaces (see the overlapping blue and green lines in Fig. 4(a), and the red line in Fig. 4(b)).

Event B: Also the neighboring WiFi resources (e.g., BS_2) saturate (see, e.g., blue and green lines in Fig. 4(a)), and users in L_1 start accessing the Internet through the local LTE resources (purple line). Also, since at BS_2 all WiFi resources are allocated to users in L_1, the users in L_2 migrate to the neighboring (clock-wise) WiFi interface, i.e., BS_3 (green line in Fig. 4(b), and red line in Fig. 4(c)).

Event C: In L_1 the local LTE saturates (purple line in Fig. 4(a)) and users start using LTE resources at neighboring BSs (light blue and gray lines).

Event D: All resources at BS_1, as well as those in neighboring BSs that can be accessed by users in L_1, are saturated. As a consequence, requests start to be dropped (black line in Fig. 4(a)). For representation purposes, the curve representing the number of losses is truncated; it actually increases till the request burst in L_1 ends (event G), reaching a total of 135 dropped requests over the whole observation time. We remark that such high losses could be avoided if a dynamic resource allocation were implemented in the system.

Events E: All radio resources at BS_2 are allocated to serve local users and those in L_1; the WiFi resources are saturated at BS_3 too. It follows that users in L_2

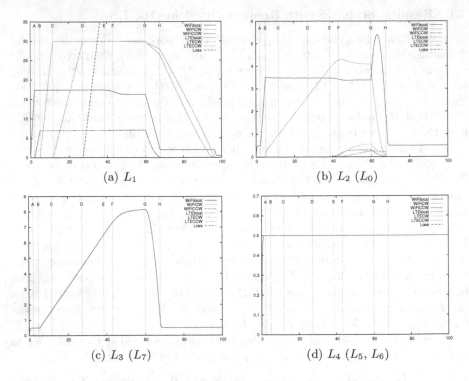

Fig. 4. Transient evolution of the user agents in locations $L_0, \ldots, L_7$. Main events are represented by vertical lines.

start accessing the Internet through LTE at BS_3 (light blue line in Fig. 4(b)); in addition, there are some oscillations in the use of the WiFi resources at BS_1 (blue line in Fig. 4(b)).

Events F: The situation arised in correspondence of event C persists. The WiFi quality of service at BS_3, however, further degrades, hence those users in L_2 that were accessing the BS_3 WiFi resources move back to their local LTE (purple line in Fig. 4(b)).

Events G: The request burst in L_1 ends, thus the traffic load at all BSs start to decrease. BS_2 continues to serve the remaining requests by using both WiFi (whose load increases, see the red line in Fig. 4(b)) and LTE resources.

Events H: The decrease of the traffic load in L_1 allows the handover from LTE (purple, light blue and gray lines, in Fig. 4(a)) to WiFi (red line). Note also that, as the backlog of requests is served by BS_1, the load of the local WiFi resources (red lines) settles at the same value as that exhbited by the other BSs (namely, 0.5, which is consistent with the generation rate of 2 requests per second).

5   Conclusions and Future Work

We presented an analytical framework based on the Markovian agent formalism, which models next generation cellular networks. We envisioned a system where base stations can provide Internet connectivity through the LTE as well as the WiFi technology, and we highlighted how our framework can model the different dynamics of the system reflecting the user traffic. By solving the model through a mean-field based methodology, we also showed that the framework can well capture the system behavior in flash-crowd scenarios.

Future work will focus on the model validation through simulation, as well as on the study of dynamic resource allocation strategies and of user connectivity policies. Metrics such as throughput, latency, user energy consumption and number of served requests will be evaluated. Furthermore, the model will be extended to account for user mobility during data transfers.

Acknowledgment. C.F. Chiasserini was supported by NPRP grant # NPRP 5 - 782 - 2 - 322 from the Qatar National Research Fund (a member of Qatar Foundation). M. Gribaudo was supported by ForgeSDK project sponsored by Reply S.R.L. The statements made herein are solely the responsibility of the authors.

References

1. Architecture for mobile data offload over Wi-Fi access networks, http://www.cisco.com/en/US/solutions/collateral/ns341/ns524/ns673/white_paper_c11-701018.html (accessed February 2013)
2. Mobile data offloading for 3g and lte networks, http://www.alvarion.it/applications/mobile-data-offloading (accessed February 2013)
3. Economy + internet trends: Web 2.0 summit (2009), http://www.morganstanley.com/institutional/techresearch/pdfs/MS_Economy_Internet_Trends_102009_FINAL.pdf
4. Cisco Visual Networking Index: Global mobile data traffic forecast update, 2011–2016, Cisco White Paper (February 2012)
5. Balasubramanian, A., Mahajan, R., Venkataramani, A.: Augmenting mobile 3g using wifi. In: Proceedings of the 8th International Conference on Mobile Systems, Applications, and Services, MobiSys 2010, pp. 209–222. ACM (2010)
6. Benaim, M., Boudec, J.Y.L.: A class of mean field interaction models for computer and communication systems. Performance Evaluation 65(11-12), 823–838 (2008)
7. Bobbio, A., Gribaudo, M., Telek, M.: Analysis of large scale interacting systems by mean field method. In: 5th International Conference on Quantitative Evaluation of Systems - QEST 2008, St. Malo (2008)
8. Cordero, F., Manini, D., Gribaudo, M.: Modeling biological pathways: an object-oriented like methodology based on mean field analysis. In: The Third International Conference on Advanced Engineering Computing and Applications in Sciences (ADVCOM), pp. 193–211. IEEE Computer Society Press (2009)
9. Dimatteo, S., Hui, P., Han, B., Li, V.: Cellular traffic offloading through wifi networks. In: 2011 IEEE 8th International Conference on Mobile Adhoc and Sensor Systems (MASS), pp. 192–201 (October 2011)

10. Gribaudo, M., Cerotti, D., Bobbio, A.: Analysis of on-off policies in sensor networks using interacting markovian agents. In: 4th International Workshop on Sensor Networks and Systems for Pervasive Computing - PerSens 2008, Hong Kong (2008)
11. Guenther, M., Bradley, J.: Higher moment analysis of a spatial stochastic process algebra. In: Thomas, N. (ed.) EPEW 2011. LNCS, vol. 6977, pp. 87–101. Springer, Heidelberg (2011), http://dx.doi.org/10.1007/978-3-642-24749-1_8
12. Hadaller, D., Keshav, S., Brecht, T., Agarwal, S.: Vehicular opportunistic communication under the microscope. In: Proceedings of the 5th International Conference on Mobile Systems, Applications and Services, MobiSys 2007, pp. 206–219. ACM (2007)
13. Han, B., Hui, P., Kumar, V., Marathe, M., Shao, J., Srinivasan, A.: Mobile data offloading through opportunistic communications and social participation. IEEE Transactions on Mobile Computing 11(5), 821–834 (2012)
14. Heusse, M., Rousseau, F., Berger-Sabbatel, G., Duda, A.: Performance anomaly of 802.11b. In: Twenty-Second Annual Joint Conference of the IEEE Computer and Communications, INFOCOM 2003, March 30-April 3, vol. 2, pp. 836–843. IEEE Society (2003)
15. Hull, B., Bychkovsky, V., Zhang, Y., Chen, K., Goraczko, M., Miu, A., Shih, E., Balakrishnan, H., Madden, S.: Cartel: a distributed mobile sensor computing system. In: Proceedings of the 4th International Conference on Embedded Networked Sensor Systems, SenSys 2006, pp. 125–138. ACM (2006)
16. Press, W.H., Teukolsky, S.A., Vetterling, W.T., Flannery, B.P.: Numerical Recipes, 3rd edn. The Art of Scientific Computing. Cambridge University Press, New York (2007)

iSWoM: The Incremental Storage Workload Model Based on Hidden Markov Models

Tiberiu Chis and Peter G. Harrison

Department of Computing, Imperial College London,
Huxley Building, 180 Queens Gate, London SW7 2RH, UK
{tc207,pgh}@doc.ic.ac.uk

Abstract. We propose a storage workload model able to process discrete time series incrementally, continually updating its parameters with the availability of new data. More specifically, a Hidden Markov Model (HMM) with an adaptive Baum-Welch algorithm is trained on two raw traces: a NetApp network trace consisting of timestamped I/O commands and a Microsoft trace also with timestamped entries containing reads and writes. Each of these traces is analyzed statistically and HMM parameters are inferred, from which a fluid input model with rates modulated by a Markov chain is derived. We generate new data traces using this Markovian fluid, workload model. To validate our parsimonious model, we compare statistics of the raw and generated traces and use the Viterbi algorithm to produce representative sequences of the hidden states. The incremental model is measured against both the standard model (parameterized on the whole dataset) and the raw data trace.

1 Introduction

In modern, large-scale storage environments, workload arises from multiple, time-varying, correlated traffic streams that may create different resource bottlenecks in the system at different times. It is therefore important to categorize and model workload in a portable and efficient way for at least three purposes:

- to produce workload traces for live systems, on which quantitative measurements can then be made;
- to generate similar, representative traces for system simulation;
- to provide input parameters for analytical performance models.

We focus on the first purpose, with the emphasis on live systems. It takes considerable time, training-data and computing power to produce a reliably parameterized model and our incremental approach, by which a model's parameters are progressively updated rather than periodically re-calculated, is appealing in terms of its run-time performance. The question is, therefore, is it accurate?

Primarily, we use a Hidden Markov Model (HMM), which is a bivariate Markov chain pairing one observable state sequence with a hidden state sequence (first developed by Baum and Petrie [1]). Each hidden state produces an observation based on a probability given in a so-called observation matrix. The transition probability from one hidden state to another is based on the state transition matrix of

A. Dudin and K. De Turck (Eds.): ASMTA 2013, LNCS 7984, pp. 127–141, 2013.

the Markov chain. This encodes information on the evolution of a time series and uses its parameter estimation algorithm to help infer the behaviour of the given series. As HMMs can efficiently represent workload dynamics, acting as parsimonious models that obtain trace characteristics, their applications include storage workload modelling [7], speech recognition [6,10] and genome sequence prediction [4,5]. HMMs act as portable benchmarks to explain and predict the complex behaviour of multiapplication workloads and therefore optimize storage and access times. One aim of this paper is to create such a benchmark, which can be applicable to a number of discrete processes, by adapting slightly the HMM mechanisms that represent these processes.

We utilize statistical algorithms to investigate the three fundamental problems associated with HMMs – here, as they pertain to workload traces. The first problem is determining $P(O; \lambda)$, which is the probability of the observed sequence O given some model λ; the second is to maximize $P(O; \lambda)$ by adjusting the model parameters λ for a given observation sequence O; and the third problem is finding the most likely hidden state sequence associated with a given observed sequence. The three fundamental problems are solved by using the Forward-Backward algorithm, the Baum-Welch algorithm[1] [2] and the Viterbi algorithm [3], respectively. Initially, this paper focuses on creating an adaptive Baum-Welch algorithm by using a forward recurrence formula to approximate certain parameters in the Forward-Backward algorithm. The Viterbi algorithm is briefly used in the results section to further validate the model.

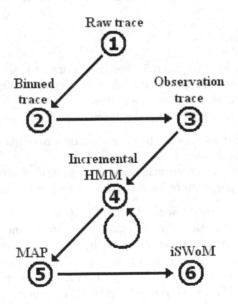

Fig. 1. A map showing our steps needed to obtain the iSWoM

---

[1] This algorithm uses the Forward-Backward algorithm iteratively.

Most of the paper is concerned with the development of an incremental Storage Workload Model (iSWoM), which consists of various procedures and submodels. The journey to form the iSWoM is summarised in Fig. 1. Firstly, we gather two distinct raw traces, which are transformed into binned traces using partitioning, and then into observation traces using K-means clustering. Secondly, our incremental HMM (i.e. a HMM that uses an adaptive Baum-Welch algorithm) trains on each observation trace to provide estimates for the model parameters. Note in Fig. 1 the circular arrow pointing to the incremental HMM, signifying the iterative process of training until no new data is made available. On completion, our incremental HMM defines (a special case of) a discrete Markov Arrival Process (MAP). The MAP and a random distribution (that chooses probabilistically particular input values from a specified cluster) form the iSWoM, which is capable of processing incoming data incrementally and updates its parameters on-the-fly as new trace data is available. The iSWoM can generate unlimited discrete traces, corresponding to and representative of the respective observation traces that have been inputted into the model. These generated traces can be validated using statistical comparisons (mean, standard deviation, confidence intervals) against raw traces. The Viterbi algorithm processes both the raw and iSWoM-generated traces and produces hidden state sequences for each, where accuracy is obtained by comparing correspondingly similar state patterns. Further applications can characterize workloads according to their HMM parameters and also be used to model quantitatively *a priori* similar workload environments – for example as inputs to simulation or analytical models. This paper aims for a storage workload model, formed by HMMs, capable of incremental learning of discrete data. Investigating other storage workload system benchmarks [19], we compare the iSWoM with our peers' models. Xhang et al [12] conducted measurements on three benchmarks (namely TPC-W [16], TPC-C [14] and RUBiS [13]) with the aim of understanding behaviour of e-commerce storage systems. Their findings consisted of workloads dominated by transactions (i.e. writes) requiring more storage than workloads dominated by browsing (i.e. reads). Similarities can be found between the read-dominated NetApp trace and the e-commerce writing workload, in terms of increased work which they demand. Regardless, the iSWoM provides on-line data characterization for its workloads (reads or writes), which Xhang et al have not attempted to fit on their model. Another workload characterization was presented by Kurmas et al [15], more specifically on an open mail trace using an FC-60 disk array as the storage system. The authors formed a cumulative distribution function (CDF) of read latency using a workload generator which reads values from a list. The I/O requests used in their workload included read or write commands and were replicated in a synthetic trace, much like the iSWoM-generated traces, but no incremental learning was attempted. We conclude the paper with evaluating our results in the context of other work and discuss improvements and continued research for the iSWoM. In the next section, we present the adaptive Baum-Welch algorithm as the key to the formation of the iSWoM.

2   Adaptive Baum-Welch Algorithm

The creation of an adaptive Baum-Welch algorithm is the foundation of the iSWoM and can be perceived as an updated Baum-Welch algorithm with a new Forward-Backward algorithm. More specifically, the backward part of the Forward-Backward algorithm will be updated with a forward-recurrence formula such that the probabilities for the new observation set can be easily calculated. We begin this section by defining both the Forward-Backward and Baum-Welch algorithms, and then describe the modifications to each one, along with a mathematical approximation for the backward variables. In the definitions, we bear in mind the need for normalization to avoid underflow with large data sets. However, to simplify notation, the standard algorithmic terms are used in this section. The approximation is followed by the definition of the incremental HMM.

2.1   Forward-Backward Algorithm

The Forward-Backward algorithm, aims to find $P(O; \lambda)$, the probability of the given sequence of observations $O = (O_1, O_2, \ldots, O_T)$ given the model $\lambda = (A, B, \pi)$, where T is the number of observations, A is the state transition matrix, B is the observation matrix and π is the initial state distribution. This is equivalent to determining the likelihood of the observed sequence O occuring. We use Rabiner's solution [9] to aid our calculations, focusing initially on the α-pass, which is the "forward" part of the Forward-Backward algorithm. Then, we shift our attention to the corresponding β-pass, aka. the "backward" part of the algorithm.

To begin with, we define $\alpha_t(i)$ as the probability of obtaining the observation sequence up to time t together with the state q_i at time t, given our model λ. Using N as the number of states and T as the number of observations, the mathematical notation is

$$\alpha_t(i) = P(O_1, O_2, \ldots, O_t, s_t = q_i; \lambda) \tag{1}$$

where $i = 1, 2, \ldots, N$, $t = 1, 2, \ldots, T$, and s_t is the state at time t.

Proceeding inductively, we write the solution for $\alpha_t(i)$ as follows:

1. For $i = 1, 2, \ldots, N$,

$$\alpha_1(i) = \pi_i b_i(O_1).$$

2. For $i = 1, 2, \ldots, N$ and $t = 1, 2, \ldots, T - 1$,

$$\alpha_{t+1}(i) = [\sum_{j=1}^{N} \alpha_t(j) a_{ji}] b_i(O_{t+1})$$

where $\alpha_t(j) a_{ji}$ is the probability of the joint event that $O_1, O_2, \ldots O_t$ are observed and we move from state q_j at time t to state q_i at $t + 1$.

3. It follows that,

$$P(O; \lambda) = \sum_{i=1}^{N} \alpha_T(i)$$

where $\alpha_T(i) = P(O_1, O_2, \ldots, O_T, s_T = q_i; \lambda)$

The backward variable, $\beta_t(i)$, is defined as the probability of obtaining the observation sequence from time $t + 1$ to T, given state q_i at time t and the model λ. So we have,

$$\beta_t(i) = P(O_{t+1}, O_{t+2}, \ldots, O_T; s_t = q_i, \lambda) \tag{2}$$

and the solution of $\beta_t(i)$ is given by

1. For $i = 1, 2, \ldots, N$,

$$\beta_T(i) = 1$$

2. For $i = 1, 2, \ldots, N$ and $t = T - 1, T - 2, \ldots, 1$,

$$\beta_t(i) = \sum_{j=1}^{N} a_{ij} b_j(O_{t+1}) \beta_{t+1}(j)$$

where we note that O_{t+1} can be observed from any state q_j.

2.2 Baum-Welch Algorithm

The Baum-Welch algorithm attempts to maximize $P(O; \lambda)$ by adjusting the parameters A, B, π given the model $\lambda = (A, B, \pi)$ and the observation sequence $O = (O_1, O_2, \ldots, O_T)$. This is done as an iterative process. We first define the probability of making a transition from state q_i at time t to state q_j at time $t + 1$, given O and λ, as

$$\xi_t(i, j) = P(s_t = q_i, s_{t+1} = q_j; O, \lambda) \tag{3}$$

Computing $\xi_t(i, j)$ can be described as a three-step process. Firstly, the observations $O_1, O_2, \ldots, O_t$ finishing in state q_i at time t will be covered by $\alpha_t(i)$. Secondly, the transition from q_i to q_j, where O_{t+1} was observed at time $t + 1$, is represented by the term $a_{ij} b_j(O_{t+1})$. Thirdly, the remaining observations $O_{t+2}, O_{t+3} \ldots O_T$ beginning in state q_j at time $t+1$ are covered by β_{t+1}. Putting those together, and dividing by a normalizing term $(P(O; \lambda))$ we have

$$\xi_t(i, j) = \frac{\alpha_t(i) a_{ij} b_j(O_{t+1}) \beta_{t+1}(j)}{P(O; \lambda)} \tag{4}$$

We now sum the terms in (4) over j and notice that this gives the probability of being in state q_i at time t, given the observation sequence O and model λ. This probability is defined as

$$\gamma_t(i) = P(s_t = q_i; O, \lambda) = \sum_{j=1}^{N} \xi_t(i, j)$$

Summing $\gamma_t(i)$ over time t up to T, we get the number of times we expect to visit state q_i. Similarly, summing up to $T - 1$ gives the expected number of transitions made from q_i. Thus:

$$\sum_{t=1}^{T} \gamma_t(i) = \text{Expected times state } q_i \text{ is visited.}$$

$$\sum_{t=1}^{T-1} \gamma_t(i) = \text{Expected transitions from } q_i.$$

Similarly, we sum $\xi_t(i,j)$ over t as follows:

$$\sum_{t=1}^{T} \xi_t(i,j) = \text{Expected visits of } q_i \text{ then } q_j.$$

$$\sum_{t=1}^{T-1} \xi_t(i,j) = \text{Expected transitions } q_i \text{ to } q_j.$$

Using these terms, the re-estimation formulas for our HMM parameters are:

$$\pi_i' = \gamma_1(i)$$

$$a_{ij}' = \frac{\sum_{t=1}^{T-1} \xi_t(i,j)}{\sum_{j=1}^{N} \sum_{t=1}^{T-1} \xi_t(i,j)}$$

$$b_j(k)' = \frac{\sum_{t=1, O_t=k}^{T} \gamma_t(j)}{\sum_{t=1}^{T} \gamma_t(j)}$$

Using these re-estimation formulas, we can update our model $\lambda' = (A', B', \pi')$, where $A' = \{a_{ij}'\}$, $B' = \{b_j(k)'\}$ and $\pi' = \{\pi_i'\}$. Our model will have fixed parameters once $P(O; \lambda') > P(O; \lambda)$, which means the iterative process to find the optimal model λ' ends.

2.3 Incremental HMM

The re-estimation of the model $\lambda' = (A', B', \pi')$ only works on a fixed set of observations. The aim of the adaptive Baum-Welch algorithm is to continually read in new trace data and update its parameters on-the-fly. This will form an incremental HMM (IncHMM) capable of efficiently supporting real time workload data, so demonstrating that infrequent, higher density, additional loads can be handled, mainly for on-line characterization of workloads as in [7]. The IncHMM initially starts as a standard HMM reading a set of observations and updating its current parameters A, B, π according to new incoming data. Therefore, after the standard HMM has finished training on its observation set, we calculate the revised α, β, ξ and γ variables based on the new set of observations.

To achieve an efficient IncHMM, we first update the terms of the Forward-Backward algorithm, which are the α and β values. Firstly, a HMM is trained on a trace of T observations and then M new observations are added. To update our model incrementally, we notice that the next α value is given by $\alpha_{T+1}(i) = [\sum_{j=1}^{N} \alpha_T(j) a_{ji}] b_i(O_T)$. The knowledge of the terms $\alpha_T(j)$, a_{ji} and $b_i(O_T)$ allows the new α variables to be computed easily using the forward recurrence formula. However, to find $\beta_{T+1}(i)$ is more difficult because it is dependent on a one step lookahead $\beta_{T+1}(i) = \sum_{j=1}^{N} a_{ij} b_j(O_{T+2}) \beta_{T+2}(j)$ and unfortunately we do not have $\beta_{T+2}(j)$. Therefore an approximation for the β variables is needed, preferrably a forward recurrence formula similar to the α formula. This β approximation will be adapted from [17], which was a preliminary attempt on a single data trace. The new ξ and γ variables (and thus the entries a'_{ij} and $b_j(k)'$) are calculated easily once the α and β sets are complete.

The β approximation will assume that, at time t and for state i, we have that $\beta_t(i) = \delta(t, i)$ is a continuous function with parameters t and i. For any state i, the function $\delta(t, i)$, w.r.t. t, is increasing from 0 to 1. Equivalently, $\delta(t, i)$ tends to 0 as $t \to 0$. The logic of this assumption comes from the backward recurrence formula in (2), which calculates the β values with a one-step lookahead. All β terms from $t = T - 1$ to $t = 1$ are less than 1, and with every step that t decreases, $\beta_t(i)$ gets closer to 0 through the computations of the backward formula. Therefore, for a sufficiently large observation set, we obtain the approximate equality $\delta(t, i) \approx 0 \approx \delta(t, j)$, where i and j are different states. We write the β approximation as:

$$\beta_t(i) \approx \beta_t(j) \tag{5}$$

Let us now transform the β backward-recurrence formula into a forward-recurrence version, with simplified notation $b_j = b_j(O_{t+1})$. Since there are two states in our model, let $N = 2$. It then follows that

$$\begin{pmatrix} \beta_t(1) \\ \beta_t(2) \end{pmatrix} = \begin{pmatrix} \sum_{j=1}^{2} a_{1j} b_j(O_{t+1}) \beta_{t+1}(j) \\ \sum_{j=1}^{2} a_{2j} b_j(O_{t+1}) \beta_{t+1}(j) \end{pmatrix}$$

Taking $\beta_t(1)$ and expanding the summation on the RHS, we obtain:

$$\beta_t(1) = a_{11} b_1(O_{t+1}) \beta_{t+1}(1) + a_{12} b_2(O_{t+1}) \beta_{t+1}(2)$$

Assuming that $t + 1$ is sufficiently small and using (5) we can deduce that $\beta_{t+1}(1) \approx \beta_{t+1}(2)$, giving us:

$$\beta_t(1) = \beta_{t+1}(1)(a_{11} b_1(O_{t+1}) + a_{12} b_2(O_{t+1}))$$

Making $\beta_{t+1}(1)$ the subject results in:

$$\beta_{t+1}(1) = \frac{\beta_t(1)}{a_{11} b_1(O_{t+1}) + a_{12} b_2(O_{t+1})} \tag{6}$$

Generalising for state i yields our forward-recurrence β approximation:

$$\beta_{t+1}(i) \approx \frac{\beta_t(i)}{\sum_{j=1}^{N} a_{ij} b_j(O_{t+1})} \tag{7}$$

With the β approximation defined in the new backward formula (7), we run the Forward-Backward algorithm and execute the α-pass, followed by the β-pass. Following the calculation of both α and β value sets on the new observations $\{O_{T+1}, O_{T+2}, \ldots, O_{T+M}\}$, the ξ and γ values are defined as follows:

For $T + 1 \leq t \leq T + M - 1$,

$$\xi_t(i, j) = \frac{\alpha_t(i) a_{ij} b_j(O_{t+1}) \beta_{t+1}(i)}{P(O; \lambda)}$$

and for $T + 1 \leq t \leq T + M$,

$$\gamma_t = \frac{\alpha_t(i)\beta_t(i)}{P(O;\lambda)}$$

We can now define the IncHMM (based on methods seen in [11]) using the updated Forward-Backward algorithm in the adaptive Baum-Welch algorithm. For each new observation, we define the modified re-estimation formulas for our IncHMM parameters $(\hat{\pi}, \hat{A}, \hat{B})$ as follows:

$$\hat{\pi}'_i = \gamma_1(i).$$

$$\hat{a}_{ij}^{T+1} = \frac{\sum_{t=1}^{T} \xi_t(i,j) + \xi_{T+1}(i,j)}{\sum_{j=1}^{N}\sum_{t=1}^{T} \xi_t(i,j) + \sum_{j=1}^{N} \xi_{T+1}(i,j)}$$

$$= \frac{\sum_{t=1}^{T} \gamma_t(i)}{\sum_{t=1}^{T+1} \gamma_t(i)} \frac{\sum_{t=1}^{T} \xi_t(i,j)}{\sum_{t=1}^{T} \gamma_t(i)} + \frac{\xi_{T+1}(i,j)}{\sum_{t=1}^{T+1} \gamma_t(i)}$$

$$= \frac{\sum_{t=1}^{T} \gamma_t(i)}{\sum_{t=1}^{T+1} \gamma_t(i)} \hat{a}_{ij}^{T} + \frac{\xi_{T+1}(i,j)}{\sum_{t=1}^{T+1} \gamma_t(i)}$$

where we only compute the new $\xi_{T+1}(i,j)$ and $\gamma_{T+1}(i)$ for each new observation (note the $\xi_t(i,j)$ values for $1 \leq t \leq T$ are already stored in the $\hat{a}_{ij}^{T}$ entry).

$$\hat{b}_j(k)^{T+1} = \frac{\sum_{t=1, O_t=k}^{T} \gamma_t(j) + \sum_{t=T+1, O_t=k}^{T+1} \gamma_t(j)}{\sum_{t=1}^{T} \gamma_t(j) + \gamma_{T+1}(j)}$$

$$= \frac{\sum_{t=1}^{T} \gamma_t(j)}{\sum_{t=1}^{T+1} \gamma_t(j)} \hat{b}_j(k)^{T} + \frac{\sum_{t=T+1, O_t=k}^{T+1} \gamma_t(j)}{\sum_{t=1}^{T+1} \gamma_t(j)}$$

where we only update $\gamma_{T+1}(j)$ (such that $O_{T+1} = k$) after storing all previous γ values in the $\hat{b}_j(k)^{T}$ entries.

This concludes the definition of the IncHMM. To achieve our iSWoM, we first transform the raw traces into observation traces and use these more concise forms in the IncHMM to form a discrete Markov Arrival Process.

3 Transformation of Traces

The transformation of the raw traces into binned traces and then into observation traces, acting as input for HMM training, is summarised briefly as follows:

In this section, we discuss, in the order presented in Fig. 2, the processes which evolve the discrete time series data. These include binning and clustering raw traces to transform data into a more manageable form for HMM processing.

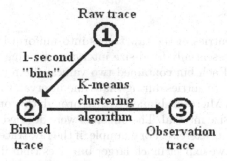

Fig. 2. The transformation of traces by various processes

3.1   Raw Traces

The first step of the process is analysing the raw NetApp trace, which contains hundreds of thousands of entries collected from NetApp storage servers. A CIFS (Common Internet File System) network trace (of about 750 GB) was gathered from file servers at the NetApp headquarters, where the servers were accessed mainly by Windows desktops and laptops using various applications. In this paper, the aforementioned trace will be known simply as the NetApp trace.

This NetApp trace is a part of a larger network trace (about 12 GB in size) consisting of I/O commands (single CIFS reads and writes). Each entry of the collected trace is of the form:

```
Cmd: Write
Timestamp: 3.4051253054
PID: 2520
IP: 10.58.48.58
Filename: 3G314EG822563KF312D4G4F
Size: 72
Offset: 0
```

For each of these entries, the command (Cmd), either "Read" or "Write," and its "Timestamp" (i.e. the time in seconds when the command was made) were collected and stored in separate read and write arrays using an *InputStreamReader*. Similarly, the Microsoft trace was collected over 15-minute periods and captures the complete build process. The trace events were primarily Disk IO (block level) and File IO, where we focused on reads and writes. Other trace specifications include a Windows Server 2008 OS and two socket, quad-core processors. Analysing the primary fields of interest for our storage workload research, we used the timestamp of the event (measured in microseconds from the start of the trace) and the type of trace event (i.e. FileIoRead and FileIoWrite). Other data, which were not considered for this paper, included size of the request in bytes and elapsed time from start of event to completion event for a specific IO (also in microseconds). The next stage of the transformation process is assigning "bins" to our traces, thus creating two distinct, binned traces.

3.2   Binned Traces

We partitioned the entries of the raw traces into uniform bins of a pre-defined size. These bins are essentially fixed-size intervals, dividing the raw data into a discrete time series. Each bin contained two values: the number of read entries and the number of write entries during that time interval. Consistency is kept in both the NetApp and Microsoft binned traces through the formation of reads and writes and the same size interval. The bin sizes were decided by the timescale required for the modelling exercise. For example, if the raw trace spans a time period of several days, then we expect much larger bin sizes than if we had a raw trace spanning several hours. Also, the level of detail at which the raw trace is operating (e.g. at the Application level) is also an important factor in determining the bin size. After experimenting with the NetApp raw trace, we found the best bin size was one second. Having tried 100 milliseconds resulted in too many empty time intervals, whilst with a larger time interval (i.e. five seconds), there were issues of missing out low-level, operation sequence characteristics such as mode transitions. Using one second bin sizes allowed us to represent each index in an array as a second. Counting the number of commands occuring each second (separately for reads and writes) resulted in filling our arrays easily. For example, reads[7] = 85 and writes[7] = 20 represents 85 reads and 20 writes in the seventh second. A vector list, holding a pair of read and write values as each data point, was formed from the arrays. Similarly to the NetApp trace, we binned the Microsoft trace in one-second intervals. After transforming the timestamped microseconds into seconds, we filtered through the raw trace and chose only the read and write events. The result was two arrays of length 9000 (i.e. 9000 seconds is 150 minutes), for reads and for writes. Both NetApp and Microsoft binned traces (acting as vectors with (read,write) tuples as data points) are inputted into a K-means clustering algorithm to obtain observation traces.

3.3   K-Means Clustering

The next step of the transformation was to apply a clustering algorithm to the binned traces and further reduce them to more manageable format (i.e. the observation trace). We implemented the K-means clustering algorithm, which grouped data into K clusters. Each cluster contains a pair of values representing the centroid (i.e. the mean number of reads and mean number of writes for that group of points) and all the data points belonging to that cluster (i.e. all the reads and writes in the group). A Euclidean-distance itervative algorithm calculates the cluster centroids over and over again until they became fixed. As we inputted K manually, we chose a value of seven clusters for the NetApp trace and four clusters for the Microsoft trace. These values were not too large (giving surplus or even empty clusters) nor too little (missing out significant differences among clusters) for our data traces. The NetApp trace lists seven clusters as vectors, with each centroid written as a pair of values, reads and writes, respectively.

$$\begin{pmatrix} 964.53 & 2.18 \\ 221.87 & 0.35 \\ 1.49 & 0.69 \\ 160.92 & 0.78 \\ 637.5 & 0.37 \\ 394.08 & 0.2 \\ 77.35 & 2.95 \end{pmatrix} \tag{8}$$

As shown in (8), observation values represent low writes with increasing reads. Some read values are low and progress to medium and high values. It is expected that there are not many varying writes (i.e. medium or high writes) in this read-dominated trace. The Microsoft observation trace was given four clusters to assign its data points to (i.e. $1, 4, 2, 3, 1, 3, \ldots$). Finalising with two discrete data traces, we proceed to simulating the iSWoM. Model validation includes comparing statistical averages of raw, HMM and iSWoM-generated traces, and also using Viterbi algorithm to produce hidden state sequences.

4   iSWoM Simulation and Results

4.1   Simulating the iSWoM Using NetApp and Microsoft Traces

To achieve the first simulation, the NetApp observation trace is inputted into the Baum-Welch algorithm as a training set of 8000 points (i.e. 8000 seconds). The IncHMM is trained on this set until parameter convergence (i.e. A, B, π converge). Afterwards, 2000 new observations are added and are evaluated using the new β approximation from the Forward-Backward algorithm. Thus, an incremental MAP (IncMAP) with fixed parameters is formed, which stores information on 10000 consecutive observation points. Our IncMAP then generates its own synthetic NetApp trace using its parameters (A, B, π). The IncMAP reproduces the observed values using random generation sampling, and simulated 1000 times, thus forming the iSWoM. In fact, the iSWoM possesses its own distribution of NetApp reads and writes defined by mean and standard deviation, where 95% intervals are performed on both statistics. We compare iSWoM-generated results with mean and standard deviation for raw and HMM-generated traces. Note that the HMM-generated trace is a result of a traditional HMM trained on an observation trace of length 10000, with no incremental learning. The second simulation involves a Microsoft data trace as input and follows the same process as that of the NetApp simulation, except that the original training set holds 7000 points. A new set of 2000 unseen Microsoft data points is added to this training set and the process follows identically. In the next section, we present the results of our simulation.

4.2   Results of iSWoM

Tables 1 and 3 present statistics on Reads/bin and Tables 2 and 4 represent Writes/bin, where the "bin" is a 1-second interval. For example, a "Raw Mean of 111.350 Reads/bin" implies that the raw NetApp trace produces, on average, 111.350 read commands per second. The "iSWoM Mean" and "iSWoM Std Dev" are the averages of the iSWoM-generated trace. The "HMM"-prefixed averages are calculated from a standard HMM-generated trace with no incremental activity. The results for the NetApp trace are summarised as follows:

Table 1. Reads/bin statistics on the raw, HMM and iSWoM NetApp traces

Trace	Mean	Std Dev
Raw	111.350	254.904
HMM	111.26 ± 0.66	254.38 ± 0.65
iSWoM	113.32 ± 0.60	253.18 ± 0.58

Table 1 shows very similar results between raw and HMM-generated means and standard deviations. The iSWoM produces a mean of 113.32 with a 95% confidence interval of 0.6, which is pleasing after 1000 simulations, but this mean is less accurate than the traditional HMM. The standard deviation of the iSWoM-generated trace (253.18) matches the raw trace well, but is outperformed slightly by the value of the HMM trace. Table 2 presents good results for the iSWoM-generated mean and standard deviation, which again slightly underperform compared to the values produced by the HMM trace.

Table 2. Writes/bin statistics on the raw, HMM and iSWoM NetApp traces

Trace	Mean	Std Dev
Raw	0.382	0.208
HMM	0.38 ± 0.0005	0.21 ± 0.001
iSWoM	0.41 ± 0.0005	0.24 ± 0.001

Table 3. Reads/bin statistics on the raw, HMM and iSWoM Microsoft traces

Trace	Mean	Std Dev
Raw	1.153	20.6
HMM	1.14 ± 0.01	20.33 ± 0.16
iSWoM	1.15 ± 0.01	20.42 ± 0.15

Table 3 summarises the statistics for the raw, HMM and iSWoM-generated Microsoft reads. The iSWoM mean and standard deviation have outperformed the traditional HMM-generated traces.

Table 4. Writes/bin statistics on the raw, HMM and iSWoM Microsoft traces

Trace	Mean	Std Dev
Raw	0.242	0.719
HMM	0.243 ± 0.0005	0.719 ± 0.001
iSWoM	0.242 ± 0.0005	0.718 ± 0.001

The statistics for the Microsoft Writes/bin reveal very similar results for raw, HMM and iSWoM traces. Both sets of confidence intervals at 95% level are identical for the HMM and iSWoM traces, as a result of the small mean and standard deviation values and also setting the population size equal to 1000 (i.e. 1000 simulations). The next section focuses on using the Viterbi algorithm as a means of validating the iSWoM.

4.3 Viterbi Hidden State Sequence Analysis

The Viterbi algorithm is used to return a sequence of hidden states, corresponding to a sequence of observations. Viterbi uses the parameters A, B and π to calculate its hidden state sequence. Therefore, Viterbi signifies, through its hidden state sequence, if the parameters for two different HMMs are similar. We compare the hidden state sequence based on the raw observation trace (i.e. the sequence of observations originally inputted into HMMs) with that of the iSWoM-generated synthetic observation trace. This process is applied to both NetApp and Microsoft traces, in turn. Unlike the simulation of 1000 runs which used the Baum-Welch algorithm, the Viterbi uses only one observation trace per hidden state sequence. The results of the Viterbi algorithm based on this raw NetApp trace and its iSWoM equivalent are as follows:

The sequences seen in Table 5 match very well, with an approximate 4:1 ratio supported by both raw and iSWoM traces. Thus, it proves, most importantly, that the HMM and iSWoM each produce a set of parameters which are in agreement.

Table 5. Viterbi state sequence ratio on the raw and iSWoM NetApp trace

Trace	State 1	State 2
Raw	7938	2062
iSWoM	8182	1818

Table 6. Viterbi state sequence ratio on the raw and iSWoM Microsoft trace

Trace	State 1	State 2
Raw	1	8999
iSWoM	1	8999

The following Viterbi results were obtained for the raw Microsoft trace and iSWoM-generated trace, and show an identical ratio of hidden states:

From Table 6, it is clear both raw and iSWoM traces have the majority of their observations from one state. This is due to the transition matrix heavily favouring one state, with a very small probability of exiting. The Viterbi algorithm has sustained the validity of the iSWoM given agreeable hidden state sequences and thus similar model parameters to a traditional HMM.

5   Conclusions and Further Work

HMMs, combined with the supporting clustering analysis and appropriate choice of bins, is able to provide a concise, parsimonious and portable synthetic workload. This has already been established, in [7], but the deficiency of such models is their heavy computing resource requirement, which essentially precludes them from any form of on-line analysis. The incremental model we have developed, iSWoM, has a vastly reduced computing requirement making it ideal for modelling workload data in real-time. In fact, with the availability of decoding new data, the iSWoM avoids re-training on "old data" like the traditional HMM. Additionally, compared with both the resource-costly HMM and raw traces, the iSWoM provides excellent accuracy of training data. Such mathematical descriptions of workload should be measured quantitatively against independent data (i.e. traces not used in model construction) that they represent, and more extensive tests are planned for our incremental model. Nonetheless, the iSWoM β approximation has been successful after statistical comparisons between raw and iSWoM-generated traces. Other related work, for example, the incremental model from [8], used a backward formula in its learning that was not recursive in terms of previous β values. The iSWoM backward formula, however, stores all information on the complete β set, unlike the formula used in [18] where all β variables were equal to 1. In fact, our forward-recurrence β formula used in the iSWoM is validated by NetApp and Microsoft data traces and produces satisfying simulated results. Extensions to this paper arise from other possible validation techniques of the iSWoM. The sensitive autocorrelation function would indeed be a useful validation method for our model, mainly because it allows time series comparison through lagged versions of the original data trace. Hidden trends can be exposed in both raw and iSWoM-generated autocorrelated data. Another extension might be a cumulative distribution function (CDF) for the iSWoM workload distribution. Given any processed trace, a CDF can be used to obtain specific time series probabilities, such as observing k reads or writes within a time t, etc. This will highlight the iSWoM as a more transparent probabilistic model.

References

1. Baum, L.E., Petrie, T.: Stastical Inference for Probabilistic Functions of Finite Markov Chains. The Annals of Mathematical Statistics 37, 1554–1563 (1966)
2. Baum, L.E., Petrie, T., Soules, G., Weiss, N.: A maximization technique occurring in the statistical analysis of probabilistic functions of Markov chains. The Annals of Mathematical Statistics 41, 164–171 (1970)
3. Viterbi, A.J.: Error bounds for convolutional codes and an asymptotically optimum decoding algorithm. IEEE Transactions on Information Theory 13, 260–269 (1967)
4. Krough, A., Brown, M., Mian, S., Sjolander, K., Haussler, D.: Hidden Markov Models in Computational Biology. Journal of Molecular Biology, 1501–1531 (1994)
5. Burge, C., Karlin, S.: Prediction of complete gene structures in human genomic DNA. Journal of Molecular Biology, 78–94 (1997)
6. Ashraf, J., Iqbal, N., Khattak, N.S., Zaidi, A.M.: Speaker Independent Urdu Speech Recognition Using HMM (2010)
7. Harrison, P.G., Harrison, S.K., Patel, N.M., Zertal, S.: Storage Workload Modelling by Hidden Markov Models: Application to Flash Memory. Performance Evaluation 69, 17–40 (2012)
8. Florez-Larrahondo, G., Bridges, S., Hansen, E.A.: Incremental Estimation of Discrete Hidden Markov Models on a New Backward Procedure (2005)
9. Rabiner, L.R., Juang, B.H.: An Introduction to Hidden Markov Models. IEEE ASSP Magazine 3, 4–16 (1986)
10. Rabiner, L.R.: A Tutorial on Hidden Markov Models and Selected Applications in Speech Recognition. IEEE 77, 257–286 (1989)
11. Chis, T.: Hidden Markov Models: Applications to Flash Memory Data and Hospital Arrival Times (2011)
12. Zhang, X., Riska, A., Riedel, E.: Characterization of the E-commerce Storage Subsystem Workload. In: Proceedings of Quantitative Evaluation of SysTems (2008)
13. RUBiS Implementation: http://rubis.objectweb.org/
14. On-Line Transaction Processing (OLTP) Benchmark: http://www.tpc.org/tpcc
15. Kumas, Z., Keeton, K., Becker-Szendy, R.: I/O Workload Characterization. In: CAECW 2001, before HPCA-7 (2001)
16. Transactional Web e-Commerce Benchmark: http://www.tpc.org/tpcw
17. Chis, T., Harrison, P.G.: Incremental HMM with an improved Baum-Welch Algorithm. In: Proceedings of Imperial College Computing Student Workshop (2012)
18. Stenger, B., Ramesh, V., Paragois, N., Coetzee, F., Buhmann, J.M.: Topology free Hidden Markov Models: Application to background modeling. In: Proceedings of the International Conference on Computer Vision, pp. 297–301 (2001)
19. Keaton, K., Veitch, A., Obal, D., Wilkes, J.: I/O characterization of commercial workloads. In: CAECW 2000 (2000)

Maximizing the Probability of Arriving on Time

Ananya Christman[1] and Joao Cassamano[2]

[1] Middlebury College, Middlebury VT 05753
achristman@middlebury.edu
[2] Lake Forest College, Lake Forest IL 60045
cassamanojc@lakeforest.edu

Abstract. We study the problem of maximizing the probability of arriving on time in a stochastic network. Nodes and links in the network may be congested or uncongested, and their states change over time and are based on states of adjacent nodes. Given a source, destination, and time limit, the goal is to adaptively choose the next node to visit to maximize the probability of arriving to the destination on time. We present a dynamic programming solution to solve this problem. We also consider a variation of this problem where the traveler is allowed the option to wait at a node rather than visit the next node. For this setting, we identify properties of networks for which the optimal solution does not require revisiting nodes.

Keywords: stochastic networks, routing, dynamic programming.

1 Introduction

Transportation networks are inherently uncertain with random events such as accidents, vehicle failure, inclement weather, and road closures often causing congestion and delays. These events often affect a group of locations and roads in a region of the network, rather than just one location. Congestion at one location is likely to cause congestion at nearby locations and roads. Fortunately, new technologies make it increasingly easy for travelers to obtain real-time information about traffic conditions, allowing them to make potentially better route decisions. This setting can be modeled with a stochastic network where a node can be in one of two states: congested or uncongested. Congestion at one node is likely to cause congestion at nearby nodes and congestion at a node causes congestion at its incident edges. Traversing an edge in the congested state requires more time than traversing the edge when it is uncongested. Since the state of a node may change over time and cannot be determined until the node is reached, there is uncertainty with both node and link states.

For this setting, we consider the problem of maximizing the probability of arriving at a specified destination node within a given amount of time. Since the network is stochastic, we cannot solve this problem *a-priori*. Instead, we solve the problem using a step-by-step approach where for each step, we make a decision based on the state of the current node, the remaining time, and the conditional congestion probabilities of adjacent nodes.

A. Dudin and K. De Turck (Eds.): ASMTA 2013, LNCS 7984, pp. 142–157, 2013.

Specifically, given source and destination locations, and a maximum time limit, we propose an efficient dynamic programming solution to find a path from the source to the destination that maximizes the probability of arriving by the time limit. We find that some optimal routes require revisiting nodes, which may not be reflective of realistic networks. Therefore, we also consider a variation of this problem where the traveler has the option to wait at a node rather than visit another node. We identify properties of networks for which the optimal solution requires waiting at a node rather than revisiting nodes.

2   Related Work

The problem of finding an optimal path in a stochastic network has been studied extensively and numerous variations of the problem have been considered. Some early examples include [5] and [12] which consider networks where link costs are random variables following known probability distributions. The work in [5] studies the problem of finding the path that maximizes the probability of arriving by a predefined time whereas [12] finds the path that has the highest probability of being the shortest path. The authors of [11] assume a network where link travel times evolve based on an independent Markov process. They give solutions for finding a path with minimal expected travel time. The authors of [6] consider a setting where link travel times depend on the time of day and propose heuristics to estimate the mean and variance of arrival times for a source-destination node pair. In [13], the authors study a network setting in which the state of each link is dependent on the predecessor link, and is independent of the states of nodes. They develop heuristics to determine the sequence of links to traverse such that the expected travel time is minimized. For different settings, several works consider the problem of maximizing the expected value of various utility functions ([2], [8], [9]). However, the authors of [8] show that such utility-based models can be solved efficiently only if an affine or exponential utility function is employed.

More recently, the authors of [3] considered a network where link costs are based on conditional probabilities of adjacent nodes. They consider the problem of finding the path with the minimum expected travel time. The authors of [4] consider a network where link costs follow independent probability distributions. They give an approximate solution to the problem of finding the path that maximizes the probability of arriving to a destination by a predetermined time. The authors of [10] also assume link travel times are defined by a probability distribution and propose an algorithm to address a similar problem: finding the shortest paths to guarantee a given probability of arriving on-time. Recently, the authors of [7] consider a network setting where the traveler has information about several link travel times (not just adjacent links) and give heuristics for maximizing a general utility function.

In this work we consider the stochastic setting studied in [3] where the congestion probability for nodes and links depend on nearby nodes. Whereas they focus on minimizing expected travel time, we focus on a different goal: given a source, destination, and desired arrival time, find the sequence of nodes to visit that maximizes the probability of arriving on time. Whereas previous studies proposed approximate solutions to routing problems in stochastic networks ([4]), we provide a dynamic programming solution that, given a maximum time limit, yields an exact solution to the problem. We also consider the cycling policy, described in [13], where the traveler has the option to revisit locations; and the waiting policy, described in [1], where the traveler has the option to wait at a location. Both policies may help to improve the objective function. We compare these policies by identifying networks for which the waiting policy will always be more beneficial than the cycling policy.

3   Problem Statement

We consider the Arriving on Time Problem. The input is a network with n nodes, a source node s, a destination node d, and a time limit t. The goal is to find the path that maximizes the probability of arriving to d from s within time t. Since the states of nodes can change, we cannot determine the optimal path *a-priori*. Instead, once we arrive at a node, we must adaptively determine the next node to visit that maximizes the probability of reaching d given the remaining time. We assume that there is a time limit T_{max} such that for every pair of vertices i and j, t cannot exceed T_{max}.

For adjacent nodes i and j, let $P_c(j|i)$ denote the conditional probability that j is congested given that i is congested and let $P_u(j|i)$ denote the conditional probability that j is uncongested given that i is uncongested. Then if i is congested, j is uncongested with probability $1 - P_c(j|i)$ and if i is uncongested then j is congested with probability $1 - P_u(j|i)$. To reach node j directly from i, we must traverse the edge (i,j). It takes $t_c(i,j) > 0$ time units to traverse (i,j) if i is congested and $t_u(i,j) > 0$ time units if i is uncongested. We assume the uncongested travel time is no more than the congested travel time so $t_u(i,j) \leq t_c(i,j)$. For simplicity, we assume that if i is congested, we can reach an adjacent node j within t time units with probability 0 if $t < t_c(i,j)$ and with probability 1 if $t \geq t_c(i,j)$. Similarly if i is uncongested, we can reach an adjacent node j within t time units with probability 0 if $t < t_u(i,j)$ and with probability 1 if $t \geq t_u(i,j)$. Note that extending this model so that these probabilities are instead determined by a distribution is straightforward. Let $N_c^t(i)$ denote the set of nodes, j, such that j is adjacent to i and $t \geq t_c(i,j)$ and let $N_u^t(i)$ denote the set of nodes, j, such that j is adjacent to i and $t \geq t_u(i,j)$. Then if node i is congested or uncongested, we consider all nodes in $N_c^t(i)$ or $N_u^t(i)$, respectively, for the next possible node to visit.

In Fig. 1(a), suppose we determine that j is the optimal node to visit from i. Since i is congested, traversing link (i,j) will take time $t_c(i,j)$ and in Fig. 1(b), since i is uncongested, traversing link (i,j) will take time $t_u(i,j)$. In either case, once we arrive at node j, if j is congested then traversing any link from j will take time equal to the congested time of the link. If j is uncongested then traversing any link from j will take time equal to the uncongested time of the link. Once we

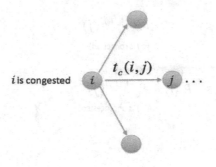

(a)Node i is congested.

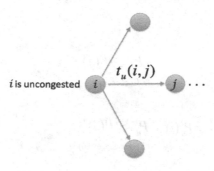

(b)Node i is uncongested.

Fig. 1. Traversing link (i, j). (a) Since i is congested, traversing link (i, j) takes time $t_c(i, j)$. (b) Since i is uncongested, traversing link (i, j) takes time $t_u(i, j)$. When we arrive at j, it will be either congested or uncongested. In either case we must find the next node to visit that maximizes the on-time arrival probability from j.

arrive at j we find the optimal next node to visit as we did for node i. Specifically, we find the node that yields the maximum on-time arrival probability from j given the amount of time remaining.

For all nodes i and some destination, let $P_c^t(i)$ denote the maximum probability of arriving to the destination on time given t time units when i is congested; and let $P_u^t(i)$ denote the maximum on-time arrival probability given t time units when i is uncongested. Suppose node i is currently congested. If j is the optimal next node to visit, then when we move from i to j, j will either be congested (with probability $P_c(j|i)$) or uncongested (with probability $1 - P_c(j|i)$). In either case, we will have $t - t_c(i, j)$ time units remaining and we would like to maximize the probability of arriving on time from j to the destination with the remaining time. In Fig. 2(a), if j is congested, the probability of arriving to the destination on time from j is $P_c^{t-t_c(i,j)}(j)$. If j is uncongested, the probability is $P_u^{t-t_c(i,j)}(j)$. If instead, i was uncongested (see Fig. 2(b)), then the maximum on-time arrival probability from j is $P_c^{t-t_u(i,j)}(j)$ (if j is congested) or $P_u^{t-t_u(i,j)}(j)$ (if j is uncongested).

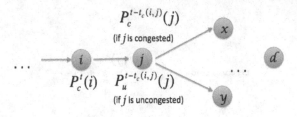

(a)Node i is congested.

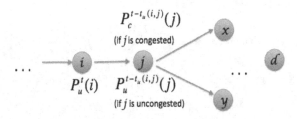

(b)Node i is uncongested.

Fig. 2. Traversing a graph for the Arriving on Time Problem. (a) Since i is congested, link (i, j) costs $t_c(i, j)$ so we have $t - t_c(i, j)$ time units remaining when we arrive at j. (b) Since i is uncongested, link (i, j) costs $t_u(i, j)$ so we have $t - t_u(i, j)$ time units remaining. When we arrive at j, it will be either congested or uncongested. In either case we must find the next node to visit that maximizes the on-time arrival probability from j.

In this manner, we can determine the optimal sequence of nodes to visit from the source to the destination and the maximum probability of arriving on time. Notice that we cannot determine the optimal node to visit from j until we arrive at j and observe its state. For example, if j is congested, the optimal next node may be x and if j is uncongested the optimal next node may be y.

Figure 3 shows a simple example of the problem. Assume the source is s and the destination is d. For the conditional probabilities, assume $P_c(a|s) = .9$, $P_c(b|s) = .6$, $P_u(a|s) = .7$, and $P_u(b|s) = .6$. For the travel times, assume $t_c(s, a) = 6$, $t_c(a, d) = 7$, $t_c(s, b) = 5$, $t_c(b, d) = 9$, $t_u(s, a) = 5$, $t_u(a, d) = 4$, $t_u(s, b) = 4$, and $t_u(b, d) = 3$. The optimal path depends on both the state of s and the time limit. If s is congested, then with a time limit of $t = 10$ units, the optimal path is s-b-d with probability $P_c^{10}(s) = .4$, whereas path s-a-d yields probability .1. However, if s is uncongested, then the optimal path is s-a-d with

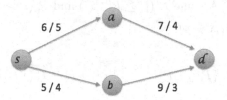

Fig. 3. Example of the Arriving On-Time Problem. Congested/uncongested travel times are shown beside each edge. In this example, the optimal path depends on the state of the source node and the time limit.

probability $P_u^{10}(s) = .7$, whereas path s-b-d yields probability .6. If we have a time limit of $t = 7$ and s is uncongested, then the optimal path is s-b-d with probability $P_c^7(s) = .6$, whereas s-a-d yields probability 0. If s is congested, then both paths yield probability 0.

The general problem can be solved with the following dynamic programming formulation:

$$P_c^0(i) = 0 \text{ if } i \text{ is not the destination} \tag{1}$$
$$P_u^0(i) = 0 \text{ if } i \text{ is not the destination} \tag{2}$$
$$P_c^t(i) = 1 \text{ for all } t \geq 0 \text{ if } i \text{ is the destination} \tag{3}$$
$$P_u^t(i) = 1 \text{ for all } t \geq 0 \text{ if } i \text{ is the destination} \tag{4}$$

$P_c^t(i) = 0$ if i is not the destination and $t < t_c(i,j)$ for all j adjacent to i
$$\max_{\forall j \in N_c^t(i)} \{P_c(j|i)P_c^{t-t_c(i,j)}(j) + (1 - P_c(j|i))P_u^{t-t_c(i,j)}(j)\} \text{ otherwise}$$

$P_u^t(i) = 0$ if i is not the destination and $t < t_u(i,j)$ for all j adjacent to i
$$\max_{\forall j \in N_u^t(i)} \{P_u(j|i)P_u^{t-t_u(i,j)}(j) + (1 - P_u(j|i))P_c^{t-t_u(i,j)}(j)\} \text{ otherwise}$$

The algorithm can be implemented in a manner similar to Dijkstra's shortest path algorithm. For a graph with n nodes and maximum time limit T_{max}, it requires $O(n^2 T_{max})$ space and time.

Proposition 1. *The quantities $P_c^t(i)$ and $P_u^t(i)$ are non-decreasing in t.*

Proof. We will prove that $P_c^t(i) \geq P_c^{t-1}(i)$ for all $t \geq 1$. The proof also holds for $P_u^t(i)$.

<u>Base Case:</u> If i is the destination then for all $t \geq 1$, $P_c^t(i) = 1$ so $P_c^t(i) = P_c^{t-1}(i)$. If i is not the destination, then for $t = 1$, $P_c^{t-1}(i) = P_c^0(i) = 0$, so $P_c^t(i) = P_c^1 \geq P_c^0(i)$.

Inductive Hypothesis. Assume $P_c^t(i) \geq P_c^{t-1}(i)$ and $P_u^t(i) \geq P_u^{t-1}(i)$.

We will show $P_c^{t+1}(i) \geq P_c^t(i)$.

$$P_c^{t+1}(i) = \max_{\forall j \in N_c^t(i)} \{P_c(j|i)P_c^{t+1-t_c(i,j)}(j) + (1 - P_c(j|i))P_u^{t+1-t_c(i,j)}(j)\}$$

By the inductive hypothesis, we know $P_c^{t+1-t_c(i,j)}(j) \geq P_c^{t-t_c(i,j)}(j)$ and $P_u^{t+1-t_c(i,j)}(j) \geq P_u^{t-t_c(i,j)}(j)$.

$$P_c^{t+1}(i) \geq \max_{\forall j \in N_c^t(i)} \{P_c(j|i)P_c^{t-t_c(i,j)}(j) + (1 - P_c(j|i))P_u^{t-t_c(i,j)}(j)\}$$
$$= P_c^t(i)$$

Similarly, $P_u^{t+1}(i) \geq P_u^t(i)$.

3.1 Node Revisiting

In the current problem formulation, the optimal route from the source to the destination may include revisiting one or more nodes. For example, suppose we arrive at a congested node i and there is a short path from i to the destination that yields a low on-time arrival probability if taken when i is congested (see Fig. 4). Suppose from i, we can also traverse a cycle such that the probability of returning to i in the congested state is small. Then, the optimal solution may be to traverse this cycle if doing so yields a higher on-time arrival probability than directly heading towards the destination from i in the congested state. The subgraph in Fig. 4 shows an example where continuously revisiting a node improves the probability of arriving on time. Suppose we would like to arrive at d from s in the congested state. Assume $P_c(i|j) = P_c(j|i) = .6$, and $P_c(i|s) = P_u(i|j) = P_u(j|i) = .9$. Assume $t_c(i,d) = 10$ and $t_u(i,d) = 1$. Assume that both the congested and uncongested travel times of all other edges are one time unit, so $t_c(x,y) = t_u(x,y) = 1$ for all edges (x,y) except (i,d). In this example, traveling directly from s to d without traversing the cycle containing nodes i and j yields probability $P_c^2(s) = 0.1$. However, with two additional time units, traversing the cycle containing i and j, specifically with the path $s - i - j - i - d$, yields probability $P_c^4(s) = .625$. Similarly, traversing the cycle twice and thrice yields probabilities $P_c^6(s) = .756$ and $P_c^8(s) = .789$, respectively. The example shows that revisiting a node may help to improve the probability of arriving to the destination within the time limit.

However, this scenario does not reflect realistic situations. Specifically, if we are at a congested location, traversing a cycle to revisit the location would never yield the highest probability of arriving on time. If heading directly towards the destination from a congested location is unlikely to get us to the destination on time, then a better option would be to simply *wait* at the current location (for a short time) for the congestion to clear up.

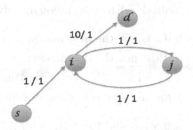

Fig. 4. Example where revisiting node i improves the on-time arrival probability. Congested/uncongested travel times are shown beside each edge.

It is possible that after waiting for some time, the location will become uncongested and therefore yield a higher probability of arriving on time. When the location becomes uncongested, we can then decide which route will yield the highest on-time arrival probability. In the following section we consider a variation of the problem that models this more realistic setting.

4   Waiting Option

We now present a version of the Arriving on Time Problem where the traveler has the option to wait at a congested node (in the hopes that the node will become uncongested). As described above, in real-world situations, we may decide to wait at a congested location for traffic to clear up and the location to become uncongested. However, we would find no reason to wait at an uncongested location for it to become congested. Therefore, we assume that the option to wait only applies to a node in the congested state.

As before, $P_c(j|i)$ denotes the conditional probability that j is congested given that i is congested, $P_u(j|i)$ denotes the conditional probability that j is uncongested given that i is uncongested, $t_c(i, j)$ and $t_u(i, j)$ denote the times to traverse edge (i, j) when i is congested and uncongested, respectively. We also define a new term $P_c^*(i, w)$, which is the conditional probability that i is congested after waiting at congested i for w time units. As in the original problem, we limit t and w to T_{max}.

As before, $P_c^t(i)$ denotes the maximum probability of arriving to the destination from i in the congested state, within at most t time units and $P_u^t(i)$ denotes the probability of arriving to the destination from i in the uncongested state within at most t time units. Given a source s and a destination, our goal is find the find the sequence of nodes to visit that maximizes $P_c^t(s)$ and $P_u^t(s)$.

Let us consider the previous example in Fig. 3. If s was congested and the time limit was $t = 10$, then paths s-b-d and s-a-d yielded on-time arrival probabilities of .4 and .1, respectively. Now, suppose we have the option to wait at a node. Assume $P_c^*(a, 1) = .1$, so with probability .1, node a stays congested if we wait at a for one time unit. For simplicity, assume this probability is zero for all other nodes. Then if s is congested, the optimal solution is to wait at a for one time unit, i.e. by using path s-a-a-d, with probability $P_c^{10}(s) = .91$.

This problem can be solved with the following dynamic programming formulation:

$$P_c^0(i) = 0 \text{ if } i \text{ is not the destination} \tag{5}$$

$$P_u^0(i) = 0 \text{ if } i \text{ is not the destination} \tag{6}$$

$$P_c^t(i) = 1 \text{ for all } t \geq 0 \text{ if } i \text{ is the destination} \tag{7}$$

$$P_u^t(i) = 1 \text{ for all } t \geq 0 \text{ if } i \text{ is the destination} \tag{8}$$

$$P_c^t(i) = 0 \text{ if } i \text{ is not the destination and } t < t_c(i,j) \text{ for all } j \text{ adjacent to } i \tag{9}$$

$$\max \begin{cases} \max_{\forall j \in N_c^t(i)} \{ P_c(j|i) P_c^{t-t_c(i,j)}(j) & \text{(Not waiting at } i\text{)} \\ \quad + (1 - P_c(j|i)) P_u^{t-t_c(i,j)}(j) \} & \\ & \\ \max_{\forall w \leq T_{max}} \{ P_c^*(i,w) P_c^{t-w}(i) & \\ \quad + (1 - P_c^*(i,w)) P_u^{t-w}(i) \} & \text{(Waiting at } i\text{)} \end{cases} \tag{10}$$

$$P_u^t(i) = 0 \text{ if } i \text{ is not the destination and } t < t_c(i,j) \text{ for all } j \text{ adjacent to } i \tag{11}$$

$$\max_{\forall j \in N_u^t(i)} \{ P_u(j|i) P_u^{t-t_u(i,j)}(j) + (1 - P_u(j|i)) P_c^{t-t_u(i,j)}(j) \} \tag{12}$$

Proposition 2. *The quantities $P_c^t(i)$ and $P_u^t(i)$ are non-decreasing in t.*

Proof. We first prove the proposition for $P_c^t(i)$. We will prove that $P_c^t(i) \geq P_c^{t-1}(i)$ for all $t \geq 1$.

Base Case: If i is the destination then for all $t \geq 1$, $P_c^t(i) = 1$ so $P_c^t(i) = P_c^{t-1}(i)$. If i is not the destination, then for $t = 1$, $P_c^{t-1}(i) = P_c^0(i) = 0$, so $P_c^t(i) = P_c^1(i) \geq P_c^0(i)$.

Inductive Hypothesis. Assume $P_c^t(i) \geq P_c^{t-1}(i)$ and $P_u^t(i) \geq P_u^{t-1}(i)$.

We will show $P_c^{t+1}(i) \geq P_c^t(i)$. From 10, we have:

$$P_c^{t+1}(i) = \max \begin{cases} \max_{\forall j \in N_c^t(i)} \{ P_c(j|i) P_c^{t+1-t_c(i,j)}(j) & \\ \quad + (1 - P_c(j|i)) P_u^{t+1-t_c(i,j)}(j) \} & \text{(Not waiting at } i\text{)} \\ & \\ \max_{\forall w \leq T_{max}} \{ P_c^*(i,w) P_c^{t+1-w}(i) & \\ \quad + (1 - P_c^*(i,w)) P_u^{t+1-w}(i) \} & \text{(Waiting at } i\text{)} \end{cases}$$

$$\geq \max \begin{cases} \max_{\forall j \in N_c^t(i)} \{ P_c(j|i) P_c^{t-t_c(i,j)}(j) & \\ \quad + (1 - P_c(j|i)) P_u^{t-t_c(i,j)}(j) \} & \\ & \text{(By Ind. Hyp.)} \\ \max_{\forall w \leq T_{max}} \{ P_c^*(i,w) P_c^{t-w}(i) & \\ \quad + (1 - P_c^*(i,w)) P_u^{t-w}(i) \} & \end{cases}$$

$$= P_c^t(i)$$

Proof. We now prove that $P_u^t(i)$ is non-decreasing in t. We show that $P_u^t(i) \geq P_u^{t-1}(i)$ for all $t \geq 1$.

Base Case: If i is the destination then for all $t \geq 1$, $P_u^t(i) = 1$ so $P_u^t(i) = P_u^{t-1}(i)$. If i is not the destination, then for $t = 1$, $P_u^{t-1}(i) = P_u^0(i) = 0$, so $P_u^t(i) = P_u^1 \geq P_u^0(i)$.

Inductive Hypothesis. Assume $P_c^t(i) \geq P_c^{t-1}(i)$ and $P_u^t(i) \geq P_u^{t-1}(i)$

We will show $P_u^{t+1}(i) \geq P_u^t(i)$. From 12, we have:

$$P_u^{t+1}(i) = \max_{\forall j \in N_u^t(i)} \{P_u(j|i)P_u^{t+1-t_u(i,j)}(j) + (1 - P_u(j|i))P_c^{t+1-t_u(i,j)}(j)\}$$

$$\geq \max_{\forall j \in N_u^t(i)} \{P_u(j|i)P_u^{t-t_u(i,j)}(j) + (1 - P_u(j|i))P_c^{t-t_u(i,j)}(j)\} \quad \text{(By Ind. Hyp.)}$$

$$= P_u^t(i)$$

4.1 Revisiting vs. Waiting

Recall that in the first version of the problem, where there is no option to wait at a node, revisiting a node may improve the probability of arriving on time. In this section we focus on the new version of the problem where the traveler has the option to wait at a node. We identify a class of networks for which revisiting nodes cannot improve the on-time arrival probability.

We will show that in this new model, revisiting a node cannot improve the on-time arrival probability for networks that exhibit both of the following two properties:

Property 1. Given some destination, for any node i and some fixed t, $P_u^t(i) > P_c^t(i)$. In other words, there is a higher chance of reaching the destination on time from location i if i is uncongested rather than congested. This property reflects realistic settings since the presence of traffic at the current location does not improve the probability of arriving to the destination on time.

Property 2. For any two neighbors i and j, and any w, $1 - P_c^*(i, w) > P_u(i|j)$ and $1 - P_c^*(i, w) > 1 - P_c(i|j)$. In other words, it is more probable for location i to become uncongested after waiting at i for w time units than by visiting i from a neighboring node j.

We now prove that for networks that exhibit Properties 1 and 2, revisiting a node cannot improve the on-time arrival probability. Specifically, we prove that for all nodes i, the arrival probability achieved by *revisiting* i after k time units will be no more than the on-time arrival probability achieved by *waiting* at a node i for k time units.

For this proof, we use the following notion of a *Tree of Paths*.

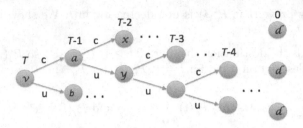

Fig. 5. $Tree_{v,T}$

Definition 1. *Tree of Paths: Given a destination d, for every node v and all time limits T > 0, we can construct a binary tree of paths, $Tree_{v,T}$ that has its root labeled as v and all leaves labeled as d (see Fig. 5). $Tree_{v,T}$ is built as follows. Suppose node a is the optimal node to visit from the root v when v is congested and b is the optimal node to visit from v when v is uncongested. Then in $Tree_{v,T}$, the children of v will be a and b. Similarly, if x and y are the optimal nodes to visit from a when a is congested and uncongested, respectively, then the children of a will be x and y. If some node i has a child who is also i, this means the optimal decision at the parent i is to wait at i for one time unit. Every path in $Tree_{v,T}$ is the optimal path to visit from v, given a sequence of node states and using at most T time units.*

Consider $Tree_{s,T}$ for some source s, destination d, and time limit T. There are two ways that a path from s to d on $Tree_{s,T}$ may contain a repeating node i (see Fig. 6). We now describe these two types of paths. Type 1: For some $k > 0$, i first appears when there are $t+k$ time units remaining, reappears when there are t time units remaining, and each node between the first and second appearance is not i. Paths of Type 1 indicate that the optimal solution requires visiting i when there will be $t + k$ time units remaining and then *revisiting* i when there will be t time units remaining (see Fig. 6(a)). Type 2: For some $k > 0$, i appears in a consecutive sequence for k time units. Paths of Type 2 indicate that the optimal solution requires *waiting* at i for k time units (see Fig. 6(b)). We will show that for networks that exhibit Properties 1 and 2, only paths of Type 2 exist in $Tree_{s,T}$. In other words, a repeating node, i, in an optimal path indicates the decision to wait at i, rather than revisit i.

Theorem 1. *For any network that exhibits Properties 1 and 2, the on-time arrival probability achieved from a path that requires waiting at a node i will be greater than the on-time arrival probability achieved from a path that requires revisiting i.*

Proof. Consider a path P from $Tree_{s,T}$. P represents the optimal path from s to the destination given T time units and a sequence of node states. There are four possible cases where a path P may contain a repeating node i: Case 1) i is uncongested in the first appearance and uncongested in the last appearance; Case 2) i is uncongested in the first appearance and congested in the last appearance; Case 3) i is congested in the first appearance and uncongested in the last

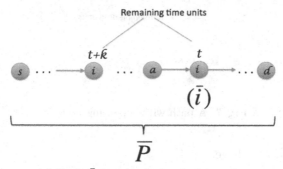

(a) Path $\bar{P}$ revisits i after k time units.

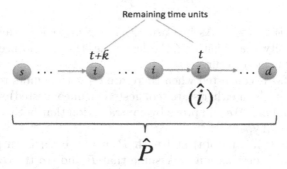

(b) Path $\hat{P}$ waits at i for k time units.

Fig. 6. Two paths with repeating node i

appearance; and Case 4) i is congested in the first appearance and congested in the last appearance. We will first show that if P is an optimal path then Cases 1) and 2) cannot occur. We will then show that Cases 3) and 4) can occur only if P is a path of Type 2. In other words, the repeating i must occur from the decision to wait at i rather than revisit i.

Let $t + k$ denote the number of time units remaining after the first occurrence of i and let t denote the number of time units remaining after the last occurrence of i (see Fig. 7).

Case 1: i is uncongested when there are $t + k$ time units remaining and uncongested when there are t time units remaining. From Proposition 2, we know $P_u^{t+k}(i) \geq P_u^t(i)$. Since visiting i the first time (when there will be $t + k$ time units remaining) yields a higher probability than visiting i the last time (when there will be t time units remaining), there is no need to revisit i, therefore Case 1 cannot occur.

Case 2: i is uncongested when there are $t + k$ time units remaining and congested when there are t time units remaining. By Property 1 we have:

$$P_u^{t+k}(i) > P_c^{t+k}(i) \tag{13}$$

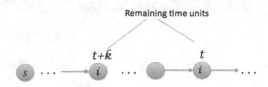

Fig. 7. A path with repeating node i

From Proposition 2 we have:

$$P_c^{t+k}(i) > P_c^t(i) \tag{14}$$

Therefore $P_u^{t+k}(i) > P_c^t(i)$. As in Case 1, since visiting i the first time yields a higher probability than visiting i the last time, there is no need to revisit i, therefore Case 2 cannot occur.

<u>Cases 3 and 4</u>: i is congested when there are $t + k$ time units remaining. The following is true if i is in either state (congested or uncongested) when there are t time units remaining. We will prove by contradiction that both cases can occur only with paths of Type 2.

Consider path $\bar{P}$ in Fig. 6(a) and path $\hat{P}$ in Fig. 6(b). Both paths start at source s and end at destination d. Assume that $\bar{P}$ and (to the contrary) $\hat{P}$ are the optimal paths given a time limit and sequence of node states. Suppose that $\bar{P}$ and $\hat{P}$ are identical up to the first appearance of node i. In path $\bar{P}$, we first visit i when there will be $t + k$ time units remaining, we then visit a sequence of nodes (none of which are i), and finally, revisit i when there will be t time units remaining. In other words, path $\bar{P}$ contains a cycle starting and ending at i (see Fig. 6(a)). In path $\hat{P}$, we wait at i for k time units, starting when there will be $t + k$ time units remaining and ending when there will be t time units remaining (see Fig. 6(b)). Let $\bar{i}_t$ denote the last occurrence of i on path $\bar{P}$ and let $\hat{i}_t$ denote the last occurrence of i on path $\hat{P}$. We will show that the on-time arrival probability from $\hat{i}_t$ will always be more than that of $\bar{i}_t$. Since replacing the subpath from i to $\bar{i}_t$ in $\bar{P}$ with the subpath from i to $\hat{i}_t$ in $\hat{P}$ will always yield a higher on-time arrival probability, $\bar{P}$ cannot be optimal. In other words, waiting at a node will always yield a higher on-time arrival probability than revisiting a node.

We first consider $\bar{P}$ (i.e. paths of Type 1) and the on-time probability achieved from $\bar{i}$. Recall that we reach $\bar{i}$ after revisiting i after k time units. Let a denote the node immediately before $\bar{i}_t$ in $\bar{P}$ (see Fig. 6(a)).

Sub-Case 1: a is congested. We first consider the case where a is congested. Node i will be congested with probability $P_c(i|a)$ and uncongested with probability $1 - P_c(i|a)$. If i is congested, then with probability $P_c^t(i)$, we will reach d on time; if node i is uncongested, then with probability $P_u^t(i)$ we will reach d on time. Therefore, we can express the on-time arrival probability from $\bar{i}_t$ given that the previously visited node was a in the congested state as follows:

$$P_c(i|a)P_c^t(i) + (1 - P_c(i|a))P_u^t(i) \tag{15}$$

We now consider $\hat{P}$ (i.e. paths of Type 2) and the on-time probability achieved from $\hat{i}$. Recall that we reach $\hat{i}$ after waiting at i for k time units. Node i will be congested with probability $P_c^*(i,k)$ and uncongested with probability $1 - P_c^*(i,k)$. If there are t time units remaining and i is congested, then with probability $P_c^t(i)$, we will reach d on time; if node i is uncongested, then with probability $P_u^t(i)$ we will reach d on time. Note that since $\hat{i}$ represents node i during the *last* time unit of waiting at i, the node following $\hat{i}$ cannot be i. Therefore the arrival probability from $\hat{i}$ is:

$$P_c^*(i,k)P_c^t(i) + (1 - P_c^*(i,k))P_u^t(i) \tag{16}$$

We will prove by contradiction that for networks exhibiting Properties 1 and 2, (15) always yields a lower probability than (waiteq).

For networks exhibiting Properties 1 and 2, assume the contrary:

$$P_c(i|a)P_c^t(i) + (1 - P_c(i|a))P_u^t(i) > P_c^*(i,k)P_c^t(i) + (1 - P_c^*(i,k))P_u^t(i)$$
$$P_c(i|a)P_c^t(i) + P_u^t(i) - P_c(i|a)P_u^t(i) > P_c^*(i,k)P_c^t(i) + P_u^t(i) - P_c^*(i,k)P_u^t(i)$$
$$P_c(i|a)[P_c^t(i) - P_u^t(i)] > P_c^*(i,k)[P_c^t(i) - P_u^t(i)]$$

By Property 1 we know $P_c^t(i) < P_u^t(i)$, so we have:

$$P_c(i|a) < P_c^*(i,k) \tag{17}$$

which violates Property 2 (i.e. $1 - P_c(i|a) < 1 - P_c^*(i,k)$), which contradicts our assumption. Therefore, if a is congested, we achieve a higher arrival probability from waiting at i for k time units than by revisiting i after a.

Sub-Case 2: a is uncongested We now consider the case where a is uncongested. Then the arrival probability from $\bar{i}_t$ is:

$$(1 - P_u(i|a))P_c^t(i) + P_u(i|a)P_u^t(i) \tag{18}$$

Again, the arrival probability achieved from waiting at i for k time units is:

$$P_c^*(i,k)P_c^t(i) + (1 - P_c^*(i,k))P_u^t(i) \tag{19}$$

Again, we will prove by contradiction that for networks exhibiting Properties 1 and 2, (18) always yields a lower probability than that of (19).

For networks exhibiting Properties 1 and 2, assume the contrary:

$$P_c^*(i,k)P_c^t(i) + (1 - P_c^*(i,k))P_u^t(i) < (1 - P_u(i|a))P_c^t(i) + P_u(i|a)P_u^t(i)$$
$$P_c^*(i,k)P_c^t(i) + P_u^t(i) - P_c^*(i,k)P_u^t(i) < P_c^t(i) - P_u(i|a)P_c^t(i) + P_u(i|a)P_u^t(i)$$
$$P_c^*(i,k)[P_c^t(i) - P_u^t(i)] + P_u^t(i) < P_u(i|a)[P_u^t(i) - P_c^t(i)] + P_c^t(i)$$
$$P_c^*(i,k)[P_c^t(i) - P_u^t(i)] + P_u^t(i) - P_c^t(i) < P_u(i|a)[P_u^t(i) - P_c^t(i)]$$
$$1 - P_c^*(i,k) < P_u(i|a)$$

which violates Property 2, which contradicts our assumption. Therefore, if a is uncongested, we achieve a higher arrival probability by waiting at i for k time units than by revisiting i k time units after a.

We have shown that regardless of the state of a, waiting at node i will yield a higher on-time probability than revisiting i.

We have shown that Cases 3 and 4 are the only situations where an optimal path may contain a reappearing node. We have also shown that in both of these cases, for networks that exhibit Properties 1 and 2, only paths of Type 2 may occur (i.e. paths where the reappearing node occurs from the decision to wait at the node) and not paths of Type 1 (i.e. paths where the reappearing node occurs from the decision to revisit the node). This affirms our claim that for these networks, the on-time arrival probability achieved from waiting at a node will be greater than the on-time arrival probability achieved from revisiting a node.

5   Conclusion

We study the problem of finding the path that maximizes the probability of arriving to a destination from a source within a specified time limit. We present an efficient dynamic programming solution that solves this adaptive routing problem. We prove that as the time limit increases, the maximum probability also increases. We also show that for some networks, revisiting a node is required to achieve the optimal probability. Since this is contrary to most realistic settings, we consider a variation of the problem where the traveler is always given the option to wait at a node rather than visit another node. We identify two important properties of networks in this setting: (1) The on-time arrival probability from any node is always higher if the node is uncongested rather than congested; and (2) a node is more likely to become uncongested after waiting at the node for any amount of time than from visiting it from an adjacent node. For networks with these properties, we prove that the optimal solution will not require revisiting nodes, which is reflective of most realistic settings.

References

1. Dean, B.C.: Algorithms for Minimum-Cost Path in Time-Dependent Networks with Waiting Policies. Networks 44, 41–46 (2004)
2. Eiger, A., Mirchandani, P.B., Soroush, H.: Path Preferences and Optimal Paths in Probabilistic Networks. Transportation Science 19, 75–84 (1985)
3. Fan, Y., Kalaba, R., Moore, J.: Shortest Paths in Stochastic Networks with Correlated Link Costs. Computers and Mathematics with Applications 49, 1549–1564 (2005)
4. Fan, Y., Kalaba, R., Moore, J.: Arriving on Time. Journal of Optimization Theory and Applications 127(3), 497–513 (2005)
5. Frank, H.: Shortest Paths in Probabilistic Graphs. Operations Research 17, 583–599 (1969)
6. Fu, L., Rillet, L.: Expected Shortest Paths in Dynamic and Stochastic Traffic Networks. Transportation Research 32(7), 499–516 (1998)
7. Gao, S., Huang, H.: Real-Time Traveler Information for Optimal Adaptive Routing in Stochastic Time-Dependent Networks. Transportation Research Part C 21, 196–213 (2011)
8. Loui, R.P.: Optimal Paths in Graphs with Stochastic or Multidimensional Weights. Communications of the ACM 26, 670–676 (1983)
9. Murthy, I., Sarkar, S.: Stochastic Shortest Path Problems with Piecewise Linear Concave Utility Functions. Management Science 44, 125–136 (1998)
10. Nie, Y., Wu, X.: Shortest Path Problem Considering On-Time Arrival Probability. Transportation Research Part B 43, 597–613 (2009)
11. Polychronopoulos, G.H., Tsitsiklis, J.N.: Stochastic Shortest Path Problems with Recourse. Operations Research Letters 10, 329–334 (1991)
12. Sigal, E.H., Pritsker, A.A.B., Solberg, J.J.: The Stochastic Shortest Route Problem. Operations Research 28, 1122–1129 (1980)
13. Waller, S., Ziliaskopoulos, A.: On the Online Shortest Path Problem with Limited Arc Cost Dependencies. Networks 40(4), 216–227 (2002)

Optimal Inventory Management in a Fluctuating Market

Eline De Cuypere, Koen De Turck, Herwig Bruneel, and Dieter Fiems

Department of Telecommunications and Information Processing, Ghent University,
St-Pietersnieuwstraat 41, B-9000, Belgium
(eline.decuypere,koen.deturck,herwig.bruneel,dieter.fiems)@telin.ugent.be

Abstract. In the highly competitive environment in which companies operate today, it is crucial that the supporting processes such as inventory management are as efficient as possible. In particular, a trade-off between inventory costs and service levels needs to be assessed. In this paper, we determine an optimal batch ordering policy accounting for both demand and market price fluctuations such that the long-term discounted cost is minimised. This means that future costs are reduced by a constant factor as we need to take inflation and other factors into account. To this end, the inventory system is modelled as a Markovian queueing system with finite capacity in a random environment. Assuming phase-type distributed lead times, Markovian demand and price fluctuations, the optimal ordering strategy is determined by a Markov decision process (MDP) approach. To illustrate our results, we analyse the ordering policy under several price fluctuation scenarios by some numerical examples.

1 Introduction

The impact of demand fluctuations has been extensively studied as an extension of the classical EOQ model [3]. Yan K. and Kulkarni V. have developed an inventory system in which the stochastic demand is assumed to be a continuous-time Markov chain [11]. Moreover, they assume an immediate delivery, i.e. the lead time equals zero, and the ordering is only possible when the inventory is empty. In this paper, we analyse an inventory system with a stochastic lead time and demand in discrete time.

In contrast to fluctuations in demand, the impact of price fluctuations is a rather unexplored scientific domain. J.-T. Teng et. al have established an algorithm assuming both fluctuating demand and price but this algorithm relates to products with a continuously decreasing price, such as high-technology products [9]. In our work, we focus on short- or medium-term price fluctuations around a long-term average. With regard to demand fluctuations, the model is relevant for products in their maturity phase where customer demand levels off.

Our paper is organised as follows. Section 2 describes the inventory model at hand. In particular, the inventory system is modelled as a three-dimensional Markov chain and the optimal ordering policy is determined by the theory of the

A. Dudin and K. De Turck (Eds.): ASMTA 2013, LNCS 7984, pp. 158–170, 2013.

Markov decision process. Also, the solution methods are explained. To illustrate our approach, section 3 considers some numerical examples. In particular, we determine the optimal ordering policy under several price fluctuation scenarios. Finally, conclusions are drawn in section 4.

2   Model Description

We consider the inventory system of a product subject to demand as well as price fluctuations. Furthermore, the lead time of this product is stochastic and demand is lost when the inventory is empty, i.e. there is no backlogging. Finally, there is at most one ongoing order and a customer demand always consists of one item. In this section, the assumptions and the research approach are explained.

2.1   Assumptions

To determine an optimal ordering policy we need to minimise the total cost. In this work, we define the following costs:

- inventory cost: $C_h \frac{Q}{2}$
- order cost: C_o
- purchase cost: QP_t
- cost due to lost demand: C_v

This give us the total cost function:

$$C = C_h \frac{Q}{2} + C_0 + QP_t + C_v$$

where C_o and C_h are the cost per order and the inventory cost per unit respectively, Q is equal to the order quantity and P_t is the price at the time of ordering t. The first cost includes the opportunity cost of fixed capital and the cost for storage and internal transport of the goods. This cost is often the largest proportion of the total cost and have varied between 6 and 18 % during the last 25 years [5]. We assume that the inventory cost is proportional to the inventory level. The second is only relevant when an order is placed and includes a fixed order cost C_o and a variable purchase cost per order QP_t. In contrast with the inventory costs, the purchase cost depends on the price and the order quantity but not on the current inventory level. The last cost is the cost due to lost demand. This is the case when the demand is not satisfied because the inventory is empty. This cost can be calculated as the sum of several factors, such as the cost due to lost sales and the damage on the industry's reputation.

Concerning the lead time and the market fluctuations, we assume that:

- the time between two consecutive order deliveries is geometrically distributed with parameter κ,
- the demand is Bernoulli distributed with parameter p. This means that at each discrete time unit the inventory level decreases by one or remains the same with probability p and $1 - p$ respectively,

– the price fluctuates around a long-term average with frequency α and β and with amplitude v. The parameters α and β denote the probability that the price decreases and increases respectively and v is the rate at which α and β decrease as the price moves further from the average price. The probability that the price level remains unchanged at the next time unit is thus equal to $1 - (\alpha + \beta)$ and is therefore independent of the price level.

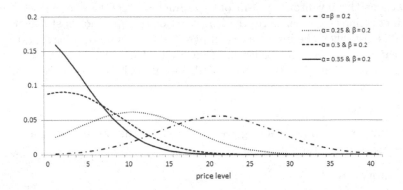

Fig. 1. Price distributions in function of α and β

Figure 1 depicts the price distribution in function of α and β, in the case that there are $K = 40$ possible price levels. Note that the absolute values of α and β have no effect on the distribution, only on the volatility. Indeed, if these probabilities increase, the probability to remain in the same level $1 - (\alpha + \beta)$ decreases and the volatility of the price is consequently much higher. Concerning the distribution, only the relative size is relevant. As the parameter α increases and β remains constant, the average price level decreases and the steady-state probability to have a relatively low price level increases.

Figure 2 depicts the steady-state price probability versus the price level for different values of v. As the figure shows, the smaller the rate v, the higher the probability that the price level is in the vicinity of the long-term average and the smaller the probability that an extreme (high or low) price level is reached. However, if the rate is high, the probability of an extreme price level is relatively high. Finally, if the probability of a price decrease and increase are independent of the current price level, i.e. $v = 1$, the probability distribution of the price is represented by a straight line.

Out of the results of figure 1 and 2 we may mention that, the frequency refers to the short-term volatility of the price level while the amplitude has an influence on the fluctuations in the long term.

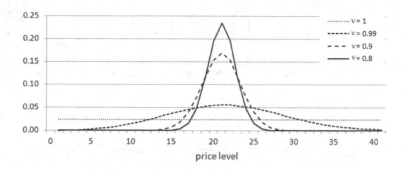

Fig. 2. Price distributions in function of v

2.2   Research Approach

Three-Dimensional Discrete-Time Markov Chain. The assumptions made above allow us to model the inventory system as a stationary Markov chain. In particular, we describe the inventory system as a three-dimensional discrete-time Markov chain with finite state space $\mathcal{S} = (\mathcal{C} + 1) \times \mathcal{K} \times \mathcal{C}$ with $\mathcal{C} = \{0, \ldots, C\}$ where C is the buffer capacity of the inventory system and $\mathcal{K} = \{1, 2 \ldots, K\}$ where K is the number of possible price levels. The state of the inventory system S_t at a discrete time t is described by the triplet $[n, m, i]$, n being the inventory level, m being the price level and i being the ordering in process that indicates whether or not an order is placed but not yet delivered, and also how many units are ordered. Note that as we analyse a stationary Markov chain, the transition matrices are homogeneous such that S_t at time t equals s and S_{t+1} at time $t + 1$ equals s'.

Note that the assumption of a cost due to lost demand leads to the introduction of a fictional inventory level -1 in the model. If the inventory is empty, a demand at time t leads to a system in inventory -1 at time $t + 1$ and the cost is added to the total cost. An inventory system in this fictional level behaves in the same way as a system with an empty inventory. At time $t + 2$, the system returns to an inventory level equal to zero if there is no demand and remains at level -1 when there is a new demand.

Figure 3 depicts the state diagram of the price level with an average price P^*. As previously mentioned, we assume a range of K possible price levels $P = (P_1, P_2, \ldots, P_K)$ and a decrease of the probability of a price increase β and a price decrease α with rate v as the price level moves further from the price average P^*. This means that the probability that the price further deviates from the average price decreases at each step with rate v. If this was not the case, the price probabilities would be independent of the state in which the system is and the steady-state probabilities of all the price levels would be equal, this being an unrealistic assumption according to us. The parameters in figure 3 are defined as follows:

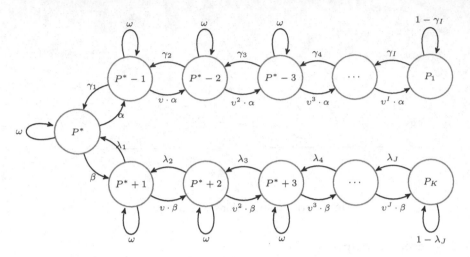

Fig. 3. State diagram of the price level

- $\omega = 1 - (\alpha + \beta)$
- $\gamma_i = \beta + (\alpha - \upsilon^i \alpha)$
- $\lambda_j = \alpha + (\beta - \upsilon^j \beta)$

where $i = 1, \ldots, I$ and $j = 1, \ldots, J$. The parameter ω is the probability that the price remains at the same price level and the parameter γ_i and λ_j represent the probability of a price increase and decrease after i price decreases and j price increases, respectively.

Markov Decision Process. A Markov decision process (MDP) is very similar to a Markov chain in the sense that the evolution of a system is described by a series of transitions in a set of predefined states. However, the decision process differs from a Markov chain as the transition matrix now depends on the action taken by the decision maker. A cost is attached at each combination of state and action. The goal is to find an optimal policy which take both immediate and future costs into account. A commonly used cost criterion is the minimisation of the (long-term) expected average cost per unit time. Another way is to minimise the expected total discounted cost with discount factor d. Both criteria are suitable for our purposes since we work with an infinite horizon. However, we choose to minimise the long-term discounted cost. The motivation for this choice is given below.

By means of the theory of a Markov decision process, we define the tuple $(\mathcal{S}, \mathcal{A}, P_a, R_a)$ being:

- the finite state space $\mathcal{S} = (\mathcal{C} + 1) \times \mathcal{K} \times \mathcal{C}$ of our discrete-time Markov chain with $\mathcal{C} = \{0, \ldots, C\}$ and $\mathcal{K} = \{1, 2 \ldots, K\}$,
- the set of actions $d_s \in \mathcal{A} = \{0, 1, \ldots, C\}$,

- the transition matrix $P_a(s, s') = \Pr[S_t + 1 = s'|S_t = s, d_t = a]$,
- the cost matrix $R_a(s, s') = R_a(s) = \mathrm{E}[R_{t+1}|S_t = s, d_t = a]$.

In a Markov decision process, a decision $d_s \in \mathcal{A}$ is taken at each possible state s. Action $d_s = 0$ means that no order is placed and action $d_s = x = \{1, \ldots, C\}$ means that an order of x units is placed. The chosen action influences the transition probabilities, the immediate and the future costs as well as the behaviour of the system in the future.

Then, there are two types of matrices of interest. The first one, the transition matrices $P_a(s, s')$, give the probability to reach s' at time $t+1$ when the system is in state s and action a has been undertaken by the decision maker. The second one, the cost matrices $R_a(s, s')$ give the cost obtained when the system is in state s' at time $t + 1$ after choosing action a in state s. The cost matrix can be written as $R_a(s)$ because the cost is independent of the state in which the system ends up at time $t + 1$. In particular, the inventory cost depends on the current inventory level and the purchase cost depends on the chosen action (and consequently on the order quantity) and on the current price.

Finally, after having defined our Markov decision process, we aim at finding an optimal action at each state such that the total discounted cost is minimised. The discounted cost is defined as,

$$\sum_{t=0}^{\infty} d^t R_{a_t}(s_t), \ d = \frac{1}{1 + r}, \ 0 < d \leq 1, \tag{1}$$

where r is the current interest rate and d^t equal to the current value of one unit-cost after t time periods. The value of this factor, typically close to 1, plays an important role in the model [2]. If the costs are greatly discounted, i.e. the discount factor has a low value, the immediate ordering costs have much more weight than the future inventory costs. Hence, the order quantity will be larger such that the ordering cost can be assigned to a larger number of units. Furthermore, the value of the discount rate depends on the specific situation of the company and of the size of the time interval. Indeed, the cost will be much more discounted for a time interval of one month than of a day. Finally, the advantage of applying a discounted cost as cost criterion is that the model can easily be adapted to a Markov decision process with a finite-time horizon. For example, this can be important for products that are sold for only several seasons.

Although it is theoretically possible to solve a Markov decision process by calculating the total cost for each possible policy, it becomes infeasible for larger problems. Several algorithms are available to find the optimal ordering policy in a more efficient way. In this work, five different solution methods are used and compared in order to address the most efficient method for our specific model. These methods are linear programming, policy iteration, value iteration and a modified algorithm of both policy iteration and value iteration.

Solution Methods

Linear Programming. It is possible to solve a Markov decision process with the aid of linear programming. This method uses a mathematical model for describing a problem and determining the best outcome by optimising a linear objective function subject to linear constraints. We use the simplex method to find our optimal deterministic policy. Further details of this method can be found in [2].

Dynamic Programming. Besides linear programming, Markov decision processes can also be solved with the aid of dynamic programming. To find the optimal policy of a Markov decision process, there are two main functions that are solved iteratively for each state s. First, the policy $\phi(s)$ for each state s represents the order quantity when the system is in state s. Second, the variable $V^\phi(s)$ is the 'value' of a given state s if a well-defined policy $\phi(s)$ is followed. This variable is calculated for each state s, which leads to a system of equations. The step in which this system is solved is called 'value determination'. The optimum policy is denoted as ϕ^* and is thus the policy for which the values $V^{\phi^*}(s)$ are smaller than or equal to any other values of the ordering policies in every possible state s. Note that we aim at finding the smallest value for each state s as the value is calculated based on costs and not on rewards. In the formulas below, the parameter k represents the successive iterations [7,10]. In other words, $k+1$ is an indication for the next iteration.

$$\phi(s) = \operatorname*{argmin}_a \{\sum_{s'} P_a(s, s')(R_a(s, s') + dV(s'))\} \tag{2}$$

$$V_{k+1}(s) = \sum_{s'} P_{\phi(s)}(s, s')(R_{\phi(s)}(s, s') + dV_k(s')) \tag{3}$$

The value function $V(s)$ is the expected total discounted cost when the system starts in state s where policy ϕ is applied and evolves to infinity. The formula (3) includes as many equations as there are states being the unknowns.

Several algorithms make use of these two steps. The order in which these steps are carried out depends on the particular variant of the algorithm. The different solution methods are now explained further in details.

Policy iteration. A popular method of dynamic programming is the 'policy improvement algorithm' or 'policy iteration', in which the policy iteratively improves [4]. The main advantage of this method is the high efficiency. In particular, a much smaller number of iterations is required to reach the optimal solution of a Markov decision process compared to the simplex method in linear programming.

The first step in this algorithm is to choose an arbitrary policy ϕ_0. In other words, an arbitrary action d_s is selected for each state s. Then, in the value determination — also called the value evaluation — step, a series of equations are solved for each state s as in formula (2). In other words, a value V^{ϕ_0} is calculated for each state s. Based on this result an improved policy ϕ_1 is set in the policy improvement step as shown in formula (3). This process corresponds to one iteration. The equations $V_1(s)$ of the valuation determination step is set and developed, after which a new improved policy ϕ_2 is given. This iterative process is repeated until two successive iterations yield an identical policy. The policy ϕ^* is defined as the optimal ordering policy. Figure 4 shows the interaction between the policy and the value function graphically. Here, E, represents the value evaluation step and I represents the policy improvement step.

In this solution method, there is a stop criterion. The algorithm is complete if ϕ does not change when using the formula (2) in the policy improvement step. Moreover, the algorithm guarantees an optimal solution in a finite number of iterations since the number of possibilities for the ordering policy is finite. Also note that each policy evaluation $V_{k+1}(s)$ starts from the value function $V_k(s')$ of the previous iteration, resulting in a very fast convergence speed. Further details on this algorithm can be found in [8], [10] and [2].

$$\phi_0 \xrightarrow{\ E\ } V^{\phi_0} \xrightarrow{\ I\ } \phi_1 \xrightarrow{\ E\ } V^{\phi_1} \xrightarrow{\ I\ } \phi_2 \xrightarrow{\ E\ } \cdots \xrightarrow{\ I\ } \phi^* \xrightarrow{\ E\ } V^*$$

Fig. 4. Policy iteration

Value iteration. A disadvantage of policy iteration is that each iteration in the policy evaluation step requires iterative calculations in itself. Moreover, convergence takes place only in the limit, i.e. if the policy is the same in two successive iterations. The question is whether we can stop the iteration process earlier and still find the optimal ordering policy. A possibility is to perform the policy evaluation step in each iteration only once for each state s and not until convergence occurs [8]. This popular solution method, developed by Bellman, is called 'value iteration' [1]. This is a method based on the principle of backward induction. In this algorithm, the so-called 'Bellman equation' is used instead of the ϕ-sequence, which is obtained by substituting the formula for $\phi(s)$ in the formula for $V(s)$.

$$V_k(s) = \min_a \{ E[r_{t+1} + dV_k(s_{t+1}) | s_t = s, a_t = a] \} \tag{4}$$

$$= \min_a \{ \sum_{s'} P_a(s, s')(R_a(s, s') + dV_k(s')) \} \tag{5}$$

In other words, the value of a certain state is defined as the sum of the minimum expected cost in that state and the expected discounted value of all possible states s' that can be achieved from state s. The value iteration algorithm consists of iteratively solving the Bellman equation for each state s. In theory, an infinite number of iterations is needed to find the optimal values V^*. In practice, the algorithm is completed when an epsilon-optimal solution is found. A disadvantage of the value iteration algorithm is that a solution is found only at the end of the iterations, in contrast to policy iteration where the solution is systematically improved, which is more transparent. For very large problems, a decent, but not optimal, solution can be found in a reasonable time with policy iteration while this is impossible with value iteration. A more detailed explanation on this algorithm can also be found in [10] and [2].

Modified policy iteration. The 'modified policy iteration' is, as the name suggests, a slightly modified version of the policy iteration algorithm. In the original algorithm, the first step, i.e. the policy improvement, iterates once, while the second step, i.e. the policy evaluation, is repeated until it converges.

In the modified version, the first step is the same and the second step is repeated only a limited number of times and thus not until convergence occurs. In particular, a number of value iteration steps are performed such that a reasonably good approximation of the solution is given whereas an exact solution is given in the original algorithm. In addition, the algorithm is completed when an epsilon-optimal solution is found, in contrast to the original algorithm where there is a strict stop criterion for optimality, i.e. if the policy is the same in two successive iterations [6]. To summarise, the modified policy iteration algorithm is similar to the standard policy iteration algorithm except that the policy evaluation step is calculated by using the value iteration algorithm and the algorithm stops when an epsilon-solution is found.

Gauss-Seidel value iteration. The adapted version of the value iteration algorithm that will be used in this work is the 'Gauss-Seidel value iteration' algorithm. As for the original algorithm, the Bellman equation is iteratively solved, but in this variant $V_{k+1}(s)$ is used instead of $V_k(s)$.

3   Numerical Results

Markovian Demand and Price Fluctuations

The first numerical example considers an inventory system with immediate delivery, i.e. the probability to have a delivery κ equals 1. In particular, figure 5 depicts the optimal ordering policy for a combined inventory- and price level. The inventory system has a buffer capacity C equal to 15 and we assume a price range of 10 levels (from 10 to 100). The holding cost C_h equals 0.1 and the fixed ordering cost C_o equals 20. Concerning the price fluctuations, the probability of a price decrease and increase, respectively α and β, equal 0.4 and the amplitude is defined by υ equal to 0.2. The probability to have a demand of one unit p equals 0.6 and we assume a discount factor d equal to 0.9. Finally, to ensure that an order is placed when the inventory is empty, we assume a very high cost due lost demand, i.e. C_v equals 1000.

As expected, the lower the inventory- and the price level, the larger the order quantity. The order quantity as well as the threshold at which an order is placed depends on several factors such as the ordering costs, the inventory costs and

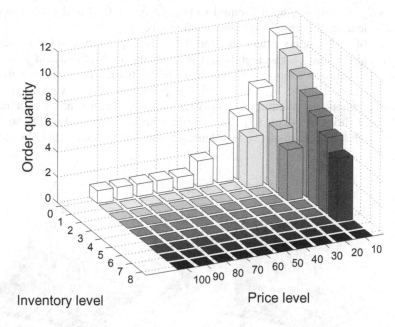

Fig. 5. Optimal ordering policy at a given inventory- and price level and with immediate delivery

the price fluctuations. Moreover, each inventory level is characterised by an exponential function with the order quantity as dependent variable and the price level as independent variable. In the higher inventory levels, this exponential function is earlier interrupted because it is possible to wait for a decreased price in the near future such that the holding costs and the purchase costs are kept low. The same is observed when the inventory level is the independent variable. For higher price levels, the decreasing function is earlier interrupted due to the combined effect of the inventory- and purchase costs. Finally, the decision maker should not order in the vast majority of the states. This is explained by the low uncertainty inherent in the model due to the absence of an uncertain lead time. If the price level is high, we will always wait because the probability of a price decrease is real and there is no risk that the demand is not satisfied due to an empty inventory. If a lead time is introduced, the risk of an unsatisfied demand is present and the decision maker will have to handle faster and will sometimes even be forced to order at a very high price. A numerical example of this scenario is given in the next paragraph.

Phase-Type Distributed Lead Times

Both figures 6(a) and 6(b) quantify the impact of the cost due to lost demand on the ordering policy. In both figures, the parameters have the same value as in the previous section and we assume that the probability to have a delivery is exponentially distributed with κ equal to 0.5. In figure 6(a) and 6(b), the cost due to lost demand C_v equals 100 and 1000 respectively.

As figure 6(a) shows, no orders are placed at a high price even if the inventory is empty. This is due to the fact that the cost due to lost demand does not outweigh the high ordering cost. However, in figure 6(b), where the cost due to lost demand is multiplied by 10, an order is placed even at a very high price and not only when the inventory is empty. In this case, an inventory shortage is avoided at all times. As expected, the combination of a low probability to have a delivery with a high cost due to lost demand would enhance further this effect.

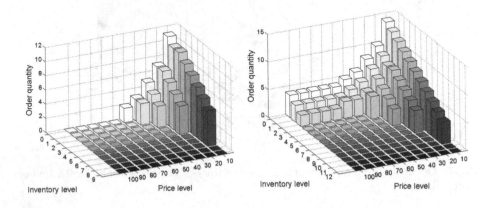

Fig. 6. Impact of the cost due to lost demand on the optimal ordering policy ($C_v = 100$ and $C_v = 1000$ in the left- and right figure respectively)

Comparison between the Different Solution Methods

In this section, we aim at comparing the different solution method algorithms explained in the section 'Solution methods'. Figure 7 summarises the performance of the various solution methods as a function of the number of states.

As the figure shows, the linear programming algorithm is not suitable for our model with a large state space. The reason is that this solution method is not specifically developed to solve Markov decision processes, while it is the case for the other four methods. Furthermore, the value iteration and its derivative Gauss-Seidel algorithm rise exponentially with large state spaces. Moreover, the standard value iteration algorithm is much more efficient than the Gauss-Seidel approximation, probably due to a lower number of iterations. Furthermore, we can observe that the state space has to be very large before the value iteration algorithm performs significantly worse than the policy iteration. The modified policy iteration algorithm is in turn slightly more efficient than the standard policy iteration algorithm, but the difference is negligible. Finally, we may conclude that the modified policy iteration algorithm is the most efficient solution method for this specific model.

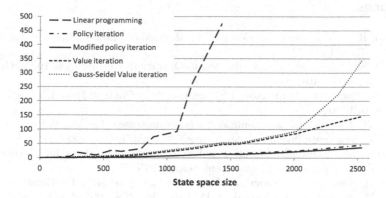

Fig. 7. A comparison between the different solution methods (CPU-time in seconds)

4   Conclusion

In this paper, we determine the optimal batch ordering policy for an inventory system with stochastic lead times, demand and price fluctuations by means of a versatile Markovian decision model. In the studied systems, there is at most one ongoing order and demand is of one unit. The optimal policy is determined under several price fluctuation scenarios such that a total cost function is minimised. The proposed cost function includes a holding cost, a fixed ordering cost, a variable purchase cost and a cost due to lost demand. The latter represents the risk that the demand is not satisfied due to an unexpected long lead time. As it becomes infeasible to solve a Markov decision process with a large state space by hand, several algorithms are used to find the optimal ordering policy.

As our numerical examples show, a relative small change in the parameters, e.g. the discount factor, the frequency and the amplitude of the price or the ordering cost, often has a major impact on the outcome of the model. This observation underlines the importance of a model customised for each company. Furthermore, a comparison between the different solution methods with respect to the state space is made. The research shows that the policy iteration algorithm, or a modified version of this algorithm, and secondly the value iteration algorithm present the best results for a system with such a large state space. However, the formulation of the problem as a linear programming problem is not appropriate. Finally, it can be stated that the required CPU time does not constitute an obstacle for solving our specific problem. The condition, however is to use a computer with enough RAM. This may also be solved by reducing the number of possible price levels or by limiting the number of possible batches.

References

1. Bellman, R.: A Markovian decision process. Journal of Mathematics and Mechanics 6(4), 679–684 (1957)
2. Hillier, F.S., Lieberman, G.: Introduction to Operations Research, ch. 19. McGraw-Hill, Singapore (2010)
3. Hopp, W., Spearman, M.: Factory Physics, Inventory Control: From EOQ to RDP. McGraw-Hill (2008)
4. Howard, R.: Dynamic programming and Markov processes. The M.I.T. Press (1960)
5. Miller, G.: Aggregate inventory management. PROACTION (2006)
6. Puterman, M., Shin, M.: Modified policy iteration algorithms for discounted Markov decision problems. Management Science 24 (1978)
7. Sutton, R.: On the significance of Markov decision processes. In: Gerstner, W., Hasler, M., Germond, A., Nicoud, J.-D. (eds.) ICANN 1997. LNCS, vol. 1327, pp. 273–282. Springer, Heidelberg (1997)
8. Sutton, R., Barto, A.: Reinforcement learning: An introduction. IEEE Transactions on Neural Networks 9(5), 1054 (1998)
9. Teng, J.T., Chern, M.S., Chan, Y.L.: Deterministic inventory lot-size models with shortages for fluctuating demand and unit purchase cost. International Transactions in Operational Research 12(1), 83–100 (2005)
10. Tijms, H.C.: Stochastic modelling and analysis: A computational approach, Markovian decision processes and their application. John Wiley & Sons Inc., New York (1986)
11. Yan, K., Kulkarni, V.: Optimal inventory policies under stochastic production and demand rates. Stochastic Models 24(2), 173–190 (2008)

Efficient Evaluation of Out-Patient Scheduling with Unpunctuality

Dieter Fiems[1] and Stijn De Vuyst[2]

[1] Department of Telecommunication and Information Processing, Ghent University
St-Pietersnieuwstraat 41, 9000 Gent, Belgium
[2] Department of Industrial Management, Ghent University
Technologiepark 903, 9052 Zwijnaarde, Belgium
{Dieter.Fiems,Stijn.DeVuyst}@UGent.be

Abstract. We assess appointment scheduling for outpatients in a hospital. A physician sees K patients during a fixed-length session. Each patient has been given an appointment time during the session in advance. Our evaluation approach aims at obtaining accurate predictions at a very low computational cost for the waiting times of the patients and the idle time of the physician. To this end, we investigate a modified Lindley recursion in a discrete-time framework. We assume general, possibly distinct, distributions for the patient's consultation times as well as for the patient unpunctuality, which allows for accounting for multiple treatment types, for patient unpunctuality as well as for patient no-shows. The moments of waiting and idle times are obtained and the computational complexity of the algorithm is discussed.

Keywords: Patient scheduling, Queueing Theory.

1 Introduction

Like any service providing industry, health care providers operate in an economic environment, and aim for improved service quality at reduced cost. Outpatient services are key in cost-effective health care, given the current emphasis on preventive medicine and shorter lengths of stay at the hospital [2]. Outpatient services constitute those medical procedures that can be performed in a medical center or hospital without an overnight stay. Full waiting rooms for outpatient services not being uncommon, the problem of scheduling a hospital's outpatients into the consultation session of a physician has received a lot of attention over the last sixty years. Many studies are motivated from a specific practical situation and aim at improving the organisational procedures in a particular (part of a) hospital [1,6,8,9]. Although, practical settings considerably differ in terms of medical practice, organisation, administrative demands or limitations and preferences of patients or medical staff, very often the underlying problem is largely the same and can be formulated as follows.

A physician receives patients during a time interval of a fixed length called a session. The physician is assisted by a nurse or secretary at the administration

A. Dudin and K. De Turck (Eds.): ASMTA 2013, LNCS 7984, pp. 171–182, 2013.

desk who takes the calls of patients who wish to consult the physician. The nurse decides whether a calling patient can be admitted to a particular session and if so, at what time during the session the patient should arrive. All appointments are fixed before the session starts. The physician arrives at some point during the session, which is not necessarily the beginning. Given the session lengths and the number of patients, a schedule consists of both the patient's appointment times and the physician's arrival time.

Uncertainty on the lengths of the consultation times makes waiting unavoidable. It is in practice very unlikely that the next patient arrives at the exact moment where the previous one leaves: either the patient waits until the physician has seen all preceding patients or the physician waits until the following patient arrives. For convenience, the waiting times of the physician are referred to as idle times. Clearly, scheduling appointments far apart results in low waiting times but long idle times and vice versa if the appointments are closely together. As a reduction in patient waiting times leads to increased idle times, a good schedule aims for an optimal trade-off.

In this paper, a numerical evaluation methodology for the evaluation of outpatient schedules is introduced. The aim of our evaluation approach is to obtain accurate performance predictions at a low computational cost. This is achieved by (1) using a modified Lindley recursion to allow for explicit expressions and (2) choosing a discrete-time (slotted) setting to make those expressions easy to compute. Our analysis requires no assumptions on the consultation times of the patients other than that their distributions are known and that they are independent. The fact that each patient can have a different consultation time distribution allows for evaluating schedules containing heterogeneous patients, which is important when making tight schedules with low cost as shown further. In addition, we account for patient unpunctuality as well as no-shows. That is, patients may arrive before or after their appointment time or may not show up at all, the only restriction being that unpunctuality cannot lead to reordering of patients. This contribution builds upon the method to evaluate patient schedules under uncertainty introduced in [3]. The method is extended here to include patient unpunctuality.

Unpunctuality complicates the analysis of appointment driven scheduling. Koeleman and Koole [7] recently noted that most papers that obtain numerical or structural results assume punctuality, [4] being a notable exception. The present study most closely relates to [4], where the scheduling problem is studied under the additional constraint that the service times are exponentially distributed. The same authors also consider multi-server appointment driven scheduling, again assuming exponential service times [5].

The remainder of the paper is organised as follows. The mathematical model is introduced in the following section. Sections 3 and 4 then present the analysis method and some numerical examples, respectively. Finally conclusions are drawn in section 5.

2   Mathematical Model

Throughout the analysis, we assume a discrete-time setting. That is, all time-related quantities in the model, such as waiting and idle times, are discretised into fixed-length intervals of length Δ. A suitable choice of Δ follows from a trade-off: whereas using small slots ensures a maximal accuracy of the performance predictions, choosing large slots results in a lower computational effort.

We consider a session of K consecutive appointments, spanning a time period $[0, t_{\max}]$. Let τ_k denote the appointment time of the kth patient ($0 \leq \tau_1 \Delta < \tau_2 \Delta < \ldots < \tau_K \Delta \leq t_{\max} \Delta$) and let $\theta \Delta$ ($0 \leq \theta \leq t_{\max} \doteq \tau_{K+1}$) denote the arrival time of the physician. Here, we introduced a virtual arrival instant τ_{K+1} at the end of the session for further use. We also introduce notation for the time between consecutive appointment times: $a_k \doteq \tau_{k+1} - \tau_k$, $k = 1, \ldots, K$. Note that, in accordance with this definition, a_K denotes the time between the appointment time of the last patient and the end of the session.

Patients introduce uncertainty in the schedule by their unpunctuality, by possibly not showing up and by the length of their consultation time.

- *Unpuctuality* is modelled by a sequence of independent random variables $\{U_k\}$ that denote the difference between actual arrival times and appointed arrival times. The unpunctuality U_k is positive if the kth patient is late, negative if this patient is early and equals 0 if the patient arrives at the appointed time. Let $u_k(n) = \Pr[U_k = n\Delta]$ denote the probability mass function of U_k. It is assumed that U_k is bounded for each patient, let $\underline{u}_k \Delta$ and $\overline{u}_k \Delta$ denote lower and upper bound, respectively and let $\Omega_k = \{\underline{u}_k, \underline{u}_k + 1, \ldots, \overline{u}_k\}$ denote the set of possible unpunctuality times: $\Pr[\underline{u}_k \Delta \leq U_k \leq \overline{u}_k \Delta] = 1$. The model at hand does not allow for changes in the arrival order of patients. To ensure this, we assume $\tau_k + \overline{u}_k \leq \tau_{k+1} + \underline{u}_{k+1}$. Note that by these bounds on the unpunctuality, the interarrival time between patient k and $k+1$ is at most $\tilde{a}_k \doteq a_k + \overline{u}_{k+1} - \underline{u}_k$. Finally, for further use, the virtual $K + 1$st patient is assumed to be punctual: $\overline{u}_{K+1} = \underline{u}_{K+1} = 0$.
- *No-shows* are modelled by a sequence of independent Bernoulli random variables $\{B_k\}$, $B_k = 0$ if the kth patient does not show up. Let $p_k = 1 - \mathrm{E}[B_k]$ denote the probability that the kth patient does not show up.
- The *consultation times* constitute a sequence of independent random variables. Let $s_k(n) = \Pr[S_k = n\Delta]$ denote the probability mass function of the consultation time S_k of the kth patient.

The former assumptions support heterogeneity in patient characteristics which allows for the evaluation of schedules that account for prior knowledge about patients. For example, for each calling patient, the administration can estimate the required consultation time distribution based on the person's characteristics like age and medical record as well as the required type of medical treatment. More information typically yields a reduction in variance of the consultation time which is beneficial for the performance of the schedule.

3   Performance Analysis

System equations. Let A_k denote the actual interarrival time (expressed in terms of the slot length Δ) between the kth and $k+1$st patient. In view of the modelling assumptions on the unpuctuality, we have,

$$A_k = a_k - U_k + U_{k+1} , \tag{1}$$

for $k = 1, \ldots, K$. Let the waiting time of the kth patient W_k be defined as the number of slots between the arrival of this patient and the start of his consultation. As patients are not allowed to change order, $A_k \geq 0$ and consecutive waiting times satisfy Lindley's recursion,

$$W_{k+1} = (W_k - A_k + S_k)^+ , \tag{2}$$

for $k = 1, \ldots, K$ and with,

$$W_1 = (\theta - \tau_1 - U_1)^+ . \tag{3}$$

From the system equations (1) and (2), the consecutive waiting times clearly do not constitute a Markov chain. However, a Markov chain is obtained if we augment the waiting time with the corresponding unpunctuality: (W_k, U_k) denotes the state of the chain for patient k. We now express the moments and probabilities of this chain for the $k + 1$st patient in terms of moments and probabilities of the preceding patient.

Waiting times. First, consider the moments of the waiting time, conditioned on the unpunctuality. By conditioning on the unpunctuality of the kth patient and by accounting for the cases where $W_k - A_k + S_k$ is negative, we find the following expression for the qth conditional moment,

$$
\begin{aligned}
\mathcal{W}_{k+1}^{(q)}(j) &\doteq \mathrm{E}\big[(W_{k+1})^q \mathbb{1}_{\{U_{k+1}=j\}}\big] \\
&= \ell_k^{(q)}(j) + u_{k+1}(j)\, \mathrm{E}[(W_k + S_k - a_k + U_k - j)^q], \\
&= \ell_k^{(q)}(j) + u_{k+1}(j) \sum_{r=0}^{q} \sum_{i=\underline{u}_k}^{\overline{u}_k} \sum_{m=0}^{q-r} \binom{q}{r}\binom{q-r}{m} \\
&\qquad \times \mathrm{E}[(S_k)^m](i - a_k - j)^{q-r-m} \mathcal{W}_k^{(r)}(i) ,
\end{aligned}
\tag{4}
$$

with,

$$
\begin{aligned}
\ell_k^{(q)}(j) &= u_{k+1}(j)\, \mathrm{E}[(W_k + S_k - a_k + U_k - j)^q \mathbb{1}_{\{W_k+S_k+U_k<j+a_k\}}] \\
&= u_{k+1}(j) \sum_{i=\underline{u}_k}^{\overline{u}_k} \mathrm{E}[(W_k + S_k - a_k + i - j)^q \mathbb{1}_{\{U_k=i, W_k+S_k<a_k+j-i\}}] \\
&= u_{k+1}(j) \sum_{i=\underline{u}_k}^{\overline{u}_k} \sum_{r=0}^{a_k+j-i-1} \sum_{m=0}^{r} s_k(r-m)(r-a_k+i-j)^q w_k(m,i) .
\end{aligned}
\tag{5}
$$

Here, $w_k(m,i) \doteq \Pr[W_k = m, U_k = i]$ denotes the joint probability of waiting time and unpunctuality of patient k. These joint probabilities of consecutive patients relate as,

$$w_{k+1}(n,j) = \sum_{i=\underline{u}_k}^{\overline{u}_k} \sum_{m=0}^{n+a_k-i+j} s_k(n + a_k - i + j - m)u_{k+1}(j)w_k(m,i) \qquad (6)$$

for $n > 0$ and $j \in \Omega_{k+1}$, whereas for the boundary case $n = 0$, we have,

$$w_{k+1}(0,j) = \sum_{i=\underline{u}_k}^{\overline{u}_k} \sum_{m=0}^{a_k-i+j} \sum_{\ell=0}^{a_k-i+j-m} u_{k+1}(j)s_k(\ell)w_k(m,i), \qquad (7)$$

for $j \in \Omega_{k+1}$. Similarly, we have for the first patient,

$$w_1(n,j) = u_1(j)\mathbb{1}_{\{\theta - \tau_1 - n - j = 0\}}$$

for $n > 0$ and

$$w_1(0,j) = u_1(j)\mathbb{1}_{\{j - \theta + \tau_1 \geq 0\}}$$

Equations (4) to (7) now allow for determining the conditional moments of the waiting times of the consecutive patients. Clearly, by equation (4), we can recursively determine the conditional moments of the waiting times provided the moments of the waiting time of the first patient and a number of probabilities $w_k(m,i)$ are known, see (5). We would like to stress that only a finite number of probabilities needs to be determined, even if the consultation time distribution is unbounded. Calculation the moments of the waiting time of the virtual patient $K + 1$ requires to calculate the probabilities $w_K(m,i)$ for $m = 0 \rightarrow \tilde{a}_K - 1$ and $i = \underline{u}_K \rightarrow \overline{u}_K$. This in turn requires calculation of $w_{K-1}(m,i)$ for $m = 0 \rightarrow \tilde{a}_K + \tilde{a}_{K-1} - 1$ and $i = \underline{u}_{K-1} \rightarrow \overline{u}_{K-1}$, calculation of $w_{K-2}(m,i)$ for $m = 0 \rightarrow \tilde{a}_K + \tilde{a}_{K-1} + \tilde{a}_{K-2} - 1$ and $i = \underline{u}_{K-2} \rightarrow \overline{u}_{K-2}$ and so on. Finally, all probabilities can be expressed in terms of $w_1(m,i)$. Expressions for the moments and probabilities of the waiting time of the first patient easily follow from (3),

$$\mathcal{W}_1^{(q)}(j) = u_1(j)((\theta - \tau_1 - j)^+)^q, \qquad (8)$$

$$w_1(n,j) = u_1(j)\mathbb{1}_{\{(\theta - \tau_1 - j)^+ = n\}}. \qquad (9)$$

Once the conditional moments of the waiting times are retrieved, we immediately find the (unconditional) moments,

$$\mathrm{E}[W_k^q] = \sum_{j=\underline{u}_k}^{\overline{u}_k} \mathcal{W}_k^{(q)}(j). \qquad (10)$$

Patient waiting times are not the only performance measures of interest. Equally important are the idle times, the modified waiting times and the session overtime. We here define these performance measures and show that these can be expressed in terms of the moments and probabilities of the waiting times as calculated above.

Idle times. The idle time I_k of the kth patient is defined as the time (expressed in terms of slots) the physician has to wait for the arrival of the next patient. By similar arguments as those for Lindley's equation (2), we find,

$$I_{k+1} = (A_k - S_k - W_k)^+ , \qquad (11)$$

such that $W_{k+1} - I_{k+1} = W_k + S_k - A_k$. Clearly, a non-zero waiting time $W_k > 0$ implies a zero idle time $I_k = 0$ and vice versa. Hence, $\mathrm{E}[(W_k)^q (I_k)^r] = 0$ for $q, r \geq 1$, which further yields,

$$\mathrm{E}[W_{k+1}^q] + (-1)^q \, \mathrm{E}[(I_{k+1})^q] = \mathrm{E}[(W_k - A_k + S_k)^q] . \qquad (12)$$

Solving for $\mathrm{E}[(I_{k+1})^q]$ and by equations (4) and (5), we find,

$$\mathrm{E}[(I_{k+1})^q] = (-1)^q \left(\mathrm{E}[(W_k - A_k + S_k)^q] - \mathrm{E}[W_{k+1}^q] \right)$$

$$= (-1)^{q+1} \sum_{j=\underline{u}_{k+1}}^{\overline{u}_{k+1}} \ell_k^{(q)}(j) . \qquad (13)$$

Modified waiting time. If a patient is early, waiting prior to its appointment time is not seen as a shortcoming of the appointment system. Therefore, the modified waiting time of a patient is defined as the time between the appointment time and the start of the consultation if the patient is early, and to the arrival instant of the patient and its consultation time if this is not the case. Clearly, the modified waiting time $\tilde{W}_k$ relates to the waiting time as follows,

$$\tilde{W}_k = W_k - (-U_k)^+ , \qquad (14)$$

such that the moments of the modified waiting time relate to those of the waiting time as,

$$\tilde{\mathcal{W}}_k^{(q)}(j) = \mathrm{E}[(\tilde{W}_k)^q \mathbb{1}_{\{U_k=j\}}]$$

$$= \mathbb{1}_{\{j<0\}} \sum_{r=0}^q \binom{q}{r} j^{q-r} \mathcal{W}_k^{(r)}(j) + \mathbb{1}_{\{j\geq 0\}} \mathcal{W}_k^{(q)}(j) .$$

Overtime. Finally, the overtime O is the amount of time by which the actual session length exceeds the previsioned session length. Recalling the introduction of the virtual punctual patient at the end of the session, it is easy to see that the waiting time of this patient exactly corresponds to the session overtime. If a patient were to be scheduled at the end of the session, this patient must wait till the overtime is completed. Hence, we directly find,

$$\mathrm{E}[O^q] = \mathcal{W}_{K+1}^{(q)}(0) . \qquad (15)$$

No-shows. The former expressions do not yet account for no-shows. These are most easily introduced by modifying the patient service time: if the patient does not show up, its consultation time is set to zero. Let $\tilde{S}_k$ denote the modified consultation time. Its probability mass function and qth moment is then given by,

$$\tilde{s}_k(n) = \Pr[\tilde{S}_k = n] = p_k \mathbb{1}_{\{n=0\}} + (1 - p_k)s_k(n)\,,$$

$$\mathrm{E}[(\tilde{S}_k)^q] = (1 - p_k)\,\mathrm{E}[(S_k)^q]\,.$$

The modified consultation time clearly incorporates the effects that a no-show has on the waiting times of the following patients. However, we can no longer define the waiting time and idle time of the kth patient as this patient may not show up. Fortunately, as no-shows are independent of the consultation times and unpunctualities, the expressions of the moments and probabilities of waiting times and idle times remain perfectly valid provided that these are interpreted as the waiting and idle times of the kth patient if this patient shows up.

Numerical complexity. We discuss the numerical complexity under the assumption that (i) all $a_k \doteq a = 2u+1$ are equal and (2) that the unpunctuality ranges from $-u$ to u for all patients. As such, the worst case scenario for adding unpunctuality to the model is considered. As the main computational effort comes from the computation of the probabilities $w_k(n, i)$, we focus on the number of floating point multiplications that are needed to calculate all probabilities required to calculate the moments of waiting and idle times of all patients. Carefully counting the number of multiplications in (6) and (7) and only accounting for the probabilities needed in (5), yields a total of $64/3\,K^3u^4 + O(u^3K^3)$ floating-point multiplications. The number of patients in a session is practically limited, hence K^3 is not overly large, posing no computational issues. Given the amount of time between patients, the value of u (or a) depends on the discretisation of time, that is, on the value of Δ. Small Δ yield large a and u: an increase of accuracy leads to considerably more computational effort.

4 Numerical Results

We now illustrate our approach by some numerical examples, with a particular focus on studying the effects of lateness on patient scheduling. We make the following assumptions:

- the time unit Δ is one minute;
- the consultation time distribution is a gamma distribution with mean 10 minutes and standard deviation 0.5 for all patients;
- patients are scheduled every 12 minutes, the physician arrives 2 minutes late;
- we consider three different unpunctuality distributions as depicted in Figure 1: (S. Normal) a discretised normal distribution with mean 0 and variance $5\mathrm{min}^2$, truncated to the interval $[-8\mathrm{min}, 8\mathrm{min}]$; (Uniform) a discrete uniform distribution over the same interval; and (A. Normal) a discretised normal distribution with mean $-3\mathrm{min}$ and variance $200\mathrm{min}^2$, again truncated to the interval $[-8\mathrm{min}, 8\mathrm{min}]$.

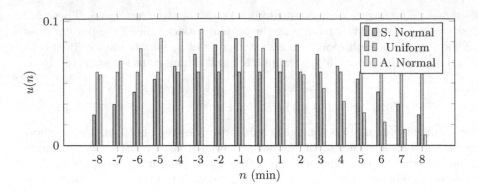

Fig. 1. Unpunctuality distribution

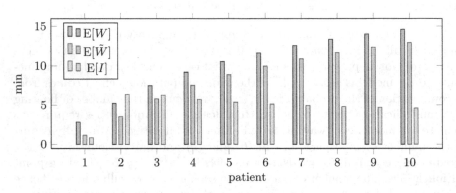

Fig. 2. Mean waiting and idle times of the patients for the symmetric normal unpunctuality distribution

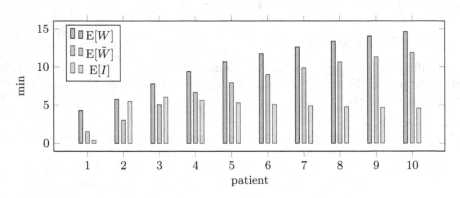

Fig. 3. Mean waiting and idle times for the asymmetric normal unpunctuality distribution

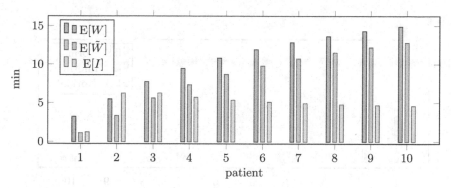

Fig. 4. Mean waiting and idle times for the uniform unpunctuality distribution

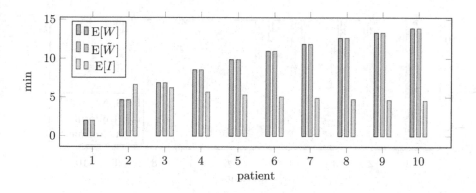

Fig. 5. Mean waiting and idle times without unpunctuality

Figures 2 till 4 show the mean waiting times and the mean modified waiting times of the consecutive patients as well as the mean idle times of the physician for the different unpunctuality distributions under consideration. In addition, Figure 5 shows the mean waiting times and the mean modified waiting times of the consecutive patients as well as the mean idle times of the physician without unpunctuality for comparison. Both the mean waiting times and mean modified waiting times increase to a limiting value in accordance with known results in queueing theory. The "steady-state" waiting time limit exists as the time between the consecutive patients is larger than the mean consultation time. Obviously, for punctual patients, the mean waiting time equals the mean modified waiting time. For the mean idle times, we first note an increase followed by a steady decrease, independent of unpunctuality which is a consequence of the interplay between the lateness of the physician, the consultation time distribution and the unpunctuality.

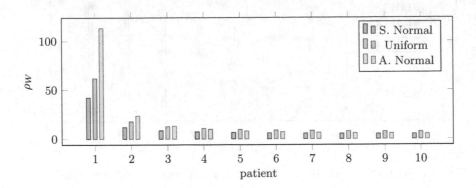

Fig. 6. Effect of the unpunctuality distribution on the mean waiting times

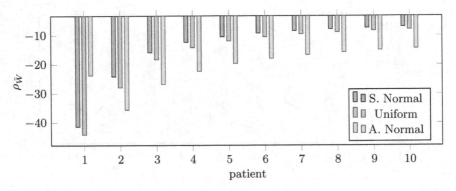

Fig. 7. Effect of the unpunctuality distribution on the modified mean waiting times

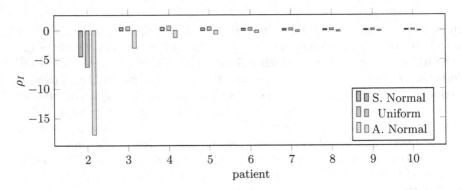

Fig. 8. Effect of the unpunctuality distribution on the mean idle times

To better illustrate the impact of the unpunctuality distribution on the waiting times and idle times, we depict the change with respect to the case of punctual patients for the different unpunctuality distributions considered in Figures 6 to 8. To this end, let ρ_X — for $X \in \{W, \tilde{W}, I\}$ — be the deviation (in percentage) with respect to the punctual case. That is,

$$\rho_X = \left(\frac{E[X^{(u)}]}{E[X^{(0)}]} - 1 \right) \times 100\% ,$$

with $X^{(0)}$ the random variable at hand assuming punctuality and with $X^{(u)}$ the random variable at hand assuming unpunctuality distribution u.

In contrast to the "queueing behaviour" discussed above, the effect of unpunctuality is quite surprising. For the mean waiting time, all patients but the first only wait a little longer when unpunctual (for all unpunctuality distributions); see Figure 6. The deviation of the mean waiting time for the first patient is large in percentage, but not large in absolute terms as the waiting time of the first patient is short. Most of the extra mean waiting time is explained by the fact that patients are early. This is clear from Figure 7 which shows the deviation from the mean modified waiting times. Patients are received considerably earlier after their appointed time if the patients are unpunctual. From the vantage point of the physician, Figure 8 shows that unpunctuality is beneficial as well: the mean idle times are smaller than when the patients arrive punctual.

5   Conclusions

This paper introduces an effective numerical method to assess the moments of the waiting times of patients and the moments of the idle times of the physician in between patients under fairly general conditions. There is no restriction on the consultation time distributions of the consecutive patients, a different distribution can be specified for each patient. The physician may arrive at any time during the session and the appointment times can be freely chosen. Finally, patients do not have to be punctual. There are no restrictions on the unpunctuality distribution apart from the fact that unpunctuality should not lead to reordering of the patients (which is possible if a first patient is late while a second patient is early).

The algorithmic approach advocated here is fast in comparison with simulations and was illustrated by numerical examples. The examples particularly focus on the effects of unpunctuality and surprisingly reveal that unpunctuality can be actually beneficial for the performance of the schedule.

References

1. Bailey, N.T.J.: A study of queues and appointment systems in hospital out-patient departments, with special reference to waiting-times. Journal of the Royal Statistical Society. Series B (Methodological) 14(2), 185–199 (1952)

2. Cayirli, T., Veral, E.: Outpatient scheduling in health care: A review of literature. Production and Operations Management 12(4), 519–549 (2003)
3. De Vuyst, S., Bruneel, H., Fiems, D.: Fast evaluation of appointment schedules for outpatients in health care. In: Al-Begain, K., Balsamo, S., Fiems, D., Marin, A. (eds.) ASMTA 2011. LNCS, vol. 6751, pp. 113–131. Springer, Heidelberg (2011)
4. Jouini, O., Benjaafar, S.: Appointment Scheduling with Non-Punctual Arrivals. In: Proceedings of the 13th IFAC Symposium on Information Control Problems in Manufacturing, Moscow, June 3-5, pp. 235–239 (2009)
5. Jouini, O., Benjaafar, S.: Queueing systems with appointment-driven arrivals, nonpunctual customers, and no-shows (2012) (under review)
6. Harper, P.R., Gamlin, H.M.: Reduced outpatient waiting times with improved appointment scheduling: a simulation modelling approach. OR Spectrum 25(2), 207–222 (2003)
7. Koeleman, P., Koole, G.: Appointment scheduling using optimisation via simulation. In: Proceedings of the 2012 Winter Simulation Conference (December 2012)
8. Reinus, W.R., Enyan, A., Flanagan, P., Pim, B., Skip Sallee, D., Segrist, J.: A proposed scheduling model to improve use of computed tomography facilities. Journal of Medical Systems 24(2), 61–76 (2000)
9. Zonderland, M.E., Boer, F., Boucherie, R.J., de Roode, A., van Kleef, J.W.: Redesign of a university hospital preanesthesia evaluation clinic using a queuing theory approach. Anesthesia and Analgesia 109(5), 1612–1621 (2009)

Interconnection of Large Scale Unstructured P2P Networks: Modeling and Analysis*

Vincenzo Ciancaglini[1], Rossano Gaeta[2], Riccardo Loti[2], and Luigi Liquori[1]

[1] LogNet Team, INRIA Sophia Antipolis Méditerranée, Sophia Antipolis, France
{firstname.lastname}@inria.fr
[2] Dipartimento di Informatica, Università di Torino, Torino, Italy
{firstname.lastname}@di.unito.it

Abstract. Interconnection of multiple P2P networks has emerged as a viable solution to increase system reliability and fault-tolerance as well as to increase resource availability. In this paper we consider interconnection of large scale unstructured P2P networks by means of special nodes (called *Synapses*) that are co-located in more than one overlay. Synapses act as *trait d'union* by forwarding a query to all the P2P networks they belong to. Modeling and analysis of the resulting interconnected system is crucial to design efficient and effective search algorithms and to control the cost of interconnection. To this end, we develop a generalized random graph based model that is validated against simulations and it is used to investigate the performance of search algorithms for different interconnection costs and to provide some insight in the characteristics of the interconnection of a large number of P2P networks.

1 Introduction

The last decade has seen the rise of peer to peer networks with a variety of applications, such as file sharing, resource lookup, real time services, up to the most recent research in SmartGrids. Common issues which affect all P2P systems, such as scalability, fault tolerance and security, arise from the different peculiarities each class of applications might expose. Increasing the locality properties of such systems, be it geographical, semantic, network, or social-based locality, is one of the most valued approaches to face such challenges: by grouping together peers representing users, and increasing their connections with one another, one can improve scalability, fault tolerance, and security (consider the possible creation of a "circle of trust" amongst nearby peers).

In this paper, we consider the interconnection of large-scale unstructured P2P networks by means of special nodes called *Synapses* [1], which are co-located in more than one network, and act as *connectors* by sending or forwarding a query to some or all the P2P networks they belong to. Modeling and analysis of the resulting interconnected system is crucial to design efficient and effective search

---

* This work was partially supported by Università degli Studi di Torino and Compagnia di San Paolo under project AMALFI (ORTO119W8J).

A. Dudin and K. De Turck (Eds.): ASMTA 2013, LNCS 7984, pp. 183–197, 2013.

algorithms and to control the cost of network interconnection. Yet, simulation and/or prototype deployment based analysis can be very difficult - if not impossible - due to the size of each component (we consider large scale systems that can be composed of millions of nodes) and to the complexity arising from the interconnection of several such complex systems.

Our Contribution

To overcome this strong limitation, we develop a generalized random graph based model to represent the topology of one unstructured P2P network, the partition of nodes into Synapses, the probabilistic flooding based search algorithms, and the resource popularity. We validate our model against simulations and prove that its predictions are reliable and accurate. We use the model to investigate the performance and the cost of different search strategies in terms of the probability of successfully locating at least one copy of the resource and the number of queries as well as the interconnection cost. We also gain interesting insights on the dependency between interconnection cost and statistical properties of the distribution of Synapses. Finally, we show that thanks to our model we can analyze the performance of a system composed of a large number of P2P networks. To the best of our knowledge, this is the first paper on model-based analysis of interconnection of large scale unstructured P2P networks[1].

The paper is organized as follows: Section 2 describes our system, Section 3 presents the mathematical derivation of the generalized random graph model we develop, Section 4 contains model validation through simulation, as well as model exploitation to study the performance of three search algorithms, Section 5 discusses related works, and in Section 6, we draw conclusions and outline ongoing activities that extend the current work.

2   System Description

In this paper, we focus on unstructured P2P networks where peers organize into an overlay network by establishing application level connections among them. The topological properties of an overlay network are represented by the number of connections of any of its participants. To this end, we describe an overlay by means of the degree distribution $\{p_k\}$ that can be interpreted as the probability that a randomly chosen peer has k connections in the overlay ($\sum_{k=1}^{\infty} p_k = 1$).

We consider a set of X unstructured P2P networks that are interconnected thanks to a subset of peers that belong to multiple overlays (these special peers are denoted as *Synapses*). Any peer may then belong to $i \in \{1, \ldots, X\}$ overlays: we denote i as the *Synapse degree* of a peer. The interconnected system is then described by $\{s_i\}$ ($i \in \{1, \ldots, X\}$) where s_i is the fraction of peers belonging to i overlays ($\sum_{i=1}^{X} s_i = 1$).

The search algorithm we consider is *flooding-based*. A peer starting a search sends *queries* to a randomly chosen subset of its one-hop neighbors. These nodes

---

[1] This paper is the full version of a two pages poster paper presented in [2].

forward the queries to a randomly chosen subset of their one-hop neighbors, excluding the query originator, and so on until the maximum number of allowed hops, i.e. the query time-to-live (TTL). A simple schema of interconnection through synapses and search is depicted in Fig. 1.

We also consider a variation of this search algorithm where a query is not forwarded by peers that own a copy of the resource. We focus on probabilistic versions of this general algorithm where any peer flips a coin before sending or forwarding a query to a specific neighbor. We allow the weight of this coin to be dependent on the Synapse degree of a peer; hence, a peer that belongs to i overlays sends/forwards a query to a particular neighbor with probability $p_f(i)$ ($i \in \{1, \ldots, X\}$). Please note that $\{p_f(i)\}$ ($i \in \{1, \ldots, X\}$) *is not* a probability distribution hence in general $\sum_{i=1}^{X} p_f(i) \neq 1$.

The goal of a search is to localize at least one resource related to the key we are looking for. There could be more replicas of the same resource hosted by different peers for two reasons: a resource is popular and/or is owned by peers located in different P2P networks. We represent resource popularity by $0 \leq \alpha \leq 1$, the average fraction of nodes that globally hold a copy of a given resource, and interpret it as the probability that a randomly chosen peer owns a copy of the resource.

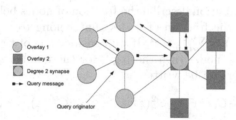

Fig. 1. Example of two P2P interconnected networks ($X = 2$) and one degree 2 synapse that belongs to both

3   System Model

This section illustrates the random graph modeling approach to represent one overlay topology, the interconnection of X P2P networks, the search algorithm, and resource popularity as described in Section 2.

3.1   One Overlay Topology

Each P2P network is organized into an overlay that we model as a generalized random graph whose degree distribution is $\{p_k\}$ that can be interpreted as the probability that a randomly chosen peer has k connections in the overlay. The random graph degree distribution is a probability distribution therefore we consider its probability generating function (henceforth denoted as p.g.f.) that is equal to $G_0(z) = \sum_{k=0}^{\infty} p_k z^k$. To correctly characterize the neighborhood of a randomly chosen peer we also need to characterize the probability distribution of the number of connections of a peer reached by randomly choosing an *edge* of the overlay. This probability is proportional to the degree of the peer (kp_k) and it can be proved that its p.g.f. is given by $\frac{\sum_k kp_k z^k}{\sum_k kp_k} = z\frac{G_0'(z)}{G_0'(1)}$ where $G_0'(z)$ denotes the first derivative of $G_0(z)$ with respect to z and $G_0'(1)$ yields the average value of distribution $\{p_k\}$. Finally, to characterize the number of connections *excluding* the edge we chose we obtain the p.g.f. $G_1(z) = \frac{G_0'(z)}{G_0'(1)}$. Starting from Equations defining $G_0(z)$ and $G_1(z)$ we can compute the p.g.f. for the number of two hops neighbors of a randomly chosen peer as $G_0(G_1(z))$. Similarly, the p.g.f. for three hops neighbor is given by $G_0(G_1(G_1(z)))$, and so on.

For a detailed overview on analyzing generalized random graphs using generating functions, we refer the reader to [3].

3.2   Interconnection of Multiple P2P Networks

To interconnect multiple overlays we consider some peers as Synapses nodes: these peers belong to multiple P2P networks hence the interconnected system can be modeled by the probability distribution $\{s_i\}$ (with $i \in \{1, \ldots X\}$). The elements of this distribution describe the fraction of nodes belonging to multiple P2P networks: s_i is the fraction of nodes that belong to k P2P networks. Its p.g.f. is given by $F(z) = \sum_{i=0}^{\infty} s_i z^i$. If we consider one of the X P2P networks including the Synapse nodes then the p.g.f. for the number of connections of a randomly chosen peer can be written as

$$M(z) = s_1 G_0(z) + s_2 G_0^2(z) + \ldots + s_X G_0^X(z) = F(G_0(z))$$

that is, if the chosen node is a degree 1 synapse (this event has probability s_1) then the number of connections is represented by $G_0(z)$. If the node is a degree 2 synapse (this event has probability s_2), then the number of connections is represented by the sum of two independent random variables whose p.g.f. is $G_0(z)$; it is well-known that the generating function of the sum of two independent random variables is equal to the product of the respective generating functions yielding the $G_0^2(z)$ factor in the equation for $M(z)$. The same reasoning is valid for synapses whose degree is greater than 2.

A similar expression can be written for the neighborhood of a node reached by following one randomly chosen edge *excluding the selected edge*:

$$N(z) = s_1 G_1(z) + s_2 G_1(z)G_0(z) + \ldots + s_X G_1(z)G_0^{X-1}(z) = \frac{G_1(z)}{G_0(z)}F(G_0(z)).$$

If we denote as $N_t(z)$ the p.g.f. for the probability distribution of the number of neighbors t hops away from a randomly chosen node we have that: $N_1(z) = M(z)$, and $N_2(z) = M(N(z))$, and $N_3(z) = M(N(N(z)))$, and so on. From these p.g.f. the average number of neighbors can be computed by evaluating their first derivative w.r.t. z in $z = 1$.

As such, each probability distribution $\{s_i\}$ induces an *interconnection cost* that we define as the average number of P2P networks a randomly chosen node belongs to:

$$f = F'(1). \tag{1}$$

3.3 Search Algorithm

To model a flooding-based search in the interconnected system, we consider the set of probabilities $\{p_f(i)\}$, where $i \in \{1, \dots X\}$. A peer belonging to i overlays sends/forwards a query to a particular neighbor with probability $p_f(i)$, where $i \in \{1, \dots X\}$). Therefore, $\{p_f(i)\}$ *is not* a probability distribution.

We denote as q_h the probability that h first hop neighbors received a query from the peer that started the search. If the peer belongs to i overlays, it sends a query to one of its neighbors with probability $p_f(i)$. Therefore, the number of neighbors that receive the query follows a binomial distribution with parameter $p_f(i)$. Therefore, it is well known that the probability distribution $\{q_h\}$ has p.g.f. given by [3] $Q(z) = \sum_{i=1}^{X} s_i G_0^i(1 + p_f(i)(z - 1))$.

Similarly, for the p.g.f. of the probability distribution describing the number of queries sent by a node reached by following a randomly chosen edge, we obtain:

$$R(z) = \sum_{i=1}^{X} s_i \, G_1(1 + p_f(i)(z - 1)) \, G_0^{i-1}(1 + p_f(i)(z - 1)). \tag{2}$$

If we denote as $Q_t(z)$ the p.g.f. for the probability distribution of the number of neighbors t hops away from a randomly chosen peer that received a query, we have that: $Q_1(z) = Q(z)$, $Q_2(z) = Q(R(z))$, and $Q_3(z) = Q(R(R(z)))$, etc. As a special case, we may consider constant forwarding probabilities, *i.e.* $p_f(i) = p_f, \forall i \in \{1, \dots X\}$. In this case, we would obtain: $Q(z) = M(1+p_f(z-1))$ and $R(z) = N(1 + p_f(z - 1))$. Since the p.g.f. of the probability distribution of the sum of independent random variables is given by the product of the corresponding p.g.f., the total number of queries generated by a search issued by a randomly chosen peer is described by: $T(z) = \prod_{t=1}^{TTL} Q_t(z)$ yielding the *average number of queries*

$$m = T'(1). \tag{3}$$

3.4 Hit Probability

We model resource popularity by $0 \leq \alpha \leq 1$ that is the average fraction of peers that globally hold the given resource. We interpret this parameter as the probability that a randomly chosen node owns a copy of the resource.

If we denote as w_h the probability that h first hop neighbors hold a copy of the requested resource *and* received a query from a peer that belongs to i overlays we note that the number of such neighbors follows a binomial distribution with parameter $\alpha p_f(i)$. If we denote as $H_t(z)$ the p.g.f. for the probability distribution of the number of neighbors t hops away from a randomly chosen peer that received a query *and* hold a copy of the requested resource then we have that: $H_1(z) = Q_1(1 + \alpha(z-1))$, $H_2(z) = Q_2(1 + \alpha(z-1))$, $H_3(z) = Q_3(1 + \alpha(z-1))$, and so on. Therefore, the total number of search hits is described by a probability distribution whose p.g.f. is given by: $H(z) = \prod_{t=1}^{TTL} H_t(z)$ yielding the search *hit probability*

$$p_{hit} = 1 - H(0). \tag{4}$$

3.5   A Variation of the Search Algorithm

To model a search algorithm where peers that own a copy of the resource do not forward a query message it suffices to redefine $R(z)$ in Equation 2. In particular, when a peer owns a copy of the resource the number of its neighbors that receive the query is equal to 0: this happens with probability α. In Equation 5 this is represented by the term α that can be written as $\alpha p_0 z^0$ with $p_0 = 1$. With probability $1 - \alpha$ Equation 2 holds, therefore we obtain the p.g.f. of the probability distribution describing the number of queries sent by a node reached by following a randomly chosen edge as:

$$R(z) = \alpha + (1-\alpha) \sum_{i=1}^{X} s_i G_1(1 + p_f(i)(z-1)) G_0^{i-1}(1 + p_f(i)(z-1)). \tag{5}$$

The definition of $Q_t(z)$, and $T(z)$, and m remains unchanged.

4   Results

In this section, we will first show the results of the model validation, performed via a heavily multi-threaded simulator, written in Erlang, that reproduces, in terms of message routing, the exact behavior of a system described by our model. Also, we will show the results of some broad system evaluations made possible by the use of our model to compute metrics that would otherwise, if performed by means of simulations, require too much in terms of simulation time and computational power.

In our analysis, we consider different routing policies that can be employed in our scenarios, modeled by defining the $p_f(i)$ mentioned in Section 2. Those are:

- $p_f(i) = \dfrac{1}{i}$, henceforth referred to as *1/i*, i.e. the probability of selecting a neighbor is inversely proportional to the number of overlays a node is connected to. This routing tends to maintain a constant number of messages, but "flattens" the interconnected topology, not allowing synapse nodes to exploit the extended neighborhood.
- $p_f(i) = min(1, \dfrac{z_{max}}{zi})$, henceforth referred to as *zmax*, where $z = E[\{p_k\}]$ is the average number of neighbors for a node based on the current degree distribution and z_{max} is a system parameter, specified upon design, indicating

the upper bound for the average number of forwarded messages. This policy allows for a better exploitation of Synapse nodes, while still finely limiting the number of messages in the system. In our evaluations, z_{max} has been set to $2z$, twice the average number of neighbors per node.

- $p_f(i) = 1$, henceforth referred to as *flood*, i.e. a routing where every node selects forwards a message to *every* neighbor, regardless of the number of connected overlays.

In both simulations and evaluations, the individual overlays have been modeled following the neighbors degree distribution measured in [4] from real world applications and used already in [5], in order to have an accurate overlay model.

4.1 Model Validation

In order to evaluate the accuracy of our model in predicting the performance indexes of a real network, we validated the obtained results by means of simulation. The simulator employs standard statistical procedures to estimate 68% and 95% confidence intervals for the p_{hit} and m indexes defined in Section 3.

Simulation Methodology. The simulator has been developed from scratch in Erlang. The choice of Erlang has been driven by its native multi-threading capabilities and inter-process communication model based on the message passing paradigm embedded in the language, thus allowing for a rapid implementation of an accurate network model made of node processes running independently and exchanging messages with one another. Each process has a list of other processes it can exchange messages with, that constitutes its neighborhood.

We consider N_s independent realizations for the interconnected overlay topologies (in our experiments $N_s = 30$); each interconnected topology is used to obtain one realization of m and p_{hit}. The hth realization is obtained as follows:

- We first generate a new topology, made of X overlays interconnected by synapse nodes, using as input parameter the number of nodes $N = 500000$, the nodes degree distribution $\{p_k\}$ [4], and the $\{s_i\}$ to be validated;
- From the generated topology file, the simulator instantiates N node processes and assigns each the corresponding list of neighbors;
- One or more resources are then seeded in the system, according to their respective popularity α, by sending a PUT(value) message to $N\alpha$ random nodes;
- Separate worker processes take care of sending a query message SEARCH(value,TTL) to each node process in the network.
- Meanwhile, a listener process receives then the responses, either the resource being found or the TTL being reached, and of computes the statistics.

Topology Generation. The generation of a network made of interconnected overlays mainly consists of generating first X individual overlay topologies, and then connecting them by "merging" nodes from different overlays in one Synapse

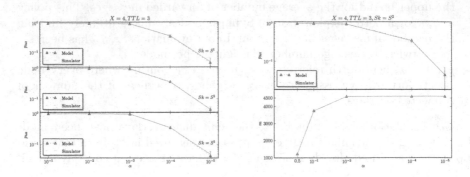

Fig. 2. p_{hit} for different α and s_i (left) and alternative search algorithm (right)

node, thus creating nodes with extended neighborhoods spanning across all the connected overlays. In order to generate X random graphs with a specified degree $\{p_k\}$ we relied on the algorithm presented in [6], that provides short generation times while guaranteeing the respect of the specified degree.

Validation Results. The first validation we performed was conducted for a system with only one overlay ($X = 1$). For the sake of brevity we only show the results for the *flood* routing strategy, $\alpha = 0.0001$, and $TTL = 3$. Table 1 shows the model is very accurate and faithfully predicts results when compared to the simulation output.

We then validated various scenarios with a higher number of interconnected overlays ($X = 4$), at $TTL = 3, 4$ and with different values of α, different routing policies and different distributions $\{s_i\}$. We considered the distribution for the degree of synapses summarized in Table 2.

Figure 2 (left) shows a comparison between the computed p_{hit} for different values of α and the corresponding simulation results, while Table 3 summarizes the same comparison for m. The results show that both performance metrics fall within the confidence interval of the simulation results.

Furthermore, we validate the system against the alternative search algorithm detailed in Section 3.5. For the sake of brevity, we are showing results only for S^2 since the same conclusions can be drawn for S^1 and S^3. Figure 2 (right) shows both p_{hit} and m against different values of α, since with this algorithm the number of message is dependent of the resource popularity. Even in this scenario, the model results fall within the confidence interval estimated by the simulator.

Table 1. m for different s_i distributions: comparison between model and simulation

	Model	Simulation (95% C.I.)
p_{hit}	0.3733	0.373552 ± 0.003852
m	4822.63	4821.57 ± 0.0498

Therefore, we can safely conclude that our model is accurate in predicting the behavior of the performance indexes in a broad range of different scenarios. Furthermore, while simulations required hours of CPU time to complete solving our model took less than one second with a solver implemented in C.

4.2 Model Exploitation

After validating the model we conducted a few analysis to show its usefulness in the design phase of the interconnection of several peer-to-peer networks.

Comparison of Different Routing Policies. A first evaluation concerns the choice of a specific routing policy in the system, i.e. the definition of different $p_f(i)$. In this case, we want to compare for values of α down to 10^{-6}, the performances in terms of p_{hit} and m for the distribution of degree of synapses S^1 (results for the other two distributions suggested similar considerations and are omitted for the sake of brevity), $X = 10$, and $TTL = 3$. Please note that to achieve a reliable measurement via simulation for $\alpha = 10^{-6}$ we would need to conduct complex simulations (at least 1000000 nodes) for a long simulation time (ideally each of them to be queried individually for multiple topology realizations).

Figure 3 show the values of p_{hit} for the 3 different policies and different resource popularities, while Figure 4 depicts the average number of messages for the 3 policies in the case of propagation of queries up to TTL hops (Figure 4b) and for the query propagation that stops when reaching a node holding a copy of the resource (Figure 4a) modeled in Section 3.5. In the former case, the number of messages is independent of the resource popularity while in the latter case we note that reduction of the number of query messages can be obtained for popular resources, i.e., for $\alpha > 0.01$.

In this case, the model allows for a simple cost/benefit evaluation, based on the expected popularity of a resource. For one, we can notice an almost tenfold increase in the number of messages between the *zmax* and the *flood* policy, to which it does not correspond a proportional increase in the p_{hit}.

f-Cost Based Evaluation. In a cost/benefit analysis of the interconnected system, we consider p_{hit} as our benefit metric whereas m and f are considered as costs. Another kind of evaluation we performed consists of fixing the f cost and analyzing which distributions $\{s_i\}$ lead to better performances (p_{hit}) and minimum cost (m).

To this end we considered all distributions $\{s_i\}$ that can be defined for $X = 5$ where the individual probabilities are non-zero multiple of 0.05. We considered

Table 2. Definition of the $\{s_i\}$ distributions used for validation

S^1	$s_1 = 0.7, s_2 = 0.1, s_3 = 0.1, s_4 = 0.1$
S^2	$s_1 = 0.4, s_2 = 0.3, s_3 = 0.2, s_4 = 0.1$
S^3	$s_1 = 0.1, s_2 = 0.2, s_3 = 0.3, s_4 = 0.4$

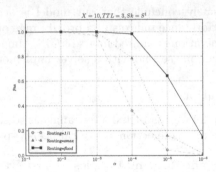

Fig. 3. Routing policies comparison: p_{hit} for different resource popularities α

3 values of f (namely, $f = 2, 3, 4$) and compared the performances of every distribution $\{s_i\}$ with given f for $TTL = 2$. Again, please note that this analysis would have required days of CPU time to be completed by means of simulation since even with a coarse granularity in the definition of $\{s_i\}$ (0.05) we tested hundreds of different distributions. This analysis required only a few seconds to complete with our model.

Figures 5a and 5b show a subset of these distributions (each point in the graph corresponds to a particular distribution $\{s_i\}$). We only plotted the ones with the highest p_{hit}; it appears that the interconnection cost f alone is not directly bound to an increase in performances. There are, as a matter of fact, different configurations with $f = 3$ that perform equally (sometimes very slightly better) than those with a $f = 4$. Furthermore, within the configuration with $f = 2$ some are better than others in terms of performance and costs. Nevertheless, a clear relation exists between message cost m and p_{hit}: the larger the average number of messages the higher the p_{hit}.

The behavior shown in the figures can be explained as following: the routing policy $zmax$ limits the number of messages that can be issued by a node to z_{max}, which is set in our evaluations to $2z$. Therefore, increasing the number connections in the interconnected system (f) beyond certain values does not lead to a significant performance increase. That is why we observe a proportionally higher increase in the p_{hit} from $f = 2$ to $f = 3$ than from $f = 3$ to $f = 4$.

Effects of Granularity. Another aspect we analyze is a performance comparison as the number of overlays to interconnect increases. In this case we chose

Table 3. m for different s_i distributions: comparison between model and simulation

	Model	Simulation (95% C.I.)
S^1	4598.02	4596.77 ± 2.38
S^2	4701.82	4700.96 ± 0.49
S^3	4449.57	4453.58 ± 3.41

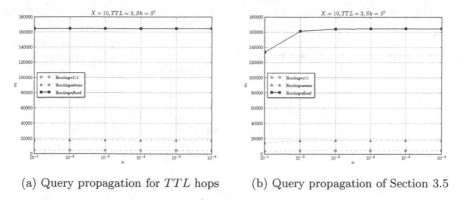

(a) Query propagation for TTL hops   (b) Query propagation of Section 3.5

Fig. 4. Average number of messages for different routing policies

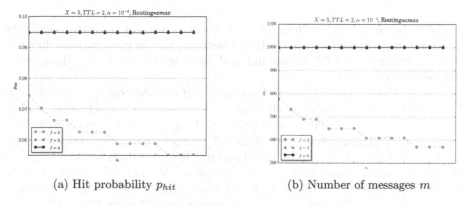

(a) Hit probability p_{hit} (b) Number of messages m

Fig. 5. s_i comparison at different f

to analyze the behavior of the *zmax* routing policy, in a system with $TTL = 3$ and $\alpha = 0.0001$, for an increasing number of overlays (X) and for different distributions $\{s_i\}$, characterized by an increasing percentage of non-synapse nodes s_1, while the remainder of the distribution is equally distributed across the remaining s_i.

Figures 6a and 6b show four different configurations, with an increasing number of non-synapse nodes in the system. The parameter s_1 indicates the share of non synapses nodes, while the remaining part $(1 - s_1)$ is equally distributed among the remaining $X - 1$ values, i.e., $s_i = \frac{1-s_1}{X-1}$ for $1 < i \le X$. It can be noted that at each given ratio of synapses vs non-synapses nodes the system behavior is roughly the same regardless the number of overlays. The efficiency is still tightly bound to the number of messages and both increase as s_1 decreases.

System Design with Minimum Requirements. Thanks to the high number of different configurations that can be evaluated with our model in a relatively short time, we conduct a further analysis to support the design of the interconnection of several peer-to-peer networks.

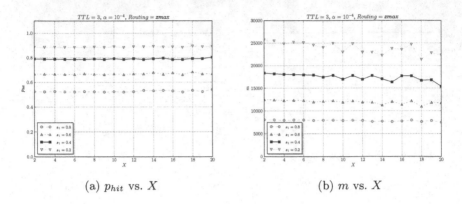

(a) p_{hit} vs. X (b) m vs. X

Fig. 6. Performance evaluation with different numbers of overlay X

For instance, we set the number of overlays X and the resource popularity α; by setting a bound for the minimum desired p_{hit}, we can compare different routing policies and TTL values and find the one that minimizes the average number of messages m.

Figures 7 and 8 (left) show a classification of distributions $\{s_i\}$ for two different routing policies and two different TTL values with respect to p_{hit} and m for $X = 10$ and $\alpha = 0.0001$ (each point in the graphs represents a particular distribution $\{s_i\}$). In the first case (Figure 7), we decided to fix a cost factor and set $f = 4$, whereas in the second case (Figure 8 left), the fixed factor is the ratio of expected non-synapse nodes in the system s_1. We are able to discriminate immediately those distributions $\{s_i\}$ that do not satisfy the imposed criteria of having $p_{hit} > 0.9$. We also discriminate among those that do the distributions $\{s_i\}$ that minimize the number of messages m, as shown in Figure 7b.

Routing without Propagation. We briefly present some evaluation results based on the model variation presented in Section 3.5. In the first version of our model, the routing of a message is assumed to continue until the TTL expires, regardless of a resource being found or not. This leads to an $H_t(z)$ able to describe different cases, such as the probability of finding *multiple* copies of a resource. However the system is not optimal message-wise. In case we are interested only in the first hit of a search query, and we want to optimize the number of messages employed, with the variant of $R(z)$ described in Section 3.5 we are able to evaluate the system under the conditions that the routing in a node stops whenever a resource is found.

Figure 8 (right) shows the trend of m for different α, and two routing policies for $X = 10$, $TTL = 3$, and distribution S^1. While the number of messages was unrelated to the resource popularity before, here we see that, as routing stops upon first hit, the more popular a resource, the lower the number of messages per query.

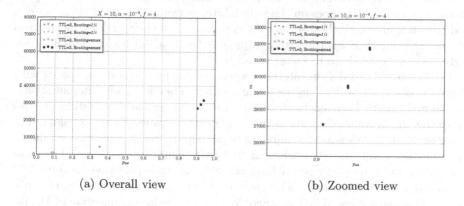

(a) Overall view   (b) Zoomed view

Fig. 7. Distribution of different routing policies with fixed f

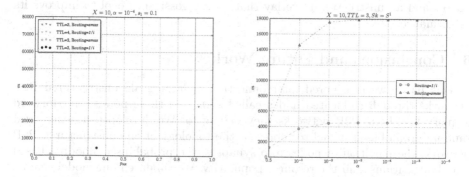

Fig. 8. Distribution of different routing policies with fixed s_1 (left) and message evaluation at different α, for different routing policies (right)

5   Related Work

Inter-cooperation of network instances has been identified in [7, 8] as one of the future trends in the current Internet architecture development. When discussing logical networks, various techniques to achieve inter-communication among them have been presented.

Synergy [9] is an architecture for the inter-cooperation of overlays which provides a cooperative forwarding mechanism of flows between networks in order to improve delay and throughput performances. Co-located nodes are, in the authors' opinion, good candidates for enabling such mechanisms and reduce traffic.

With a similar goal, authors in [10] propose algorithms tailored to file sharing applications, enabling a symbiosis between different overlays networks. They present hybrid P2P networks cooperation mechanisms and provide interesting observations on the appropriate techniques to perform network join, peer selection, network discovery, etc. Their simulations showed the effect of the popularity of a cooperative peer on the search latency evaluation, that is the more a node has neighbors, the better, as well as the effect of their caching mechanism which reduces (when appropriately adjusted) the load on nodes (but interestingly does not contribute to faster search).

Authors in [11] model an interconnected system by considering spaces with some degree of *intersection* between one another. They focus on different strategies to find a path between two overlays, and compare various routing policies analyzing which trade-offs lead to the best results. Trade-offs are considered in terms of number of messages, number of hops to find a result and state overhead. They provide a comparative analytical study of the different policies. They show that with some dynamic finger caching and with multiple gateways tactfully laid out in order to avoid bottlenecks due to the overload of a single gateway, they obtain good performances. Their protocol focuses on the interconnection of DHTs, while we focus on unstructured overlays.

Finally, [12] studies the co-existence of multiple overlay networks, namely Pastry and an unstructured overlay that uses a gossip protocol to improve its performance.

6   Conclusions and Future Work

In this paper we considered interconnection of large scale unstructured P2P networks through co-located nodes called synapses: these nodes send/forward a query to all the P2P networks they belong to. We developed a generalized random graph based model to represent the topology of one unstructured P2P network, the partition of nodes into synapses, the probabilistic flooding based search algorithms, and the resource popularity. We validated our model against simulations and proved that its predictions are reliable and accurate. The model allowed the analysis of very large and complex systems: we believe that simulation and/or prototype deployment based analysis would be unfeasible in this case.

We are currently working to further extend our model in several directions. In particular, we are generalizing equations to represent heterogeneous topologies and resource availability. As a consequence, we are also extending the analysis to more refined partition of synapses, i.e., to consider the fraction of nodes that belong to a specific set of P2P networks. Furthermore, we are extending the model to represent nodes availability due to churning. Last but not least, we are generalizing the model to represent interconnection of both unstructured and structured P2P networks.

References

1. Liquori, L., Tedeschi, C., Vanni, L., Bongiovanni, F., Ciancaglini, V., Marinković, B.: Synapse: A scalable protocol for interconnecting heterogeneous overlay networks. In: Crovella, M., Feeney, L.M., Rubenstein, D., Raghavan, S.V. (eds.) NETWORKING 2010. LNCS, vol. 6091, pp. 67–82. Springer, Heidelberg (2010)

2. Ciancaglini, V., Gaeta, R., Liquori, L., Loti, R.: Modeling and analysis of large scale interconnected unstructured p2p networks. In: Proc. of ICPADS 2012, pp. 710–711 (2012)
3. Newman, M.E.J., Strogatz, S.H., Watts, D.J.: Random graphs with arbitrary degree distributions and their applications. Phys. Rev. E 64(2), 026118 (2001)
4. Bolla, R., Gaeta, R., Magnetto, A., Sciuto, M., Sereno, M.: A measurement study supporting p2p file-sharing community models. Computer Networks 53(4), 485–500 (2009)
5. Gaeta, R., Sereno, M.: Generalized probabilistic flooding in unstructured peer-to-peer networks. IEEE Transactions on Parallel and Distributed Systems 22(12), 2055–2062 (2011)
6. Viger, F., Latapy, M.: Efficient and simple generation of random simple connected graphs with prescribed degree sequence. In: Wang, L. (ed.) COCOON 2005. LNCS, vol. 3595, pp. 440–449. Springer, Heidelberg (2005)
7. Siekkinen, M., Goebel, V., Plagemann, T., Skevik, K., Banfield, M., Brusic, I.: Beyond the Future Internet–Requirements of Autonomic Networking Architectures to Address Long Term Future Networking Challenges. In: International Workshop on Future Trends of Distributed Computing Systems, pp. 89–98 (2007)
8. Fonte, A., Curado, M., Monteiro, E.: Interdomain Quality of Service Routing: Setting the Grounds for the Way Ahead. Annals of Telecommunications 63(11), 683–695 (2008)
9. Kwon, M., Fahmy, S.: Synergy: an Overlay Internetworking Architecture. In: Proc. of International Conference on Computer Communications and Networks, pp. 401–406 (2005)
10. Junjiro, K., Naoki, W., Masayuki, M.: Design and Evaluation of a Cooperative Mechanism for Pure P2P File-Sharing Networks. IEICE Trans. Commun. (Inst. Electron. Inf. Commun. Eng.) 89(9), 2319–2326 (2006)
11. Furtado, P.: Multiple Dnamic Overlay Communities and Inter-sace Routing. In: Moro, G., Bergamaschi, S., Joseph, S., Morin, J.-H., Ouksel, A.M. (eds.) DBISP2P 2005/2006. LNCS, vol. 4125, pp. 38–49. Springer, Heidelberg (2007)
12. Maniymaran, B., Bertier, M., Kermarrec, A.: Build one, get one free: Leveraging the coexistence of multiple p2p overlay networks. In: Proc. of IEEE ICDCS 2007, pp. 33–40 (2007)

Energy-Efficient Operation of a Mobile User in a Multi-tier Cellular Network

Olga Galinina[1], Alexey Trushanin[2], Vyacheslav Shumilov[2],
Roman Maslennikov[2], Zsolt Saffer[3],
Sergey Andreev[1], and Yevgeni Koucheryavy[1]

[1] Tampere University of Technology (TUT), Finland
{olga.galinina,sergey.andreev}@tut.fi, yk@cs.tut.fi
[2] Lobachevsky State University of Nizhny Novgorod (UNN), Russia
{alexey.trushanin,vyacheslav.shumilov,roman.maslennikov}@wcc.unn.ru
[3] Budapest University of Technology and Economics (BUTE), Hungary
safferzs@hit.bme.hu

Abstract. In this paper[1], we propose a new power control scheme suitable for a multi-tier wireless network. It maximizes the energy-efficiency of a mobile device transmitting on several communication channels while at the same time ensures the required minimum quality of service. As the result, a good compromise between improving the data rate and extending the battery lifetime is provided. In order to enable energy-efficiency maximization, we formulate an optimization problem basing on the Shannon's capacity formula. The optimal transmit power is thus obtained from the direct solution of this optimization problem under several practical constraints, such as minimum bitrate and maximum transmit power. In the second part of the paper, we apply extensive simulations to calibrate the key parameters of our optimization framework. The numerical results suggest the benefit of the proposed analytical solution by comparing it against intuitive (heuristic) power control strategies.

1 Introduction and Background

Wireless cellular networks have experienced essential growth over the last decades, eventually becoming an integrated part of our daily lives [1]. As market analysts predict, this steady development is expected to continue over the following years [2]. Hence, it comes as no surprise that users are increasingly interested in extending functionality of their mobile devices to run more demanding applications. The resulting advent of the high-rate fourth generation (4G) communication technologies combined with a wide variety of new mobile devices and services brings substantial increase in the amounts of user-generated data [3]. This, in turn, implies higher power consumption when transmitting this data, which may be harmful for the battery-powered mobile devices [4]. As a result,

---

[1] Part of this work had been completed when Alexey Trushanin and Vyacheslav Shumilov were on a research visit at Tampere University of Technology, Finland.

A. Dudin and K. De Turck (Eds.): ASMTA 2013, LNCS 7984, pp. 198–213, 2013.
© Springer-Verlag Berlin Heidelberg 2013

the gap between the user's need for higher data rates and the battery lifetime restrictions of small-scale user equipment grows considerably [5].

To address this gap, wireless industry reacts with a selection of solutions ranging from device battery innovation to advanced network architecture [6]. The latter suggests the use of small cells to augment the conventional cellular layout. Such integrated multi-tier deployments offer decisive benefits to the indoor user's connectivity, as well as in the areas with limited cellular coverage [7]. For example, a user in a two-tier in-building network (see Figure 1), e.g. in a shopping mall or in an office building, may receive improved service from the infrastructure low-power nodes (LPNs). However, when traveling from a small cell to a small cell, this user may also suffer from extra signaling when selecting the best LPN to transmit to. Furthermore, frequent cell re-selections may lead to excessive power consumption and drain the device battery. This problem receives increasing attention from wireless community [8], which recognizes the need for improved device power management mechanisms that would explicitly target small cell deployments.

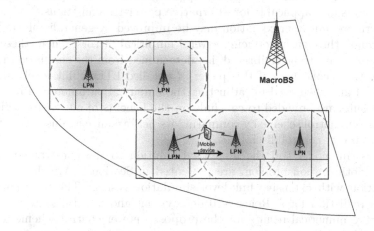

Fig. 1. Example topology of a multi-tier network

Whereas there has been much work on power control schemes for conventional cellular networks [9], it becomes crucial to address and account for the specific features of multi-tier networks. We believe that by intelligently allocating power on the available communication channels, a mobile user may considerably reduce its power consumption, while not compromising its desired quality of service. In this work, we propose a candidate power control strategy suitable for multi-tier wireless networks. We seek to maximize the energy efficiency of a user device to reach good balance between the required data rate and the resulting energy consumption.

More specifically, we consider a single user device which may *simultaneously* transmit its data to several neighboring LPNs centered at the surrounding small cells (as may be possible in future beyond-4G networks). Our goal is to advise this device on the optimal transmit power levels for each of the available LPN connections (which we term channels). We choose energy-efficiency as the optimization criterion, which is given as the relation between the data rate and the corresponding power expenses. Furthermore, we account for several important practical constraints of mobile user operation, such as the minimum required bitrate and the maximum allowed transmit power, which directly leads to an inequality-constrained optimization problem.

The energy-efficiency optimization at hand is based on the relationship between the transmit power and the resulting data rate. Instead of the actually achievable data rate, we study its theoretical maximum, that is, the capacity of a communication system. Here, the Shannon's capacity formula [10] for the channel with additive white Gaussian noise is the most widely used and popular approach. It helps abstract away the specific transmitter and receiver structures and, therefore, can be applied to most contemporary wireless networks, such as UMTS HSPA/HSPA+, WiMAX, 3GPP LTE/LTE-A, etc. In what follows, we choose 3GPP LTE-A (Long Term Evolution Advanced) as our example 4G technology [11] and demonstrate that its performance is reasonably close to the Shannon's limit, so that our capacity approximation is very precise. However, our solution is also applicable for alternative power-rate functions.

Furthermore, our approximation may be improved by generalizing the Shannon's formula through introducing several empirical factors. Then, these additional factors have to be calibrated, i.e. determined from the simulation results. Hence, in the second part of this paper, a detailed LTE-A link-level simulator is described and then used for adjusting the empirical parameters. Finally, numerical results are provided to conclude on the benefits of our energy-efficiency centric power control scheme by comparing its performance against two intuitive (heuristic) power control disciplines.

Our system model and the analytical solution to the constrained energy-efficiency optimization problem are presented in Section 2. We then augment this solution with extensive link-level simulation results. The used simulator mimics a realistic LTE-A Release-10 deployment and is detailed in Section 3. Further, the numerical results for the proposed power control scheme and the competitor heuristic strategies are given in Section 4. Finally, Section 5 concludes this paper.

2   Energy-Efficiency Optimization Problem

2.1   System Model and Assumptions

We consider an uplink data transmission of a single user device in a multi-tier wireless cellular network and study its achievable data rate, power, and energy-efficiency. This device may simultaneously use up to K available communication channels to the neighboring LPNs. In our model, every channel $i = \overline{1, K}$ may have different properties, and we only assume that the channels are mutually non-interfering. This may correspond to the practical scenario when the adjacent small cells are allocated non-overlapping radio frequencies, which is often the case in real-world deployments.

We also assume that the application-layer traffic of the considered user is saturated. The achievable data rate on the channel i is determined by the properties of the channel and by how much power is allocated by the user to transmit on it. The total user data rate can thus be obtained by aggregating the individual rates $r = \sum_{i=1}^{K} r_i$, where r_i is the data rate on the channel i. We impose a constraint on the total data rate r such that it must not drop below the minimum bitrate requirement r_0 given by e.g. a particular mobile application.

The total power consumption of the user device equals $p = p^{tx}(\mathbf{r}) + p^c = p^{tx}(\mathbf{r}) + \sum_{i=1}^{K} p_i^c$, where $p^{tx}(\mathbf{r})$ is the transmit power which is determined by a particular vector $\mathbf{r} = (r_1, ..., r_K)$ and p_i^c is the constant circuit power component (incurred by the active electronic circuitry) for the channel i. Further, we assume that the transmit power $p^{tx}(\mathbf{r})$ can also be aggregated over the individual powers $p^{tx}(\mathbf{r}) = \sum_{i=1}^{K} p_i^{tx}(r_i)$.

We introduce a variation of the Shannon's capacity formula, which would give us relationship between the transmit power and the maximum achievable data rate as:

$$p_i^{tx} = A_i(2^{B_i c_i} - 1), \tag{1}$$

where p_i^{tx} is the transmit power on the channel i, c_i is the theoretical capacity, while A_i and B_i are the additional parameters depending on wireless system implementation and configuration (including signal transmission mode, implementation-specific parameters, etc.). These can be given as:

$$B_i = \frac{1}{w_i}, \quad A_i = \frac{N_i}{g_i}, \tag{2}$$

where w_i is the channel bandwidth, $g_i = \rho_i/PL$ is the corresponding power gain, PL is path loss, ρ_i is the antenna gain and N_i is the total noise power over the given bandwidth. The coefficient B_i can also account for the overhead of the pilot signals, cyclic prefixes, and control channels occupying a portion of the system resources: $B_i = \frac{T_{total}}{w_i T_{data}}$, where T_{total} is the total amount of orthogonal (time-frequency) resources and T_{data} is the amount of resources allocated for the data transmission, w_i is the pure data channel (so-called PUSCH channel in LTE-A) bandwidth without guard bands and control channels.

The Shannon's formula relates the transmit power to the theoretical capacity, which may in reality differ from the actually achievable data rates. This may be due to the limited user knowledge about the wireless environment and thus non-optimal selection of modulation and coding schemes. In order to bring the relation (1) closer to the actual LTE-A system performance, we also introduce additional empirical factors α and β as:

$$B_i = \beta \frac{T_{total}}{w_i T_{data}}, \quad A_i = \alpha \frac{N_i}{g_i}. \tag{3}$$

Accounting for the above factors, we formally replace c_i by r_i, p_i^{tx} by p_i in (1) to summarize the considered relationship between the transmit power and the achievable data rate as:

$$p_i = A_i(2^{B_i r_i} - 1). \tag{4}$$

We name expression (4) the generalized Shannon's formula and give it as:

$$r_i = \frac{1}{B_i} log_2 \left(1 + p_i \frac{1}{A_i} \right). \tag{5}$$

We also note that the considered power-rate function (5) is bijective and monotonic with a continuous derivative, which will be used by the further analysis.

2.2 Optimization Problem

Our goal of energy-efficiency maximization can be achieved with the appropriate power control. Below, we formulate the constrained optimization problem where the argument is the achievable data rate on each available communication channel (which is equivalent to the corresponding transmit power).

We define and further optimize the energy efficiency of a user as the ratio between the total data rate r and the total power p:

$$\eta(\mathbf{r}) = \frac{r}{p} = \frac{\sum_{i=1}^{K} r_i}{\sum_{i=1}^{K} p_i(r_i) + \sum_{i=1}^{K} p_i^c}. \tag{6}$$

Further, we formulate our energy efficiency $\eta(\mathbf{r})$ optimization problem:

$$\max_{\{r_i\}_{i=1}^{K}} \eta(\mathbf{r}) = \max_{\{r_i\}_{i=1}^{K}} \frac{\sum_{i=1}^{K} r_i}{\sum_{i=1}^{K} p_i(r_i) + \sum_{i=1}^{K} p_i^c}, \text{ subject to:} \tag{7}$$

$$r = \sum_{i=1}^{K} r_i \geq r_0, \tag{8}$$

where r_0 is the minimum required bitrate. We also impose a reasonable constraint on the achievable data rate r_i (and hence, p_i), so that it cannot be negative:

$$r_i \geq 0, i = \overline{1, K}, \tag{9}$$

and, finally, account for the maximum allowed transmit power limit as:

$$p_i(r_i) \leq p_i^{max}, i = \overline{1, K}.$$

Note that since the function $p_i(r_i)$ is bijective, the above can be written as:

$$r_i \leq r_i^{max}, i = \overline{1, K}, \tag{10}$$

where $r_i^{max} = r_i(p_i^{max})$ can be calculated from (5).

For LTE-A system, the maximum throughput value can be expressed as:

$$\tilde{r}_i^{max} = w_i \frac{T_{data}}{T_{total}} N_{bits}, \tag{11}$$

where N_{bits} is number of bits per symbol for the highest supported modulation.

Therefore, the maximum data rate should actually be calculated as $r_i^{max} = min(\tilde{r}_i^{max}, r_i(p_i^{max}))$. We also take into account that $\eta(r_1, ..., r_K)$ is a non-zero function bounded on the interval $[\mathbf{0}, \infty)$, where $\mathbf{0}$ is a zero vector $(0, ..., 0)$. Now considering an equivalent form of (7) and rearranging the above inequalities (8), (9), as well as (10), our optimization problem can be summarized as:

$$\min_{\{r_i\}_{i=1}^K} U(\mathbf{r}) = \min_{\{r_i\}_{i=1}^K} \frac{1}{\eta(\mathbf{r})} = \min_{\{r_i\}_{i=1}^K} \frac{\sum_{i=1}^K p_i + \sum_{i=1}^K p_i^c}{\sum_{i=1}^K r_i} \tag{12}$$

subject to the constraints:

$$\phi(\mathbf{r}) = r_0 - \sum_{i=0}^K r_i \leq 0, \tag{13}$$

$$f_i(r_i) = -r_i \leq 0, i = \overline{1, K}, \tag{14}$$

$$g_i(r_i) = r_i - r_i^{max} \leq 0, i = \overline{1, K}. \tag{15}$$

2.3   General Way of Solving the Optimization Problem

The objective function given by (12)–(15) constitutes an inequality-constrained optimization problem. The general way of solving such optimization problem may be described by applying the Karush-Kuhn-Tucker (KKT) approach [12]. Accordingly, a system of equations and inequalities can be set up, known as regularity KKT conditions. For our optimization problem, the regularity KKT conditions are given as:

$$\frac{\partial U(\mathbf{r})}{\partial r_i} + \sum_{i=1}^K \lambda_i \frac{dg_i(r_i)}{dr_i} + \sum_{i=1}^K \mu_i \frac{df_i(r_i)}{dr_i} + \gamma \frac{d\phi(\mathbf{r})}{dr_i} = 0 \Leftrightarrow$$

$$\Leftrightarrow \frac{\frac{dp_i}{dr_i} \cdot r - (\sum_{i=1}^K p_i + \sum_{i=1}^K p_i^c)}{r^2} + \lambda_i - \mu_i - \gamma = 0, \quad i = \overline{1, K}$$

$$\lambda_i(r_i - r_i^{max}) = 0, r_i - r_i^{max} \leq 0, \lambda_i \geq 0, i = \overline{1, K}, \tag{16}$$

$$\gamma \left(\sum_{i=1}^K r_i - r_0 \right) = 0, \sum_{i=1}^K r_i - r_0 > 0, \gamma \geq 0,$$

$$\mu_i r_i = 0, r_i \geq 0, \mu_i \geq 0, i = \overline{1, K},$$

where λ_i, μ_i and γ are KKT multipliers.

Thus, in order to establish the optimal solution to the considered constrained optimization problem, the system of $3K + 1$ equations under $4K + 2$ inequalities has to be solved. We note that the search domain bounded by these inequalities has to be non-empty. Otherwise, the entire problem does not have a solution.

Noteworthy, the KKT conditions by themselves do not provide a method of finding the maximum/minimum points. Instead, they only determine the stationary points (where the gradient is zero) among which the minimum point can be found. In general, solving the system of equations and inequalities is known to be difficult. Therefore, instead of doing that, we apply a particular approach to solve the target optimization problem.

2.4   Particular Solution to the Optimization Problem

The solution to the system of equations:

$$\frac{\partial U(\mathbf{r})}{\partial r_i} = 0, i = \overline{1, K} \tag{17}$$

determines the optimum according to (12) without any inequality constraints. We begin with finding the stationary points for this unconstrained optimization problem in the domain $(-\infty, \mathbf{0}) \cup (\mathbf{0}, \infty)$. We then use this solution when dealing with the constrained optimization problem later.

Hence, we solve the target optimization problem in two steps

1. Solving the unconstrained optimization problem (12).
2. Updating the optimum by taking into account the constraints (13), (14), and (15) for each component of $\mathbf{r}$ step by step.

Optimal Solution without Constraints. In order to determine the stationary points of $U(\mathbf{r})$, we substitute the derivative $\frac{dp_i}{dr_i} = A_i B_i 2^{B_i r_i} \ln 2$ of function p_i (4) into (17). This results in:

$$\frac{\partial U(\mathbf{r})}{\partial r_i} = \frac{A_i B_i 2^{B_i r_i} \ln 2 \sum_{i=1}^{K} r_i - \left[\sum_{i=1}^{K} A_i (2^{B_i r_i} - 1) + p_c\right]}{\left(\sum_{i=1}^{K} r_i\right)^2} = 0, i = \overline{1, K}. \tag{18}$$

Therefore, we establish the following condition for the stationary points:

$$A_i B_i 2^{B_i r_i} \ln 2 \sum_{i=1}^{K} r_i - \sum_{i=1}^{K} A_i 2^{B_i r_i} + \left[\sum_{i=1}^{K} A_i - p_c\right] = 0, i = \overline{1, K}. \tag{19}$$

Rearranging (19) indicates that the term $A_i B_i 2^{B_i r_i} \ln 2$ does not depend on i:

$$A_i B_i 2^{B_i r_i} \ln 2 = \frac{\sum_{i=1}^{K} A_i 2^{B_i r_i} - \left[\sum_{i=1}^{K} A_i - p_c\right]}{\sum_{i=1}^{K} r_i}. \tag{20}$$

Further, we introduce the notation:

$$D = \frac{\sum_{i=1}^{K} A_i 2^{B_i r_i} - \left[\sum_{i=1}^{K} A_i - p_c\right]}{\sum_{i=1}^{K} r_i}. \tag{21}$$

Applying (20) and (21), the terms in (19) including the unknown values r_i can be expressed as:

$$A_i 2^{B_i r_i} = \frac{D}{B_i \ln 2}, \quad i = \overline{1, K}. \tag{22}$$

$$r_i = \frac{1}{B_i} \log_2 \frac{D}{A_i B_i \ln 2} = \frac{1}{B_i}[\log_2 D - \log_2(A_i B_i \ln 2)], \quad i = \overline{1, K} \tag{23}$$

Applying (22) and (23) in (19) and rearranging leads to:

$$\frac{D}{\ln 2} \sum_{i=1}^{K} \frac{1}{B_i} - D \sum_{i=1}^{K} \frac{1}{B_i} \log_2 D + D \sum_{i=1}^{K} \frac{1}{B_i} \log_2(A_i B_i \ln 2) = \left[\sum_{i=1}^{K} A_i - p_c \right].$$

We denote $\sum_{i=1}^{K} \frac{1}{B_i}$ and $\sum_{i=1}^{K} \frac{1}{B_i} \log_2(A_i B_i \ln 2)$ as B and G respectively. Applying these notations, we obtain:

$$DB \frac{1}{\ln 2} - DB \log_2 D + DG = \left[\sum_{i=1}^{K} A_i - p_c \right]. \tag{24}$$

Rearranging (24) yields:

$$DB \left(\ln \left[D \cdot 2^{-\left(\frac{1}{\ln 2} + \frac{G}{B}\right)} \right] \right) = \ln 2 \left[p_c - \sum_{i=1}^{K} A_i \right].$$

Let us also denote $D \cdot 2^{-\left(\frac{1}{\ln 2} + \frac{G}{B}\right)}$ as X:

$$X(\ln X) = \ln 2 \frac{\left[p_c - \sum_{i=1}^{K} A_i \right]}{B} 2^{-\left(\frac{1}{\ln 2} + \frac{G}{B}\right)}. \tag{25}$$

From equation (25), we may obtain the value of X and, consequently, the expression for D:

$$D = 2^{\left(\frac{1}{\ln 2} + \frac{G}{B}\right)} \cdot \exp \left(W \left(\ln 2 \frac{\left[p_c - \sum_{i=1}^{K} A_i \right]}{B} 2^{-\left(\frac{1}{\ln 2} + \frac{G}{B}\right)} \right) \right), \tag{26}$$

where $W(x)$ is the Lambert's function [13].

Applying (26) together with the definitions of G and B in (23) leads to the stationary point:

$$r_i^* = \frac{1}{B_i} \left[\frac{1}{\ln 2} + \frac{\sum_{i=1}^{K} \frac{1}{B_i} \log_2(A_i B_i \ln 2)}{\sum_{i=1}^{K} \frac{1}{B_i}} - \log_2(A_i B_i \ln 2) \right] +$$

$$+ \frac{1}{B_i \ln 2} W \left(\ln 2 \frac{\left[p_c - \sum_{i=1}^{K} A_i \right]}{\sum_{i=1}^{K} \frac{1}{B_i}} 2^{-\left(\frac{1}{\ln 2} + \frac{\sum_{i=1}^{K} \frac{1}{B_i} \log_2(A_i B_i \ln 2)}{\sum_{i=1}^{K} \frac{1}{B_i}} \right)} \right), \tag{27}$$

where A_i and B_i are given by (3). The power level to operate on a particular channel can finally be obtained by (4).

Optimization Under Constraints. Here we employ the solution $\mathbf{r}^* \in R^K$ of the unconstrained optimization problem (27) to the respective constrained problem in (12)–(15).

Several alternative cases are possible:

1. The argument of the function $W(x)$ in (26) is less than $-e^{-1}$, which means that the objective function of the unconstrained problem does not have a stationary point. Hence, we need to search the optimal point on the border of the domain by choosing the component r_i with the minimum contribution and setting this component to zero.

2. If $r_i^* \leq 0$ or $r_i^* \geq r_i^{max}$, then the stationary point lies outside the search domain and should be instead found on the respective plane $r_i^* = 0$ or $r_i^* = r_i^{max}$. If either of these conditions holds for several indexes, we need to choose the component r_i with the minimum contribution and set $r_i^* = 0$ or $r_i^* = r_i^{max}$.

3. If $\sum_{i=1}^{K} r_i^* < r_0$, then the optimal point lies on the plane $\sum_{i=1}^{K} r_i^* = r_0$ and we need to follow the above steps again.

Having fixed one of the components, we proceed by solving the respective optimization problem of dimension $K - 1$. It can be done in a similar way as solving the original problem $\mathbf{r}^*$ (we omit the details here due to the space constraints). The above steps are to be repeated until the set of components $\mathbf{r}^*$ is obtained which satisfy the given constraints.

3   Simulation Methodology

3.1   Description of the Simulator

After our energy-efficient power control scheme has been introduced, we aim to augment our solution with simulation results derived from a detailed model of 3GPP LTE-A system. Below we shortly describe the considered link-level simulator (LLS). In particular, we seek to apply the LLS tool for calibrating the empirical coefficients α and β in (3). Our approach is realistic modeling of (i) all the necessary transmitter operations of LTE-A Release-10 uplink, (ii) the radio channel with additive white Gaussian noise (AWGN), and (iii) all the corresponding receiver operations. The general structure of the LLS is presented in Figure 2.

A fragment of data in the form of a transport block (TB) is generated of random bits. It is then fed to the input of the transmitter part and passed through the stages of turbo encoding, rate matching [14], scrambling, and QAM mapping [15]. Turbo encoding with a fixed code rate of 1/3 is performed according to the LTE-A standard. The rate matching stage performs bit puncturing or repetition coding to match the original code rate of 1/3 to an arbitrary code rate. Puncturing and repetition patterns are implemented in full compliance with the LTE-A specifications. The scrambler performs modulus two addition of rate-matched bits with a random sequence specified by the standard, whereas

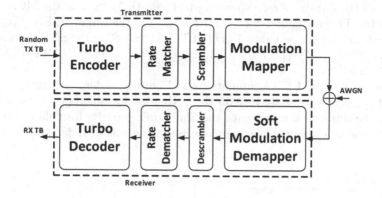

Fig. 2. Structure of the LTE-A link-level simulator

the mapper generates QPSK, 16-QAM, or 64-QAM constellations. An arbitrary modulation and coding scheme (MCS), specifying a pair of code rate and modulation type, is set as a simulation parameter and is fixed during a particular simulation run.

The AWGN channel performs addition of constellations (QAM symbols) at the output of the transmitter with randomly generated AWGN. The target is to reach the required signal-to-noise ratio (SNR) set as a simulation parameter.

Noisy constellations are fed to the input of the receiver part and pass through the soft demapping, descrambling, rate dematching, and turbo decoding stages. The soft demapper calculates logarithms of likelihood ratio using the max-log approximation. The descrambler performs the inverse transform as opposed to the scrambler, while rate dematcher performs addition of likelihood ratios corresponding to the repeated bits and/or taking the ratios corresponding to the punctured bits equal to zero. Turbo decoder realizes max-log maximum a posteriori decoding algorithm to derive a received TB with, possibly, some erroneous bits. Then, this TB is compared with the transmitted TB to detect errors and to determine the correctness of TB reception.

3.2   Experiment Plan and Results

The transmission of a particular TB is repeated multiple times for a fixed set of parameters to perform statistical averaging over the random realizations of a TB and AWGN. After the simulation is completed, the collected statistics shows the dependence of transport block error probability (BLER) versus SNR for a fixed MCS. Then, similar simulation is repeated for a set of available MCSs.

Each of the MCSs determines a possible data rate in the LTE-A system. It can be calculated as a transport block size (TBS) divided by the subframe length and multiplied by the probability of successful TB transmission:

$$r_k = (1 - BLER_{target}) \frac{TBS_k}{T_{subframe}}, \quad SNR_k = SNR^{(k)}(BLER_{target}), \quad (28)$$

where r_k is the data rate corresponding to the k-th MCS, k is the MCS index, TBS_k is the TBS of the k-th MCS, $BLER_{target}$ is the target BLER, $T_{subframe}$ is the subframe length (equal to 1 ms in LTE-A), SNR_k is the SNR required for the k-th MCS, $SNR^{(k)}(BLER)$ is the inverted dependency of BLER on SNR for the k-th MCS.

As the result of our LTE-A simulations, in Figure 3 we demonstrate the dependency of the achievable data rate on the SNR level for different MCSs. We also compare the simulation values with the Shannon's capacity formula to conclude that it serves as a reasonable approximation of the practical data rate.

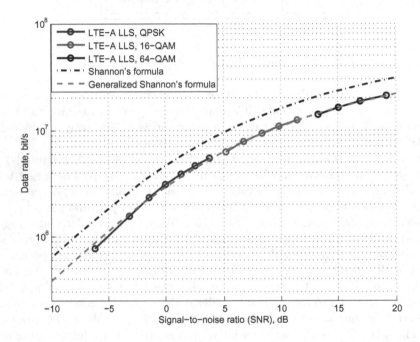

Fig. 3. Comparison of the Shannon's formula ($\alpha = 1$, $\beta = 1$) against simulation results for QPSK, 16-QAM, and 64-QAM

3.3   Calibrating the Generalized Shannon's Formula

Here we use our simulation results to calibrate the coefficients of the generalized Shannon's formula (4). Figure 3 provides dependencies corresponding to the conventional Shannon's formula ($\alpha = 1$ and $\beta = 1$), the generalized Shannon's formula (fitted coefficients $\alpha = 1.2456$ and $\beta = 1.3463$), and the simulated data rates for QPSK, 16-QAM, and 64-QAM modulation schemes. In order to properly adjust the coefficients α and β, the three sets of points corresponding to different modulations may first be converted into a single curve.

Our calibration procedure consists in the optimal selection (by e.g. least-squares technique) of the coefficients α and β in order to minimize the discrepancy between the available simulation results and the formula:

$$r_i = w \frac{T_{data}}{T_{total}} \frac{1}{\beta} \log_2 \left(1 + \frac{SNR_i}{\alpha}\right), \tag{29}$$

where $SNR = \frac{p_i g_i}{N_i}$.

4 Numerical Results

4.1 Setup Details and Parameters

In this section, we concentrate on an illustrative numerical example to evaluate the performance of the proposed power control scheme. We assume that there are $K = 2$ communication channels available to the mobile user. Correspondingly, the first recipient LPN is located at the point $x_1 = 0$, whereas the second one is at the point $x_2 = R$ (see Figure 4). The user is assumed to move all along the x-axis between the two LPNs, and its current coordinate is $x \in [0, R]$.

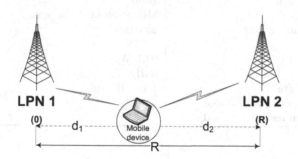

Fig. 4. Topology of our numerical setup

In order to compare our energy-efficient power control with alternative power management techniques, we introduce two simple and intuitive strategies.

1. The user transmits on both channels simultaneously by allocating a fixed amount of power to every channel.
2. The user transmits on one channel by selecting it basing on the channel quality and allocating a fixed amount of power to the best channel only.

In both cases, for the sake of simplicity, we assume that the allocated power level is equal the maximum allowed power (see Table 1). We also assume that the channels are symmetric, i.e. employ similar parameters, including the following propagation model [16]:

$$PL = 22.0 \log_{10} d + 28.0 + 20 \log_{10} f_c, 10 < d < d_{BP}, \tag{30}$$

$$PL = 40\log_{10} d + 7.8 - 18\log_{10} h'_{LPN} - 18\log_{10} h'_{user} + 2\log_{10} f_c, d_{BP} < d < 5000,$$

where PL is the path loss (dB), h'_{LPN} and h'_{user} are effective antenna heights (m), d and d_{BP} are the distance to the LPN and the break point distance (m) respectively, and f_c is the center frequency (GHz). See Table 1 for more details.

Table 1. Summary of simulation assumptions

Parameter	Value
Simulation approach	Link level simulations of LTE-A Release 10
Radio channel model	AWGN
Channel estimation and synchronization	Ideal
PUSCH bandwidth	5.4 MHz (30 resource blocks per slot)
Cyclic prefix	Normal
Turbo decoder	Max-log turbo decoder with 8 iterations
Target block error rate (BLER)	10%
Thermal noise power	-103 dB
Carrier frequency	2 GHz
User antenna height	1.5 m
LPN antenna height	10 m
Environment	Micro cell in urban area
Maximum transmit power	23 dBm
Circuit power	0.1 W
Idle power	0.01 W
Antenna gain	3 dB
Minimum data rate	1.19 Mbps
Number of bits per QAM symbol for the maximum modulation order	6

4.2   Discussion of the Results

In Figure 5, we overlay the results for the total achievable data rate, energy efficiency, and power consumption depending on the user location x. With our optimal power control, the user begins with transmitting on one channel (A). As it moves toward the center (B), the user adjusts its transmit power to compensate for the varying pathloss value [16]. The shape of the transmit power function here is determined solely by the pathloss alterations in (30).

Moving further, the user does not yet need to apply the maximum power to reach the highest energy efficiency until the point (C), when the constraint r_0 takes effect. Because of this bitrate constraint, the power rises dramatically up to the point (D), when the use of the second channel becomes reasonable. Then power gradually grows up to the maximum transmit power level (E) to stay there until (F). Further, this behavior mirrors symmetrically (as both channels are equivalent).

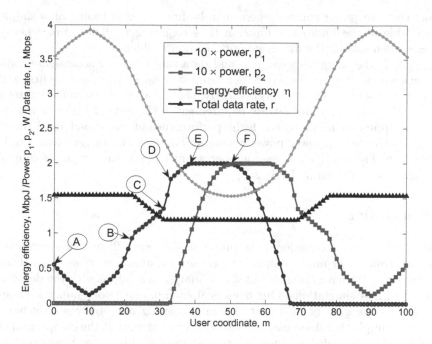

Fig. 5. Achievable data rate, transmit power, and energy efficiency of the mobile user

Table 2. Energy efficiency and total data rate comparison for base schemes, $K = 2$

Distance, m	r^*, Mbps	r_1, Mbps	r_2, Mbps	η^*, MbpJ	η_1, MbpJ	η_2, MbpJ	$\Delta\eta_1$, %	$\Delta\eta_2$, %
0	1.54	1.58	1.54	3.54	2.49	3.54	30	0
10	1.54	1.61	1.54	3.93	2.71	3.93	31	0
20	1.54	1.65	1.54	3.47	2.57	3.47	26	0
30	1.23	1.57	1.36	2.41	2.01	2.35	17	2
40	1.19	1.29	0.93	1.7	1.65	1.61	3	5
50	1.19	1.19	0.6	1.53	1.53	1.03	0	33
60	1.19	1.29	0.93	1.7	1.65	1.61	3	5
70	1.23	1.57	1.36	2.41	2.01	2.35	17	2
80	1.54	1.65	1.54	3.47	2.57	3.47	26	0
90	1.54	1.61	1.54	3.93	2.71	3.93	31	0
100	1.54	1.58	1.54	3.54	2.49	3.54	30	0

The data rate plot clearly demonstrates the upper and the lower regions of better and worse channel quality, respectively. Consequently, the lowest energy efficiency is reached in the central point, where the quality of both channels is the poorest, whereas the maximum is reached at the distance of obstruction, when channel quality is the best according to (30).

Note that our power control scheme can be implemented basing on a simple lookup table of the Lambert's function thus requiring negligible time/energy expenses when solving the discussed optimization problem.

In Table 2, the energy efficiency η^* and data rate r^* of our scheme are compared against those of two intuitive (heuristic) strategies (η_1, r_1 for the first and η_2, r_2 for the second, respectively). Clearly, the one-channel (second) strategy fails to satisfy the minimum bitrate requirement in the central region. Furthermore, our approach allows to reach the performance of the two-channel (first) strategy, but results in lower power consumption when the target bitrate r_0 is already met. The relative increase in energy efficiency with our power control ($\Delta\eta_1$, $\Delta\eta_2$) is also given in Table 2.

5 Conclusion

In this work, we have considered the problem of energy efficient power control when the mobile user may communicate on several uplink wireless channels at the same time. We have also calibrated our analytical solution with the detailed link-level LTE-A simulations. Our numerical example for two symmetric channels suggests the gain of up to 30 % when prefering our power management strategy to simpler heuristic mechanisms. Our current work is the comparison of the proposed energy efficient approach against more sophisticated power control strategies and the consideration of more practical device power models.

Acknowledgment. This work is supported by the Internet of Things program of Tivit, funded by Tekes, and Hungarian Government through the TÁMOP-4.2.2C-11/1/KONV-2012-0001 and the OTKA K101150 projects.

References

1. Ericsson. More than 50 billion connected devices (2011)
2. Cisco. Global Mobile Data Traffic Forecast Update, 2011-2016 (2012)
3. Akyildiz, I., Gutierrez-Estevez, D., Reyes, E.: The evolution to 4G cellular systems: LTE-Advanced. Physical Communication 3, 217–244 (2010)
4. Pentikousis, K.: In search of energy-efficient mobile networking. IEEE Communications Magazine 48, 95–103 (2010)
5. Tombaz, S., Vastberg, A., Zander, J.: Energy- and cost-efficient ultra-high capacity wireless access. IEEE Wireless Communications 18, 18–24 (2011)
6. Raychaudhuri, D., Mandayam, N.: Frontiers of wireless and mobile communications. Proceedings of the IEEE 100, 824–840 (2012)
7. Hu, R., Talwar, S., Zong, P.: Cooperative, Green and Mobile Heterogeneous Wireless Networks. IEEE 802.16x (2011)
8. R2-130478. Mobility in HetNet scenarios. Nokia Siemens Networks (2013)
9. Miao, G., Himayat, N., Li, G.: Energy-efficient link adaptation in frequency-selective channels. IEEE Transactions on Communications 58, 545–554 (2010)
10. Rappaport, T.: Wireless Communications: Principles and Practice. Pearson Education (2009)

11. 3GPP LTE Release 10 & beyond (LTE-Advanced)
12. Kuhn, H., Tucker, A.: Nonlinear programming. In: Proc. of 2nd Berkeley Symposium (1951)
13. Corless, R., Gonnet, G., Hare, D., Jeffrey, D.: On Lambert's W function. Technical report, University of Waterloo (1993)
14. 3GPP TS 36.212. Multiplexing and channel coding (2012)
15. 3GPP TS 36.211. Physical Channels and Modulation (2012)
16. 3GPP TR 36.814. Further advancements for E-UTRA physical layer aspects (2010)

Journey Data Based Arrival Forecasting
for Bicycle Hire Schemes

Marcel C. Guenther and Jeremy T. Bradley

Imperial College London,
180 Queen's Gate,
SW7 2RH, London, United Kingdom
{mcg05,jb}@doc.ic.ac.uk

Abstract. The global emergence of city bicycle hire schemes has recently received a lot of attention in the performance and modelling research community. A particularly important challenge is the accurate forecast of future bicycle migration trends, as these assist service providers to ensure availability of bicycles and parking spaces at docking stations, which is vital to match customer expectations. This study looks at how historic information about individual journeys could be used to improve interval arrival forecasts for small groups of docking stations. Specifically, we compare the performance of small area arrival predictions for two types of models, a mean-field analysable time-inhomogeneous population CTMC model (IPCTMC) and a multiple linear regression model with ARIMA error (LRA). The models are validated using historical rush hour journey data from the London Barclays Cycle Hire scheme, which is used to train the models and to test their prediction accuracy.

Keywords: IPCTMC, Time-inhomogeneous population models, Mean-field analysis, Multiple linear regression with ARIMA error, Bicycle sharing schemes, Spatial modelling.

1 Introduction

On the 30th of July 2010, the Barclays Cycle Hire scheme launched in London, after similar schemes had proved to be popular in metropoles such as Paris, Barcelona and Vienna. Bike hire schemes generally feature a number of docking stations where bikes can either be rented or returned. Stations are installed all over the city so that the maximum distance between neighbouring stations is at most 500 metres. Moreover, stations vary in the numbers of parking slots they provide, the largest stations being close to transport hubs. The schemes are aimed to provide a cost-effective, green solution to the last-mile transport problem in large cities for both tourists and commuters. While tourists can purchase day memberships for the hire scheme, commuters can also opt for a discounted annual membership. Naturally, the growing popularity of cycle hire as well as the abundance of publicly available data from various operational hire schemes has attracted interest in the performance research community.

A. Dudin and K. De Turck (Eds.): ASMTA 2013, LNCS 7984, pp. 214–231, 2013.

Much like traditional transport providers, cycle hire operators face classical problems such as infrastructure planning [1], pricing [2] and policy improvement [3, 4]. However, our focus in this paper is related to the challenge of being able to forecast the number of available bikes and parking slots at different docking stations [5–7]. Being able to make such forecasts is of vital interest to both operators and customers. With the growing availability of mobile internet access, users of transport systems nowadays expect the availability of real-time transport information. Hence, multi-modal end-to-end routing applications that consider bicycle hire as a possible mode of transport, need to be able to accurately forecast the availability of bikes and parking spaces at suggested origin and destination docking stations [7, 8]. Moreover, operators further require good future estimates as to when stations become empty or full in order to redistribute bikes and antagonise such trends [9]. Aside from purely quantitative performance evaluation, efforts have also been made to visualise migration trends of bicycle schemes. In particular [10] and [11] show that a lot can be learnt about the dynamics of bike sharing systems by using an appropriate visualisation.

In this paper we present a novel time-inhomogeneous Population CTMC (IPCTMC) model that can provide forecasts for the number of interval arrivals for small sets of docking stations at rush hour[1]. Moreover, we compare the accuracy of the model with a more traditional multiple linear regression with ARIMA error (LRA) approach that is more akin to the time series techniques used in [6, 7]. Our results indicate that despite the time-inhomogeneous nature of bicycle arrivals and departures, the LRA model produces better forecasts under realistic circumstances. It would, however, be premature to discard IPCTMCs as a modelling formalism for similar systems, as the IPCTMC model helps us to understand the important dynamics of the system better and further offers a lot of expressiveness and flexibility, which makes it very suitable for other applications such as model based provisioning [12]. Irrespective of the modelling technique, our results further indicate that at least for short-term predictions of 15 minutes or less, the availability of historic origin to destination journey data, can help to produce better accuracy than mere time and location dependent departure and arrival time series information. This motivates further study on how to combine individual journey data with station occupancy time series data in order to improve forecasts of the latter.

The remainder of this paper is organised as follows; In Section 1.1 we briefly review the literature on bicycle hire schemes. Subsequently Section 2 introduces the reader to Population Continuous-Time Markov Chains and presents a time-inhomogeneous extension. In Section 3 and Section 4 we develop an IPCTMC as well as an LRA model for interval arrival forecasts and compare their forecast quality in Section 5. Section 5.1 discusses other aspects of the two different modelling approaches.

---

[1] By interval arrivals we mean the total number of future arrivals for a set of neighbouring docking stations over a fixed interval of time.

1.1   Related Work

Froehlich *et al.* [5] were the first to propose a station occupancy forecast model based on historic time series data describing the number of bikes docked at a particular station. To make predictions about individual stations, they trained a Bayesian Network model for each station, using time, prediction window and the current proportion of occupied parking slots as regression parameters. Given the current state of the system, their model could then be used to forecast future station capacity as either $0 - 20\%$ full, $20 - 40\%$ full and so on. However, judging from their error analysis, their model only marginally outperformed a simple historical trend predictor.

Kaltenbrunner *et al.* [6] suggested an ARMA model, which was trained on similar time series data. Their most important contribution was the observation that the prediction error can be vastly reduced by incorporating information about the occupancy of positively correlated neighbouring stations. Although their ARMA model produces significantly better results than comparable historical trend models, their decision to use an ARMA process to predict a highly time-inhomogeneous process is possibly sub-optimal. Furthermore, both their model fitting and their error analysis is performed on smoothed time series data, making it harder to judge how accurate their model truly is.

Yoon *et al.* [7] recently addressed some of these shortcomings. By fitting an ARIMA process they were able to capture the inhomogeneous nature of bike hires better. Furthermore, instead of only looking at neighbouring stations for extra exogenous variables for the regression model, the authors computed a detailed time-lag dependent cross-correlation metric for all pairs of stations. This approach improves the long-term forecast quality, as it captures both positively correlated neighbouring stations that experience similar traffic as well as negatively correlated stations that have opposite migration trends. Interestingly, while their ARIMA model provides the best forecast in a benchmark carried out on data from the Dublin cycle hire scheme, it does not appear to outperform Kaltenbrunner's ARMA model by much.

Aside from research on out-of-sample forecasts of future station occupancy, other notable contributions are the cluster analysis provided in [13] and the departure forecasting model discussed in [14].

2   PCTMCs

Population Continuous-Time Markov Chains (PCTMCs) consist of a finite set of populations S, $n = |S|$ and a set E of transition classes [15]. States are represented as an integer vector $\boldsymbol{P}(t) = (P_1(t), \ldots, P_n(t)) \in \mathbb{Z}^n$, with the ith component being the current population level of species $S_i \in S$ at time t. A transition class $(r_e, \boldsymbol{c}_e) \in E$ for an event e describes a transition with negatively exponentially distributed delay D at rate $r_e : \mathbb{Z}^n \to \mathbb{R}$ which changes the population vector $\boldsymbol{P}(t + D)$ to $\boldsymbol{P}(t) + \boldsymbol{c}_e$. The analogue to PCTMCs in the systems biology literature are Chemical Reaction Systems, were $\boldsymbol{P}(t)$ describes a molecule count vector and transition classes represent chemical reactions between the molecules with r_e being the reaction rate function and c_e the stoichiometric vector for a specific reaction. For notational convenience we write an event/reaction e as

$$\underbrace{S_* + \cdots + S_*}_{\text{in}} \rightarrow \underbrace{S_* + \cdots + S_*}_{\text{out}} \qquad\qquad \text{at } r_e(\boldsymbol{P}(t)) \qquad (1)$$

where $S_* \in S$ represent different species that are involved in the event. The corresponding change vector is $c_e = (s_1^{\text{out}} - s_1^{\text{in}}, \ldots, s_n^{\text{out}} - s_n^{\text{in}}) \in \mathbb{Z}^n$ where s_i^{in} represents the number of occurrences of a particular species $S_i \in S$ on the left hand side of the event and s_i^{out} its number of occurrences on the right hand side. The event rate is

$$\begin{cases} r_e(\boldsymbol{P}(t)) & \text{if } P_i(t) \geq s_i^{\text{in}} \ \forall i = 1, \ldots, n \\ 0 & \text{otherwise} \end{cases} \qquad (2)$$

When used to describe spatially distributed populations, we denote a species S at location l at time t as $S@l(t)$ [16].

An important feature of PCTMC models is that approximations to the evolution of population moments of the underlying stochastic process can be represented by the following system of ODEs [17]

$$\frac{\mathrm{d}}{\mathrm{d}t}\mathbb{E}[T(\boldsymbol{P}(t))] = \sum_{e \in E} \mathbb{E}[(T(\boldsymbol{P}(t) + c_e) - T(\boldsymbol{P}(t)))r_e(\boldsymbol{P}(t))] \qquad (3)$$

To obtain the ODE describing the evolution of the mean of a population, all we need to do is to substitute $T(\boldsymbol{P}(t)) = P_i(t)$ in the above equation, where $P_i(t)$ is the random variable representing the population count of species S_i at time t. In the literature the resulting ODEs are often referred to as mean-field approximations [17].

2.1   Time-Inhomogeneous PCTMCs

While the PCTMC formalism has been applied to problems in many application areas, it would be rather inaccurate to describe the model presented in Section 3 using a time-homogeneous CTMC process, since many parameters, such as departure rates and destinations of bikes, vary with time. Hence, we present a time-inhomogeneous extension to the PCTMC formalism, which we term IPCTMC, that is going to be released in a future version of the Grouped PEPA Analyser(GPA) [18]. To our knowledge this is the only paper aside from [19], which applies mean-field analysis to IPCTMC models. In IPTMCs we allow deterministic rate and population changes that occur at deterministic times. This implies that any reaction rate $r_e(\boldsymbol{P}(t))$ (cf. Section 2) is now time dependent, i.e. $r_e(\boldsymbol{P}(t), t)$ becomes

$$\begin{cases} r_e(\boldsymbol{P}(t), t_1) & \text{if } P_i(t) \geq s_i^{\text{in}} \ \forall i = 1, \ldots, n \wedge t < t_1 \\ r_e(\boldsymbol{P}(t), t_2) & \text{if } P_i(t) \geq s_i^{\text{in}} \ \forall i = 1, \ldots, n \wedge t_1 \leq t < t_2 \\ \ldots & \\ 0 & \text{otherwise} \end{cases} \qquad (4)$$

where $t_1, t_2, \ldots$ are deterministic times at which reaction rate changes occur. Similarly we allow deterministically timed events that result in an affine transformation of the population vector $\boldsymbol{P}(t)$. In the following we informally assert that if a deterministic population change occurs at time t_d then no population changes occur due to random PCTMC events between $t_d - \delta t$ and t_d. Should no such interval exist, then we assume that the deterministic event is triggered immediately after the random event. Let $\mathcal{D}$ denote the set of all deterministic events, s.t. $(t_d, \mathbf{M}) \in \mathcal{D}$, where

$$
\mathbf{M}_{n,n+1} = \begin{pmatrix} \lambda_1 & 0 & \cdots & 0 & d_1 \\ 0 & \lambda_2 & \cdots & 0 & d_2 \\ \vdots & \vdots & \ddots & \vdots & \vdots \\ 0 & 0 & \cdots & \lambda_n & d_n \end{pmatrix} \tag{5}
$$

and the updated population count vector becomes

$$
\boldsymbol{P}(t_d) = \mathbf{M} \left(\begin{pmatrix} 1 & 0 & \cdots & 0 \\ 0 & 1 & \cdots & 0 \\ \vdots & \vdots & \ddots & \vdots \\ 0 & 0 & \cdots & 1 \\ 0 & 0 & \cdots & 0 \end{pmatrix} \boldsymbol{P}(t_d - \delta t) + \begin{pmatrix} 0 \\ 0 \\ \vdots \\ 0 \\ 1 \end{pmatrix} \right) \tag{6}
$$

As an example we can now describe the reset of population p_1 to d_1 and the population jump of population p_1 by d_1 individuals as

$$
\mathbf{Reset}_{n,n+1} = \begin{pmatrix} 0 & 0 & \cdots & 0 & d_1 \\ 0 & 1 & \cdots & 0 & 0 \\ \vdots & \vdots & \ddots & \vdots & \vdots \\ 0 & 0 & \cdots & 1 & 0 \end{pmatrix} \quad \mathbf{Jump}_{n,n+1} = \begin{pmatrix} 1 & 0 & \cdots & 0 & d_1 \\ 0 & 1 & \cdots & 0 & 0 \\ \vdots & \vdots & \ddots & \vdots & \vdots \\ 0 & 0 & \cdots & 1 & 0 \end{pmatrix} \tag{7}
$$

Similarly we could vary the λ_i, for instance to double a population. In both IPCTMC simulation runs and mean-field analysis, rate changes that are part of a deterministic event, simply involve an update of that rate as the event occurs. Moreover, when simulating an IPCTMC model, any population change can be immediately applied to the population count vector using Eq. (6). However, mean-field ODE analysis of IPCTMCs is more complicated with regards to the population vector updates, since we keep track of population moments rather than the values of actual random variables. As a consequence we need to expand the affine transformation inside the expectation expressions. For instance assuming that we a have two populations X and Y and that the following deterministic population change occurs at t_d; $X(t_d) = \lambda_x X(t_d - \delta t) + d_1$ and $Y(t_d) = Y(t_d - \delta t)$ then

$$
\begin{aligned}
\mathbb{E}[X(t_d)Y(t_d)] &= \mathbb{E}[(\lambda_x X(t_d - \delta t) + d_1)Y(t_d - \delta t)] \\
&= \lambda_x \mathbb{E}[X(t_d - \delta t)Y(t_d - \delta t)] + d_1 \mathbb{E}[Y(t_d - \delta t)]
\end{aligned} \tag{8}
$$

and similarly for $\mathbb{E}[X(t_d)^2]$ or any other moments. Fortunately, so long as the transformation of the population vector remains affine, we do not have to use moment closures [15] in order to compute the new values for the moments at time t_d. More complex variants of mean-field analysable IPCTMC models, such as ones with non-linear population vector transformations or population transformations that are subject to boundary conditions, are also possible. However, these require further treatment, which are beyond the scope of this work.

3   An IPCTMC Interval Arrival Forecasting Model

Figure 1 shows the species and parameters of our IPCTMC forecasting model. In the following we assume that our model is estimating the number of future arrivals for a set of neighbouring stations denoted in a small area A. In Section 5 we will then look at forecasts for different areas A at different times of the day, e.g. for areas depicted in Figure 2. Since we model arrivals, we are particularly interested in areas around stations that are in danger of running out of available parking spaces.

All states shown in Figure 1 correspond to species in the formal definition of the underlying IPCTMC. Each journey is assumed to start in one of the states $DeparturesCl@1, \ldots, DeparturesCl@n$. Those which end in A, will finish in $Arrivals@A$ after experiencing a delay represented by $PhCl@i$. All other journeys are assumed to end $Elsewhere$. Naturally, for the purpose of forecasting, we are most interested in the population of agents that transit to state $Arrivals@A$ during the forecast interval $[t_0, t_{fcast}]$. $DeparturesCl@1, \ldots, DeparturesCl@n$ encapsulate all stations that are starting points for journeys that terminate in A, including stations that lie inside A. However, while the $Arrivals@A$ state captures arrivals at stations within a specific geographic area, departure cluster membership of stations is based on similarity in journey time distribution. In other words for each station within a cluster, the distribution of time it takes a journey that starts from this station and ends at any station in A has to be similar. Since journey durations are generally not exponentially distributed, we use additional states (cf. $PhCl@i$) to represent the clusters' characteristic journey time distribution as phase-type approximations.

Having described the states of the IPCTMC model, we now need to explain its parameters. In the following we assume that we can initialise our model using historical information for the interval $[t_{-warmup}, \ldots, t_0]$ and that we make a forecast for the number of agents that reach state $Arrivals@A$ during the interval $[t_0, t_{fcast}]$. $t_{-warmup}$ is chosen to be large enough such that it represents the 99% quantile of the clusters' joint journey time distribution, i.e. departures before $t_{-warmup}$ are unlikely to affect our forecast. $\delta_i(t)$ is the time-inhomogeneous component that describes departures from cluster i. During the interval $[t_{-warmup}, t_0]$, we know all journeys that have departed from any station in the cluster and therefore we can use population jumps to increase the departure population of a given cluster. This is done every minute, i.e. at $t_{-warmup}$ we increase the population of $DeparturesCl@i$ by the number of departures that were observed for this cluster

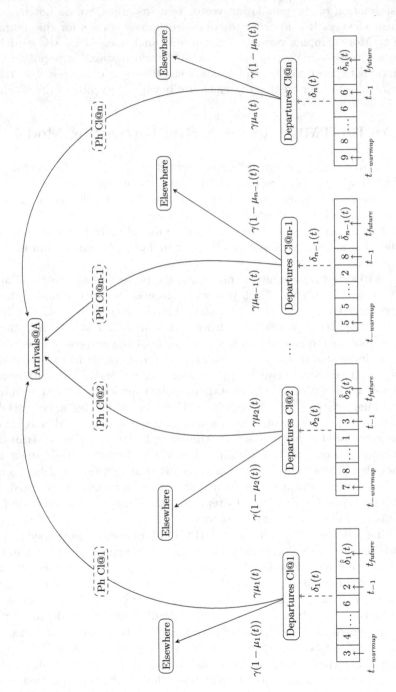

Fig. 1. The IPCTMC arrival forecasting model

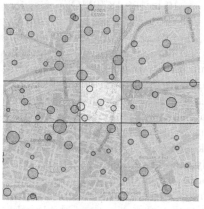

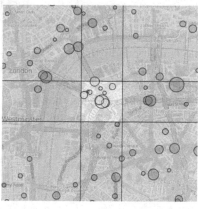

(a) City of London (Stn175)   (b) Waterloo (Stn361)

Fig. 2. Two examples of areas that lack parking spaces during rush hour (cf. Table 1). Each circle represents a docking station, the larger the circle the more docks it has.

during the interval $[t_{-warmup}, t_{-warmup+1}]$. To ensure that the additional journeys actually leave state $DeparturesCl@i$ by $t_{-warmup+1}$, we choose a sufficiently large exponential rate γ. During the actual forecast period $[t_0, t_{fcast}]$, we do no longer use population jumps, but instead treat $\delta_i(t)$ as exponential rates at which new journeys start from cluster i. As we cannot look into the future, these have to be predicted from previous departure rates. Last but not least the $\mu_i(t)$ parameters reflect the proportion of departures from cluster i that is heading for stations in A.

3.1   IPCTMC Model Parameter Estimation

To create clusters we first chose a forecast period and assumed journey times to be iid. Weekday rush hour journeys are particularly suitable for this, since commuters, unlike tourists, are less likely to cycle in groups. To train a model for a particular area A, we first measured the journey time samples for all training journeys that ended in A. Subsequently we grouped these observations by their start station and computed the 10% and the 90% quantiles of the resulting station journey time distributions. After that, a k-means clustering algorithm was run on the coordinates. To ensure that stations with a larger number of outgoing journeys have a larger impact than less frequently used ones, we created n identical (10%, 90%) cluster points for a station with n observations in the training data set. We varied the number of clusters k to get a decent trade-off between cluster compactness and clusters size. Although more clusters should generally produce convoluted cluster journey time distributions that are more similar to the distributions of the individual stations in the clusters, too many clusters generate large state-spaces and make cluster parameter estimation harder. To fit the cluster journey time distributions we generated Hyper-Erlang approximations using HyperStar [20]. We discuss the impact of the number of clusters on the forecast accuracy in Section 5. Using between 10 to 100 PCTMC species for each $PhCl@i$, we generally obtained low relative errors of less than 5% up to the 4th uncentred distribution moment. In addition to generating the states and the transitions in the IPCTMC model using the phase-type fits, we also generated the time series data describing the minutely departures from all clusters for all training and test dates. $\hat{\mu}_i(t)$ estimates were computed using 5-minute window daily averages that were then further averaged for identical times on different days, thus producing a single estimate for each time point in the observed interval. Moreover, for all minutely cluster departures we also recorded the precise $\mu_i(t)$ value, and assumed perfect knowledge of journey destinations up to t_0 when computing a forecast. Naturally, this is an idealised assumption, however, as Come et al. note in [13], the majority of rush hour journeys is made by subscribed cyclists, which would allow service providers to accurately predict journey destinations from their history, especially when we consider the destination to be an area rather than a specific station.

4   Linear Regression ARIMA Error Forecasting Model

Our motivation for comparing the IPCTMC model to a time series model (cf. [6, 7]), was to investigate whether an inhomogeneous process would yield better results than a time series error model fitted to arrival and departure data, which was made stationary by means of seasonal normalising. After various experiments with different time series model classes, we found that a simple multiple linear regression model with ARIMA error (LRA) worked best. Like Yoon et al. [7] we decided to choose an observation frequency of 5 minutes for departure and arrival observations. The model can then be expressed as follows

$$
\begin{aligned}
Y(t+h) = \alpha_{11}X_1(t+h) + \ldots + \alpha_{1m}X_1(t+h-(m-1)*5) + \ldots + \\
\alpha_{n1}X_n(t+h) + \ldots + \alpha_{nm}X_n(t+h-(m-1)*5) + N(t+h)
\end{aligned}
\tag{9}
$$

where $h = 5i, i \in \mathbb{N}$ and $Y(t+h)$ is the number of arrivals for a set of chosen stations during the interval $[t, t+h]$. $X_c(f)$ is the number of departures from cluster c during the interval $[f-5, f]$. $m \geq 1$ determines how many of the departure intervals X_c we fit the multivariate regression model to, for example when $h = 15$ we may want to have $m = 7$ to ensure we take into account all departures that occurred in the interval $[t-20, t+15]$. Note, when forecasting $Y(t+h)$ we also need to forecast any interval departures $X_c(f)$ for which $f > t$. $N(t+h)$ is an ARIMA(p,d,q) process fitted to the regression residuals. We used the R forecast package [21] to estimate the α_{ci} parameters from the training data. However, since we had to fit a single LRA model to a number of non-consecutive time series traces for the same interval on different days, e.g. an LRA model for the interval arrivals of a specific set of stations during the afternoon rush hour, we had to use the interleaving technique described in [22]^2. Furthermore, instead of

---

^{2} While the actual error is assumed to be ARIMA(p,d,q), the interleaving technique requires us to fit a SARIMA($p \cdot \#days, 0, q \cdot \#days)(0, d, 0)_{\#days}$, where $\#days$ is the number of days in our rush hour time series training data set.

fitting the observed $Y, \boldsymbol{X}$ directly, we first normalised each observation at time t by its corresponding time dependent sample mean and standard deviation, which were estimated from the training data. For genuine out-of-sample forecasts, future values of the regression variables $\boldsymbol{X}$ were estimated using ARIMA models, which we fitted for each departure cluster using training data.

5   Model Analysis

In this section we compare the forecasting accuracy of models described in Sections 3.1 and 4 for different areas A during the morning and the afternoon rush hour, 6-9am and 4-7pm respectively. Each area is a $500m \times 500m$ square (cf. Figure 2) centred around the stations listed in Table 1, which usually run out of bikes during one of the two rush hours (cf. *Rush hour* in Table 1) and out of bikes during the other. We consider an area rather than a single station as we assume that cyclists, who arrive at a full station, are likely to seek alternate parking spaces in the surrounding area so long as these lie within $250m$ of their initial destination. From a user perspective area forecasts should therefore be equally useful as single station forecasts, so long as the radius of the area is small enough.

We used May 2012 journey data from the Barclays Cycle Hire scheme to train our models and June 2012 data to test out-of-sample forecast accuracy. As mentioned earlier, our aim is twofold, namely to investigate the impact of additional origin destination journey information on prediction accuracy and a comparison between the IPCTMC and the LRA model. In the following we distinguish between *journey based* models that use origin destination information and *normal* models which only use information about departures and arrivals. Additionally, we discriminate between *realistic* out-of-sample forecasts, which use forecasting techniques for both regression variable and interval arrival predictions and *perfect* information models, where future regression variables are assumed to be known at the time of a forecast. Recall from Section 3.1 that even for realistic forecasts we assume that journey destinations of departures up to t_0 are assumed to be known.

We start by looking at different ways to incorporate journey information in our models. One way to achieve this is journey time dependent clustering that we described in Section 3.1. Figure 3a shows the Root Mean Squared Error (RMSE) for the interval arrival forecasts of IPCTMC models with one departure cluster

Table 1. Each $500m \times 500m$ arrival area *Ref* is centred around a single station *Station name* that tends to lack free parking slots during *Rush hour*

Ref	Station name	Rush hour	#Bike slots (station / ttl area)
Stn66 am	Holborn Circus	morning	39 / 155
Stn175 am	Appold Street	morning	26 / 130
Stn354 pm	Northumberland Avenue	afternoon	36 / 134
Stn361 pm	Waterloo Station 2	afternoon	55 / 280

divided by the RMSE of the corresponding IPCTMC models with $4-6$ clusters for different areas and forecast interval lengths. A similar comparison is made in Figure 3b for the corresponding LRA models. In both cases we compare the accuracy of realistic forecasts. The IPCTMC graph indicates that more clusters appear to produce significantly better fits for realistic short-term forecasts, whereas for longer intervals, extra clusters do not outperform single cluster predictions. For the LRA model on the other hand clusters do not seem to have a significant positive impact on the RMSE, although when using log-likelihood instead of RMSE, it appears that extra clusters do produce better short-term forecasts as in the IPCTMC case. However, as Figure 4a shows, IPCTMC models fitted on more departure clusters still outperform single cluster predictions when perfect information about future departures and journey destinations is available after t_0. Again for LRA models (cf. Figure 4b) no such trend is visible. This suggests that the extra regression variables for additional clusters result in a worse fit, making the model insensitive to better future departure information. In view of this finding we will use the more parsimonious single cluster models from this point onwards. Nevertheless, when using IPCTMC models for other purposes such as model based provisioning, our observations emphasise that additional clusters do indeed ensure higher accuracy. In the above examples the IPCTMC model uses journey data for clustering as well as for estimating the $\hat{\mu}_i(t)$ parameters, whereas for the LRA only the cluster information is added. In a second case we look at what happens if we train a single cluster LRA model on journey data. To do this, we train the model considering only journeys that are heading to the arrival area which we are monitoring. As before, future LRA departure regression variables are estimated using ARIMA models. In Figure 5 we show the RMSE of normal single cluster LRA model forecasts divided by the RMSE of a journey information based single cluster LRA model. Under realistic conditions (cf. Figure 5a) the forecast accuracy of journey based LRA models decreases as the forecast interval increases. On the other hand when provided

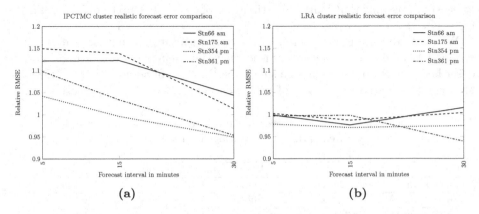

Fig. 3. Comparing realistic forecast error of multi-cluster departure models against single departure cluster models for both IPCTMC and LRA

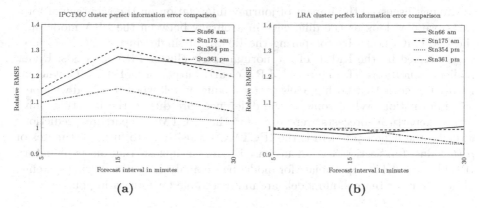

Fig. 4. Same error metric as in Figure 3 assuming perfect information

with perfect information about future departures and their destinations (cf. Figure 5b), journey based LRA models visibly outperform normal LRA models, no matter what the forecast length is. Comparing the log-likelihood gave similar results. Further experiments in which we compared the quality of the regression variable forecast accuracy let us to the conclusion that the journey data trained LRA model is much more sensitive to the quality of regression variable forecasts. Moreover, it is much harder to train ARIMA models to predict the number of departures that head for a certain area than it is to merely predict the total number of departures. The combination of these two facts explains the effects seen in Figure 5.

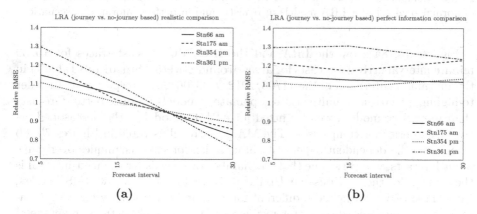

Fig. 5. Comparing the accuracy of normal single cluster LRA models with that of journey information based single cluster LRA models in realistic and in perfect information scenarios

Having discussed the impact of journey information on the accuracy of forecasts, we now look at the difference in accuracy between the LRA model and the IPCTMC model by comparing the RMSE of single cluster IPCTMC forecasts divided by the RMSE of the normal single cluster LRA forecasts. Under realistic conditions (cf. Figure 6a), LRA clearly outperforms IPCTMC, whereas Figure 6b shows that both models perform equally well when provided with perfect information. While some of the differences are due to the fact that naive future departure forecasts were used for the IPCTMC departure predictions after t_0, the bigger problem is the IPCTMC's sensitivity to future estimates of $\mu_i(t)$ values (cf. Section 3.1). In any case Figure 6b suggests that our explanatory IPCTMC model is approapiate for model based analysis of bike sharing systems, although linear regression models are more suitable for forecasting purposes.

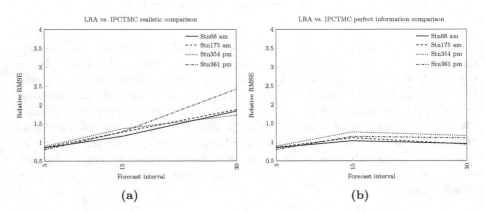

Fig. 6. Comparing the accuracy of normal single cluster IPCTMC models with that of normal single cluster LRA models in realistic and in perfect information scenarios

Figures 7a, 8a show the LRA and the IPCTMC forecast traces for 15/30 minute interval arrival forecasts for areas around Stn175 / Stn361 (cf. Table 1) in the morning / afternoon of the 11/06/2012 & 06/06/2012, respectively. In order to highlight that these multivariate explanatory models produce better forecasts than univariate models, we also fitted an ARIMA model to the deseasoned arrival time series for comparison. The MASE[3] error illustrated in Figures 7b, 8b visualises time dependent superiority of complex forecast techniques over naive point forecasts, which assume that the number of arrivals for the next interval is the same as for the previous one. Unlike [6, 7] we do not show raw RMSE errors, since these vary a lot between different forecast areas. Ultimately for actual station occupancy forecasts a better metric would be to look at the percentage of incorrect predictions that would cause users of the system to wait for available bikes or parking slots.

---

[3] The MASE [21] statistic is the average model forecast error divided by the average naive forecast error and thus provides a good benchmark for the prediction accuracy.

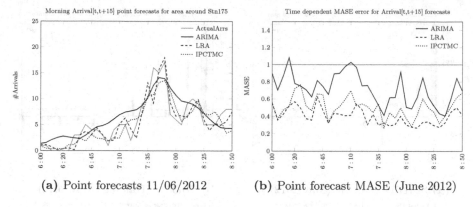

(a) Point forecasts 11/06/2012   **(b)** Point forecast MASE (June 2012)

Fig. 7. Comparing different morning arrival forecasts for area around Stn175

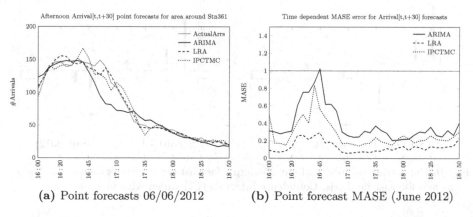

(a) Point forecasts 06/06/2012   **(b)** Point forecast MASE (June 2012)

Fig. 8. Comparing different afternoon arrival forecasts for area around Stn361

The graphs in Figures 7b, 8b study the time dependent MASE statistic for different times in the morning and the afternoon rush hour in different areas over the entire month of June 2012. In both examples the LRA model performs better than the IPCTMC, although for 15 minute interval arrival predictions in the area around Stn175 the difference is less significant than for the longer forecasts around Stn361.

The comparison between auto-correlation values of forecast errors obtained from our models when testing them on the training data Figures 9a, 9b and the correlograms of genuine out-of-sample forecast errors in Figures 9c, 9d suggests that some of the errors might be systematic, especially for the IPCTMC model. However, since we are dealing with time-inhomogeneous processes it requires further investigation to see whether errors are sufficiently time-homogeneous to apply any time series based error correction.

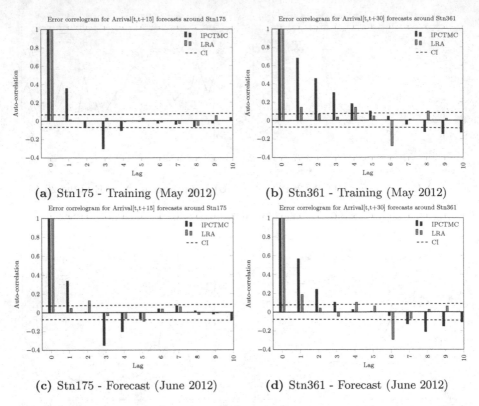

(a) Stn175 - Training (May 2012) **(b)** Stn361 - Training (May 2012)

(c) Stn175 - Forecast (June 2012) **(d)** Stn361 - Forecast (June 2012)

Fig. 9. Comparing training and out-of-sample forecast error correlograms for different areas and forecast methods. Confidence intervals (CI) widen with increasing lag.

5.1 Model Discussion

The examples above indicate that LRA model forecasts generally outperform IPCMTC ones under realistic conditions. However, while this might be of importance to many practical applications, it cannot gloss over the fact that the LRA model remains, to some extend, a blackbox model, whereas the corresponding IPCTMC models are fully transparent. In particular there are two opaque aspects to the LRA model. The first one is the procedure used to transform the time-inhomogeneous series into a homogeneous one (cf. Section 4). While constructing the LRA model, we tried various different ways to achieve this, until we found a method that gave sufficiently good fits for our training data. Even though the chosen transformation works for our purposes, it is hard to judge whether the resulting model is truly time-homogeneous. Secondly, the order of the ARIMA error model cannot immediately be deduced from the forecast interval length. Needless to say, the longer the arrival forecast interval, the more regression variables $X_c(\cdot)$ need to be predicted and hence we would expect the need for a higher auto-regressive order in our model. However, the LRA generally performs better with additional moving average terms, even though it is not entirely clear as to why this is the case. Given that the IPCTMC model transparently describes the relationship between departures and arrivals, it is not possible to interpret the relationship between journey time, origin and destination from LRA parameters. This could potentially be a problem if we were to extended the model for other purposes such as model provisioning or policy testing. For example, for future work we would like investigate whether sensors on bicycles can be used to collect enough data to provide sufficiently timely information about different geographical areas, using similar mechanisms as described in [23]. Clearly, in this case IPCTMC models are likely to produce better, easier to understand results than regression models.

Aside from modelling challenges, two other important factors are the time required to train the model and to compute a forecast. Parameter fitting is significantly faster for the IPCTMC model than for the LRA model. Although the fitting procedure described in Section 3.1 sounds work intensive, clustering and phase-type fitting can generally be done quickly. For the purpose of this study we fitted phase-type distributions by hand, but there is no reason why this cannot be automatised. The time required to fit the LRA model in R on the other hand depends on the amount of training data available, the number of ARIMA configurations that we are prepared to test and on how large we choose m (cf. Section 4) to be. In general, if models can be fitted offline, either IPCMTC or LRA fitting is sufficiently fast as it can usually be done within minutes. Note, however, that we do need to fit an LRA model for each forecast interval length, whereas the same IPCTMC model can be used for different interval lengths.

The subsequent forecast speed for either LRA and IPCTMC models is negligably small, e.g. point forecasts can be computed in less than 10 seconds using either model. Moreover, due to the linear nature of the model, it is also feasible to use simulation in order to analyse the IPCTMC models. However, if we were to add any non-linear rates to our model, for instance in order to investigate gossip like behaviour like in [23], then mean-field analysis would become significantly faster than simulation. In addition to this, non-linear relationships would make regression model fitting harder and also more computationally expensive.

6   Conclusions

In this paper we investigated the benefit of using extra journey data on the accuracy of future station arrival forecasts and compared the accuracy of time series models with those of a strictly inhomogeneous population model. It was surprising to see that the LRA models outperformed IPCTMCs for realistic predictions, since we thought that the time-inhomogeneous nature of the process would make it hard to fit such a model. However, we also showed that given perfect information on future departures, both models produce similar accuracy, which encourages further research into IPCTMC models, maybe not for forecasting purposes, but for tasks such as model based provisioning or policy impact prediction. In addition to this, IPCTMC models have the advantage that they can easily incorporate non-linear effects such as mass-action kinetics, which cannot be represented in linear models. Furthermore, our research on the accuracy of arrival forecasts shows that the origin destination information can be exploited to produce better short-term forecasts of 15 minutes or less. Moreover, the experiments with perfect information indicate that a better prediction accuracy of future departures is key to using this information to improve long-term predictions. In this work we only used ARIMA and naive prediction methods to estimate future departures, but other time dependent forecasting techniques, cf. [14], should be explored. In the future we would like to investigate, whether our small area arrival predictions can be successfully applied to improve station and area occupancy predictions similar to Yoon's and Kaltenbrunner's work and whether we can apply IPCTMC models to investigate phenomena such as geographical crowd data sourcing in a system of mobile agents.

Acknowledgements. Jeremy Bradley is supported in part by EPSRC on the AMPS project, the Analysis of Massively Parallel Stochastic Systems, ref. EP/G011737/1.

References

1. Stannard, A., Wolfenden, G.: Putting in Place a New Public Transport System in London. In: 18th ITS World Congress, Orlando, Florida (2011)
2. Le Masurier, P., Shore, F., Hiett, J.: Cycle-hire-The New Travel Option for Central London. In: European Transport Conference, Glasgow (2010)
3. Lin, J.-R., Yang, T.-H.: Strategic design of public bicycle sharing systems with service level constraints. Transportation Research Part E: Logistics and Transportation Review 47, 284–294 (2011)
4. Lathia, N., Ahmed, S., Capra, L.: Measuring the impact of opening the London shared bicycle scheme to casual users. Transportation Research Part C: Emerging Technologies 22, 88–102 (2012)
5. Froehlich, J., Neumann, J., Oliver, N.: Sensing and Predicting the Pulse of the City through Shared Bicycling. In: Twenty-First International Joint Conference on Artificial Intelligence, Pasadena (2009)
6. Kaltenbrunner, A., Meza, R., Grivolla, J., Codina, J., Banchs, R.: Urban cycles and mobility patterns: Exploring and predicting trends in a bicycle-based public transport system. Pervasive and Mobile Computing 6(4), 455–466 (2010)
7. Yoon, J.W., Pinelli, F., Calabrese, F.: Cityride: A Predictive Bike Sharing Journey Advisor. In: 2012 IEEE 13th International Conference on Mobile Data Management, pp. 306–311. IEEE (July 2012)
8. Kaleta, R.: An Integrated London Journey Planner. Master's thesis, Imperial College London (2012)
9. Li, J., Ren, C., Shao, B., Wang, Q., He, M., Dong, J., Chu, F.: A solution for reallocating public bike among bike stations. In: Proceedings of 2012 9th IEEE International Conference on Networking, Sensing and Control, pp. 352–355. IEEE (April 2012)
10. Slingsby, A., Dykes, J., Wood, J.: Visualizing the dynamics of Londons bicycle hire scheme. Cartographica the International Journal for Geographic Information and Geovisualization 46(4), 239–251 (2011)
11. O'Brien, O.: Bike Share Map (2010), http://bikes.oobrien.com

12. Stefanek, A., Hayden, R.A., Bradley, J.T.: Fluid analysis of energy consumption using rewards in massively parallel Markov models. In: 2nd ACMSPEC International Conference on Performance Engineering, ICPE, pp. 121–132 (2011)
13. Côme, E., Oukhellou, L.: Model-based count series clustering for Bike-sharing system usage mining, a case study with the V'lib' system of Paris. Submitted to ACM TIST (2012)
14. Borgnat, P., Abry, P., Flandrin, P., Robardet, C., Rouquier, J.-B., Fleury, E.: Shared Bicycle in a City: A Signal Processing and Data Analysis Perspective. Advances in Complex Systems 14, 415–438 (2011)
15. Guenther, M.C., Stefanek, A., Bradley, J.T.: Moment closures for performance models with highly non-linear rates. In: Tribastone, M., Gilmore, S. (eds.) EPEW/UKPEW 2012. LNCS, vol. 7587, pp. 32–47. Springer, Heidelberg (2013)
16. Galpin, V.: Towards a spatial stochastic process algebra. In: Proceedings of the 7th Workshop on Process Algebra and Stochastically Timed Activities (PASTA), Edinburgh (2008)
17. Hayden, R.A., Bradley, J.T.: A fluid analysis framework for a Markovian process algebra. Theoretical Computer Science 411(22-24), 2260–2297 (2010)
18. Stefanek, A., Hayden, R.A., Bradley, J.T.: A new tool for the performance analysis of massively parallel computer systems. In: Eighth Workshop on Quantitative Aspects of Programming Languages, QAPL 2010, Paphos, Cyprus, March 27-28 (2010)
19. Stefanek, A., Hayden, R.A., Bradley, J.T.: Mean-field Analysis of Large Scale Markov Fluid Models with Fluid Dependent and Time-Inhomogeneous Rates. Technical report (2013)
20. Reinecke, P., Krauss, T., Wolter, K.: HyperStar: Phase-Type Fitting Made Easy. In: 2012 Ninth International Conference on Quantitative Evaluation of Systems, pp. 201–202. IEEE (September 2012)
21. Hyndman, R.J.: Automatic Time Series Forecasting: The forecast Package for R. Journal of Statistical Software 27(3), 1–22 (2008)
22. Bowden, R.S., Clarke, B.R.: A single series representation of multiple independent ARMA processes. Journal of Time Series Analysis 33, 304–311 (2012)
23. Chaintreau, A., Le Boudec, J.Y., Ristanovic, N.: The age of gossip: spatial mean field regime. In: Evolution, pp. 109–120 (2009)

Moment Matching-Based Distribution Fitting with Generalized Hyper-Erlang Distributions

Gábor Horváth[1,2,3]

[1] Budapest University of Technology and Economics
Department of Networked Systems and Services
[2] MTA-BME Information Systems Research Group
Magyar Tudósok krt. 2, 1117 Budapest, Hungary
[3] Inter-University Center of Telecommunications and Informatics
ghorvath@hit.bme.hu

Abstract. This paper describes a novel moment matching based fitting method for phase-type (PH) distributions. A special sub-class of phase-type distributions is introduced for the fitting, called generalized hyper-Erlang distributions. The user has to provide only two parameters: the number of moments to match, and the upper bound for the sum of the multiplicities of the eigenvalues of the distribution, which is related to the maximal size of the resulting PH distribution. Given these two parameters, our method obtains *all* PH distributions that match the target moments and have a Markovian representation up to the given size. From this set of PH distributions the best one can be selected according to any distance function.

1 Introduction

Since their introduction, phase-type (PH) distributions have played an important role in performance and reliability modeling. PH distributions are simple, numerically tractable and easy to integrate into complex stochastic models.

However, the applicability of PH distributions for modeling real systems relies on efficient *fitting procedures*. A fitting procedure constructs a PH distribution based on empirical samples or based on an other known distribution.

A large number of PH fitting procedures have been published in the literature. This paper presents a fitting procedure that is based on moment matching. The moment matching problem of PH distributions can be formulated as the solution of a system of polynomial equations. It is possible, however, that none of the solutions is a valid PH distribution. To overcome this problem we introduce a sub-class of PH distributions, called generalized hyper-Erlang distributions, that can grow in size if the set of moments can not be realized with a given size. It is guaranteed that above a given size the generalized hyper-Erlang distributions can realize any valid moment set, thus any moment set that belongs to a positive distribution.

The rest of the paper is organized as follows. Section 2 introduces phase-type distributions, and shows how a Markovian representation is obtained from a

A. Dudin and K. De Turck (Eds.): ASMTA 2013, LNCS 7984, pp. 232–246, 2013.

non-Markovian one. Section 3 provides an overview on the PH fitting methods published in the literature, with an emphasis on the moment matching-based solutions. Section 4 describes generalized hyper-Erlang distributions, and how they are used to fit distributions. Some numerical examples are given in Section 5, and Section 6 concludes the paper.

2   Phase-Type Distributions

Phase-type distributions are given by two parameters: an initial vector $\alpha = \{\alpha_i, i = 1, \ldots, K\}, \alpha\mathbb{1} = 1$, and a matrix $\boldsymbol{A} = \{q_{ij}, i, j = 1, \ldots, K\}$. The vector-matrix pair $(\alpha, \boldsymbol{A})$ is called the *representation* of the PH distribution, and K is the size of the representation. The probability density function (pdf, denoted by $f(x)$), the cumulative distribution function (cdf, denoted by $F(x)$) and the kth moment of a PH($\alpha,\boldsymbol{A}$) distributed random variable $\mathcal{X}$ are

$$f(x) = \alpha e^{\boldsymbol{A}x}(-\boldsymbol{A})\mathbb{1}, \tag{1}$$

$$F(x) = P(\mathcal{X} < x) = 1 - \alpha e^{\boldsymbol{A}x}\mathbb{1}, \tag{2}$$

$$\mu_k = E(\mathcal{X}^k) = \int_0^\infty x^k f(x)dx = k!\alpha(-\boldsymbol{A})^{-k}\mathbb{1}, \tag{3}$$

where $\mathbb{1}$ is a column vector of appropriate size.

A representation is called a *Markovian representation*, if the entries of the initial vector are probabilities ($0 \leq \alpha_i \leq 1, i = 1, \ldots, K$) and matrix $\boldsymbol{A}$ is a generator of a transient continuous time Markov chain, thus, $q_{ii} < 0$, $q_{ij} \geq 0, \forall i \neq j$, and for the row sum we have that $\sum_{j=1}^K q_{ij} \leq 0, i = 1, \ldots, K$, with at least one state where the row sum is strictly negative. If $(\alpha,\boldsymbol{A})$ is a Markovian representation, then the corresponding PH distribution has a probabilistic interpretation as well: $\mathcal{X}$ is the absorption time of the transient Markov chain with sub-generator $\boldsymbol{A}$ and initial state probability vector α of the non-absorbing states.

The pdf can be expressed in a spectral form as well. Suppose the number of distinct eigenvalues is n_d. Let us denote the eigenvalues by $-\lambda_i$, and their multiplicity by r_i ($\sum_{i=1}^{n_d} r_i = N \leq K$). From (1) we have

$$f(x) = \sum_{i=1}^{n_d} \sum_{j=1}^{r_i} b_{ij} \frac{(\lambda_i x)^{j-1}}{(j-1)!} \lambda_i e^{-\lambda_i x}. \tag{4}$$

Note that if $\lambda_i \in \mathbb{C}\backslash\mathbb{R}$ then $\exists j \neq i : \lambda_j = \bar{\lambda}_i$ ($\bar{\lambda}_i$ denotes the complex conjugate of λ_i). To define a valid distribution re$\langle\lambda_i\rangle > 0, i = 1, \ldots, n_r$ must hold. Furthermore, as a consequence of the Perron-Frobenius theorem, the dominant eigenvalue (i.e., the eigenvalue with the largest real part) must be real.

When $N = K$, the $(\alpha,\boldsymbol{A})$ representation is called *minimal*. If $N < K$ then matrix $\boldsymbol{A}$ has at least one eigenvalue that does not play a role in the pdf, because the corresponding coefficients are zero.

2.1   Similarity Transformation of the Representations

The (α, A) representation of PH distributions is not unique. According to the following Theorem, different similarity transformations generate different representations of the same distribution.

Theorem 1. *(From [5], Theorem 3.) If there exists a matrix $W \in \mathbb{R}^{N,M}, M \geq N$, such that $W\mathbb{1} = \mathbb{1}$, $AW = WB$ and $\alpha W = \beta$, then (α, A) and (β, B) define the same distribution.*

Transforming a representation (α, A) with matrix W can destroy the Markovian property of the representation. In this paper, however, we are using similarity transforms to achieve the opposite effect, thus to find a Markovian representation starting from a non-Markovian one, which may require to inflate to size of the representation $(M > N)$.

2.2   Obtaining a Markovian Representation from a Non-Markovian One

In [12] a special representation, called *monocyclic representation* is defined, which has an important feature phrased by the following Theorem.

Theorem 2. *[12] Every PH distribution has a Markovian monocyclic representation.*

A monocyclic representation consists of Feedback-Erlang Blocks (FEB) arranged in a row. FEB_i is characterized by a rate parameter ν_i, a size parameter k_i and a feedback probability z_i (see Figure 1, where $k_1 = 1, k_2 = 4, k_3 = 1$).

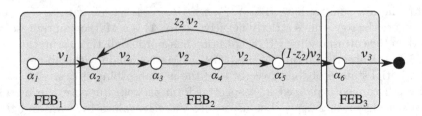

Fig. 1. Monocyclic representation of a PH distribution

[12] provides a constructive algorithm to obtain the Markovian monocyclic representation. We will rely on this result heavily in this paper, as the equations solved to match the moments typically result in a non-Markovian representation.

The transformation of a non-Markovian representation (α, A) to a Markovian (monocyclic) representation (β, B) consists of the following steps ([12]).

1. In the first step matrix B is constructed. Each FEB implements one real eigenvalue or a conjugate complex eigenvalue pair of A. Let us denote the jth eigenvalue of A by $-\lambda_j$, or, if it is a complex conjugate eigenvalue pair, by $-\lambda_j = a_j + b_j\mathrm{i}$ and $-\bar{\lambda}_j = a_j - b_j\mathrm{i}$.

- If λ_j is real, the parameters of the jth FEB are $\nu_j = \lambda_j, k_j = 1, z_j = 0$.
- If λ_j is complex, the parameters of the corresponding FEB are determined as

$$k_j = \text{the smallest integer for which } a_j/b_j > \tan(\pi/k_j), \qquad (5)$$

$$\nu_j = \frac{1}{2}\left(2a_j - b_j \tan\frac{\pi}{k_j} + b_j \cot\frac{\pi}{k_j}\right), \qquad (6)$$

$$z_j = \left(1 - \left(a_j - b_j \tan\frac{\pi}{k_j}\right)\right)^{k_j}. \qquad (7)$$

With these parameters matrix B is Markovian by construction and contains all eigenvalues of A with the proper multiplicities. However, the size of matrix B can be larger than the size of matrix A, meaning that new eigenvalues are introduced. These extra eigenvalues do not play a role in the pdf, as vector β will be such that they have zero coefficients.

2. The second step of the procedure is calculating the initial vector β. To this end, we need to obtain the transformation matrix W that transforms matrix A to matrix B.

   According to Theorem 1 matrix W is the solution of $AW = WB, W\mathbb{1} = \mathbb{1}$, which is a linear system of equations with regards to the entries of W. With the presented construction of B, this linear system always has a unique solution. If the size of A is N, and the size of B is M, equation $AW = WB$ has $N \times M$ unknowns and defines $N \times M$ equations. However, only $N \times M - N$ equations will be independent, since N eigenvalues of A and B are the same. With $W\mathbb{1} = \mathbb{1}$, however, we get $N \times M$ independent equations and obtain a unique solution.

   The initial vector is then given by $\beta = \alpha W$, which may not be a probability vector.

3. The third, last step is necessary only if vector β is not a proper probability vector. In this case, an Erlang tail needs to be appended to the row of FEBs. This Erlang tail is added to matrix B, the corresponding transformation matrix W is calculated, and we get a new initial vector. It is proven in [12] that an appropriate Erlang tail always makes the representation Markovian, if (α, A) defines a PH distribution. Unfortunately there is no explicit way to obtain the size and the rate parameter of the Erlang tail. We apply a simple heuristic that increases the size of the Erlang tail successively and applies the secant method to find its rate parameter such that β is a probability vector.

3 Fitting Methods for Phase-Type Distributions

In this section we give a short overview on some PH fitting procedures published in the past. This overview will not be exhaustive, and concentrates on the moment matching based solutions, as the fitting method presented in this paper belongs to this category as well.

Generally speaking, PH distribution fitting means to find a vector α and a matrix A such that the corresponding PH distribution is as close to the target distribution as possible. The most popular methods can be classified as follows.

- *Expectation-maximization (EM) based methods.* These iterative methods provide a PH distribution from a measurement trace (given by samples) by increasing the likelihood in each iteration. Some variants operate on the whole class of PH distributions ([1] and [13]), some others restrict themselves to a specific sub-class like hyper-exponential distributions ([7]) or hyper-Erlang distributions ([15]). The EM algorithm based methods have the advantage that they are very user friendly, the only input parameter required for the fitting is usually only the number of states of the resulting PH distribution. The execution time, however, increases with the length of the measurement trace, furthermore, these methods perform a local maximization and possibly can not find the global optimum.
- *Other optimization based methods.* Some PH fitting methods formulate and solve the PH fitting as a non-linear optimization problem. For instance, [9] applies sequential linear programming to optimize the parameters of an acyclic PH distribution according to a user-selectable distance function (some possible distance functions can be found in [3]). The optimization can take a long time, and there is no guarantee that the global optimum is found.
- *Matching based methods.* Matching based methods intend to match some statistical quantities of the target distribution exactly. In [8] a hyper-exponential distribution is constructed such that its pdf matches the pdf of the target distribution at selected points. The point selection needs to be done by the user by manual tuning, and the resulting distribution might not be a valid distribution in some cases. The moment matching based methods fall into this category as well, that we discuss in more details soon.

In the practice, the selection between the large number of PH fitting methods depends on many factors. Typically several methods are used and the one that gives the best result is selected.

3.1 PH Distribution Fitting Based on Moment Matching

Fitting distributions based on the moments has some distinct advantages:

- The moments of random variables are easy to obtain. This holds for measurement and for simulation as well. When only the moments are required, there is no need to collect and store all the samples of the observed variable, moments can be calculated in a progressive way with negligible memory and computation effort.
- Some performance measures depend only on the moments of some random variables, thus it makes sense to capture them accurately. For instance, the mean waiting time in an M/G/1 queue depends only on the first two moments of the service time distribution.

On the other hand, there are some drawbacks as well:

- Solving the equations to match the moments explicitly can be difficult if the number of moments to match is high, or if the structure of the PH is not simple enough.
- There are heavy tailed distributions with infinite moments. Moment matching can not be applied to fit such distributions.

There are a number of moment matching methods available in the literature. Based on how flexible the structure of the PH distribution is, we can distinguish between *fixed structure* and *flexible structure* methods.

- *Fixed structure* methods use a PH representation in which the number of parameters equals the number of moments to match. In [14] a PH of size 2 is used to match 3 moments, while [10] is able to match 5 moments with 3 states (it is known that the canonical form of 2-state PH distributions have 3, the one of 3-state PH distributions have 5 parameters). In [6] a hyper-exponential distribution of size N is obtained based on $2N - 1$ moments. Methods belonging to this category have the drawback that only those moments can be matched that fall into the region of feasible moments of the applied PH distribution. The region of feasible moments can be quite restrictive in the practice. Suppose we have a distribution whose first 3 moments can not be matched with a 2-state PH distribution. One would think that increasing the size of the PH can help, and there will be a better chance that a 3-state PH distribution will match the required moments. Actually the opposite is true. The contradiction is, that more (namely 5) moments need to be matched to obtain a 3-state PH distribution. The more moments we have to match, the larger the chance is that at least one of them lies outside of the feasible region.
- *Flexible structure* methods use PH representations that have *more* parameters than the number of moments to match. A set of parameters are set to ensure the matching of moments, while the remaining parameters add some extra degrees of freedom to obtain a valid PH distribution to match any set of moments. Such a method is described in [11], which uses a representation called *mixture of Erlang distributions of common order* (MECO). Mixing N Erlang distributions of common order has $2N$ parameters: $N - 1$ initial probabilities (which Erlang to choose, sums up to 1), the intensity parameters of the Erlang distributions (there are N of them), and the common order of the Erlang distributions (1 parameter). With these $2N$ parameters this procedure can match $2N - 1$ moments. The free parameter not involved in moment matching is the order parameter. The moment matching is performed with order=1, order=2, etc., the order is increased till a Markovian solution is found. It is guaranteed that, with appropriately large order, this procedure is able to match any set of $2N - 1$ moments belonging to a positive distribution. The other method operating on a flexible structure has been published in [2]. It matches 3 moments with an exponential and an Erlang distribution connected after each other in a row. The degree of freedom is the order of the Erlang component again. It is proven that, by choosing the

order of the Erlang component appropriately large, it is possible to match any 3 moments with this structure.

As this quick summary suggests, moment matching methods having a *flexible structure* are very convenient to use in the practice. The user just has to enter the moments, and these procedures find the appropriate PH distribution of appropriate size automatically. The method presented in this paper falls into this category as well.

4   Moment Matching with Generalized Hyper-Erlang Distributions

In this section we propose a new sub-class of PH distributions for matching moments.

Definition 1. *Random variable* $\mathcal{X}$ *has order-N generalized hyper-Erlang distribution iff its density function is*

$$f(x) = \sum_{i=1}^{N} \alpha_i \frac{(\lambda_i x)^{r_i - 1}}{(r_i - 1)!} \lambda_i e^{-\lambda_i x}, \tag{8}$$

with $f(x) \geq 0$ *and* $\int_0^\infty f(x)dx = 1$. *For the parameters we have that* $\lambda_i \in \mathbb{C}, re\langle \lambda_j \rangle \geq 0, \alpha_j \in \mathbb{C}, \sum_{i=1}^N \alpha_i = 1$ *and* $r_i \in \mathbb{N}$ *for* $i = 1, \ldots, N$.

Thus, generalized hyper-Erlang distributions (GHErD) are similar to hyper-Erlang distributions, the difference is that coefficients α_i do not need to be valid probabilities, and that λ_i can be complex as well. The kth moment of generalized hyper-Erlang distributions is calculated as

$$\mu_k = \int_0^\infty x^k f(x)dx = \sum_{i=1}^N \alpha_i \frac{(k + r_i - 1)!}{(r_i - 1)!} \frac{1}{\lambda_i^k}, \quad k \geq 0. \tag{9}$$

(Note that $\mu_0 = 1$).

4.1   Solution of the Moment Matching Problem When the r_i Parameters Are Fixed

For matching moments $\mu_1, \ldots, \mu_{2N-1}$ with order-N GHErD having the r_i parameters fixed we have to solve a system of polynomial equations defined by (9) for $k = 0, \ldots, 2N - 1$, such that the unknown variables are $\lambda = \{\lambda_i, i = 1, \ldots, N\}$ and $\sigma = \{\alpha_i, i = 1, \ldots, N\}$, which give $2N$ unknowns in total.

Due to the structure of the system of polynomial equations we were not able to derive an explicit solution for arbitrary N. However, there are excellent tools available that are able to solve polynomial systems numerically. For this purpose, we are using PHCpack (see [16]), which is a multi-platform open-source tool and is under continuous development and refinement. We would emphasize that this does not mean that we are applying a non-linear programming or other optimization methods to find the solution of the moment matching problem (as it is done in [4]). What we are doing is the numerical solution of the polynomial system, that is able to provide *all* the solutions of the system of polynomial equations.

This polynomial system has typically several solutions, and it is also possible that it has no solutions at all (it is inconsistent). If it does have solutions, each solution either defines a valid PH distribution, or it does not. To decide if a solution is valid, we try to obtain a Markovian representation by using the method described in Section 2.2. From a solution given by vectors λ and σ the initial (non-Markovian) representation is obtained in a direct way as

$$\alpha = [\alpha_1 \underbrace{0 \ldots 0}_{r_1-1} \alpha_2 \underbrace{0 \ldots 0}_{r_2-1} \alpha_N \underbrace{0 \ldots 0}_{r_N-1}],$$

$$A = \begin{bmatrix} -\lambda_1 & \lambda_1 & & & & \\ & \ddots & \ddots & & & \\ & & -\lambda_1 & & & \\ & & & -\lambda_2 & \lambda_2 & \\ & & & & \ddots & \ddots \\ & & & & & -\lambda_2 \\ & & & & & & \ddots \\ & & & & & & & -\lambda_N & \lambda_N \\ & & & & & & & & \ddots & \ddots \\ & & & & & & & & & -\lambda_N \end{bmatrix} \begin{matrix} \left. \vphantom{\begin{matrix}a\\a\end{matrix}} \right\} r_1 \\ \\ \left. \vphantom{\begin{matrix}a\\a\end{matrix}} \right\} r_2 \\ \\ \\ \left. \vphantom{\begin{matrix}a\\a\end{matrix}} \right\} r_N \end{matrix} \qquad (10)$$

If the output (β, B) of the algorithm of Section 2.2 is Markovian, then we found a valid solution. Given vector r it may happen that several different PH distributions are found, but it is also possible that no solutions exist. In the latter case the entries of vector r need to be increased to obtain a valid solution.

4.2   Optimizing the r_i Parameters

Finding the appropriate vector r can be made automatic as well. In this case the user just has to enter a single parameter, R, and the algorithm repeats the moment matching with all vectors r satisfying $\sum_{i=1}^{N} r_i \leq R$.

As the MECO is a sub-class of GHErD, furthermore, according to [11] it is always possible to find a MECO for any set of moments, our procedure always finds a Markovian solution (with an appropriately large R parameter).

Our algorithm is depicted in Figure 2. First, the algorithm solves the moment matching problem with different r vectors up to $\sum_{i=1}^{N} r_i \leq R$ and collects all solutions that have a Markovian representation in set *res*. Notice that all solutions in *res* match the first $2N - 1$ moments of the target distribution. In the second step (last line of the algorithm) the best solution is selected by a distance

```
 1: procedure FITGHERD(μ₁,...,μ₂ₙ₋₁, R, D(·))
 2:     res ← {}
 3:     for all r = [r₁ ... rₙ] with ∑ᵢ₌₁ᴺ rᵢ ≤ R do
 4:         {(λ,σ)ᵢ} ← Solve polynomial equations of (9) for k = 0,...,2N − 1
 5:         for each solution (λ,σ) do
 6:             (α, A) ← Create non-Markovian representation based on (10)
 7:             (β, B) ← Transform (α, A) to a Markovian representation (Section 2.2)
 8:             if (β, B) is Markovian then
 9:                 Add (β, B) to set res
10:             end if
11:         end for
12:     end for
13:     if res is {} then
14:         error "No solutions found up to size R. Parameter R has to be increased."
15:     end if
16:     (γ, G) ← arg min₍β,B₎∈res D(β, B)
17:     return (γ, G)
18: end procedure
```

Fig. 2. GHErD fitting algorithm based on moment matching

function $D(\cdot)$. Any distance function can be used that quantifies the distance between two distributions. Two possible distance functions are:

- Moment distance (MD): the sum of the squared relative difference of the moments up to moment K. Denoting the kth moment of the target distribution by $\hat{\mu}_k$ this means

$$D(\beta, \boldsymbol{B}) = \sum_{k=1}^{K} \left(\frac{\mu_k^{(\beta,\boldsymbol{B})} - \hat{\mu}_k}{\hat{\mu}_k} \right)^2. \tag{11}$$

(Note that for $k \leq 2N - 1$ we have $\mu_k^{(\beta,\boldsymbol{B})} = \hat{\mu}_k$).

- Relative entropy (RE): the relative entropy has been introduced in [3] as a measure of goodness of the approximation. By denoting the pdf of the target distribution by $\hat{f}(x)$ it is

$$D(\beta, \boldsymbol{B}) = \int_0^\infty f(x) |\log \frac{f(x)}{\hat{f}(x)}| dx, \tag{12}$$

where $f(x) = \beta e^{\boldsymbol{B}x}(-\boldsymbol{B})\mathbb{1}$.

Both distance functions have their advantage. The moment distance still relies only on the moments, thus the exact shape of the target distribution (eg. the pdf) is not required. The relative entropy, however, may quantify the similarity of the shape of the density functions better. Both distance functions will be evaluated in the subsequent numerical examples.

5   Numerical Examples

In this section we apply the presented fitting method on two well known traffic measurement traces, the BC-pAug89 and the LBL-TCP-3 traces[1]. The former one records one million packet arrivals on an Ethernet network, while the latter one captures two hours of wide-area TCP traffic. These traces are too old to be representative for the network traffic in these days, the only reason we selected them is that they both are frequently used as demonstration of various traffic fitting algorithms, thus they serve as a kind of benchmark.

Our method has been implemented in MATLAB. To solve the system of polynomial equations we used PHCpack v2.3.76[2], that has a MATLAB interface as well. All the results have been calculated on an average PC with a CPU clocked at 3.4 GHz and 4 GB of memory.

In general, our experience with PHCpack is positive. We got some warnings that the polynomial system is ill-conditioned initially, but normalizing the trace (such that the mean value is one) solved this issue. The solution time of the polynomial system depends on N heavily. We got prompt results in case of $N = 2$, it took about $0.5s$ in case of $N = 3$, and $25s$ in case of $N = 4$. As our method relies on the iterative solution of a large number of polynomial equations, we did not increase parameter N above 4, which means that we are going to match at most 7 moments in the numerical examples. When N is large, the number of solutions is large as well, however trying to create the corresponding Markovian representations is such fast that is has negligible effect on the execution time. The total execution time of the algorithm depends on parameter R as well. If R is large, a large number of r vectors are considered, and a large number of polynomial systems are solved.

In the subsequent case studies we compare the following 4 moment matching-based PH fitting methods:

1. Our method, where the PH distribution with the smallest moment distance is selected from the set of all PH distributions matching the moments of the trace;
2. Our method, where the PH distribution with the smallest relative entropy is selected from the set of all PH distributions matching the moments of the trace;
3. Moment matching with a mixture of Erlang distributions of common order (MECO, [11]);
4. The method of [2], which is able to match the first three moments only.

5.1   The LBL Trace

In this example the R parameter (the sum of the multiplicities of the eigenvalues) is set to 20. According to the algorithm in Figure 2 this means that the moment

---

[1] Downloaded from `http://ita.ee.lbl.gov/html/contrib/BC.html` and from `http://ita.ee.lbl.gov/html/contrib/LBL-TCP-3.html`, respectively.

[2] It can be obtained from `http://homepages.math.uic.edu/~jan/download.html`

matching is performed with all vectors r satisfying $\sum_{i=1}^{N} r_i \leq R$. With a given vector r, the moment matching problem can result in several different valid PH distributions. At the end we have a large number of PH distributions from which we can select the best according to some distance function. Table 1 shows how many r vectors and valid solutions there are, and how long the execution time of the algorithm is.

Table 1. The number of different r vectors, the number of valid solutions, and the total execution time in the LBL example

# of moments to match	# of different r vectors	# of valid solutions	Execution speed
3	100	88	28 sec
5	237	688	337 sec
7	408	3920	810 min

The numerical results are shown in Table 2. When the first three moments are matched, all methods found the same solution. Even if a large number of r vectors have been checked by our method, the best solution has been found to be the same according to both distance functions.

When 5 moments are matched, the MECO matching method returned a hyper-exponential distribution that was found to be optimal by our method as well according to the distance of moments. However, our method was able to find a PH distribution that has lower relative entropy. This PH distribution has 5 states and the corresponding r vector is $r = \begin{bmatrix} 1 & 1 & 2 \end{bmatrix}$. The sum of the elements of vector r is only 4, which means that a new eigenvalue has been introduced and the size of the PH has been increased by 1 to obtain a Markovian representation (this new eigenvalue cancels out, it has no effect on the pdf).

The advantage and the flexibility of our method can be seen the best when 7 moments are matched. Our method was able to find a PH distribution with significantly lower moment distance, and an other one with significantly lower relative entropy as well.

Figures 3, 4 and 5 plot the density functions belonging to the methods discussed, both on linear and on logarithmic scale. While the tail of the pdf is fitted well by all methods, the plots differ significantly in the body of the pdf. Based on a visual comparison, the solution found by our method by matching 7 moments and selecting the best according to the RE distance seems to capture the shape of the pdf best.

5.2 The BC Trace

The numerical results corresponding to the fitting of the BC trace are summarized in Table 3. When fitting the first 3 moments, the same hyper-exponential

Table 2. Results of the fitting of the LBL trace

Num. of moments	Method	MD	RE	Num. of states
3	Our method (MD)	1.786	0.3024	$2\ (r = \begin{bmatrix}1\ 1\end{bmatrix})$
	Our method (RE)	1.786	0.3024	$2\ (r = \begin{bmatrix}1\ 1\end{bmatrix})$
	MECO [11]	1.786	0.3024	$2\ (r = \begin{bmatrix}1\ 1\end{bmatrix})$
	ErlExp [2]	1.786	0.3024	2
5	Our method (MD)	0.0072	0.0984	$3\ (r = \begin{bmatrix}1\ 1\ 1\end{bmatrix})$
	Our method (RE)	0.0386	0.0953	$5\ (r = \begin{bmatrix}1\ 1\ 2\end{bmatrix})$
	MECO [11]	0.0072	0.0984	$3\ (r = \begin{bmatrix}1\ 1\ 1\end{bmatrix})$
7	Our method (MD)	8.26×10^{-6}	3.9727	$20\ (r = \begin{bmatrix}2\ 2\ 3\ 13\end{bmatrix})$
	Our method (RE)	0.00499	0.0727	$8\ (r = \begin{bmatrix}1\ 1\ 3\ 3\end{bmatrix})$
	MECO [11]	0.00475	0.1339	$8\ (r = \begin{bmatrix}2\ 2\ 2\ 2\end{bmatrix})$

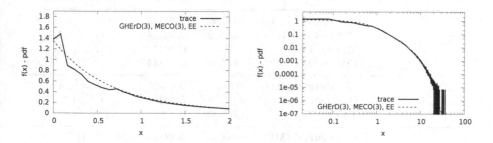

Fig. 3. Comparison of the density functions matching 3 moments of the LBL trace

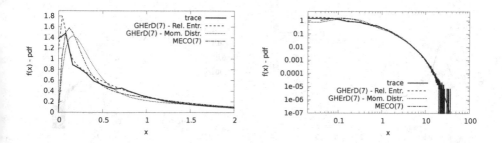

Fig. 4. Comparison of the density functions matching 7 moments of the LBL trace

distribution turned out to be optimal by all the methods involved into the comparison. When fitting 5 phases, our method has found a PH distribution with slightly lower relative entropy. In the case when 7 moments are matched, we have a PH distribution with better moment distance, and an other one with significantly better relative entropy than the MECO based method.

The density functions corresponding to the investigated cases are depicted in Figure 6, 7 and 8. Figure 7 demonstrates how different the shapes of the density

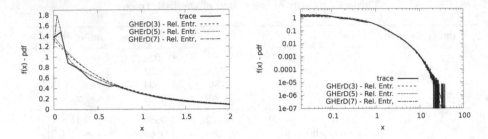

Fig. 5. Comparison of the results of our method, matching 3, 5 and 7 moments

Table 3. Results of the fitting of the BC trace

Num. of moments	Method	MD	RE	Num. of states
3	Our method (MD)	3.0509	0.30244	$2\ (r = \begin{bmatrix}1\ 1\end{bmatrix})$
	Our method (RE)	3.0509	0.30244	$2\ (r = \begin{bmatrix}1\ 1\end{bmatrix})$
	MECO [11]	3.0509	0.30244	$2\ (r = \begin{bmatrix}1\ 1\end{bmatrix})$
	ErlExp [2]	3.0509	0.30244	2
5	Our method (MD)	0.00198	0.30521	$14\ (r = \begin{bmatrix}1\ 4\ 8\end{bmatrix})$
	Our method (RE)	55.4699	0.30212	$5\ (r = \begin{bmatrix}1\ 1\ 2\end{bmatrix})$
	MECO [11]	55.4699	0.30212	$3\ (r = \begin{bmatrix}1\ 1\ 1\end{bmatrix})$
7	Our method (MD)	0.0056	0.48178	$20\ (r = \begin{bmatrix}1\ 2\ 2\ 15\end{bmatrix})$
	Our method (RE)	0.0072	0.185	$19\ (r = \begin{bmatrix}1\ 1\ 2\ 15\end{bmatrix})$
	MECO [11]	0.0391	0.3536	$16\ (r = \begin{bmatrix}4\ 4\ 4\ 4\end{bmatrix})$

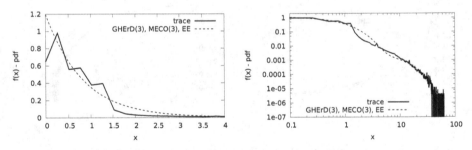

Fig. 6. Comparison of the density functions matching 3 moments of the BC trace

functions can be even if the first 7 moments are the same. The MECO-based method looks to be the least successful in this example, while our method with the relative entropy based selection managed to capture the characteristics of the density function reasonably well.

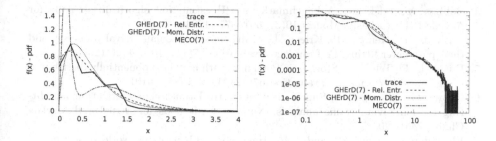

Fig. 7. Comparison of the density functions matching 7 moments of the BC trace

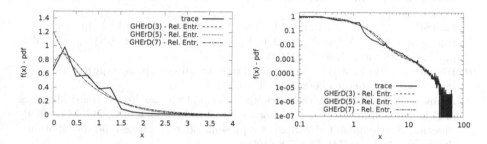

Fig. 8. Comparison of the results of our method, matching 3, 5 and 7 moments

6   Conclusion

This paper describes a unique approach to PH distribution fitting. The presented method is based on moment matching, and allows to use an arbitrary distance function to select the best solution from the ones matching the target moments.

Two case studies are presented to demonstrate the behavior and the capabilities of the procedure when fitting real measurement traces.

The weak point of the procedure is the numerical solution of the polynomial system, which limits the number of moments to match.

Acknowledgment. This work was supported by the Hungarian Government through the TAMOP-4.2.2C-11/1/KONV-2012- 0001 and the OTKA K101150 projects, and by the János Bolyai Research Scholarship of the Hungarian Academy of Sciences.

References

1. Asmussen, S., Nerman, O., Olsson, M.: Fitting phase-type distributions via the EM algorithm. Scandinavian Journal of Statistics, 419–441 (1996)
2. Bobbio, A., Horváth, A., Telek, M.: Matching three moments with minimal acyclic phase type distributions. Stochastic Models 21(2-3), 303–326 (2005)

3. Bobbio, A., Telek, M.: A benchmark for PH estimation algorithms: results for Acyclic-PH. Stochastic Models 10(3), 661–677 (1994)
4. Buchholz, P., Kemper, P., Kriege, J.: Multi-class Markovian arrival processes and their parameter fitting. Performance Evaluation 67(11), 1092–1106 (2010)
5. Buchholz, P., Telek, M.: Stochastic Petri nets with matrix exponentially distributed firing times. Performance Evaluation 67(12), 1373–1385 (2010)
6. Casale, G., Zhang, E.Z., Smirni, E.: Interarrival times characterization and fitting for markovian traffic analysis. Numerical Methods for Structured Markov Chains 7461 (2008)
7. El Abdouni Khayari, R., Sadre, R., Haverkort, B.R.: Fitting world-wide web request traces with the EM-algorithm. Performance Evaluation 52(2), 175–191 (2003)
8. Feldmann, A., Whitt, W.: Fitting mixtures of exponentials to long-tail distributions to analyze network performance models. Performance Evaluation 31(3), 245–279 (1998)
9. Horváth, A., Telek, M.: PhFit: A general phase-type fitting tool. In: Field, T., Harrison, P.G., Bradley, J., Harder, U. (eds.) TOOLS 2002. LNCS, vol. 2324, pp. 82–91. Springer, Heidelberg (2002)
10. Horváth, G., Telek, M.: On the canonical representation of phase type distributions. Performance Evaluation 66(8), 396–409 (2009)
11. Johnson, M.A., Taaffe, M.R.: Matching moments to phase distributions: Mixtures of Erlang distributions of common order. Stochastic Models 5(4), 711–743 (1989)
12. Mocanu, Ş., Commault, C.: Sparse representations of phase-type distributions. Stochastic Models 15(4), 759–778 (1999)
13. Okamura, H., Dohi, T., Trivedi, K.S.: A refined EM algorithm for PH distributions. Performance Evaluation 68(10), 938–954 (2011)
14. Telek, M., Heindl, A.: Matching moments for acyclic discrete and continuous phase-type distributions of second order. International Journal of Simulation Systems, Science & Technology 3(3-4) (2002)
15. Thummler, A., Buchholz, P., Telek, M.: A novel approach for fitting probability distributions to real trace data with the EM algorithm. In: Proceedings of the International Conference on Dependable Systems and Networks, DSN 2005, pp. 712–721. IEEE (2005)
16. Verschelde, J.: Algorithm 795: PHCpack: A general-purpose solver for polynomial systems by homotopy continuation. ACM Transactions on Mathematical Software (TOMS) 25(2), 251–276 (1999)

Output Process in Batch-Arrival Queue with N-Policy and Multiple Vacations

Wojciech M. Kempa

Silesian University of Technology, Institute of Mathematics,
ul. Kaszubska 23, 44-100 Gliwice, Poland
wojciech.kempa@polsl.pl

Abstract. In the paper the $M^X/G/1$-type queueing system with the N-policy and multiple vacations is considered. The output process, counting successive departures, is studied using the approach consisting of two main stages. Firstly, introducing an auxiliary model with the N-policy and multiple vacations, and applying the formula of total probability, the analysis is brought to the case of the corresponding system without restrictions in the service process, on its first busy cycle. Next, defining a delayed renewal process of successive vacation cycles, the general results are obtained. The explicit formula for the probability generating function of the Laplace transform of the distribution of the number of packets completely served before a fixed moment t is derived and written using transforms of "input" distributions of the system, and components of the Wiener-Hopf-type factorization identity connected with them. Moreover, illustrative numerical results are presented.

Keywords: Batch-arrival queueing system, multiple vacations, N-policy, output process, transient state.

1 Introduction

Queueing models are used in almost all fields of modern technical sciences and economics. Their application, however, is particularly evident in computer sciences and teletraffic engineering, where they are used in modelling of the operation of computer or telecommunication networks, like e.g. the Internet. In practice it is rare to observe the situation in which the incoming packets, calls, customers etc. are being served without any disruptions. Temporary service suspensions may be caused by different factors, e.g. by the requirement of server's maintenance or introducing the new service discipline, random server's breakdowns, or, finally, by the necessity of reducing the costs of system's operation.

According to the last possibility, different-type queueing models with server vacations are considered, in which the service process is blocked for a random time period, usually when the system is idle (free of packets). One of the main vacation disciplines is the multiple vacation policy in which the server takes on successive single vacations as far as, at the end of one of them, at least one packet is present in the system and waiting for service. The multiple vacation policy often occurs in

A. Dudin and K. De Turck (Eds.): ASMTA 2013, LNCS 7984, pp. 247–261, 2013.

real-life situations, especially when the costs of the system operation are high, and it is desired to reduce to minimum the time in which the server is in standby mode but is idle. Thus, evidently, it can be useful in modelling of computer networks, manufacturing processes, transportation etc. For example, queueing models with multiple vacations are being used in modelling the operation in the sleep mode of Mobile Stations in mobile WiMAX systems (see e.g. [13]).

Of course, we can affect the efficiency of the system's operation by manipulating the parameters of a random (or non-random) period of server's unavailability. Even greater control, from the perspective of optimization of the system operation, can be achieved by introducing additionally the so-called N-policy into the service discipline. In the system with multiple vacations and the N-policy the server takes on successive periods of unavailability as far as, at the end of one of them, at least N packets (calls, customers etc.) are acumulated in the queue. The N−policy can be used in situations in which the costs of keeping the server in standby and control it during a short period of time only (e.g. in the case of low arrival rate and relatively fast service intensity), including times for the setup and closedown, are disproportionately high in relation to the potential benefits. Indeed, we can change the threshold N in dependence on the service rate to provide more efficient using of the server. One of important applications of queueing systems with the N-policy is a "queued wake up" power saving mechanism in wireless sensor networks (see e.g. [4]).

One of the main stochastic characteristics helpful in the performance evaluation of each queueing model is the output process (departure process) $h(t)$, that at any fixed time t takes on a random value equal to the number of packets (calls etc.) completely served before the time t. In this paper we study this characteristic in the single-server queueing system with infinite buffer, batch Poisson arrivals and individual generally-distributed services, in which the multiple vacation with the N-policy discipline is implemented.

One can find the first review of results concerning the systems with server vacations in [3]. Besides, there are two important monographs [14] and [17] devoted to vacation queueing models. It is easy to note that majority of results obtained for systems with vacations relate to the stationary state of the system i.e. to the case as the time t tends to infinity. The formulae derived for main stochastic characteristics in the transient state of the system (at the fixed time t) are rare and often given in the forms which are difficult in numerical treatment.

Results obtained for the output process in systems with vacations one can find mainly in papers of Tang (see e.g. [15], [16]). In [15] the stationary departure process was investigated in the system of the $M/G/1$ type with vacations. The representation for the mean of departures before the fixed time t was obtained there. In [16] those results were generalized for the case of the system with batch arrivals.

Some results derived for the $M^X/G/1$ system with the N-policy and multiple vacations can be found e.g. in [11] and [12]. Moreover, in [6] and [8] the departure process was studied for the inifinite-buffer system without the threshold N, with single and multiple vacations, respectively. The compact formulae for the double transform of the number of packets completely served before the fixed time t were obtained there. Transient results for the vacation models with infinite buffers can also be found e.g. in [5] and [7], where the joint distribution of the components of the first busy cycle, and the queue-size distribution were studied, respectively. In [9] and [10] the output process was analyzed in the case of finite-buffer system.

Results obtained for the output process in the model with multiple vacations and the N-policy are almost invisible in the literature. This fact provides an additional incentive to address this problem.

In the paper we propose an approach that is, in fact, a certain step-by-step procedure and can be described as follows. After the description of the model (Section 2) we consider the departure process in the "ordinary" system, without vacations and the threshold N, corresponding to the original one, and present some results obtained for such a system (Section 3). In Section 4 we study the output process in a certain modified model, using the formula of total probability and applying results obtained for the "ordinary" model. In Section 5 we get the explicit formula for the PGF (probability generating function) of the LT (Laplace transform) of the number of packets served before t in the original system, applying the renewal-theory approach. In the last Section 6 we give numerical results illustrating the theoretical formulae for the $M/M/1$-type queueing model with single arrivals, exponentially distributed successive vacations and $N = 2$.

2   Queueing Models

In the paper we consider, in fact, three queueing models. Let us start with the original (main) one. That is the $M^X/G/1$-type system with multiple vacations and the N-policy. We assume that batches of packets occur according to a Poisson process with rate λ, while the batch sizes equal k with probability p_k. Packets are being served individually with a general-type d.f. (distribution function) $F(\cdot)$ of the service time, and the service discipline is supposed to be of FIFO type. The system starts working in the "standard" way: it is empty before the initial moment and the first batch of packets arrives at $t = 0$. After each busy period the server begins a multiple vacation period consisting of a number of i.i.d. (independent and identically distributed) single vacations which are generally distributed random variables with a d.f. $V(\cdot)$. A multiple vacation period consists of exactly n single vacations if and only if during the nth single vacation the number of packets present in the system reaches N for the first time. After finishing each multiple vacation period a new busy period starts immediately.

One can investigate the evolution of the system on successive vacation cycles C_i, $i = 0, 1, ...$, described as follows:

$$C_0 = \tau_0, \quad C_i = V_i + \tau_i = \sum_{j=1}^{n_i} v_{ij} + \tau_i, \ i = 1, 2, ..., \tag{1}$$

where τ_i denotes the ith busy period duration, V_i is the ith multiple vacation period duration, v_{ij} denotes the jth single vacation duration "inside" V_i, and n_i stands for the number of single vacations in the ith multiple vacation period. In fact, we often identify a particular period with its duration. According to the order of successive summands on the right side of (1), we mean that each vacation cycle (except for the "zero" one, consisting only of a busy period) consists of a multiple vacation period followed by a busy period. Besides, we assume total independence of all random variables describing successive interarrival times, service times, batch sizes and single vacations.

In addition to the original system, we consider the $M^X/G/1$ system without vacations and the N-policy, corresponding to the original one. Let us call this system the "ordinary" one. The "ordinary" system begins a new busy period whenever a group of packets arrives into the empty system. All remaining features of the "ordinary" system are the same as in the original one. We distinguish two different initial conditions for the "ordinary" system: we denote by $\mathbf{P}_O^n\{\cdot\}$ and $\mathbf{E}_O^n\{\cdot\}$, respectively, the probability and mean on condition that the system contains exactly n packets just after the opening (at $t = 0+$). Similarly, $\mathbf{P}_O^{std}\{\cdot\}$ and $\mathbf{E}_O^{std}\{\cdot\}$ denote, respectively, the probability and mean on the "standard" initial condition of the system (as described above).

Lastly, we introduce a modified (auxiliary) system equivalent to the original one but with $C_0 = 0$. By $\mathbf{P}_M\{\cdot\}$ and $\mathbf{E}_M\{\cdot\}$ we denote the probability and mean for the modified system.

We end this section with introducing necessary notations. So, let us put

- $h(t)$ – for the number of departures occurring in the time period $(0, t)$ (output process);

- $f(\cdot)$ – for the LST (Laplace-Stieltjes transform) of the d.f. $F(\cdot)$;

- $p(z) = \sum_{k=1}^{\infty} p_k z^k$, $|z| \leq 1$ – for the PGF of the sequence (p_k);

- $I\{\mathbb{A}\}$ – for the indicator of a random event $\mathbb{A}$;

- $I_+\left[\int_{-\infty}^{\infty} e^{-sx} g(x)dx\right] = \int_0^{\infty} e^{-sx} g(x)dx$ – for the positive projection of the LT of any function $g(\cdot)$, if only $\int_{-\infty}^{\infty} e^{-\mathrm{Re}(s)x}|g(x)|dx < \infty$;

- $G^{j*}(\cdot)$ – for the jth Stieltjes convolution of arbitrary distribution function $G(\cdot)$ with itself, i.e. $G^{j*}(t) = \int_0^t G(t - y)dG^{(j-1)*}(y)$, $j \geq 1$, and $G^{0*}(t) = 1$ for $t > 0$;

- p_k^{j*} – for the kth term of the jth convolution of the sequence (p_k) with itself, i.e. $p_k^{j*} = \sum_{i=1}^k p_i^{(j-1)*} p_{k-i}$, $j \geq 1$, and $p_0^{0*} = 1$.

3   Output Process in the "Ordinary" System

Let us consider the evolution of the "ordinary" system, corresponding to the original one, during the "zero" busy period τ_0. Introduce the following notation for the PGF of the LT of the number of packets served before t, under two initial conditions introduced in the previous section:

$$D_O^n(z,\mu) = \sum_{m=0}^{\infty} z^m \int_0^{\infty} e^{-\mu t} \mathbf{P}_O^n\{h(t) = m, \, t \in \tau_0\}dt, \tag{2}$$

$$D_O^{std}(z,\mu) = \sum_{m=0}^{\infty} z^m \int_0^{\infty} e^{-\mu t} \mathbf{P}_O^{std}\{h(t) = m, \, t \in \tau_0\}dt, \tag{3}$$

where $\mu > 0$ and $|z| < 1$.

In the formulae stated below we use the fact (see [1]) that the function

$$1 - \frac{\lambda}{\lambda + s}p\big(zf(\mu - s)\big) \tag{4}$$

can be factorized in the strip $0 \le \mathrm{Re}(s) \le \mu$ as follows:

$$1 - \frac{\lambda}{\lambda + s}p\big(zf(\mu - s)\big) = f_+(z,\mu,s)f_-(z,\mu,s), \tag{5}$$

where $|z| \le 1$, and $f_\pm(z,\mu,s)$ are analytic and non-zero for $\mathrm{Re}(s) > 0$ and $\mathrm{Re}(s) < \mu$, respectively. The condition $f_\pm(z,\mu,\infty) = 1$ guarantees the uniqueness of the factorization (5). Besides, let us denote

$$P_+^{(0)}(z,\mu,x) = I\{x > 0\} + P_+(z,\mu,x), \tag{6}$$

where the LST (Laplace-Stieltjes transform) of $P_+(z,\mu,x)$ is given as

$$\int_0^{\infty} e^{-sx}dP_+(z,\mu,x) = f_+^{-1}(z,\mu,s) - 1, \quad \mathrm{Re}(s) \ge 0. \tag{7}$$

The representation for $P_+(z,\mu,x)$ cannot be found explicitly since the component $f_+(z,\mu,s)$ of the factorization identity (5) depends essentially on mathematical shapes of particular distributions $F(\cdot)$ and (p_k).

The following theorem one can find in [6]:

Theorem 1. *For any $\mu > 0$ and $|z| < 1$ we have*

$$D_O^n(z,\mu) = \frac{z\big(1 - f(\mu)\big)}{\mu\big(1 - zf(\mu)\big)}\left[1 - \big(zf(\mu)\big)^n\right.$$

$$\left. + z^n f_+(z,\mu,0) \int_0^{\infty} e^{-\mu y} \int_{0-}^{y} \big(1 - e^{-\lambda(y-v)}\big)dP_+^{(0)}(z,\mu,v)dF^{n*}(y)\right]. \tag{8}$$

Taking into consideration all possible "volumes" of the first arriving batch we can write

$$\mathbf{P}_O^{std}\{\cdot\} = \sum_{n=1}^{\infty} p_n \mathbf{P}_O^n\{\cdot\}, \tag{9}$$

and then from (8) we obtain the following corollary:

Corollary 1. *For any* $\mu > 0$ *and* $|z| < 1$ *the following representation holds true:*

$$D_O^{std}(z,\mu) = \frac{z(1 - f(\mu))}{\mu(1 - zf(\mu))}\left[1 - p(zf(\mu)) + f_+(z,\mu,0)\right.$$

$$\left. \times \int_0^\infty e^{-\mu y}\int_{0-}^y (1 - e^{-\lambda(y-v)})dP_+^{(0)}(z,\mu,v)\sum_{n=1}^\infty p_n z^n dF^{n*}(y)\right]. \quad (10)$$

Let $h(\tau_0)$ be the number of departures occurring during the "zero" busy period τ_0. In [1] one can find the following representation for the joint transform of τ_0 and $h(\tau_0)$ in the "ordinary" system, working under the "standard" initial condition:

Theorem 2. *For any* $\mu \geq 0$ *and* $|z| \leq 1$ *the following formula is true:*

$$\mathbf{E}_O^{std}\{e^{-\mu\tau_0}z^{h(\tau_0)}\} = z(1 - f_+(z,\mu,0)). \quad (11)$$

Besides, in [2] one can find the following result for the "ordinary" system which starts working with a fixed number n of packets just after the opening:

Theorem 3. *For* $n \geq 1$ *and* $\mu > 0$ *we have*

$$\mathbf{E}_O^n\{e^{-\mu\tau_0}\} = f^n(\mu) - f_+(1,\mu,0)$$

$$\times \int_0^\infty e^{-\mu y}\int_{0-}^y (1 - e^{-\lambda(y-v)})dP_+^{(0)}(1,\mu,v)dF^{n*}(y). \quad (12)$$

Lastly, let us state another result obtained in [1] (see also [8]):

Theorem 4. *For* $n \geq 1$ *and* $|z| < 1$ *the following formula holds true:*

$$\mathbf{E}_O^n\{z^{h(\tau_0)}\}$$

$$= z^{n+1}\int_0^\infty \int_{0-}^y Q(z,0,y-v)dP_+^{(0)}(z,0,v)dF^{n*}(y), \quad (13)$$

where

$$Q(z,\mu,x) = f_+(z,\mu,0)e^{-\lambda x} - \int_x^\infty dF_+(z,\mu,y) \quad (14)$$

and the LST of $F_+(z,\mu,x)$ *is defined as*

$$\int_0^\infty e^{-sx}dF_+(z,\mu,x) = f_+(z,\mu,s) - 1, \quad \operatorname{Re}(s) \geq 0. \quad (15)$$

4   Departure Process in the Modified System

Let us take into consideration the modified $M^X/G/1$-type queueing system with multiple vacations and the N-policy. Recall that the modified system starts working at $t = 0$ with the first vacation cycle C_1 (thus, in the modified system we have $C_0 = 0$). Let us restrict our considerations to the first cycle C_1 only. The formula of total probability allows to bring the analysis of the output process $h(t)$ in the modified system to the analysis of the same characteristic in the corresponding "ordinary" system (without vacations and the threshold N) on its first "zero" busy cycle. Let us note that

$$\mathbf{P}_M\{h(t) = m, t \in C_1\} = \sum_{k=1}^{3} \mathbf{P}_M\{(h(t) = m, t \in C_1) \cap A_k\}, \qquad (16)$$

where A_k, for $k = 1, 2, 3$, are the following random events:

- A_1 – before t the number of packets present in the system reaches N but the multiple vacation period V_1 ends after t (and hence the first busy period τ_1 begins after t);
- A_2 – the number N of packets present is reached for the first time after t;
- A_3 – the first multiple vacation period V_1 ends before t, thus t is "inside" the first busy period τ_1.

For the individual summands on the right side of (16) we can write the following representations:

$$\mathbf{P}_M\{(h(t) = m, t \in C_1) \cap A_1\} = I\{m = 0\} \sum_{i=0}^{\infty} \sum_{j=0}^{N-1} \sum_{k=j}^{N-1} p_k^{j*}$$

$$\times \int_0^t \frac{(\lambda x)^j}{j!} e^{-\lambda x} dV^{i*}(x) \sum_{r=N-k}^{\infty} \sum_{l=1}^{r} p_r^{l*} \frac{[\lambda(t-x)]^l}{l!} e^{-\lambda(t-x)} (1 - V(t-x)), \quad (17)$$

$$\mathbf{P}_M\{(h(t) = m, t \in C_1) \cap A_2\} = I\{m = 0\} \sum_{i=0}^{N-1} \sum_{j=i}^{N-1} p_j^{i*} \frac{(\lambda t)^i}{i!} e^{-\lambda t}, \qquad (18)$$

$$\mathbf{P}_M\{(h(t) = m, t \in C_1) \cap A_3\} = \sum_{i=0}^{\infty} \sum_{j=0}^{N-1} \sum_{k=j}^{N-1} p_k^{j*} \int_0^t \frac{(\lambda x)^j}{j!} e^{-\lambda x} dV^{i*}(x)$$

$$\times \sum_{r=N-k}^{\infty} \sum_{l=1}^{r} p_r^{l*} \int_0^{t-x} \frac{(\lambda y)^l}{l!} e^{-\lambda y} \mathbf{P}_O^{k+r}\{h(t-x-y) = m, t-x-y \in \tau_1\} dV(y), \tag*{(19)}$$

where we take the agreement $p_0^{0*} = 1$.

Let us comment (17)–(19) in few words. In (17) the completion epoch of the penultimate single vacation during the first multiple vacation period V_1 is denoted by x. The next (last one) single vacation ends after t. In (18) the number of packets arriving before t (according to a Poisson process with intensity λ) equals at most $N-1$. In (19) the number of packets at the end of the penultimate single vacation during the first vacation cycle C_1 (at the time denoted by x) equals k and next r packets ($r \geq N-k$) occur during the last vacation (duration of which is denoted by y) that ends before t. Thus, at the time $t - x - y$ the modified system begins its operation as the "ordinary" one (on the "zero" busy period) with $k + r$ packets present just after the opening.

Let us put

$$D_M(m, \mu) = \int_0^\infty e^{-\mu t} \mathbf{P}_M\{h(t) = m, t \in C_1\}dt, \quad \mu > 0. \tag{20}$$

Collecting (17)–(19) we obtain

$$D_M(m, \mu) = \sum_{i=0}^\infty \sum_{j=0}^{N-1} \beta_{i,j}(\mu) \sum_{k=j}^{N-1} p_k^{j*} \sum_{r=N-k}^\infty \sum_{l=1}^r p_r^{l*}$$

$$\times \left(I\{m = 0\}\gamma_l(\mu) + \beta_{1,l}(\mu)D_O^{k+r}(m, \mu)\right) + I\{m = 0\} \sum_{i=0}^{N-1}\sum_{j=i}^{N-1} p_j^{i*}\alpha_i(\mu), \tag{21}$$

where we have applied the following notations:

$$\alpha_i(\mu) = \int_0^\infty e^{-(\lambda+\mu)t} \frac{(\lambda t)^i}{i!} dt, \tag{22}$$

$$\beta_{i,j}(\mu) = \int_0^\infty e^{-(\lambda+\mu)t} \frac{(\lambda t)^j}{j!} dV^{i*}(t), \tag{23}$$

$$\gamma_k(\mu) = \int_0^\infty e^{-(\lambda+\mu)t} \frac{(\lambda t)^k}{k!} \left(1 - V(t)\right) dt \tag{24}$$

and

$$D_O^{k+r}(m, \mu) = \int_0^\infty e^{-\mu t} \mathbf{P}_O^{k+r}\{h(t) = m, t \in \tau_0\}dt. \tag{25}$$

Now, denoting

$$D_M(z, \mu) = \sum_{m=0}^\infty D_M(m, \mu)z^m, \quad |z| < 1, \tag{26}$$

we obtain from (21) the following theorem:

Theorem 5. *For any $|z| < 1$ the representation for the PGF of the LT of the output process $h(t)$ in the modified system on its first vacation cycle C_1 has the form*

$$D_M(z, \mu) = \sum_{m=0}^\infty z^m \int_0^\infty e^{-\mu t} \mathbf{P}_M\{h(t) = m, t \in C_1\}dt = \sum_{i=0}^\infty \sum_{j=0}^{N-1} \beta_{i,j}(\mu)$$

$$\times \sum_{k=j}^{N-1} p_k^{j*} \sum_{r=N-k}^\infty \sum_{l=1}^r p_r^{l*} \left(\gamma_l(\mu) + \beta_{1,l}(\mu)D_O^{k+r}(z, \mu)\right) + \sum_{i=0}^{N-1}\sum_{j=i}^{N-1} p_j^{i*}\alpha_i(\mu). \tag{27}$$

Taking $p_1 = 1$ and $p_k = 0$ for $k \geq 2$, we obtain the system with single arrivals. Then from Theorem 5 we get

Corollary 2. *The representation for the PGF of the LT of the output process $h(t)$ in the modified system with single arrivals, multiple vacations and the N-policy is following:*

$$D_M(z, \mu) = \sum_{i=0}^{\infty} \sum_{j=0}^{N-1} \beta_{i,j}(\mu) \sum_{r=N-j}^{\infty} \left(\gamma_r(\mu) + \beta_{1,r}(\mu) D_O^{j+r}(z, \mu) \right) + \sum_{i=0}^{N-1} \alpha_i(\mu).$$
(28)

Taking $N = 1$ we get from (27) the representation for the case of the system with multiple vacations but without the threshold.

Corollary 3. *The representation for the PGF of the LT of the output process $h(t)$ in the modified system with batch arrivals, multiple vacations without the N-policy is following:*

$$D_M(z, \mu) = \sum_{i=0}^{\infty} \beta_{i,0}(\mu) \sum_{r=1}^{\infty} \sum_{l=1}^{r} p_r^{l*} \left(\gamma_l(\mu) + \beta_{1,l}(\mu) D_O^r(z, \mu) \right) + \alpha_0(\mu). \quad (29)$$

Note that the last result is equivalent to the formula obtained in [8].

5   Output Process in the Original System

From the memoryless property of the exponential distribution of interarrival times follows that successive vacation cycles $C_i, i \geq 0$, are independent random variables and for $i \geq 1$ have the same distributions. Thus, successive vacation cycles form a delayed renewal process. Let us denote by $B_0(\cdot)$ and $B_1(\cdot)$ d.fs of C_0 and C_1, respectively and, similarly, let $b_0(\cdot)$ and $b_1(\cdot)$ denote the LSTs of $B_0(\cdot)$ and $B_1(\cdot)$, respectively.

Since the zero vacation cycle C_0 consists only of the "zero" busy period τ_0 then, substituting $z = 1$ in (11), we get for $\mu \geq 0$

$$b_0(\mu) = \mathbf{E}\{e^{-\mu C_0}\} = \mathbf{E}_O^{std}\{e^{-\mu \tau_0}\} = 1 - f_+(1, \mu, 0). \quad (30)$$

Moreover, from the formula of total probability the following representation follows (compare (17)–(19)):

$$b_1(\mu) = \sum_{i=0}^{\infty} \sum_{j=0}^{N-1} \sum_{k=j}^{N-1} p_k^{j*} \int_0^{\infty} \frac{(\lambda x)^j}{j!} e^{-(\lambda+\mu)x} dV^{i*}(x)$$

$$\times \sum_{r=N-k}^{\infty} \sum_{l=1}^{r} p_r^{l*} \int_0^{\infty} \frac{(\lambda y)^l}{l!} e^{-(\lambda+\mu)y} dV(y) \mathbf{E}_O^{k+r}\{e^{-\mu \tau_0}\}$$

$$= \sum_{i=0}^{\infty} \sum_{j=0}^{N-1} \beta_{i,j}(\mu) \sum_{k=j}^{N-1} p_k^{j*} \sum_{r=N-k}^{\infty} \sum_{l=1}^{r} p_r^{l*} \beta_{1,l}(\mu) \mathbf{E}_O^{k+r}\{e^{-\mu \tau_0}\}, \quad (31)$$

where $\mathbf{E}_O^{k+r}\{e^{-\mu\tau_0}\}$ was defined in (12).

Now let us denote by $h(C_i)$ the number of packets completely served before the completion epoch of the ith vacation cycle. Thus, $h(C_i) - h(C_{i-1})$ represents the number of packets served during C_i for $i = 1, 2, \ldots$ Let us put

$$q_0(z) = \sum_{n=1}^{\infty} z^n q_0(n) = \sum_{n=1}^{\infty} z^n \mathbf{P}\{h(C_0) = n\}, \tag{32}$$

$$q_1(z) = \sum_{n=1}^{\infty} z^n q_1(n) = \sum_{n=1}^{\infty} z^n \mathbf{P}\{h(C_1) - h(C_0) = n\}, \tag{33}$$

where $|z| < 1$. Of course, probabilities $\mathbf{P}\{h(C_i) - h(C_{i-1}) = n\}$ are equal for $i \geq 1$ and fixed n. Since $\mathbf{E}\{z^{h(C_0)}\} = \mathbf{E}_O^{std}\{z^{h(\tau_0)}\}$ then, substituting $\mu = 0$ in (11), we get

$$q_0(z) = z(1 - f_+(z, 0, 0)), \quad |z| \leq 1. \tag{34}$$

Next, similarly as in (31), we can write the following representation:

$$q_1(z) = \sum_{i=0}^{\infty} \sum_{j=0}^{N-1} \sum_{k=j}^{N-1} p_k^{j*} \int_0^{\infty} \frac{(\lambda x)^j}{j!} e^{-\lambda x} dV^{i*}(x)$$

$$\times \sum_{r=N-k}^{\infty} \sum_{l=1}^{r} p_r^{l*} \int_0^{\infty} \frac{(\lambda y)^l}{l!} e^{-\lambda y} dV(y) \mathbf{E}_O^{k+r}\{z^{h(\tau_0)}\}$$

$$= \sum_{i=0}^{\infty} \sum_{j=0}^{N-1} \beta_{i,j}(0) \sum_{k=j}^{N-1} p_k^{j*} \sum_{r=N-k}^{\infty} \sum_{l=1}^{r} p_r^{l*} \beta_{1,l}(0) \mathbf{E}_O^{k+r}\{z^{h(\tau_0)}\}, \tag{35}$$

where the formula for $\mathbf{E}_O^{k+r}\{z^{h(\tau_0)}\}$ was given in (13).

Summing $\mathbf{P}\{h(t) = m, t \in C_k\}$ over all cycles we obtain

$$\mathbf{P}\{h(t) = m\} = \sum_{k=0}^{\infty} \mathbf{P}\{h(t) = m, t \in C_k\}. \tag{36}$$

Moreover

$$\mathbf{P}\{h(t) = m, t \in C_0\} = \mathbf{P}_O^{std}\{h(t) = m, t \in \tau_0\} \tag{37}$$

and besides, for $k \geq 1$ and $m \geq k$, we get

$$\mathbf{P}\{h(t) = m, t \in C_k\} = \sum_{n=k}^{m} (q_0 * q_1^{(k-1)*})(n)$$

$$\times \int_0^t \mathbf{P}_M\{h(t - y) = m - n, t - y \in C_1\} d(B_0 * B_1^{(k-1)*})(y), \tag{38}$$

where

$$(q_0 * q_1^{(k-1)*})(n) = \sum_{i=1}^{n-k+1} q_0(i) q_1^{(k-1)*}(n - i)$$

and

$$(B_0 * B_1^{(k-1)*})(y) = \int_0^y B_0(y - u)dB_1^{(k-1)*}(u).$$

The well-known properties of transforms, for $\mu > 0$, $|z| < 1$ and $k \geq 1$, lead to the following equation:

$$\sum_{m=0}^{\infty} z^m \int_0^{\infty} e^{-\mu t}\mathbf{P}\{h(t) = m, t \in C_k\}dt$$
$$= D_M(z,\mu)b_0(\mu)b_1^{k-1}(\mu)q_0(z)q_1^{k-1}(z). \tag{39}$$

Now, finally, as a consequence of (36), (37) and (39), we can state the following theorem that gives the formula for the PGF of the LT of the output process $h(t)$ in the original system with batch arrivals, multiple vacations and the N-policy:

Theorem 6. *For any $\mu > 0$ and $|z| < 1$ the following formula holds true:*

$$\sum_{m=0}^{\infty} z^m \int_0^{\infty} e^{-\mu t}\mathbf{P}\{h(t) = m\}dt = D_O^{std}(z,\mu) + \frac{D_M(z,\mu)b_0(\mu)q_0(z)}{1 - b_1(\mu)q_1(z)}, \tag{40}$$

where the representations for $D_O^{std}(z,\mu)$, $D_M(z,\mu)$, $b_0(\mu)$, $b_1(\mu)$, $q_0(z)$ and $q_1(z)$ are given in (10), (27), (30), (31), (34) and (35) respectively.

Of course, taking $p_1 = 1$ or $N = 1$ on the right side of (40), we can easily obtain the corresponding formulae for the systems with single arrivals and without the N-policy, respectively.

Let us note that the proposed analytical approach (Sections 4 and 5) can be succesfully used in the analysis of the $M^X/G/1$-type queue with some other vacation-type restrictions in service process, e.g. for models with additional setup (closedown) times (see [7]) or with single vacation policy (see [6]). However, in the method the assumption that the batches of packets occur according to a Poisson process is essential, in consequence the method cannot be applied e.g. in the case of general independent (GI) arrival process.

6   Numerical Results

Let us note that all the results obtained above are written by means of transforms of d.fs of "input" characteristics describing the evolution of the system: interarrival times, batch sizes, service times and single vacation durations. They also depend on components $f_\pm(z, \mu, s)$ of the factorization identity (5) and functions $P_+(z, \mu, x)$ (see (7)) and $F_+(z, \mu, x)$ (see (15)) connected with this factorization. Thus, in numerical treatment of the results the most essential is to obtain these functions explicitly. Of course, it is feasible for some special distributions only. For the remaining cases some approximations are necessary.

As an example let us consider the $M/M/1$-type system with single arrivals, exponentially distributed vacations and with the threshold $N = 2$. Let

$$F(t) = 1 - e^{-\sigma t}, \qquad V(t) = 1 - e^{-\theta t}, \quad t > 0, \sigma > 0, \theta > 0. \qquad (41)$$

The components of the factorization identity (5) are following:

$$f_+(z, \mu, s) = \frac{s - \frac{1}{2}(\sigma + \mu - \lambda - \sqrt{(\sigma + \mu + \lambda)^2 - 4\sigma\lambda z})}{s + \lambda}, \qquad (42)$$

$$f_-(z, \mu, s) = \frac{-[s - \frac{1}{2}(\sigma + \mu - \lambda + \sqrt{(\sigma + \mu + \lambda)^2 - 4\sigma\lambda z})]}{\sigma + \mu - s}. \qquad (43)$$

From the formulae (30)–(31) one can find the means of the vacation cycles as follows:

$$\mathbf{E}C_i = -\frac{d}{d\mu} b_i(\mu)\Big|_{\mu=0}, \quad i = 0, 1. \qquad (44)$$

Similarly, we have from (34)–(35)

$$\mathbf{E}L_i = \frac{d}{dz} q_i(z)\Big|_{z=1}, \quad i = 0, 1, \qquad (45)$$

where we denote $L_0 = h(C_0)$ and $L_1 = h(C_1) - h(C_0)$.

In Table 1 we present results for $\mathbf{E}L_0$, $\mathbf{E}C_0$ and their quotients, for three different sets of "input" parameters of the system satisfying the stability condition $\rho = \frac{\lambda}{\sigma} < 1$.

Table 1. Values of $\mathbf{E}L_0$ and $\mathbf{E}C_0$ for three different sets of system parameters

Arrival rate λ	Service rate σ	Traffic load ρ	$\mathbf{E}L_0$	$\mathbf{E}C_0$	$\mathbf{E}L_0/\mathbf{E}C_0$
2	8	0.25	0.16667	2.33333	0.07143
1	2	0.50	1.00000	3.00000	0.33333
3	4	0.75	1.00000	5.00000	0.20000

In particular, for $\lambda = 2$, $\sigma = 8$ (and, in consequence, for the traffic load $\rho = 0.25$) we have from (30) and (34), applying (42)–(43),

$$b_0(\mu) = 1 + \frac{1}{4}\left(6 + \mu - \sqrt{36 + 20\mu + \mu^2}\right),$$

$$q_0(z) = \left(1 + \frac{1}{2}(3 - \sqrt{25 - 16z})\right)z.$$

Similarly, for $\lambda = 1$ and $\sigma = 2$ (hence $\rho = 0.5$) we get

$$b_0(\mu) = 1 + \frac{1}{2}\left(1 + \mu - \sqrt{1 + 6\mu + \mu^2}\right),$$

$$q_0(z) = \left(1 + \frac{1}{2}(1 - \sqrt{9 - 8z})\right)z.$$

Finally, for $\lambda = 3$ and $\sigma = 4$ (so, for $\rho = 0.75$) we obtain

$$b_0(\mu) = 1 + \frac{1}{6}\left(1 + \mu - \sqrt{1 + 14\mu + \mu^2}\right),$$

$$q_0(z) = \left(1 + \frac{1}{6}\left(1 - \sqrt{49 - 48z}\right)\right)z.$$

Differentiating the above formulae as in (44)–(45) we obtain results given in Table 1.

In Tables 2 and 3 we present results for $\mathbf{EL}_1$ and $\mathbf{EC}_1$ obtained applying the formulae (31) and (35), for the same three sets of "input" system parameters, and for ten different values of the parameter θ of the exponentially distributed single vacation duration (for $\theta = 1, 2, ..., 10$).

Table 2. Values of $\mathbf{EL}_1$ in a function of θ for different sets of system parameters

Parameter θ	$\mathbf{EL}_1$ for $\rho = 0.25$	$\mathbf{EL}_1$ for $\rho = 0.50$	$\mathbf{EL}_1$ for $\rho = 0.75$
1	6.59744	10.42710	15.05246
2	5.58331	7.67547	14.01076
3	5.01013	6.18579	12.89251
4	4.58056	5.08817	12.08466
5	4.19880	4.24080	11.42016
6	3.84538	3.57896	10.81270
7	3.51893	3.05671	10.23144
8	3.22049	2.63982	9.66922
9	2.95008	2.30299	9.12721
10	2.70657	2.02759	8.60856

Table 3. Values of $\mathbf{EC}_1$ in a function of θ for different sets of system parameters

Parameter θ	$\mathbf{EC}_1$ for $\rho = 0.25$	$\mathbf{EC}_1$ for $\rho = 0.50$	$\mathbf{EC}_1$ for $\rho = 0.75$
1	3.08171	9.42489	4.96494
2	2.41421	6.44899	4.50434
3	2.04510	4.87075	4.07335
4	1.76961	3.78950	3.76307
5	1.53648	3.01754	3.50801
6	1.33548	2.45445	3.27769
7	1.16330	2.03489	3.06188
8	1.01691	1.71561	2.85816
9	0.89291	1.46776	2.66655
10	0.78795	1.27184	2.48749

As one can observe, as the values of the parameter θ increase (the means of the single vacation durations decrease), the lengths of the cycle decrease and vice versa. Indeed, for a relatively very short single vacation duration (comparing to

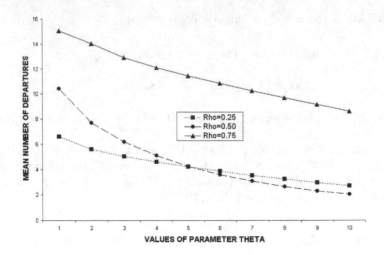

Fig. 1. Values of $\mathbf{E}L_1$ for different sets of system parameters and different values of θ

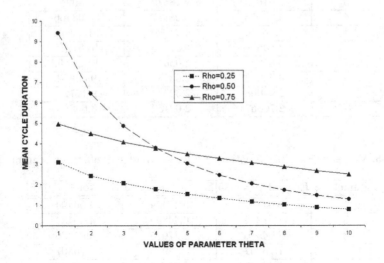

Fig. 2. Values of $\mathbf{E}C_1$ for different sets of system parameters and different values of θ

the arrival rate), a new busy period begins at the moment which is close to the instant of the second arrival, so the probability that the busy period will begin with more that two packets present (such a situation "extends" the busy period duration) is small. The same phenomenon can be noted for the number of packets completely served during a cycle: L_1 decreases as the values of the parameter θ increase, so as the mean duration of a single vacation period decreases.

The results from Tables 2 and 3 are visualized in Figures 1 and 2, respectively.

References

1. Bratiichuk, M.S., Kempa, W.M.: Application of the superposition of renewal processes to the study of batch arrival queues. Queueing Syst. 44, 51–67 (2003)
2. Bratiichuk, M.S., Kempa, W.M.: Explicit formulae for the queue length distribution of batch arrival systems. Stoch. Models 20(4), 457–472 (2004)
3. Doshi, B.T.: Queueing systems with vacations - a survey. Queueing Syst. 1(1), 29–66 (1986)
4. Jiang, F.-C.: Mitigation techniques for the energy hole problem in sernsor networks using N-policy $M/G/1$ queuing models. In: Proc. of Frontier Computing. Theory, Technologies and Applications, IET International Conference, Taichung, Taiwan, August 4-6, pp. 281–286 (2010)
5. Kempa, W.M.: $GI/G/1/\infty$ batch arrival queueing system with a single exponential vacation. Math. Methods Oper. Res. 69(1), 91–97 (2009)
6. Kempa, W.M.: Some new results for departure process in the $M^X/G/1$ queueing system with a single vacation and exhaustive service. Stoch. Anal. Appl. 28(1), 26–43 (2010)
7. Kempa, W.M.: The transient analysis of the queue-length distribution in the batch arrival system with N-policy, multiple vacations and setup times. In: AIP Conf. Proc., vol. 1293, pp. 235–242 (2010)
8. Kempa, W.M.: Analysis of departure process in batch arrival queue with multiple vacations and exhaustive service. Commun. Stat. - Theor. M. 40, 1–10 (2011)
9. Kempa, W.M.: Departure process in finite-buffer queue with batch arrivals. In: Al-Begain, K., Balsamo, S., Fiems, D., Marin, A. (eds.) ASMTA 2011. LNCS, vol. 6751, pp. 1–13. Springer, Heidelberg (2011)
10. Kempa, W.M.: Transient analysis of the output process in the $GI/M/1$-type queue with finite buffer. In: AIP Conf. Proc., vol. 1487, pp. 193–200 (2012)
11. Lee, H.W., Lee, S.S., Park, J.O., Chae, K.C.: Analysis of the $M^x/G/1$ queue with N-policy and multiple vacations. J. Appl. Probab. 31, 476–496 (1994)
12. Lee, H.W., Lee, S.S., Chae, K.C.: Operating characteristics of $M^X/G/1$ queue with N-policy. Queueing Syst. 15(1-4), 387–399 (1994)
13. Mancuso, V., Alouf, S.: Analysis of power saving with continuous connectivity. Comput. Netw. 56, 2481–2493 (2012)
14. Takagi, H.: Queueing analysis: A foundation of performance evaluation. Vacation and priority systems, vol. 1, Part 1. Elsevier, Amsterdam (1991)
15. Tang, Y.: The departure process of the $M/G/1$ queueing model with server vacation and exhaustive service descipline. J. Appl. Probab. 31(4), 1070–1082 (1994)
16. Tang, Y.: The departure process for the $M^x/G/1$ queueing system with server vacations. Math. Appl. 20(3), 478–484 (2007)
17. Tian, N., Zhang, Z.G.: Vacation queueing models: theory and applications. Springer, New York (2006)

Retrial Queueing System with Correlated Input, Finite Buffer, and Impatient Customers

Che Soong Kim[2], Valentina Klimenok[1], and Alexander Dudin[1]

[1] Department of Applied Mathematics and Computer Science
Belarusian State University
Minsk 220030, Belarus
{klimenok,dudin}@bsu.by

[2] Department of Industrial Engineering Sangji University
Wonju, Kangwon, Korea 220-702
dowoo@sangji.ac.kr

Abstract. Multi-server queue with a finite buffer and *Batch Markov Arrival Process (BMAP)* is considered. Customers, which do not succeed to enter the system upon arrival (due to unavailability of servers and buffer space), move to orbit to make repeated attempts in exponentially distributed time intervals. Customers in a buffer are impatient. After exponentially distributed amount of time they may leave the system without a service or go to the orbit. Stability condition of the system is derived, steady state distribution is computed, expressions for key performance measures and for waiting time distribution are given. Numerical illustrations are presented.

Keywords: Multi-Server Queueing System, Batch Markov Arrival Process, Retrials, Impatience.

1 Introduction

Retrial queueing models play an important role in performance evaluation and capacity planning of many telecommunication networks and they are extensively studied in the queueing literature, for references see, e.g., the books [1], [7] and survey paper [8]. The overwhelming majority of results obtained for retrial queues relate to the systems with the arrival flow described by the stationary Poisson process. However, such an arrival process is a poor descriptor of information flows in modern telecommunication networks. The Batch Markovian Arrival Process (*BMAP*) is recommended in literature for description of such flows. First papers where the retrial queues with the *BMAP* arrival process and arbitrary distribution of service time were analyzed are [5] and [6]. There, the $BMAP/G/1$ and the $BMAP/SM/1$ retrial queues, respectively, were under study.

The most general up to now well-studied model of multi-server retrial queueing system is given in [2] where the arrival process was described by the *BMAP*, service times were assumed having *PH* (phase type) distribution and arbitrary

A. Dudin and K. De Turck (Eds.): ASMTA 2013, LNCS 7984, pp. 262–276, 2013.

dependence of the total retrial rate from the orbit on the number of customers in the orbit is suggested. In [2], stability and instability conditions for this model were derived and the numerically stable algorithm for computing the steady state distribution of the system was elaborated.

In the paper [3], the results of [2] were applied to the problem of analyzing performance of hot spots in airports. This application showed some shortcoming of the software based directly on the results from [2] if the service time has more general distribution than the exponential one. This shortcoming consists of large dimension of blocks of infinite-size generator of the Markov chain describing behavior of the system. In the paper [9], the $BMAP/PH/N$ retrial queue was described by the multi-dimensional Markov chain having much smaller, than in [2], dimension of blocks of infinite-size generator. Such a description is based on the idea formulated in [15]. Instead of keeping track on the states of the service process at each busy server, it is enough to count how many servers have the respecting state of the underlying Markov chain.

In [11], the model of [2] was generalized to the case when each retrying customer may leave the system forever without getting the service after the each unsuccessful attempt, i.e., the customers are non-persistent. Also, the paper [11] essentially supplements results presented in [2] by providing extensive numerical study of the system behavior.

All mentioned above papers deal with pure retrial models, i.e., models that do not have any buffer for waiting customers. In this paper, we extend the results of [2] to the case when the system has a finite buffer which is visited by the customers that do not succeed to start the service immediately upon arrival due to the lack of available servers. Only if the buffer is full, the customers go to orbit. Similar model for different disciplines of customers access from the orbit was analyzed in [12]. Distinction of the model under study in the present paper consists of suggestion that customers in the buffer are impatient (they may leave the system forever or go to orbit after exponentially distributed sojourn time in the buffer).

2 Model Description

We consider the system with N identical independent servers and a finite buffer of capacity $R < \infty$. Service time distribution is exponential with parameter μ.

The arrivals to the system are described by a $BMAP$. The $BMAP$ is defined by the underlying process ν_t, $t \geq 0$, which is an irreducible continuous time Markov chain with state space $\{0, ..., W\}$, and by the matrices D_k, $k \geq 0$, with matrix generating function $D(z) = \sum_{k=0}^{\infty} D_k z^k$, $|z| \leq 1$. Arrivals occur only at epochs of the transitions in the process ν_t, $t \geq 0$. The intensities of transitions accompanied by arrival of a batch of size k are defined by the matrices D_k, $k \geq 1$. The intensities of the transitions of the process ν_t without generation of customers are given by non-diagonal entries of the matrix D_0. Diagonal entries of this matrix define, up to the sign, intensities of exit of the process ν_t from its states.

The matrix $D(1) = \sum_{k=0}^{\infty} D_k$ is an infinitesimal generator of the process ν_t. Thus, the stationary distribution vector θ of this process is defined as the unique solution of the system $\theta D(1) = 0, \theta e = 1$, where e is a column vector consisting of 1's, and 0 is a row vector of 0's. The intensity (fundamental rate) of the $BMAP$ is given by $\lambda = \theta D'(z)|_{z=1} e$. We assume that $\lambda < \infty$. The intensity of batch arrivals is defined by $\lambda_b = \theta(-D_0)e$.

Assumption that the arrival process is defined by the $BMAP$ allows to take into account variance and possible correlation of successive inter-arrival times. The coefficient of variation, c_{var}, of intervals between successive group arrivals is calculated by $c_{var}^2 = 2\lambda_b \theta(-D_0)^{-1} e - 1$. The coefficient of correlation, c_{cor}, of the successive intervals is given by $c_{cor} = (\lambda_b \theta(-D_0)^{-1}(D(1)-D_0)(-D_0)^{-1} e - 1)/c_{var}^2$.

For more information about the $BMAP$, its special cases and properties and related literature see, e.g., [4] [13].

If the number of available servers is greater or equal to the size of arriving batch, all customers of the batch enter the service. Otherwise, the part of a batch occupies all free servers and starts service while the rest moves to the buffer. If the number of customers is greater than the available capacity R of the buffer, the part of customers join the buffer while the rest move to the virtual place called orbit. Customers staying in the orbit repeat their attempts to reach a free server, if any, or the place in the buffer afterward, independently of each other. Inter-retrial times of a customer from the orbit have exponential distribution with intensity $\alpha > 0$. Thus, if there are i customers in the orbit, the total intensity of retrials is equal to $i\alpha$, $i \geq 1$.

Customers in a buffer wait for picking-up for the service according to $FIFO$ discipline. These customers are impatient. Impatience time of a customer is exponentially distributed with intensity γ. If this time expires before the customer is picked up for the service, with probability q, $0 \leq q \leq 1$, the customer leaves the system forever. With complementary probability $(1-q)$, the customer moves to the orbit.

The structure of the system is presented in Figure 1.

Our aim is to derive the condition for stable operation of the system, calculate the steady state distribution of the system states and the main performance measures of the system.

3   Process of the System States

The process of the system states is described in terms of the irreducible three-dimensional continuous-time Markov chain

$$\xi_t = \{i_t, r_t, \nu_t\}, \ t \geq 0,$$

where

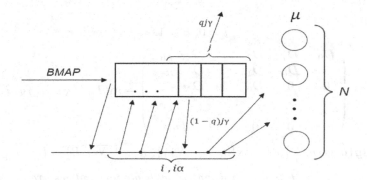

Fig. 1. Structure of the system

- i_t is the number of customers in the orbit,
- r_t is the total number of customers in the service and in the buffer,
- ν_t is the state of the directing process ν_t of the $BMAP$ at epoch t.

The state space of the chain is defined as

$$\{(i,\, r,\, \nu):\ i \geq 0,\ 0 \leq r \leq N + R,\ 0 \leq \nu \leq W\}.$$

Enumerate the states of the Markov chain ξ_t, $t \geq 0$, in the lexicographic order. In the following, the set of the states of the chain having the value i of the first component is referred as the level **i**.

Introduce the square matrices $\boldsymbol{Q}_{i,j}$, $i,\, j \geq 0$, of order $K = (N+R+1)(W+1)$ consisting of intensities of transition from the level **i** to the level **j**.

Lemma 1. *The infinitesimal generator $\boldsymbol{Q}$ of the Markov chain ξ_t, $t \geq 1$, is of the following block structure:*

$$\boldsymbol{Q} = \begin{pmatrix} \boldsymbol{Q}_{0,0} & \boldsymbol{Q}_{0,1} & \boldsymbol{Q}_{0,2} & \cdots \\ \boldsymbol{Q}_{1,0} & \boldsymbol{Q}_{1,1} & \boldsymbol{Q}_{1,2} & \cdots \\ 0 & \boldsymbol{Q}_{2,1} & \boldsymbol{Q}_{2,2} & \cdots \\ 0 & 0 & \boldsymbol{Q}_{3,2} & \cdots \\ \vdots & \vdots & \vdots & \ddots \end{pmatrix},$$

where non-zero blocks $\boldsymbol{Q}_{i,j}$, $i,\, j \geq 0$, are defined as follows:

$$\boldsymbol{Q}_{i,i-1} = i\alpha \tilde{\boldsymbol{I}}_K,\ i \geq 1,$$

$$\boldsymbol{Q}_{i,i+k} =$$

$$\begin{pmatrix} & | & 0 & 0 & \cdots 0 & 0 & \boldsymbol{D}_{N+R+k} \\ & | & \vdots & \vdots & \ddots \vdots & \vdots & \vdots \\ & | & \delta_{1k}(1-q)\gamma \boldsymbol{I}_{\overline{W}} & 0 & \cdots 0 & 0 & \boldsymbol{D}_{R+k-1} \\ \boldsymbol{0}_{N+R+1,\,N} & | & 0 & \delta_{1k}2(1-q)\gamma \boldsymbol{I}_{\overline{W}} & \cdots 0 & 0 & \boldsymbol{D}_{R+k-2} \\ & | & \vdots & \vdots & \ddots \vdots & \vdots & \vdots \\ & | & 0 & 0 & \cdots 0 & \delta_{1k}R(1-q)\gamma \boldsymbol{I}_{\overline{W}} & \boldsymbol{D}_k \end{pmatrix},$$

$$i \geq 0, \ k > 1,$$

$$Q_{i,i} = \begin{pmatrix} D_0 & D_1 & D_2 & \cdots & D_{N+R} \\ 0 & D_0 & D_1 & \cdots & D_{N+R-1} \\ 0 & 0 & D_0 & \cdots & D_{N+R-2} \\ \vdots & \vdots & \vdots & \ddots & \vdots \\ 0 & 0 & 0 & \cdots & D_0 \end{pmatrix} - i\alpha(I - \hat{I}_{N+R+1}) \otimes I_{\overline{W}}$$

$$- diag\{\max\{n, N\}\mu + \max\{0, (n-N)\gamma\}, \ n = \overline{0, N+R}\} \otimes I_{\overline{W}}, \ i \geq 0.$$

Here $\overline{W} = W + 1$; $\tilde{I}$ is a square matrix with all zero entries except the ones located on the first off-diagonal; $I_{\overline{W}}$ is an identity matrix of size $\overline{W}$; $\delta_{i,k}$ is the Kronecker delta; $\hat{I}$ is a square matrix with all zero entries except the ones located on the first sub-diagonal; $diag\{a_m, \ m = 0, \ldots, M\}$ stands for the diagonal matrix defined by the diagonal entries listed in the brackets; $\otimes$ is a symbol of Kronecher product of matrices.

Proof of the lemma is implemented by analyzing the intensities of transition of the multi-dimensional Markov chain $\xi_t, t \geq 0$.

In the further investigation of the stationary behavior of the Markov chain $\xi_t, t \geq 0$, we will use the results for continuous time Asymptotically Quasi-Toeplitz Markov Chains ($AQTMC$) presented in [10].

Corollary 1. *The Markov chain $\xi_t, t \geq 0$, belongs to the class of continuous time AQTMC.*

Proof. According to the definition given in [10], the chain $\xi_t, t \geq 0$, belongs to the class of continuous time $AQTMC$ if there exist the limits

$$Y_k = \lim_{i \to \infty} C_i^{-1} Q_{i,i+k-1}, \ k = 0, 2, 3 \ldots, \ Y_1 = \lim_{i \to \infty} C_i^{-1} Q_{i,i} + I, \qquad (1)$$

and the matrix $\sum_{k=0}^{\infty} Y_k$ is a stochastic one.

Here C_i is a diagonal matrix defined by modules of diagonal entries of the matrix $Q_{i,i}, i \geq 0$. It is easy to see that the diagonal entries of the matrix $C_i, \ i \geq 1$, corresponding to the first $(N+R)\overline{W}$ rows of the matrix $Q_{i,i}$ include the term $i\alpha$ while the rest of diagonal does not depend on i. Taking into account dependence (or independence) of the entries of the matrices $Q_{i,i+k}, \ k \geq -1$, on i, we calculate the limits defining the matrices Y_k as follows:

$$Y_0 = \tilde{I}_K, \quad Y_2 = \begin{pmatrix} 0 & 0 & 0 \\ 0 & C^{-1}(1-q)R\gamma & C^{-1}D_1 \end{pmatrix}, \qquad (2)$$

$$Y_1 = \begin{pmatrix} 0 & 0 & 0 \\ 0 & C^{-1}(N\mu + qR\gamma) & C^{-1}[D_0 - (N\mu + R\gamma)I] + I \end{pmatrix}, \qquad (3)$$

$$Y_k = \begin{pmatrix} 0 & 0 \\ 0 & C^{-1}D_{k-1} \end{pmatrix}, \ k > 2, \qquad (4)$$

where
$$C = diag\{(-D_0)_{l,l} + N\mu + R\gamma, l = \overline{0, W}\}. \tag{5}$$

It is evident that the matrices given by equations (2)-(5) satisfy conditions (1). This proves the corollary.

Let $\xi_n = \{i_n, r_n, \nu_n\}$, $n \geq 1$, denotes discrete time Markov chain embedded in continuous time $AQTMC$ $\xi_t = \{i_t, r_t, \nu_t\}, t \geq 0$, over all moments of its jumps. Then the matrices Y_k, $k \geq 0$, can be interpreted as transition probability matrices of ξ_n, $n \geq 1$, under assumption that the component i_t (the number of customers in the orbit) tends to infinity. Such a chain is called as *limiting* chain with respect to the chain ξ_n, $n \geq 1$.

Let $Y(z) = \sum_{k=0}^{\infty} Y_k z^k$, $|z| \leq 1$, be the generating function of the matrices Y_k, $k \geq 0$. This matrix function can be represented in the following block form:
$$Y(z) = \begin{pmatrix} Y_{11}(z) & Y_{12}(z) \\ 0 & Y_{22}(z) \end{pmatrix} \tag{6}$$

where
$$Y_{11}(z) = \tilde{I}_{(N+R-1)\overline{W}}, \qquad Y_{12}(z) = \begin{pmatrix} 0 & 0 \\ I_{\overline{W}} & 0 \end{pmatrix}, \tag{7}$$

$$Y_{22}(z) = \begin{pmatrix} 0 & I_{\overline{W}} \\ C^{-1}z[N\mu + qR\gamma + (1-q)R\gamma z] & C^{-1}z[D(z) - (N\mu + R\gamma)I] + zI \end{pmatrix}.$$

The matrix generating function $Y(z)$ plays an important role in steady state analysis of $AQTMC$s since it contains information about the asymptotic properties of these chains.

4   Stationary Distribution

In this section, we obtain the ergodicity condition and calculate the stationary distribution of the chain under consideration using matrix $Y(z)$ defined by formulas (6)-(7).

Theorem 2. *The sufficient condition for ergodicity of the $AQTMC$ ξ_t, $t \geq 0$, is the fulfillment of the inequality*

$$\rho = \frac{\lambda}{N\mu + R\gamma q} < 1, \tag{8}$$

where λ is fundamental rate of the $BMAP$, $\lambda = \theta D'(z)|_{z=1}\mathbf{e}$.

Proof. Analyzing the matrix $Y(z)$, one can see that this matrix is a reducible one and a structure of this matrix satisfies the condition of theorem 6 in [10]. Then, as follows from [10], the sufficient condition for ergodicity of the chain ξ_n, $n \geq 1$, is defined in terms of the matrix $Y_{22}(z)$. The chain is ergodic if the inequality $[\det(zI - Y_{22}(z))]'_{z=1} > 0$ holds. This inequality is equivalent to the following one
$$[\det(zI - \hat{Y}_{22}(z))]'_{z=1} > 0 \tag{9}$$

where $\hat{\boldsymbol{Y}}_{22}(z)$ is a normal form of the matrix $\boldsymbol{Y}_{22}(z)$,

$$\hat{\boldsymbol{Y}}_{22}(z) = \begin{pmatrix} \boldsymbol{C}^{-1}z[\boldsymbol{D}(z) - (N\mu + R\gamma)\boldsymbol{I}] + z\boldsymbol{I} & \boldsymbol{C}^{-1}z[N\mu + qR\gamma + (1-q)R\gamma z] \\ \boldsymbol{I}_{\overline{W}} & \boldsymbol{0} \end{pmatrix}.$$

Using the block structure of the determinant in (9) and the well-known formula $\det \begin{pmatrix} \boldsymbol{A} \ \boldsymbol{B} \\ \boldsymbol{F} \ \boldsymbol{H} \end{pmatrix} = \det(\boldsymbol{A} - \boldsymbol{B}\boldsymbol{H}^{-1}\boldsymbol{F})\det \boldsymbol{H}$, we reduce inequality (9) to the form

$$\det[-\boldsymbol{C}^{-1}][\det \boldsymbol{\Omega}(z)\det(z\boldsymbol{I}_{\overline{W}})]'_{z=1} > 0 \qquad (10)$$

where

$$\boldsymbol{\Omega}(z) = z\boldsymbol{D}(z) + (1-z)(N\mu + qR\gamma)\boldsymbol{I}.$$

Note that the matrix $\boldsymbol{\Omega}(1)$ is an infinitesimal generator, so $\det \boldsymbol{\Omega}(1) = 0$.

Calculating the derivative in the left-hand side of inequality (10) at the point $z = 1$ and taking into account that $\det[-\boldsymbol{C}^{-1}] < 0$, $\det(z\boldsymbol{I}_{\overline{W}})]'_{z=1} > 0$, $\det \boldsymbol{\Omega}(1) = 0$, we arrive to the equivalent inequality

$$[\det \boldsymbol{\Omega}(z)]'_{z=1} < 0. \qquad (11)$$

Since the matrix $\boldsymbol{\Omega}(1)$ is an infinitesimal generator, we can apply the results of [10] to reduce inequality (11) to the equivalent inequality

$$\mathbf{x}\boldsymbol{\Omega}'(1)\mathbf{e} < 0 \qquad (12)$$

where the vector $\mathbf{x}$ is the unique solution of the system

$$\mathbf{x}\boldsymbol{\Omega}(1) = \mathbf{0}, \ \mathbf{x}\mathbf{e} = 1.$$

It is easy to see that $\mathbf{x} = \boldsymbol{\theta}$. Substituting such a vector into inequality (12), we reduce this inequality to form (8) using obvious transformations. Theorem is proved.

In what follows we suppose that inequality (8) is fulfilled. Denote the stationary state probabilities of the Markov chain ξ_t, $t \geq 0$, by

$$p(i, r, \nu) = \lim_{t \to \infty} P(i_t = i, r_t = r, \nu_t = \nu), \ i \geq 0, \ r = \overline{0, N+R}, \ \nu = \overline{0, W}.$$

Define the vectors of these probabilities as

$$\mathbf{p}_{i,r} = (p(i, r, 0), p(i, r, 1), \dots, p(i, r, W)), \ i \geq 0, \ r = \overline{0, N+R};$$

$$\mathbf{p}_i = (\mathbf{p}_{i,0}, \mathbf{p}_{i,1}, \dots, \mathbf{p}_{i,N+R}), \ i \geq 0.$$

The vectors $\mathbf{p}_i$, $i \geq 0$, are the unique solution of the system of equilibrium equations supplemented by a normalization condition

$$\sum_{i=0}^{\infty} \mathbf{p}_i Q_{i,j} = \mathbf{0}, \ j \geq 0, \ \sum_{i=0}^{\infty} \mathbf{p}_i \mathbf{e} = 1.$$

It is well known that, in case of level-dependent Markov chains with countable state space, it is not possible to solve this system by a straightforward way. To compute the vectors $\mathbf{p}_i$, $i \geq 0$, we use the numerically stable algorithm which has been elaborated in [10] for calculating the stationary distribution of the multi-dimensional continuous time $AQTMCs$. The algorithm is based on censoring technique and consists of the next principal steps.

1. Calculate the matrix $\boldsymbol{G}$ as the minimal nonnegative solution of the matrix equation

$$\boldsymbol{G} = \boldsymbol{Y}(\boldsymbol{G}).$$

This equation is Neuts' equation, see, e.g., [14], for the discrete time Markov chain having $\boldsymbol{Y}(z)$ as the generating function of its transition probability matrices. It follows from this equation that, in our case, the matrix $\boldsymbol{G}$ is of the following structure:

$$\boldsymbol{G} = \begin{pmatrix} 0 & \boldsymbol{I}_{\overline{W}} & & & \\ & 0 & \boldsymbol{I}_{\overline{W}} & & \\ & & \ddots & & \\ & & & 0 & \boldsymbol{I}_{\overline{W}} \\ & & & \boldsymbol{G}^{(1)} & \boldsymbol{G}^{(2)} \end{pmatrix}$$

where $\boldsymbol{G}^{(1)}, \boldsymbol{G}^{(2)}$ are the square matrices of size $\overline{W}$, whose entries are calculated using the iterative method.

2. For preassigned sufficiently large integer i_0, calculate the matrices $\boldsymbol{G}_{i_0-1}$, $\boldsymbol{G}_{i_0-2}, \ldots, \boldsymbol{G}_0$ using the equation of the backward recursion

$$\boldsymbol{Q}_{i+1,i} + \sum_{n=i+1}^{\infty} \boldsymbol{Q}_{i+1,n} \boldsymbol{G}_{n-1} \boldsymbol{G}_{n-2} \ldots \boldsymbol{G}_i = 0,$$

$i = i_0 - 1, i_0 - 2, \ldots, 0$, with the boundary condition $\boldsymbol{G}_i = \boldsymbol{G}$, $i \geq i_0$.

The issues concerning the proper choice of i_0 are discussed in [10].

3. Calculate the matrices

$$\bar{\boldsymbol{Q}}_{i,l} = \boldsymbol{Q}_{i,l} + \sum_{n=l+1}^{\infty} \boldsymbol{Q}_{i,n} \boldsymbol{G}_{n-1} \boldsymbol{G}_{n-2} \ldots \boldsymbol{G}_l, \ l \geq i, \ i \geq 0.$$

4. Calculate the matrices $\boldsymbol{F}_l$ using the recurrent formulas

$$\boldsymbol{F}_l = (\bar{\boldsymbol{Q}}_{0,l} + \sum_{i=1}^{l-1} \boldsymbol{F}_i \bar{\boldsymbol{Q}}_{i,l})(-\bar{\boldsymbol{Q}}_{l,l})^{-1}, \ l \geq 1.$$

5. Calculate the vector $\mathbf{p}_0$ as the unique solution of the system

$$\mathbf{p}_0 \bar{\boldsymbol{Q}}_{0,0} = \mathbf{0}, \quad \mathbf{p}_0 (\mathbf{e} + \sum_{l=1}^{\infty} \boldsymbol{F}_l \mathbf{e}) = 1.$$

6. Calculate the vectors $\mathbf{p}_l$ as $\mathbf{p}_l = \mathbf{p}_0 \boldsymbol{F}_l$, $l \geq 1$.

5   Performance Measures

As soon as the vectors $\mathbf{p}_i$, $i \geq 0$, have been computed, we are able to calculate different performance measures of the system. Below we present some of them.

The mean number of customers in the orbit, L_{orb}, and in the buffer L_{buf}, are calculated in a trivial way by

$$L_{orb} = \sum_{l=0}^{\infty} l \mathbf{p}_l \mathbf{e}, \quad L_{buf} = \sum_{r=N+1}^{N+R} (r-N)\boldsymbol{\pi}(I^{(r)} \otimes I_{\overline{W}})\mathbf{e}.$$

Hereinafter $\boldsymbol{\pi} = \sum_{l=0}^{\infty} \mathbf{p}_l$, $I^{(r)}$ is a square matrix of size $N+R+1$ whose entries are zeroes except rth diagonal entry which is equal to 1.

Note that the expression $\boldsymbol{\pi}(I^{(r)} \otimes I_{\overline{W}})\mathbf{e}$ appearing in the formula for L_{buf} gives the probability that, at an arbitrary time, there are r customers in the system (on the servers and in the buffer).

Now we will derive some nontrivial performance measures of the system.

Theorem 3. *Probability that an arbitrary customer reaches a server immediately upon arrival is calculated as*

$$P_{imm} = \lambda^{-1}\boldsymbol{\pi}[\sum_{r=0}^{N-1} I^{(r)} \otimes \sum_{k=0}^{N-r}(k-N+r)D_k]\mathbf{e}. \tag{13}$$

Proof. Under the natural assumption that position of a call in an arriving batch is uniformly distributed, the rate of primary customers that had a luck to occupy a server immediately upon arrival is calculated as $\boldsymbol{\pi}[\sum_{r=0}^{N-1} I^{(r)} \otimes (\sum_{k=1}^{N-r} kD_k + \sum_{k=N-r+1}^{\infty}(N-r)D_k)]\mathbf{e}$. Dividing this rate by the rate λ of the $BMAP$ and taking into account the relation $\sum_{k=N-r+1}^{\infty} D_k \mathbf{e} = -\sum_{k=0}^{N-r} D_k \mathbf{e}$, we get (13).

Let $W(t)$ be the probability that an arbitrary primary customer is accepted into the buffer upon arrival, reaches service without visiting the orbit, and the waiting time of this customer is less than t. Let $w(s) = \int_0^{\infty} e^{-st}dW(t)$ denote the Laplace-Stieltjes transform (LST) of the function $W(t)$.

Theorem 4. *The LST $w(s)$ is given by expression*

$$w(s) = \lambda^{-1}\boldsymbol{\pi} \sum_{r=0}^{N+R-1} I^{(r)}\mathbf{e} \otimes \left[\sum_{k=\max\{1,N-r+1\}}^{N+R-r} D_k\mathbf{e} \sum_{j=\max\{1,r-N+1\}}^{k-N+1} w_j(s) \right.$$

$$\left. - \sum_{k=0}^{N+R-r} D_k\mathbf{e} \sum_{j=\max\{1,r-N+1\}}^{R} w_j(s) \right] \tag{14}$$

where

$$w_j(s) = \prod_{l=0}^{j-1} \frac{N\mu + l\gamma}{s + N\mu + (l+1)\gamma}. \tag{15}$$

Proof. Let $W_j(t)$ be the probability that a customer does not leave the buffer due to the impatience and his/her waiting time is less than t under condition that, upon arrival, this customer has been placed to the jth position in the buffer; $\Phi_l(t)$ be a distribution function of the time during which the l – size queue $(l = \overline{1,R})$ in the buffer reduces by one. Such a reduction occurs when one of the l customers staying in the queue whether leaves the buffer due to impatience or enters the service. By analogy, denote by $\Phi_0(t)$ a distribution function of the time during which one of the N busy servers becomes free.

Introduce the notation

$$w_j(s) = \int_0^\infty e^{-st} dW_j(t), \ j = \overline{1, R-1}, \ \phi_l(s) = \int_0^\infty e^{-st} e^{-\gamma t} d\Phi_l(t), \ l = \overline{0, R}.$$

It is easy to see that $\Phi_l(t) = 1 - e^{-(N\mu + l\gamma)t}$, $l = \overline{0, R}$. This yields

$$\phi_l(s) = \int_0^\infty e^{-st} e^{-\gamma t} dW_l(t) = \frac{N\mu + l\gamma}{s + N\mu + (l+1)\gamma}, \ l = \overline{0, R}.$$

The LST $w_j(s)$ can be represented as $w_j(s) = \prod_{l=0}^{j-1} \phi_l(s)$. Therefore, this LST is calculated by formula (15).

Let $W(t, r)$ be the probability that an arbitrary primary customer meets r, $r = \overline{0, R-1}$, customers in the system, is accepted into the buffer upon arrival, gets service without visiting the orbit and the waiting time of this customer is less than t. Let also $w(s, r) = \int_0^\infty e^{-st} dW(t, r)$, $r = \overline{0, R-1}$. The functions $w(s, r)$ are derived using the probabilistic arguments. The reader can easily follow these arguments by reading the following transparent formulas:

$$w(s, r) = \boldsymbol{\pi} \left[\boldsymbol{I}^{(r)} \otimes \left(\sum_{k=N-r+1}^{N+R-r} \frac{k D_k}{\lambda} \frac{k - N + r}{k} \frac{1}{k - N + r} \sum_{j=1}^{k-N+r} w_j(s) \right. \right.$$

$$\left. \left. + \sum_{k=N+R-r+1}^\infty \frac{k D_k}{\lambda} \frac{R}{k} \frac{1}{R} \sum_{j=1}^R w_j(s) \right) \right] \mathbf{e}, \ r = \overline{0, N}, \tag{16}$$

$$w(s, r)) = \boldsymbol{\pi} \left[\boldsymbol{I}^{(r)} \otimes \left(\sum_{k=1}^{N+R-r} \frac{k D_k}{\lambda} \frac{1}{k} \sum_{j=r-N+1}^{r-N+k} w_j(s) \right. \right.$$

$$\left. \left. + \sum_{k=N+R-r+1}^\infty \frac{k D_k}{\lambda} \frac{N+R-r}{k} \frac{1}{N+R-r} \sum_{j=r-N+1}^R w_j(s) \right) \right] \mathbf{e}, \tag{17}$$

$$r = \overline{N+1, N+R-1}.$$

Summing equations (16)-(17) over j and using the relation $\sum_{k=m+1}^{\infty} D_k e = -\sum_{k=0}^{m} D_k e$, after some algebra we get (14).

Corollary 1. *The function $W(t)$ is calculated as*

$$W(t) = \lambda^{-1}\pi \sum_{r=0}^{N+R-1} I^{(r)} e \otimes \left[\sum_{k=\max\{1,N-r+1\}}^{N+R-r} D_k e \sum_{j=\max\{1,r-N+1\}}^{k-N+1} \frac{1}{N\mu + j\gamma} E_j^{(gen)}(t) \right.$$

$$\left. - \sum_{k=0}^{N+R-r} D_k e \sum_{j=\max\{1,r-N+1\}}^{R} \frac{1}{N\mu + j\gamma} E_j^{(gen)}(t) \right] \tag{18}$$

where $E_j^{(gen)}(t)$ is a generalized Erlang distribution function with the parameters $(j; N\mu + \gamma, \ldots, N\mu + j\gamma)$.

Proof is implemented by noting that the inverse of the function $w_j(s)$ given by formula (15) is $(N\mu + j\gamma)^{-1} E_j^{(gen)}(t)$.

Corollary 2. *Probability, P_{buf}, that an arbitrary arriving customer is accepted into the buffer upon arrival and reaches service without visiting the orbit is defined by*

$$P_{buf} = \frac{N\mu}{\lambda}\pi \sum_{r=0}^{N+R-1} I^{(r)} e \otimes \left[\sum_{k=\max\{1,N-r+1\}}^{N+R-r} D_k e \sum_{j=\max\{1,r-N+1\}}^{k-N+1} \frac{1}{N\mu + j\gamma} \right.$$

$$\left. - \sum_{k=0}^{N+R-r} D_k e \sum_{j=\max\{1,r-N+1\}}^{R} \frac{1}{N\mu + j\gamma} \right].$$

Proof is immediate from (14) by taking $s = 0$.

Based on the obtained results, we are able to find one more important performance measure of the system. It is the probability that an arbitrary primary customer will reach the server immediately or after waiting in the buffer without visiting the orbit. This probability is calculated by

$$P_{succ} = P_{imm} + P_{buf}.$$

6   Numerical Example

Having the main performance measures of the system been calculated, we can analyze their behavior numerically. Here we restrict ourselves by presenting the results of only one numerical experiment where we investigate dependence of the system performance measures on the system load ρ under different values of impatience rate γ.

The arrival flow is defined by the matrices

$$D_0 = \begin{pmatrix} -4.32575 & 0.32725 \\ 0.11675 & -0.477 \end{pmatrix}, \quad D_k = h^{k-1}\frac{1-h}{1-h^k}\begin{pmatrix} 3.89125 & 0.10725 \\ 0.077 & 0.28325 \end{pmatrix}, \quad k = \overline{1,3},$$

where $h = 0.8$. This MAP has the squared coefficient of variation $c_{var}^2 = 3.86$ and the coefficient of correlation $c_{corr} = 0.2$.

In the experiment, we vary the system load by means of changing the value of $BMAP$ fundamental rate λ. To this end, we normalize the matrices $D_k, k = \overline{0,3}$, by multiplying them by some positive number.

The other parameters of the system are assumed to be as follows:

$$N = 10, \ R = 20, \ \mu = 5, \ q = 0, \ \alpha = 3.$$

Figures 2-6 depict the system performance measures as functions of the load ρ under different values of impatience rate γ.

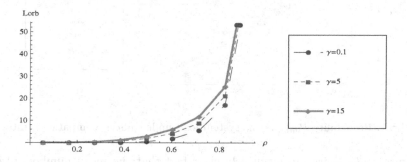

Fig. 2. Mean number of customers in the orbit vs the system load under different impatience rates

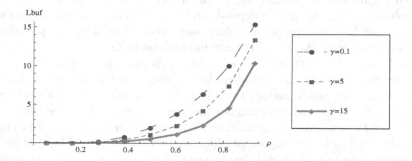

Fig. 3. Mean number of customers in the buffer vs the system load under different impatience rates

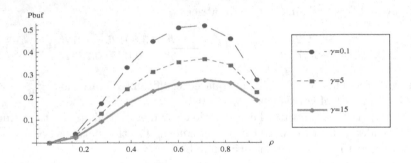

Fig. 4. Probability P_{buf} vs the system load under different impatience rates

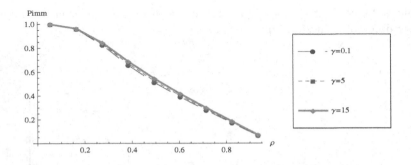

Fig. 5. Probability P_{imm} vs the system load under different impatience rates

Figures 2-3 confirm an intuitively clear fact that the mean numbers of customers in the buffer, L_{buf}, and in the orbit, L_{orb}, increase with increase in the load ρ. Under the fixed value of ρ, L_{orb} increases and L_{buf} decreases when the impatience rate γ increases. Not so evident observation is that the relative difference in the values of each of these performance measures reduces when ρ becomes large.

From Figure 4, it is evidently seen that, under the fixed value of the load ρ, the probability P_{buf} essentially depends on the impatience rate γ. As expected, with increase in γ this probability decreases. The behavior of the probability P_{buf} as a function of the load ρ is more unpredictable. One can see that, under the fixed value of γ, with increase in ρ this probability goes up to some value and then begins to decrease. Explanation of such a behavior can be done as follows. Under the small load, ρ, there is a good probability that an arbitrary primary customer immediately reaches the service, not entering the buffer. So, the value of P_{buf} increases insignificantly. With increasing ρ, the chance to meet a free server reduces. This implies considerable increase in P_{buf}. This probability increases to some maximum and then begins to decrease since arriving customers frequently meet the buffer full and are forced to go to the orbit.

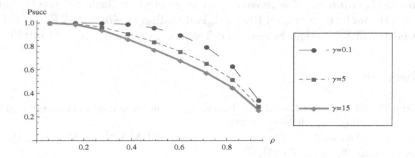

Fig. 6. Probability P_{succ} vs the system load under different impatience rates

The graphs shown in Figure 5 confirm the obvious fact that, under the fixed value of the impatience rate γ, the probability to reach a server immediately upon arrival, P_{imm}, decreases with increase in the load ρ. In the region of small values of ρ, the probability under consideration is not sensitive to the change in the value of the impatience rate γ because there is a good probability that an arbitrary arriving customer will be accepted for service immediately. In the region of average values of ρ, the probability P_{imm} increases with increase in γ because more customers leave the system due to impatience and, therefore, an arriving customer more likely will reach a server immediately upon arrival. However, the dependence of P_{imm} on γ is not very significant because, with increase in γ, the rate of leaving the system due to impatience increases along with increase of the total retrial rate. With further increase in ρ, the difference in the values of the probability P_{imm} for different γ vanishes because the total retrial rate becomes large and the probability to see a free server becomes very small.

One of the most important performance measures of the system under consideration is the introduced above probability P_{succ} that an arbitrary primary customer will be served without visiting the orbit. It is obvious that the behavior of this probability depends on the behavior of probabilities P_{imm} and P_{buf}. The behavior of P_{succ} is shown in figure 6. As it is seen from the figure, under the fixed value of γ, the probability P_{succ} decreases as the load ρ increases. However, as it can be observed from comparison of figures 5 and 6, the function P_{succ} decreases more slowly than function P_{imm} due to possibility for arriving customer to reach service after waiting in the buffer. Under the fixed value of ρ, the impatience phenomenon has a greater effect on P_{succ} in the region when the load ρ is neither light nor heavy. This effect is caused by behavior of the probabilities P_{imm} and P_{buf} that was discussed above.

7 Conclusion

Multi-server retrial queue with a finite buffer and impatient customers is analyzed. Impact of impatience and system load on performance measures is illustrated.

Acknowledgements. This research was supported by Basic Science Research Program through the National Research Foundation of Korea (NRF) funded by the Ministry of Education, Science and Technology (Grant No. 2011-0015214).

References

1. Artalejo, J.R., Comez-Corral, A.: Retrial queueing systems: a computational approach. Springer, Heidelberg (2008)
2. Breuer, L., Dudin, A.N., Klimenok, V.I.: A retrial $BMAP/PH/N$ system. Queueing Systems 40, 433–457 (2002)
3. Breuer, L., Klimenok, V.I., Birukov, A.A., Dudin, A.N., Krieger, U.: Modeling the access to a wireless network at hot spots. European Transactions on Telecommunications 16, 309–316 (2005)
4. Chakravarthy, S.R.: The batch Markovian arrival process: a review and future work. In: Krishnamoorthy, A., et al. (eds.) Proc. of Advances in Probability Theory and Stochastic Process: Proc., pp. 21–49. Notable Publications, NJ (2001)
5. Dudin, A.N., Klimenok, V.I.: Queueing System $BMAP/G/1$ with repeated calls. Mathematical and Computer Modelling 30, 115–128 (1999)
6. Dudin, A.N., Klimenok, V.I.: A retrial $BMAP/SM/1$ system with linear repeated requests. Queueing Systems 34, 47–66 (2000)
7. Falin, G.I., Templeton, J.G.C.: Retrial queues. Chapman and Hall, London (1997)
8. Gomez-Corral, A.: A bibliographical guide to the analysis of retrial queues through matrix analytic techniques. Annals of Operations Research 141, 163–191 (2006)
9. Kim, C.S., Mushko, V.V., Dudin, A.N.: Computation of the steady state distribution for multi-server retrial queues with phase type service process. Annals of Operations Research 201, 307–323 (2012)
10. Klimenok, V.I., Dudin, A.N.: Multi-dimensional asymptotically quasi-toeplitz Markov chains and their application in queueing theory. Queueing Systems 54, 245–259 (2006)
11. Klimenok, V.I., Orlovsky, D.S., Dudin, A.N.: A $BMAP/PH/N$ system with impatient repeated calls. Asia-Pacific Journal of Operational Research 24, 293–312 (2007)
12. Klimenok, V.I., Orlovsky, D.S., Kim, C.S.: The $BMAP/PH/N/N + R$ retrial queueing system with different disciplines of retrials. In: Proceedings of 11th International Conference on Analytical and Stochastic Modelling Techniques and Applications (ASMTA 2004), Magdeburg, Germany, June 13-16, pp. 93–98 (2004)
13. Lucantoni, D.M.: New results on the single server queue with a batch Markovian arrival process. Communications in Statistics-Stochastic Models 7, 1–46 (1991)
14. Neuts, M.F.: Structured stochastic matrices of $M/G/1$ type and their applications. Marcel Dekker, New York (1989)
15. Ramaswami, V., Lucantoni, D.: Algorithm for the multi-server queue with phase-type service. Communications in Statistics-Stochastic Models 1, 393–417 (1985)

Representation Transformations
for Finding Markovian Representations

András Mészáros[1], Gábor Horváth[1,3], and Miklós Telek[1,2]

[1] Budapest University of Technology and Economics
[2] MTA-BME Information Systems Research Group,
[3] Inter-University Center of Telecommunications and Informatics
{meszarosa,ghorvath,telek}@hit.bme.hu

Abstract. In this paper we consider existing and new representation transformation methods for non-Markovian generalizations of Markov chain driven stochastic models which intend transforming non-Markovian representations into Markovian ones and evaluate their efficiency through numerical experiments. One of the new features of the considered methods is the ability to obtain a Markovian representation of larger size.

Keywords: Markov arrival process, Rational arrival process, representation transformation, Markovian representation.

1 Introduction

Background Markov chain driven stochastic models, like PH distributions [1], MAPs [2], and MMAPs [3], are efficiently used for describing practical systems of various application fields, including computer networks and telecommunication systems. These models are described with a set of vectors and/or matrices, which are referred to as representation. In some cases, e.g., when the coefficient of variation of inter-event times is low, the non-Markovian generalizations of these Markov chain based models are more efficient with respect to the size of the representation. Another motivation for dealing with non-Markovian generalizations comes from the fact that there are moments based fitting methods which generate non-Markovian representations based on a set of moments and joint moments.

Stochastic processes with Markovian representations have essentially important nice features. There is always a valid stochastic process associated with a Markovian representation, and it is easy to check if a representation is Markovian or not. The main drawback of using non-Markovian representations is that non-Markovian representations might or might not represent valid stochastic processes. A non-Markovian representation is non-valid the joint density function of consecutive inter-arrival times defined by the representation, see eg. (1), is negative at some point. In this case there is no stochastic process associated with the non-Markovian representation. Additionally, it is not obvious how to determine if a matrix exponential distribution (defined by a non-Markovian representation) is a valid distribution (e.g. its density function is non-negative) and

A. Dudin and K. De Turck (Eds.): ASMTA 2013, LNCS 7984, pp. 277–291, 2013.
© Springer-Verlag Berlin Heidelberg 2013

for more complex processes, like RAPs and MRAPs, we do not have the methodology for checking if a non-Markovian representation defines a valid process or not.

In this paper we cannot present a general solution for checking if a non-Markovian representation is associated with a stochastic process or not, we only propose an elementary step. If a non-Markovian representation can be transformed to a Markovian representation such that the representation defines the same process, then the non-Markovian representation surely defines a valid stochastic process. Our proposed elementary step is through numerical procedures for transforming non-Markovian representations to Markovian ones. We survey previously available methods and present new ones. The main new feature in this work is the representation transformation procedure which searches for a Markovian representation of larger size.

The rest of the paper is organized as follows. Section 2 surveys the considered set of stochastic models and summarizes the properties which are utilized later on. Existing and new representation transformation methods are introduced for same size in Section 3 and for extended size in Section 4. Section 5 is devoted to the numerical experimentation with the different representation methods. The paper is concluded in Section 6.

2   Markov Chain Driven Stochastic Models and Their Non-markovian Generalizations

2.1   Phase Type and Matrix Exponential Distributions

We start with the basic definition of PH [1] and ME [4] distributions.

Definition 1. *Let $\mathcal{X}$ be a random variable with cumulative distribution function (cdf)*

$$F_X(x) = Pr(\mathcal{X} < x) = 1 - \alpha e^{Ax}\mathbb{1},$$

where α is an initial row vector of size n, A is a square matrix of size $n \times n$, and $\mathbb{1}$ is the column vector of ones of size n. In this case, we say that $\mathcal{X}$ is matrix exponentially distributed with representation α, A, or shortly, $ME(\alpha, A)$ distributed.

Definition 2. *If $\mathcal{X}$ is an $ME(\alpha, A)$ distributed random variable, where α and A have the following properties:*

- *$\alpha_i \geq 0$, $\alpha\mathbb{1} = 1$ (there is no probability mass at $x = 0$),*
- *$A_{ii} < 0$, $A_{ij} \geq 0$ for $i \neq j$, $A\mathbb{1} \leq 0$,*

then we say that $\mathcal{X}$ is phase type distributed with representation α, A, or shortly, $PH(\alpha, A)$ distributed.

The vector-matrix representations satisfying the conditions of Definition 2 are called Markovian.

2.2 Marked MAPs and Marked RAPs

MAPs and RAPs model point processes with a single type of events. Extension of these processes to multiple event types are referred to as marked MAPs (MMAPs) [3] and marked RAPs (MRAPs) [5]. Due to space limitations we summarize the properties of these more general processes only.

Let $\{X(t), Y(t)\}$ be a point process supplemented with the type of the events $\{1, \ldots, K\}$, with joint probability density function of inter-event times and associated types $f(t_1, k_1, \ldots, t_j, k_j)$ for $j = 1, 2, \ldots$, and $k_j \in \{1, \ldots, K\}$.

Definition 3. *$\{\mathcal{X}(t), \mathcal{Y}(t)\}$ is called a marked rational arrival process if there exists a set of $K + 1$ finite matrices $(\boldsymbol{H_0}, \ldots, \boldsymbol{H_K})$, such that*

$$f_{(\boldsymbol{H_0}, \ldots, \boldsymbol{H_K})}(t_1, k_1, \ldots, t_j, k_j) = \underline{\pi} e^{\boldsymbol{H_0} t_1} \boldsymbol{H_{k_1}} e^{\boldsymbol{H_0} t_2} \boldsymbol{H_{k_2}} \ldots e^{\boldsymbol{H_0} t_j} \boldsymbol{H_{k_j}} \mathbb{1} \quad (1)$$

where $\sum_{i=0}^{K} \boldsymbol{H_i} \mathbb{1} = 0$, and π is the solution of

$$\underline{\pi}(-\boldsymbol{H_0})^{-1} \sum_{i=1}^{K} \boldsymbol{H_i} = \underline{\pi}, \quad \underline{\pi}\mathbb{1} = \mathbb{1} . \quad (2)$$

In this case, we say that $\{\mathcal{X}(t), \mathcal{Y}(t)\}$ is a marked rational arrival process with representation $(\boldsymbol{H_0}, \ldots, \boldsymbol{H_K})$, or shortly, MRAP$(\boldsymbol{H_0}, \ldots, \boldsymbol{H_K})$.

Definition 4. *If $\{\mathcal{X}(t), \mathcal{Y}(t)\}$ is a MRAP$(\boldsymbol{H_0}, \ldots, \boldsymbol{H_K})$, such that*

- $\boldsymbol{H_{kij}} \geq 0$ for $k \geq 1$,
- $\boldsymbol{H_{0ii}} < 0$, $\boldsymbol{H_{0ij}} \geq 0$ for $i \neq j$, $\boldsymbol{H_0}\mathbb{1} \leq 0$,

then we say that $\{\mathcal{X}(t), \mathcal{Y}(t)\}$ is a marked MAP with representation $(\boldsymbol{H_0}, \ldots, \boldsymbol{H_K})$, that is, MMAP$(\boldsymbol{H_0}, \ldots, \boldsymbol{H_K})$.

The representations satisfying the conditions of Definition 4 are called Markovian. Later we are going to use the following property of MRAPs from [6]

Definition 5. *The rank of the $(\boldsymbol{H_0}, \ldots, \boldsymbol{H_K})$ representation, with respect to the initial vector, is the number of linear independent vectors of the form $\underline{\pi} \boldsymbol{H_0}^{a_1} \boldsymbol{H_{k_1}} \boldsymbol{H_0}^{a_2} \boldsymbol{H_{k_2}} \ldots$, with $a_i \in \{1, 2, \ldots\}$ and $k_i \in \{1, 2, \ldots, K\}$.*

An efficient computational method for computing this rank is provided in [6].

Obviously, MRAPs and MMAPs with $K = 1$ are RAPs and MAPs. Based on this property in the rest of the paper we are going to follow a unified treatment of MRAPs/MMAPs and RAPs/MAPs with K event types. This way we are going to present the existing representation transformation methods of the literature in a more general environment than it is in the original publications.

```
 1: procedure MRAP2MMAP-[7](H₀,...,H_K)
 2:     a = 0.5;
 3:     while a > ε do
 4:         for n = 1 to maxIter do
 5:             (i*, j*) =    arg min    G(Tr(H₀,...,H_K,T(i,j,a)))
                           i,j∈[1,...,n],i≠j
 6:             (H₀^(new),...,H_K^(new)) = Tr(H₀,...,H_K,T(i*,j*,a))
 7:             if G(H₀^(new),...,H_K^(new)) < G(H₀,...,H_K) then
 8:                 (H₀,...,H_K)=(H₀^(new),...,H_K^(new))
 9:             end if
10:             if H₀,...,H_K is Markovian then
11:                 return (H₀,...,H_K)
12:             end if
13:         end for
14:         a = a/2
15:     end while
16: return (H₀,...,H_K)
17: end procedure
```

Fig. 1. The representation transformation method of [7]

3   Representation Transformation Methods with the Same Size

In this section we first present the previous research related to RAP/MRAP to MAP/MMAP transformation. After that we introduce the new algorithms. The past work only dealt with transformation between representations of the same size. In this section we also introduce a representation transformation method with size extension.

3.1   Previous Works with Single Element Modifications

The development of efficient numerical representation transformation methods for finding a Markovian representation from a non-Markovian one dates back to 2007 [7]. This algorithm, generalized to MRAPs, is shown in Figure 1.

The algorithm applies a series of elementary transformations improving the representation in each step until it reaches a Markovian representation or a local optimum of the goal function. The matrix representing the elementary transformations is given by

$$T(i,j,x) = I + xE_{ij} - xE_{ii}, \qquad (x \neq 1), \tag{3}$$

where I is the identity matrix and E_{ij} is the matrix whose only nonzero entry is the i,j element which equals to 1. The procedure determines the best possible elementary transformation matrix and transforms the representation in each step. The transformation step in the algorithm is represented by $Tr(H_0,\ldots,H_K,T) = (T^{-1}H_0T,\ldots,T^{-1}H_KT)$.

```
 1: procedure MRAP2MMAP-[8](H_0,...,H_K)
 2:     r ← 1
 3:     repeat
 4:         (i*,j*,x*) =   arg min   G_4(Tr(H_0,...,H_K,T(i,j,x)),r)
                        i,j∈[1,...,n],i≠j,
                             x∈[0,1]
 5:         (H_0,...,H_K) = Tr(H_0,...,H_K,T(i*,j*,x*))
 6:         if H_0,...,H_K is Markovian then
 7:             return (H_0,...,H_K)
 8:         end if
 9:         r ← 1.05r
10:     until x* < ε
11: return (H_0,...,H_K)
12: end procedure
```

Fig. 2. The representation transformation method of [8]

In [7] the following three goal functions are considered (denoted by $G_1(\cdot), G_2(\cdot)$ and $G_3(\cdot)$):

$$G_1(H_0,...,H_K) = \sum_{i,j,i\neq j} \exp(-2[H_0]_{i,j}) + \sum_{k=1}^{K}\sum_{i,j} \exp(-2[H_k]_{i,j}),$$

$$G_2(H_0,...,H_K) = \sum_{i,j,i\neq j} \exp(-1000[H_0]_{i,j}) + \sum_{k=1}^{K}\sum_{i,j} \exp(-1000[H_k]_{i,j}),$$

$$G_3(H_0,...,H_K) = \sum_{i,j,i\neq j} \exp(-([H_0]_{i,j}-1)^3) + \sum_{k=1}^{K}\sum_{i,j} \exp(-([H_k]_{i,j}-1)^3).$$

The algorithm switches goal functions from time to time, which, for simplicity, is not reflected in Figure 1. The multiple goal functions help to leave a local optimum and let the optimization go for a better solution.

While the algorithm is successful in finding the Markovian representation in many cases, our current intuition is that the rigidity of the transformation matrix (3) might pose a problem in some cases. Buchholz et al. proposed a modified version of this method in [8] by applying a different goal function and parameter x in (3) is included in the optimization as well. The proposed procedure is summarized in Figure 2. The goal function used in this algorithm is given by

$$G_4(H_0,...,H_K,r) = r\sum_{\substack{i,j\\i\neq j}}((-[H_0]_{i,j})^+)^2 - \sum_{\substack{i,j\\i\neq j}}(([H_0]_{i,j})^+)^2$$

$$+ r\sum_{k=1}^{K}\sum_{i,j}((-[H_k]_{i,j})^+)^2 - \sum_{k=1}^{K}\sum_{i,j}(([H_k]_{i,j})^+)^2, \quad (4)$$

where $(x)^+$ is the positive part of x, i.e., $(x)^+ = \max\{0,x\}$.

The intuitive motivation of such a goal function is the following. If we do not reward the higher non-negative elements, the algorithm will only try to increase the negative elements, while possibly decreasing the other elements to zero. When these elements reach zero, the algorithm can get stuck in a non-Markovian representation. Therefore, this algorithm tries to make a "provision" for the non-negative elements in the beginning. Later the weight of these elements decreases, and the procedure focuses more and more on increasing the negative elements of the representation.

Apart from the different goal function, the main difference between MRAP2MMAP-[7] and MRAP2MMAP-[8] is that in each iteration the former one performs an exhaustive search for the optimal i, j parameters of the elementary transformation matrix, while the latter one incorporates a continuous variable x into the optimization as well, needing a more involved solution method.

3.2   A New Family of Representation Transformation Methods

Based on our experience with the past algorithms discussed in Section 3.1 we found that the overly simple structure of the elementary transformation matrix they are using limits their efficiency considerably.

The structure of the elementary matrix plays an important role during the representation transformation. A too simple elementary transformation matrix can restrict the basic movements of the optimization method. Along the restricted path potentially no Markovian representations may be reachable. On the other hand, a too general elementary transformation matrix introduces too many variables to optimize. This makes the solution of the corresponding non-linear optimization problem less effective. It will be slow and might give a local optimum that is far from the global one.

In this section we introduce a family of representation transformation procedures. All these procedures follow the same basic idea as Algorithms 1 and 2, that is, they still use consecutive transformations to find a Markovian representation of a RAP. However, we propose enhancements in two principal details of the algorithm.

1. *Elementary modifications of the transformation matrix* The simple elementary transformation matrix given by (3) is basically an identity matrix (acting as an initial transformation matrix) modified such that a single entry is put in an off-diagonal position. It is possible to generalize this idea by allowing more general modifications of the initial transformation matrix.

   In the family of algorithms introduced in this paper we are making use of three kinds of such elementary modifications.

   – *Single scalar modification.* The initial transformation matrix T_0 is modified by a single scalar x as

   $$M_S(T_0, i, j, x) = T_0 + xE_{ij} - xE_{ii}. \qquad (5)$$

   In this case, only a single variable has to be optimized.

- *Single vector modification.* The initial transformation matrix is increased by a vector as

$$M_V(T_0, i, \underline{x}) = T_0 + \underline{e}_i^T \cdot \underline{x}, \tag{6}$$

where $\underline{e}_i$ is the ith unit row vector. Vector $\underline{x}$ is such that $M_V(T_0, i, \underline{x})\mathbb{1} = \mathbb{1}$, thus number of elements to be optimized is $n - 1$, where n is the size of the representaiton.
- *Dyadic product modification.* The most general modification of the initial transformation matrix used in this paper is given by a dyadic product of two vectors (where the i, j element of matrix $\underline{u}^T \cdot \underline{v}$ is $u_i v_j$), yielding

$$M_D(T_0, \underline{u}, \underline{v}) = T_0 + \underline{u}^T \cdot \underline{v}, \tag{7}$$

with $M_D(T_0, \underline{u}, \underline{v})\mathbb{1} = \mathbb{1}$, which means that $2n - 1$ elements need to be optimized.

2. *The way the transformation matrix is constructed.* We investigate two cases on how the transformation matrix is obtained in each step of the algorithm.
   - *Transformation with a single elementary modification.* In this case, the transformation matrix used to transform the current representation in each step of the algorithm is an identity matrix with an elementary modification applied. Thus, in each step the optimal matrix $T = M_S(I, i, j, x), T = M_V(I, i, \underline{x})$ or $T = M_D(I, \underline{u}, \underline{v})$ is determined (depending on which variant of the algorithm we are using) and the corresponding transformation is applied.
   - *Transformation with cumulative elementary modifications.* According to this approach the transformation is not applied in each step of the algorithm. The algorithm starts with $T = I$, and makes several elementary modifications one after the other in a cumulative way (thus $T = M_S(T, i, j, x), T = M_V(T, i, \underline{x})$ or $T = M_D(T, \underline{u}, \underline{v})$), as long as there is improvement according to the goal function. The transformation is then applied with matrix T. Observe that the matrix T obtained this way is much more general then the one obtained by a single modification.

The three options according to the elementary transformations and the two options according to the way the transformation matrix is obtained gives 6 possible algorithms in total. These algorithms will be identified as Algorithm MRAP2MMAP-xy, where x can be

- x=S, if the algorithm uses single scalar modifications,
- x=V, if it uses single vector modifications,
- and x=D, if dyadic product modifications are used.

Similarly, the letter at position y determines the way the transformation matrix is constructed

- y=S, if the algorithm uses single elementary modifications,
- y=C, if it uses cumulative elementary modifications.

```
1: procedure MRAP2MMAP-DC (H₀,...,H_K)
2:     iter ← 1
3:     while H₀,...,H_K not Markovian and iter < maxIter do
4:         T ← I
5:         for i = 1 to N do
6:             (u*, v*) ← arg min G(Tr(H₀,...,H_K, M_D(T, u, v)))
                           u,v
7:             T ← M_D(T, u*, v*)
8:         end for
9:         (H₀,...,H_K) ← Tr(H₀,...,H_K, T)
10:        iter ← iter + 1
11:    end while
12: return (H₀,...,H_K)
13: end procedure
```

Fig. 3. Representation transformation method with single ($N = 1$) and cumulative ($N > 1$) dyadic elementary modifications

Figure 3 with $N = 1$ and $N > 1$ depicts two possible variants of this family of methods. The goal function is the sum of the square of the negative elements of the representation, thus

$$G(\boldsymbol{H_0},\ldots,\boldsymbol{H_K}) = \sum_{\substack{i,j \\ i \neq j}} ((-[\boldsymbol{H_0}]_{i,j})^+)^2 + \sum_{k=1}^{K} \sum_{i,j} ((-[\boldsymbol{H_k}]_{i,j})^+)^2. \qquad (8)$$

4 Representation Transformation Methods with Size Extension

If a size n non-Markovian MRAP has no Markovian representation of size n, then the above discussed procedures are not applicable. There are cases, however, when a MRAP of order n has a Markovian representation of order $m > n$. With a simple modification, our algorithms can be applied to the problem of finding a larger Markovian representation as well. The idea is based on the results of [6].

Theorem 1. ([6], Theorem 4) If, for an MRAP $(\boldsymbol{D_0},\ldots,\boldsymbol{D_K})$ of size m and initial vector $\underline{\pi}$ (where $\underline{\pi}$ is the solution of $\underline{\pi} = \underline{\pi}(\boldsymbol{D_0})^{-1} \sum_{k=1}^{k} \boldsymbol{D_k}$ and $\underline{\pi}\mathbb{1} = 1$), the rank of the representation with respect to the initial vector (see Definition 5) is n, and $n < m$, then there exists a non singular $m \times m$ transformation matrix $\boldsymbol{T}$, such that the transformed representation has the following block structure (with block sizes n and $m - n$)

$$\boldsymbol{T}\boldsymbol{D_k}\boldsymbol{T}^{-1} = \begin{bmatrix} \boldsymbol{H_k} & \boldsymbol{0} \\ \star & \star \end{bmatrix}, \; for \; k = 0, 1, \ldots, K \; and \; \underline{\pi}\boldsymbol{T} = \begin{bmatrix} \underline{\gamma} & 0 \end{bmatrix}, \qquad (9)$$

where $\star$ denotes irrelevant matrix blocks with arbitrary elements, and $(\boldsymbol{H_0},\ldots,\boldsymbol{H_K})$ is an equivalent representation of the same process with size n and initial vector $\underline{\gamma}$.

This theorem together with the further results of [6] provide a way to obtain a smaller representation of a RAP, if it is possible. In this paper, however, we have the opposite problem. We are investigating RAPs not having a Markovian representation of the same size, thus we need to find a way to extend the size of the representation.

Theorem 2. *If representation $(H_0, \ldots, H_K)$ of size n is non-Markovian, but the process has a $(D_0, \ldots, D_K)$ Markovian representation of size $n + 1$ whose rank with respect to the initial vector is n, then such a representation can be obtained in the form $D_k = T^{-1} H_k' T$, $k = 0, 1, \ldots, K$, where*

$$
H_k' = \left[
\begin{array}{c|c}
H_k & \begin{matrix} 0 \\ \vdots \\ 0 \end{matrix} \\
\hline
[x_k]_1 \cdots \cdots [x_k]_{n+1}
\end{array}
\right],
$$

and matrix T, vectors x_k, $k = 0, 1, \ldots, K$, are such that $T\mathbb{1} = \mathbb{1}$, $\sum_{k=0}^{K} x_k \mathbb{1} = 0$, and $[x_0]_{n+1} / \sum_{k=1}^{K} [x_k]_{n+1} \neq -1$.

Proof. By construction we have that T transforms matrices D_k, $k = 0, 1, \ldots, K$, into the required block structure of Theorem 1.

It remains to be proven that the initial vector also has the required block structure. That is, the solution of $\underline{\gamma}' = \underline{\gamma}'(-H_0')^{-1} \sum_{k=1}^{K} H_k'$ has the form $\underline{\gamma}' = [\gamma\ 0]$. Focusing on the last element of the vector equation and utilizing the block structure of matrices H_k' we have $[\gamma']_{n+1} = [\gamma']_{n+1} \frac{-1}{[x_0]_{n+1}} \sum_{k=1}^{K} [x_k]_{n+1}$. From this scalar equation we have that $[\gamma']_{n+1} = 0$ if $[x_0]_{n+1} / \sum_{k=1}^{K} [x_k]_{n+1} \neq -1$. $\square$

Based on Theorem 2 we search for a Markovian representation of size $n + 1$ with the algorithm in Figure 4. For simplicity we omit the constraint $[x_0]_{n+1} / \sum_{k=1}^{K} [x_k]_{n+1} \neq -1$ in the algorithm description, as, in practice, the procedure never generates vector elements which violate it. This algorithm is based on the alternating optimization of the elementary transformation matrix and vectors x_k, $k = 0, 1, \ldots, K$.

The $Tr()$ operator that performs the similarity transformation with size extension is defined by

$$
Tr'(H_0, \ldots, H_K, \underline{x_0}, \ldots, \underline{x_K}, T) =
$$

$$
\left(T^{-1} \left[
\begin{array}{c|c}
H_0 & \begin{matrix} 0 \\ \vdots \\ 0 \end{matrix} \\
\hline
[x_0]_1 \cdots \cdots [x_0]_{n+1}
\end{array}
\right] T, \ldots, T^{-1} \left[
\begin{array}{c|c}
H_K & \begin{matrix} 0 \\ \vdots \\ 0 \end{matrix} \\
\hline
[x_K]_1 \cdots \cdots [x_K]_{n+1}
\end{array}
\right] T \right)
\tag{10}
$$

Based on the duality of redundant representations in [6], one might miss a size extension approach with redundancy according to the closing vector. It is possible

```
1:  procedure MRAP2MMAP-SizeExt(H_0, ..., H_K)
2:      x_k ← random vector, k = 0, ..., K, such that ∑_{k=0}^{K} x_k 1 = 0
3:      T ← I
4:      while D_0, ..., D_K not Markovian and iter < maxIter do
5:          B ← I
6:          for i = 1 to N do
7:              (u*, v*) ← arg min G(Tr'(H_0, ..., H_K, x_0, ..., x_K, T · M_D(B, u, v)))
                              u,v
8:              B ← M_D(B, u*, v*)
9:          end for
10:         T ← T · B;
11:         (x_0*, ..., x_K*) ←      arg min      G(Tr'(H_0, ..., H_K, x_0, ..., x_K, T))
                              x_0,...,x_k ∈ ℝ^(n+1)
                                  ∑_{k=0}^{K} x_k 1 = 0
12:         (x_0, ..., x_k) ← (x_0*, ..., x_k*)
13:         (D_0 ..., D_K) = Tr'(H_0, ..., H_K, x_0, ..., x_K, T)
14:         iter ← iter + 1
15:     end while
16: return (D_0, ..., D_K)
17: end procedure
```

Fig. 4. Representation transformation with size extension using cumulative dyadic product modifications

to construct a counterpart procedure of similar nature that generates representation with redundancy according to the closing vector, but we have two reasons for not recommending such procedures.

- Each ME representation satisfying simple eigenvalue and density conditions (namely, the dominant eigenvalue is real and has the largest real part; the density is strictly positive on $(0, \infty)$) has a PH representation with redundant initial vector.
- We have the following conjecture:
  If $(H_0, ..., H_K)$ of size n is non-Markovian and it has a Markovian representation of size $n + 1$, then it has a Markovian representation of size $n + 1$ with a redundant initial vector and an other one with redundant closing vector.

5   Numerical Results

In this section we will provide some examples to demonstrate the behavior of the algorithms introduced.

5.1   Finding a Markovian Representation of the Same Size

To investigate how efficient the various representation transformation methods are, we define a kind of benchmark that consists of a large number of non-Markovian representations of MAPs. We are interested in how many times the various methods succeed to obtain a Markovian representation.

The MAPs belonging to the benchmark are generated as follows.

I *"Balanced" problems.* In this problem class we generate MAPs with matrices having random integer entries falling into $[0, 10]$. This random, but Markovian representation is then transformed to a non-Markovian one with a random transformation matrix. The benchmark includes

    a.) 10 representations of size 3,

    b.) 10 representations of size 4,

    c.) and 10 representations of size 5.

II *"Stiff" problems.* MAPs belonging to this class have random entries as in case of "Balanced" MAPs, but a particular matrix row or an entry is several orders of magnitudes smaller or larger than the other ones. Again, a random transformation matrix is applied to obtain a non-Markovian representation. We expect that it will be more difficult to obtain a Markovian representation for these problems. The benchmark includes

    a.) 10 representations of size 3 with random integer entries falling into $[0, 100000]$ except for the first row, where the random entries are from $[0, 10]$.

    b.) 10 representations with similar construction, but the size is 4

    c.) 10 representations of size 3 with random integer entries falling into $[0, 100000]$ except for the first row, where the random entries are from $[0, 10]$. One entry in the first row is from $[0, 100000]$.

    d.) 10 representations with similar construction, but the size is 4.

    e.) 10 representations of size 3 with random integer entries falling into $[0, 10]$ except for the first row, where the random entries are from $[0, 100000]$.

    f.) 10 representations with similar construction, but the size is 4

    g.) 10 representations of size 3 with random integer entries falling into $[0, 10]$ except for the first row, where the random entries are from $[0, 100000]$. One entry in the first row is from $[0, 10]$.

    h.) 10 representations with similar construction, but the size is 4.

III *"Sparse" problems.* In this case the matrices of the MAPs have a large number of 0 entries. The non-zero entries are random integers from $[0, 10]$, and the positions of the non-zero entries are randomly chosen as well (such that the irreducibility of the background process is ensured). The non-Markovian input for the algorithms is obtained by a random similarity transformation again. The more zero entries the matrices have, the smaller the space of Markovian representations for that particular process is, thus we expect that it is difficult to find a Markovian solution for these problems as well. The following sparse problems are included in the benchmark

    a.) 10 representations of size 3 with 1 non-zero entry in each row (including the rows of $H_k, k = 0, \ldots, K$, except the diagonal of H_0),

    b.) 10 representations of size 3 with 2 non-zero entry in each row,

    c.) 10 representations of size 3 with 3 non-zero entry in each row,

    d.) 10 representations of size 4 with 1 non-zero entry in each row,

    e.) 10 representations of size 4 with 2 non-zero entry in each row,

    f.) 10 representations of size 4 with 3 non-zero entry in each row.

IV *"Near-bound" problems.* These problems are based on the numerical example of [7]. Matrices H_0 and H_1 are given by

$$H_0 = \begin{bmatrix} -1 & 0 & 0 \\ 0 & -3 & h \\ 0 & -h & -3 \end{bmatrix}, \quad H_1 = \begin{bmatrix} 1 \\ 3-h \\ 3+h \end{bmatrix} \cdot [0.2\ 0.3\ 0.5]. \tag{11}$$

It is proven in [7] that this RAP has a Markovian representation if $h \leq 0.552748375$. In order to quantify the efficiency of the algorithms, we are trying to approach this bound as much as possible and check which algorithm finds the Markovian representation. Thus, the benchmark includes 12 MAPs defined by (11) with h values running from 0 to 0.55.

In total, this benchmark consists of 182 MAPs given by non-Markovian representations. The algorithms involved in the comparison are the MRAP2MMAP-[7] (see Figure 1) and the 6 variants of our algorithm introduced in Section 3.2.

We implemented all the algorithms in MATLAB environment. To solve the arising nonlinear optimization problems we used exhaustive search for the discrete and the built-in fminsearch function for the continuous variables. The fminsearch function is based on the Nelder-Mead simplex algorithm that can be found in most scientific toolboxes.

We are interested in the following measures:

- The number of times the algorithm finds a Markovian representation with a tolerance of -10^{-5}, denoted by N_1. All results are considered as Markovian if the smallest element (except the diagonal of H_0) is greater than the tolerance.
- The number of times the algorithm finds a Markovian representation with a tolerance of -10^{-7}, denoted by N_2.
- The average execution time of the algorithm denoted by T.

The results are summarized in Table 1. For the "balanced" problems all algorithms performed well. There are some cases when an algorithm stucks in a local optimum and fails to find a solution, but it does not occur frequently. For the "stiff" problems the results are quite mixed. Our VC variant wins in this case, with the DC and DS variants being the second. MRAP2MMAP-[7] failed to find a solution several times. The algorithms had a hard time obtaining a solution for "sparse" problems as well. Our DC variant managed to solve more problems than the other algorithms.

In general, we can conclude that the cumulative transformation matrix modifications perform much better than the single ones, and that the "dyadic product" and the "single vector" modifications give the best overall performance. Regarding the execution times, it is surprising how fast MRAP2MMAP-[7] is, it is clearly the fastest method in the comparison.

5.2   Finding a Markovian Representation with Size Extension

There is no method available to determine the MAP order of a RAP in the general case. Therefore the examples discussed here will be special RAPs: we

Table 1. Results of the comparison of the algorithms

Problem	MRAP2MMAP-[7]			MRAP2MMAP-DS			MRAP2MMAP-DC			MRAP2MMAP-VS			MRAP2MMAP-VC			MRAP2MMAP-SS			MRAP2MMAP-SC		
	N_1	N_2	$T[s]$	N_1	N_2	$T[s]$	N_1	N_2	$T[s]$	N_1	N_2	$T[s]$	N_1	N_2	$T[s]$	N_1	N_2	$T[s]$	N_1	N_2	$T[s]$
I. "Balanced" problems																					
I.a.	10	10	0.028	10	10	1.63	10	10	1.92	9	9	20.63	10	10	4.81	10	10	2.63	10	10	2.66
I.b.	10	10	0.68	10	10	6.74	10	10	4.69	10	10	49.9	10	10	17.9	10	9	40.8	10	10	11.4
I.c.	10	10	0.36	10	10	11.9	10	10	8.69	9	8	354	9	9	166	9	9	57.7	9	9	96.7
II. "Stiff" problems																					
II.a.	10	10	2.07	10	10	5.59	9	9	13.8	10	10	57.6	10	10	10.2	9	9	10.8	10	10	6.94
II.b.	10	10	3.85	9	9	48.6	10	10	8.8	7	7	219	9	9	72.9	9	9	48.8	8	8	70.8
II.c.	10	10	0.11	9	9	15.8	10	10	2.63	7	7	76.9	10	10	10.6	8	7	20.4	9	9	12.2
II.d.	10	10	0.44	10	10	21.3	10	10	6.73	8	8	182	9	9	81.5	7	7	70.3	10	10	31.6
II.e.	5	5	7.66	10	10	8.16	10	10	4.5	9	9	47.3	9	9	28.8	8	8	14.6	10	10	7.66
II.f.	2	2	23.6	9	9	82.3	8	8	79.2	7	7	270	10	10	51.5	10	9	48.9	9	8	57.2
II.g.	4	4	9	10	10	11.3	10	10	4.65	9	9	34.6	9	9	37.4	6	6	21.1	10	10	8.48
II.h.	2	2	23.6	6	6	165	7	7	107	8	7	255	10	10	80.8	5	5	86.1	8	7	65.5
III. "Sparse" problems																					
III.a.	9	9	8.06	10	10	42.6	10	10	7.07	8	8	83.8	10	10	14.9	8	7	25.4	9	9	19
III.b.	5	5	6.42	10	10	6.16	10	10	4.69	7	6	72.3	10	9	36.3	5	4	31.9	7	6	28.9
III.c.	8	8	1.69	9	9	20.8	9	9	20.4	5	5	69.6	8	8	32.8	5	5	29.4	7	7	20.8
III.d.	8	7	20	9	6	192	10	9	71.5	3	2	383	8	8	175	7	5	122	7	7	93.5
III.e.	4	2	16.4	4	2	287	8	8	88.3	2	1	364	7	6	269	2	1	151	3	2	131
III.f.	4	4	10.5	6	6	154	6	6	114	2	2	311	5	5	239	1	1	140	3	3	132
IV. "Near bound" problem																					
IV.	6	5	3.58	12	12	6.23	12	12	2.56	1	1	127	12	12	11.4	12	12	36.9	10	10	25.2
Summary																					
Average	70%	68%	7.64	90%	87%	60.4	93%	92%	30.6	66%	64%	165	91%	90%	74.5	72%	68%	53.3	82%	80%	45.6

will only examine renewal processes with ME(3) distributed inter-event times as there are results for the order of these.

Our first example is an order 3 RAP. The inter-arrival time distribution is a matrix exponential distribution discussed by Harris in [9].

$$
H_0 = \begin{bmatrix} -1 & 1 & 0 \\ 0 & -2 & 2 \\ 0 & 0 & -3 \end{bmatrix}, \quad H_1 = \begin{bmatrix} 0 & 0 & 0 \\ 0 & 0 & 0 \\ a & b & a \end{bmatrix},
$$

where $b = 3 - 2a$.

With $a > 1.5$ the process does not have an order 3 Markovian representation. With $a = 2$ the process is on the border of the representation space of order 4 MAPs because with parameter $a > 2$ it will not have a size 4 Markovian representation. All our size extension based methods find a size 4 Markovian representation for $a = 2$ with a tolerance of 10^{-7}. The fastest are the dyadic optimization methods, needing only a few minutes, while the slowest SC scalar optimization method needs more than an hour. As an example we present the output of the VS algorithm

$$
D_0 = \begin{bmatrix} -1 & 0 & 0 & 1 \\ 0 & -2.12 & 2.12 & 0 \\ 0 & 0 & -4 & 0.03 \\ 0 & 2.79 & 0 & -2.79 \end{bmatrix}, \quad D_1 = \begin{bmatrix} 0 & 0 & 0 & 0 \\ 0 & 0 & 0 & 0 \\ 1.97 & 0 & 2 & 0 \\ 0 & 0 & 0 & 0 \end{bmatrix}.
$$

In the second example we examine an order 3 RAP with a structure similar to the previous one.

$$
H_0 = \begin{bmatrix} -1 & 1 & 0 \\ 0 & -1 & 1 \\ 0 & 0 & -1 \end{bmatrix}, \quad H_1 = \begin{bmatrix} 0 & 0 & 0 \\ 0 & 0 & 0 \\ a & b & a \end{bmatrix},
$$

with $b = 1 - 2a$. By applying Theorem 7.5 in [10] one can get that this process has a size $n = 3 + \lceil \frac{\xi}{2-\xi} \rceil$ Markovian representation with triangular D_0, where $\xi = \frac{(1-2a)^2}{a^2}$ if and only if $a < \frac{1}{2-\sqrt{2}}$. According to Conjecture 7.6 in the same paper this is also the order of the MAP. If $a > 0.5$, this order is at least 4. (The confirmation can be done similar to the first example.) If $0.5 < a \leq 1$ the order is exactly 4, as can be seen from the previous formula for n. For order 5, we get the $1 < a \leq 1 + (0.25(3 + \sqrt{3}))$ boundaries. We choose $a = 1$, thus the process will be on the border of order 4 and order 5 MAP class. The results are similar to that of the first example. Each method finds a Markovian representation with 10^{-7} tolerance. The dyadic and vector based optimizations need a couple of minutes, the scalar based optimizations finish in around an hour. We give the output of the VS method in this case as an example.

$$
D_0 = \begin{bmatrix} -1 & 0.048 & 0.005 & 0.802 \\ 0 & -1.07 & 0.48 & 0 \\ 0 & 0.119 & -1.83 & 0 \\ 0 & 0.972 & 0 & -1 \end{bmatrix}, \quad D_1 = \begin{bmatrix} 0.104 & 0 & 0.041 & 0 \\ 0.38 & 0.015 & 0.194 & 0.002 \\ 1.029 & 0.069 & 0.607 & 0 \\ 0 & 0.008 & 0.02 & 0 \end{bmatrix}.
$$

6   Conclusion

We have presented a set of representation transformation methods for finding Markovian representation based on general (non-Markovian) matrix representations of various stochastic models including ME distributions, RAPs and MRAPs. The presented new methods relax two limitations of the previously applied ones: the single element based similarity transformation of the representation in each iteration cycle and the fixed size of the representations. The price of the more complex methods is the potentially increased computational complexity and the dependence on non-linear optimization tools.

Acknowledgement. This work was supported by the Hungarian Government through the TÁMOP-4.2.2C-11/1/KONV-2012-0001 and the OTKA K101150 projects, and by the János Bolyai Research Scholarship of the Hungarian Academy of Sciences.

References

1. Neuts, M.F.: Matrix-Geometric Solutions in Stochastic Models: An Algorithmic Approach. Dover (1981)
2. Neuts, M.F.: A versatile Markovian point process. Journal of Applied Probability 16(4), 764–779 (1979)
3. He, Q.M., Neuts, M.F.: Markov arrival processes with marked transitions. Stochastic Processes and their Applications 74, 37–52 (1998)
4. Bladt, M., Neuts, M.F.: Matrix-exponential distributions: Calculus and interpretations via flows. Stochastic Models 19(1), 113–124 (2003)
5. Bean, N., Nielsen, B.: Quasi-birth-and-death processes with rational arrival process components. Stochastic Models 26(3), 309–334 (2010)
6. Buchholz, P., Telek, M.: On minimal representations of rational arrival processes. Annals of Operations Research, 1–24 (2011)
7. Telek, M., Horváth, G.: A minimal representation of Markov arrival processes and a moments matching method. Performance Evaluation 64(9), 1153–1168 (2007)
8. Buchholz, P., Kemper, P., Kriege, J.: Multi-class Markovian arrival processes and their parameter fitting. Performance Evaluation 67(11), 1092–1106 (2010)
9. Harris, C.M., Marchal, W.G., Botta, R.F.: A note on generalized hyperexponential distributions. Stochastic Models 8(1), 179–191 (1992)
10. O'Cinneide, C.A.: Triangular order of triangular phase-type distributions. Stochastic Models 9(4), 507–529 (1993)

Unbiased Simultaneous Prediction Limits on Observations in Future Samples

Nicholas Nechval[1], Konstantin Nechval[2], Maris Purgailis[1], and Uldis Rozevskis[1]

[1] University of Latvia, EVF Research Institute, Statistics Department,
Raina Blvd 19, LV-1050 Riga, Latvia
Nicholas Nechval, Maris Purgailis, Uldis Rozevskis
nechval@junik.lv
[2] Transport and Telecommunication Institute, Applied Mathematics Department,
Lomonosov Street 1, LV-1019 Riga, Latvia
konstan@tsi.lv

Abstract. This paper provides procedures for constructing unbiased simultaneous prediction limits on the observations or functions of observations of all of k future samples using the results of a previous sample from the same underlying distribution belonging to invariant family. The results have direct application in reliability theory, where the time until the first failure in a group of several items in service provides a measure of assurance regarding the operation of the items. The simultaneous prediction limits are required as specifications on future life for components, as warranty limits for the future performance of a specified number of systems with standby units, and in various other applications. Prediction limit is an important statistical tool in the area of quality control. The lower simultaneous prediction limits are often used as warranty criteria by manufacturers. The initial sample and k future samples are available, and the manufacturer wants to have a high assurance that all of the k future orders will be accepted. It is assumed throughout that $k + 1$ samples are obtained by taking random samples from the same population. In other words, the manufacturing process remains constant. The results in this paper are generalizations of the usual prediction limits on observations or functions of observations of only one future sample. In the paper, attention is restricted to invariant families of distributions. The technique used here emphasizes pivotal quantities relevant for obtaining ancillary statistics and is applicable whenever the statistical problem is invariant under a group of transformations that acts transitively on the parameter space. Applications of the proposed procedures are given for the two-parameter exponential and Weibull distributions. The exact prediction limits are found and illustrated with a numerical example.

Keywords: Future samples, observations, simultaneous prediction limits.

1 Introduction

Statistical intervals used by engineers and others include confidence intervals on a population parameter, such as the mean, and tolerance intervals. Confidence intervals

A. Dudin and K. De Turck (Eds.): ASMTA 2013, LNCS 7984, pp. 292–307, 2013.

give information about parameter of the population or a function of population parameters such as a percentile; tolerance intervals give information about a region which contains a specified proportion of a population.

Often one desires to construct from the results of a previous sample an interval which will have a high probability of containing the values of all of k future observations. For example, such an interval would be required in establishing limits on the values of some performance variable for a small shipment of equipment when the satisfactory performance of all units is to be guaranteed, or in setting acceptance limits on a specific lot of material, when acceptance requires the values of all items in a future sample to fall within the limits. An interval which contains the values of a specified number of future observations with a specified probability is known as a prediction interval. Such an interval need be distinguished both from a confidence interval on an unknown distribution parameter, and from a tolerance interval to contain the values of a specified proportion of the population. Research works on prediction intervals related to a single future statistic are abundant (see Hahn and Meeker [1], Patel [2], and references therein).

In many situations of interest, it is desirable to construct lower simultaneous prediction limits that are exceeded with probability γ by observations or functions of observations of all of k future samples, each consisting of m units. The prediction limits depend upon a previously available complete or type II censored sample from the same distribution. For instance, two situations where such limits are required are:

1. A customer has placed an order for a product which has an underlying time-to-failure distribution. The terms of his purchase call for k monthly shipments. From each shipment the customer will select a random sample of m units and accept the shipment only if the smallest time to failure for this sample exceeds a specified lower limit. The manufacturer wishes to use the results of a previous sample of n units to calculate this limit so that the probability is γ that all k shipments will be accepted. It is assumed that the n past units and the km future units are random samples from the same population.

2. A system consists of n identical components whose times to failure follow an underlying distribution. Initially one component is operating and the remaining $n-1$ components are in a standby mode; a new component goes into operation as soon as the preceding component has failed. The system is said to fail when all n components have failed. Thus, the system time to failure is the total of the failure times for the n components. A simultaneous lower prediction limit to be exceeded with probability γ by the system time to failure of all of k future systems is desired. This limit is to be calculated from the times to failure of n previously tested components. Similar problems also arise in various product maintenance and servicing problems.

Prediction limits can be of several forms. Hahn [3] dealt with simultaneous prediction limits on the standard deviations of all of the k future samples from a normal population. Hahn [4] considered the problem of obtaining simultaneous prediction limits on the means of all of k future samples from an exponential distribution. In addition, Hahn and Nelson [5] discussed such limits and their applications. Mann, Schafer, and Singpurwalla [6] gave an interval that contains, with probability γ, all m

observations of a single future sample from the same population. Fertig and Mann [7] constructed prediction intervals to contain at least $m - k + 1$ out of m future observations from a normal distribution with probability $1-\beta$. They considered life-test data, and the performance variate of interest is the failure time of an item. Their lower prediction limit constitutes a "warranty period".

In this paper we give an expression for obtaining unbiased simultaneous prediction limits on order statistics of all of k future samples. In order to obtain the unbiased simultaneous prediction limits, attention is restricted to invariant families of distributions. In particular, the case is considered where a previously available complete or type II censored sample is from a continuous distribution with cumulative distribution function (cdf) $F((x-\mu)/\sigma)$ and probability density function (pdf) $1/\sigma f((x-\mu)/\sigma)$, where $F(\cdot)$ is known but both the location (μ) and scale (σ) parameters are unknown. For such family of distributions the decision problem remains invariant under a group of transformations (a subgroup of the full affine group) which takes μ (the location parameter) and σ (the scale) into $c\mu + b$ and $c\sigma$, respectively, where b lies in the range of μ, $c > 0$. This group acts transitively on the parameter space and, consequently, the risk of any equivariant estimator is a constant. Among the class of such estimators there is therefore a "best" one. The effect of imposing the principle of invariance, in this case, is to reduce the class of all possible estimators to one. In the present paper we investigate this question for the problem of constructing the unbiased simultaneous prediction limits on order statistics in future samples.

The technique used here emphasizes pivotal quantities relevant for obtaining ancillary statistics. It is a special case of the method of invariant embedding of sample statistics into a performance index [8-11] applicable whenever the statistical problem is invariant under a group of transformations which acts transitively on the parameter space (i.e., in problems where there is a unique best invariant procedure). The exact unbiased simultaneous prediction limits on order statistics of all of k future samples are obtained via the technique of invariant embedding and illustrated with numerical example.

2   Mathematical Preliminaries

The main theorem, which shows how to construct lower (upper) simultaneous prediction limit for the order statistics in all of k future samples when prediction limit for a single future sample is available, is given below.

Theorem 1. *(Lower (upper) simultaneous prediction limit under complete information).* Let $(Y_{1_j}, ..., Y_{m_j})$ be the jth random sample of m_j "future" observations from the cdf $F_\theta(.)$, where θ is the parameter (in general, vector), $j \in \{1, ..., k\}$, and let $Y_{(r_j, m_j)}$ denote the r_jth order statistic in the jth sample of size m_j.. Assume that all of k samples from the same cdf are independent. Then a lower simultaneous $(1-\alpha)$ prediction limit h on the r_jth order statistics $Y_{(r_j, m_j)}$, $j=1, ..., k$, of all of k future samples may be obtained from

$$P_\theta\left\{Y_{(r_1,m_1)} \geq h, \ldots, Y_{(r_j,m_j)} \geq h, \ldots, Y_{(r_k,m_k)} \geq h\right\}$$

$$= \sum_{i_1=0}^{r_1-1} \cdots \sum_{i_j=0}^{r_j-1} \cdots \sum_{i_k=0}^{r_k-1} \binom{m_1}{i_1} \cdots \binom{m_j}{i_j} \cdots \binom{m_k}{i_k} \frac{P_\theta\left\{Y_{(i_\Sigma+1,m_\Sigma)} \geq h\right\} - P_\theta\left\{Y_{(i_\Sigma,m_\Sigma)} \geq h\right\}}{\binom{m_\Sigma}{i_\Sigma}} = 1-\alpha, \quad (1)$$

where

$$i_\Sigma = \sum_{j=1}^{k} i_j, \quad m_\Sigma = \sum_{j=1}^{k} m_j. \tag{2}$$

(Observe that an upper simultaneous α prediction limit h may be obtained from a lower simultaneous prediction limit by replacing $1-\alpha$ by α.)

Proof

$$P_\theta\left\{Y_{(r_1,m_1)} \geq h, \ldots, Y_{(r_j,m_j)} \geq h, \ldots, Y_{(r_k,m_k)} \geq h\right\} = \prod_{j=1}^{k} P_\theta\left\{Y_{(r_j,m_j)} \geq h\right\}$$

$$= \prod_{j=1}^{k} \sum_{i_j=0}^{r_j-1} \binom{m_j}{i_j} [F_\theta(h)]^{i_j} [1-F_\theta(h)]^{m_j-i_j}$$

$$= \sum_{i_1=0}^{r_1-1} \cdots \sum_{i_j=0}^{r_j-1} \cdots \sum_{i_k=0}^{r_k-1} \binom{m_1}{i_1} \cdots \binom{m_j}{i_j} \cdots \binom{m_k}{i_k} [F_\theta(h)]^{i_\Sigma} [1-F_\theta(h)]^{m_\Sigma-i_\Sigma}. \tag{3}$$

Since

$$[F_\theta(h)]^{i_\Sigma} [1-F_\theta(h)]^{m_\Sigma-i_\Sigma}$$

$$= \binom{m_\Sigma}{i_\Sigma}^{-1} \left[\sum_{i=0}^{i_\Sigma} \binom{m_\Sigma}{i} [F_\theta(h)]^{i} [1-F_\theta(h)]^{m_\Sigma-i} - \sum_{i=0}^{i_\Sigma-1} \binom{m_\Sigma}{i} [F_\theta(h)]^{i} [1-F_\theta(h)]^{m_\Sigma-i} \right]$$

$$= \frac{P_\theta\left\{Y_{(i_\Sigma+1,m_\Sigma)} \geq h\right\} - P_\theta\left\{Y_{(i_\Sigma,m_\Sigma)} \geq h\right\}}{\binom{m_\Sigma}{i_\Sigma}}, \tag{4}$$

the joint probability can be written as

$$P_\theta\left\{Y_{(r_1,m_1)} \geq h, \ldots, Y_{(r_j,m_j)} \geq h, \ldots, Y_{(r_k,m_k)} \geq h\right\}$$

$$= \sum_{i_1=0}^{r_1-1} \cdots \sum_{i_j=0}^{r_j-1} \cdots \sum_{i_k=0}^{r_k-1} \binom{m_1}{i_1} \cdots \binom{m_j}{i_j} \cdots \binom{m_k}{i_k} \frac{P_\theta\left\{Y_{(i_\Sigma+1,m_\Sigma)} \ge h\right\} - P_\theta\left\{Y_{(i_\Sigma,m_\Sigma)} \ge h\right\}}{\binom{m_\Sigma}{i_\Sigma}}. \quad (5)$$

This ends the proof.   □

Corollary 1.1. If $r_j = 1$, $\forall j=1(1)k$, then

$$P_\theta\left\{Y_{(1,m_1)} \ge h, \ldots, Y_{(1,m_j)} \ge h, \ldots, Y_{(1,m_k)} \ge h\right\} = P_\theta\left\{Y_{(1,m_\Sigma)} \ge h\right\} = 1-\alpha. \quad (6)$$

Theorem 2. *(Lower (upper) unbiased simultaneous prediction limit under parametric uncertainty).* Let $(X_1 \le \ldots \le X_r)$ be the r smallest observations in a random sample of size n from the cdf $F_\theta(.)$, where the θ is the parameter (in general, vector), and let $(Y_{1_j}, \ldots, Y_{m_j})$ be the jth random sample of m_j "future" observations from the same cdf, $j \in \{1, \ldots, k\}$. Assume that $(k+1)$ samples are independent and the parameter θ is unknown. Let $H=H(X_1, \ldots, X_r)$ be any statistic based on the preliminary sample and let $Y_{(r_j,m_j)}$ denote the r_jth order statistic in the jth sample of size m_j. Then an unbiased lower simultaneous $(1-\alpha)$ prediction limit H on the r_jth order statistics $Y_{(r_j,m_j)}$, $j=1$, $\ldots, k$, of all of k future samples may be obtained from

$$E_\theta\left\{P_\theta\left\{Y_{(r_1,m_1)} \ge H, \ldots, Y_{(r_j,m_j)} \ge H, \ldots, Y_{(r_k,m_k)} \ge H\right\}\right\}$$

$$= \sum_{i_1=0}^{r_1-1} \cdots \sum_{i_j=0}^{r_j-1} \cdots \sum_{i_k=0}^{r_k-1} \binom{m_1}{i_1} \cdots \binom{m_j}{i_j} \cdots \binom{m_k}{i_k}$$

$$\times \frac{E_\theta\left\{P_\theta\left\{Y_{(i_\Sigma+1,m_\Sigma)} \ge H\right\}\right\} - E_\theta\left\{P_\theta\left\{Y_{(i_\Sigma,m_\Sigma)} \ge H\right\}\right\}}{\binom{m_\Sigma}{i_\Sigma}} = 1-\alpha. \quad (7)$$

Proof. For the proof we refer to Theorem 1.   □

Corollary 2.1. If $r_j = 1$, $\forall j=1(1)k$, then

$$E_\theta\left\{P_\theta\left\{Y_{(1,m_1)} \ge H, \ldots, Y_{(1,m_j)} \ge H, \ldots, Y_{(1,m_k)} \ge H\right\}\right\} = E_\theta\left\{P_\theta\left\{Y_{(1,m_\Sigma)} \ge H\right\}\right\} = 1-\alpha. (8)$$

Remark. In this paper, in order to find the unbiased lower simultaneous $(1-\alpha)$ prediction limit H on the r_jth order statistics $Y_{(r_j, m_j)}$, $j=1$, ..., k, of all of k future samples, the technique of invariant embedding [8-11] is used.

2.1   Weibull Distribution

In this paper, the two-parameter Weibull distribution with the pdf

$$f_\theta(x) = \frac{\delta}{\beta}\left(\frac{x}{\beta}\right)^{\delta-1} \exp\left[-\left(\frac{x}{\beta}\right)^\delta\right], \quad x > 0,\ \beta > 0,\ \delta > 0, \tag{9}$$

indexed by scale and shape parameters β and δ is used as the underlying distribution of a random variable X in a sample of the lifetime data, where $\theta = (\beta, \delta)$. We consider both parameters β, δ to be unknown. Let $(X_1, ..., X_n)$ be a random sample from the two-parameter Weibull distribution (9), and let $\widehat{\beta}$, $\widehat{\delta}$ be maximum likelihood estimates of β, δ computed on the basis of $(X_1, ..., X_n)$. In terms of the Weibull variates, we have that

$$V_1 = \left(\frac{\widehat{\beta}}{\beta}\right)^\delta, \quad V_2 = \frac{\delta}{\widehat{\delta}}, \quad V_3 = \left(\frac{\widehat{\beta}}{\beta}\right)^{\widehat{\delta}} \tag{10}$$

are pivotal quantities. Furthermore, let

$$Z_i = (X_i / \widehat{\beta})^{\widehat{\delta}}, \quad i=1, ..., n. \tag{11}$$

It is readily verified that any $n-2$ of the Z_i's, say Z_i, ..., Z_{n-2} form a set of $n-2$ functionally independent ancillary statistics. The appropriate conditional approach, first suggested by Fisher [12], is to consider the distributions of V_1, V_2, V_3 conditional on the observed value of $\mathbf{Z}^{(n)} = (Z_i, ..., Z_n)$. (For purposes of symmetry of notation we include all of Z_i, ..., Z_n in expressions stated here; it can be shown that Z_n, Z_{n-1}, can be determined as functions of Z_i, ..., Z_{n-2} only.)

Theorem 3. (*Joint pdf of the pivotal quantities V_1, V_2 from the two-parameter Weibull distribution*) Let $(X_1 \leq ... \leq X_r)$ be the first r ordered observations from a sample of size n from the two-parameter Weibull distribution (9). Then the joint pdf of the pivotal quantities

$$V_1 = \left(\frac{\widehat{\beta}}{\beta}\right)^\delta, \quad V_2 = \frac{\delta}{\widehat{\delta}}, \tag{12}$$

conditional on fixed

$$\mathbf{z}^{(r)} = (z_i, ..., z_r), \tag{13}$$

where

$$Z_i = \left(\frac{X_i}{\hat{\beta}}\right)^{\hat{\delta}}, \quad i = 1, \ldots, r, \tag{14}$$

are ancillary statistics, any $r-2$ of which form a functionally independent set, $\hat{\beta}$ and $\hat{\delta}$ are the maximum likelihood estimates for β and δ based on the first r ordered observations $(X_1 \leq \ldots \leq X_r)$ from a sample of size n from the two-parameter Weibull distribution (9), which can be found from solution of

$$\hat{\beta} = \left(\left[\sum_{i=1}^{r} x_i^{\hat{\delta}} + (n-r)x_r^{\hat{\delta}}\right]/r\right)^{1/\hat{\delta}}, \tag{15}$$

and

$$\hat{\delta} = \left[\left(\sum_{i=1}^{r} x_i^{\hat{\delta}} \ln x_i + (n-r)x_r^{\hat{\delta}} \ln x_r\right)\left(\sum_{i=1}^{r} x_i^{\hat{\delta}} + (n-r)x_r^{\hat{\delta}}\right)^{-1} - \frac{1}{r}\sum_{i=1}^{r} \ln x_i\right]^{-1}, \tag{16}$$

is given by

$$f(v_1, v_2 \mid \mathbf{z}^{(r)}) = \vartheta^{\bullet}(\mathbf{z}^{(r)})v_2^{r-2} \prod_{i=1}^{r} z_i^{v_2} v_1^{r-1} \exp\left(-v_1\left[\sum_{i=1}^{r} z_i^{v_2} + (n-r)z_r^{v_2}\right]\right)$$

$$= f(v_2 \mid \mathbf{z}^{(r)})f(v_1 \mid v_2, \mathbf{z}^{(r)}), \quad v_1 \in (0, \infty), \quad v_2 \in (0, \infty), \tag{17}$$

where

$$\vartheta^{\bullet}(\mathbf{z}^{(r)}) = \left[\int_0^{\infty} \Gamma(r)v_2^{r-2} \prod_{i=1}^{r} z_i^{v_2} \left(\sum_{i=1}^{r} z_i^{v_2} + (n-r)z_r^{v_2}\right)^{-r} dv_2\right]^{-1} \tag{18}$$

is the normalizing constant,

$$f(v_2 \mid \mathbf{z}^{(r)}) = \vartheta(\mathbf{z}^{(r)})v_2^{r-2} \prod_{i=1}^{r} z_i^{v_2} \left(\sum_{i=1}^{r} z_i^{v_2} + (n-r)z_r^{v_2}\right)^{-r}, \quad v_2 \in (0, \infty), \tag{19}$$

$$\vartheta(\mathbf{z}^{(r)}) = \left[\int_0^{\infty} v_2^{r-2} \prod_{i=1}^{r} z_i^{v_2} \left(\sum_{i=1}^{r} z_i^{v_2} + (n-r)z_r^{v_2}\right)^{-r} dv_2\right]^{-1}, \tag{20}$$

$$f(v_1 \mid v_2, \mathbf{z}^{(r)}) = \frac{\left[\sum_{i=1}^{r} z_i^{v_2} + (n-r)z_r^{v_2}\right]^{r}}{\Gamma(r)} v_1^{r-1} \exp\left(-v_1\left[\sum_{i=1}^{r} z_i^{v_2} + (n-r)z_r^{v_2}\right]\right)$$

$$= \frac{1}{\Gamma(r)} \left(v_1 \left[\sum_{i=1}^{r} z_i^{v_2} + (n-r) z_r^{v_2} \right] \right)^{r-1} \exp\left(-v_1 \left[\sum_{i=1}^{r} z_i^{v_2} + (n-r) z_r^{v_2} \right] \right)$$

$$\times \left[\sum_{i=1}^{r} z_i^{v_2} + (n-r) z_r^{v_2} \right], \quad v_1 \in (0, \infty). \tag{21}$$

Proof. The joint density of $X_1 \le \ldots \le X_r$ is given by

$$f_\theta(x_1, \ldots, x_r) = \frac{n!}{(n-r)!} \prod_{i=1}^{r} \frac{\delta}{\beta} \left(\frac{x_i}{\beta} \right)^{\delta-1} \exp\left(-\left(\frac{x_i}{\beta} \right)^{\delta} \right) \exp\left(-(n-r) \left(\frac{x_r}{\beta} \right)^{\delta} \right). \tag{22}$$

Using the invariant embedding technique [8-11], we transform (22) to

$$f_\theta(x_1, \ldots, x_r) \, d\hat{\beta} \, d\hat{\delta}$$

$$= \frac{n!}{(n-r)!} \prod_{i=1}^{r} x_i^{-1} \delta^r \prod_{i=1}^{r} \left(\frac{x_i}{\beta} \right)^{\delta} \exp\left(-\sum_{i=1}^{r} \left(\frac{x_i}{\beta} \right)^{\delta} - (n-r) \left(\frac{x_r}{\beta} \right)^{\delta} \right) d\hat{\beta} \, d\hat{\delta}$$

$$= -\frac{n!}{(n-r)!} \hat{\beta} \hat{\delta}^r \prod_{i=1}^{r} x_i^{-1} \left(\frac{\delta}{\hat{\delta}} \right)^{r-2} \prod_{i=1}^{r} \left(\frac{x_i}{\hat{\beta}} \right)^{\hat{\delta}\left(\frac{\delta}{\hat{\delta}} \right)} \left(\frac{\hat{\beta}}{\beta} \right)^{\delta(r-1)}$$

$$\times \exp\left(-\left(\frac{\hat{\beta}}{\beta} \right)^{\delta} \left[\sum_{i=1}^{r} \left(\frac{x_i}{\hat{\beta}} \right)^{\hat{\delta}\left(\frac{\delta}{\hat{\delta}} \right)} + (n-r) \left(\frac{x_r}{\hat{\beta}} \right)^{\hat{\delta}\left(\frac{\delta}{\hat{\delta}} \right)} \right] \right) \left(\frac{\delta}{\beta} \left(\frac{\hat{\beta}}{\beta} \right)^{\delta-1} d\hat{\beta} \right) \left(-\frac{\delta}{\hat{\delta}^2} d\hat{\delta} \right)$$

$$= -\frac{n!}{(n-r)!} \hat{\beta} \hat{\delta}^r \prod_{i=1}^{r} x_i^{-1} v_2^{r-2} \prod_{i=1}^{r} z_i^{v_2} v_1^{r-1} \exp\left(-v_1 \left[\sum_{i=1}^{r} z_i^{v_2} + (n-r) z_r^{v_2} \right] \right) dv_1 dv_2. \tag{23}$$

Normalizing (23), we obtain (17). This ends the proof. □

Theorem 4. (*Lower (upper) unbiased prediction limit H for the lth order statistic Y_l in a new (future) sample of m observations from the two-parameter Weibull distribution on the basis of the preliminary data sample*) Let $X_1 \le \ldots \le X_r$ be the first r ordered observations from the preliminary sample of size n from the two-parameter Weibull distribution (9). Then a lower unbiased $(1-\alpha)$ prediction limit H on the lth order statistic Y_l from a set of m future ordered observations $Y_1 \le \ldots \le Y_m$ also from the distribution (9) is given by

$$H = \arg\left[E_\theta\{P_\theta\{Y_l \geq H\} \mid \mathbf{z}^{(r)}\} = 1 - \alpha\right] = z_H^{1/\hat{\delta}} \hat{\beta}, \tag{24}$$

where

$$E_\theta\{P_\theta\{Y_l \geq H\} \mid \mathbf{z}^{(r)}\}$$

$$= \frac{\displaystyle\int_0^\infty v_2^{r-2} \prod_{i=1}^r z_i^{v_2} \sum_{k=0}^{l-1}\binom{m}{k}\sum_{j=0}^k\binom{k}{j}(-1)^j\left((m-k+j)z_H^{v_2} + \sum_{i=1}^r z_i^{v_2} + (n-r)z_r^{v_2}\right)^{-r} dv_2}{\displaystyle\int_0^\infty v_2^{r-2} \prod_{i=1}^r z_i^{v_2}\left(\sum_{i=1}^r z_i^{v_2} + (n-r)z_r^{v_2}\right)^{-r} dv_2}, \tag{25}$$

$$z_H = \left(\frac{H}{\hat{\beta}}\right)^{\hat{\delta}}, \tag{26}$$

$Z_i = (X_i / \hat{\beta})^{\hat{\delta}}$, $i = 1, \ldots, r$; $\hat{\beta}$ and $\hat{\delta}$ are the maximum likelihood estimates for β and β based on the first r ordered observations $(X_1 \leq \ldots \leq X_r)$ from a sample of size n from the two-parameter Weibull distribution (9).

(Observe that an upper unbiased α prediction limit H on the lth order statistic Y_l from a set of m future ordered observations $Y_1 \leq \ldots \leq Y_m$ may be obtained from a lower unbiased$(1-\alpha)$ prediction limit by replacing $1-\alpha$ by α.)

Proof. If there is a random sample of m ordered observations $Y_1 \leq \ldots \leq Y_m$ from the two-parameter Weibull distribution (9) with the pdf $f_\theta(y)$ and cdf $F_\theta(y)$, then for the lth order statistic Y_l we have

$$P_\theta\{Y_l \geq H\} = \sum_{k=0}^{l-1}\binom{m}{k}[F_\theta(H)]^k[1 - F_\theta(H)]^{m-k}$$

$$= \sum_{k=0}^{l-1}\binom{m}{k}\left[1 - \exp\left(-\left(\frac{H}{\beta}\right)^\delta\right)\right]^k\left[\exp\left(-\left(\frac{H}{\beta}\right)^\delta\right)\right]^{m-k}. \tag{27}$$

Writing (27) as

$$P_\theta\{Y_l \geq H\} = \sum_{k=0}^{l-1}\binom{m}{k}\left[1 - \exp\left(-\left(\frac{H}{\beta}\right)^\delta\right)\right]^k \exp\left(-(m-k)\left(\frac{H}{\beta}\right)^\delta\right)$$

$$= \sum_{k=0}^{l-1}\binom{m}{k}\left[1 - \exp\left(-\left(\frac{H}{\hat{\beta}}\right)^{\hat{\delta}\left(\frac{\delta}{\hat{\delta}}\right)}\left(\frac{\hat{\beta}}{\beta}\right)^\delta\right)\right]^k \exp\left(-(m-k)\left(\frac{H}{\hat{\beta}}\right)^{\hat{\delta}\left(\frac{\delta}{\hat{\delta}}\right)}\left(\frac{\hat{\beta}}{\beta}\right)^\delta\right)$$

$$= \sum_{k=0}^{l-1} \binom{m}{k} [1 - \exp(-z_H^{v_2} v_1)]^k \exp(-(m-k)z_H^{v_2} v_1)$$

$$= \sum_{k=0}^{l-1} \binom{m}{k} \sum_{j=0}^{k} \binom{k}{j} (-1)^j \exp[-v_1(m-k+j)z_H^{v_2}] = P\{Z_l > z_H \mid v_1, v_2\}, \qquad (28)$$

where

$$Z_l = \left(\frac{Y_l}{\hat{\beta}}\right)^{\hat{\delta}}, \qquad (29)$$

we have from (17) and (28) that

$$E_\theta\{P_\theta\{Y_l \geq H\} \mid \mathbf{z}^{(r)}\} = E\{P\{Z_l \geq z_H \mid v_1, v_2\} \mid \mathbf{z}^{(r)}\}$$

$$= \int_0^\infty \int_0^\infty P\{Z_l \geq z_H \mid v_1, v_2\} f(v_1, v_2 \mid \mathbf{z}^{(r)}) dv_1 dv_2. \qquad (30)$$

Now v_1 can be integrated out of (30) in a straightforward way to give (25). This completes the proof. □

Corollary 4.1. If $l=1$, then

$$H = \arg \left[\frac{\int_0^\infty v_2^{r-2} \prod_{i=1}^r z_i^{v_2} \left(m\left[\left(\frac{H}{\hat{\beta}}\right)^{\hat{\delta}}\right]^{v_2} + \sum_{i=1}^r z_i^{v_2} + (n-r)z_r^{v_2} \right)^{-r} dv_2}{\int_0^\infty v_2^{r-2} \prod_{i=1}^r z_i^{v_2} \left(\sum_{i=1}^r z_i^{v_2} + (n-r)z_r^{v_2} \right)^{-r} dv_2} = 1-\alpha \right]. \qquad (31)$$

Theorem 5. ((*Lower (upper) unbiased prediction limit H for the lth order statistic Y_l in a new (future) sample of m observations from the left-truncated Weibull distribution on the basis of the preliminary data sample*) Let $X_1 \leq \ldots \leq X_r$ be the first r ordered observations from the preliminary sample of size n from the left-truncated Weibull distribution with the pdf

$$f_\theta(x) = \frac{\delta}{\sigma} x^{\delta-1} \exp[-(x^\delta - \mu)/\sigma], \quad (x^\delta \geq \mu, \sigma, \delta > 0), \qquad (32)$$

where $\theta = (\mu, \sigma, \delta)$, δ is termed the shape parameter, σ is the scale parameter, and μ is the truncation parameter. It is assumed that the parameter δ is known. Then a lower unbiased $(1-\alpha)$ prediction limit H on the lth order statistic Y_l from a set of m future ordered observations $Y_1 \leq \ldots \leq Y_m$ also from the distribution (32) is given by

$$H = \left(X_1^\delta + w_H S\right)^{1/\delta}, \qquad (33)$$

where

$$
w_H =
\begin{cases}
\arg\left[nl \binom{m}{l} \sum_{i=0}^{l-1} \dfrac{\binom{l-1}{i}(-1)^i [1+w_H(m-l+i+1)]^{-(r-1)}}{(n+m-l+i+1)(m-l+i+1)} = 1-\alpha \right], & \text{if } \alpha \geq \dfrac{m!(n+m-l)!}{(m-l)!(n+m)!}, \\[4ex]
\arg\left(1 - \dfrac{m!(m+n-l)!}{(m-l)!(m+n)!}(1-nw_H)^{-(r-1)} = 1-\alpha \right), & \text{if } \alpha < \dfrac{m!(n+m-l)!}{(m-l)!(n+m)!},
\end{cases}
$$

$$(34)$$

$$S = \sum_{i=1}^{r}(X_i^\delta - X_1^\delta) + (n-r)(X_r^\delta - X_1^\delta). \qquad (35)$$

(Observe that an upper unbiased α prediction limit H on the lth order statistic Y_l may be obtained from a lower unbiased $(1-\alpha)$ prediction limit by replacing $1-\alpha$ by α.)

Proof. It can be justified by using the factorization theorem that (X_1^δ, S) is a sufficient statistic for (μ, σ). We wish, on the basis of the sufficient statistic (X_1^δ, S) for (μ, σ), to construct the predictive density function of the lth order statistic Y_l from a set of m future ordered observations $Y_1 \leq \dots \leq Y_m$. By using the technique of invariant embedding [8-11] of (X_1^δ, S), if $X_1 \leq Y_l$, or (Y_l^δ, S), if $X_1 \geq Y_l$, into a pivotal quantity $(Y_l^\delta - \mu)/\sigma$ or $(X_1^\delta - \mu)/\sigma$, respectively, we obtain an ancillary statistic

$$W_l = \left(Y_l^\delta - X_1^\delta\right)/S. \qquad (36)$$

It can be shown that the pdf of W_l is given by

$$
f(w_l) =
\begin{cases}
n(r-1)l \binom{m}{l} \sum_{i=0}^{l-1} \dfrac{\binom{l-1}{i}(-1)^i [1+w_l(m-l+i+1)]^{-r}}{n+m-l+i+1}, & \text{if } w_l \geq 0, \\[4ex]
n(r-1) \dfrac{m!(n+m-l)!}{(m-l)!(n+m)!}(1-nw_l)^{-r}, & \text{if } w_l < 0.
\end{cases}
$$

$$(37)$$

It follows from (37) that

$$P(W_l > w_H) = \begin{cases} nl\binom{m}{l}\displaystyle\sum_{i=0}^{l-1}\frac{\binom{l-1}{i}(-1)^i[1+w_H(m-l+i+1)]^{-(r-1)}}{(n+m-l+i+1)(m-l+i+1)}, & \text{if } w_H \geq 0, \\[2em] 1 - \dfrac{m!(m+n-l)!}{(m-l)!(m+n)!}(1-nw_H)^{-(r-1)}, & \text{if } w_H < 0. \end{cases} \tag{38}$$

where

$$w_H = \left(H^\delta - X_1^\delta\right)/S. \tag{39}$$

This ends the proof. □

Corollary 5.1. If $l = 1$, then a lower $(1-\alpha)$ prediction limit H on the minimum Y_1 of a set of m future ordered observations $Y_1 \leq \ldots \leq Y_m$ is given by

$$H = \begin{cases} \left(X_1^\delta + \dfrac{S}{m}\left[\left(\dfrac{n}{(1-\alpha)(n+m)}\right)^{\frac{1}{r-1}} - 1\right]\right)^{1/\delta}, & \text{if } \alpha \geq \dfrac{m}{n+m}, \\[2.5em] \left(X_1^\delta - \dfrac{S}{n}\left[\left(\dfrac{m}{\alpha(n+m)}\right)^{\frac{1}{r-1}} - 1\right]\right)^{1/\delta}, & \text{if } \alpha < \dfrac{m}{n+m}. \end{cases} \tag{40}$$

2.2   Two-Parameter Exponential Distribution

Theorem 6. ((*Lower (upper) unbiased prediction limit H for the lth order statistic Y_l in a new (future) sample of m observations from the two-parameter exponential distribution on the basis of the preliminary data sample*) Let $X_1 \leq \ldots \leq X_r$ be the first r ordered observations from the preliminary sample of size n from the two-parameter exponential distribution with the pdf

$$f_\theta(x) = \frac{1}{\sigma}\exp[-(x-\mu)/\sigma], \quad (x^\delta \geq \mu, \ \sigma > 0), \tag{41}$$

where $\theta = (\mu, \sigma)$, σ is the scale parameter, and μ is the shift parameter. It is assumed that these parameters are unknown. Then a lower unbiased $(1-\alpha)$ prediction limit H on the lth order statistic Y_l from a set of m future ordered observations $Y_1 \leq \ldots \leq Y_m$ also from the distribution (41) is given by

$$H = X_1 + w_H S, \tag{42}$$

where

$$
w_H = \begin{cases}
\arg\left[nl \binom{m}{l} \sum_{i=0}^{l-1} \dfrac{\binom{l-1}{i}(-1)^i [1+w_H(m-l+i+1)]^{-(r-1)}}{(n+m-l+i+1)(m-l+i+1)} = 1-\alpha \right], & \text{if } \alpha \geq \dfrac{m!(n+m-l)!}{(m-l)!(n+m)!}, \\[3em]
\arg\left[1 - \dfrac{m!(m+n-l)!}{(m-l)!(m+n)!}(1-nw_H)^{-(r-1)} = 1-\alpha \right], & \text{if } \alpha < \dfrac{m!(n+m-l)!}{(m-l)!(n+m)!},
\end{cases}
$$

(43)

$$
S = \sum_{i=1}^{r}(X_i - X_1) + (n-r)(X_r - X_1).
$$

(44)

(Observe that an upper unbiased α prediction limit H on the lth order statistic Y_l may be obtained from a lower unbiased $(1-\alpha)$ prediction limit by replacing $1-\alpha$ by α.)

Proof. For the proof we refer to Theorem 5.   □

Corollary 6.1. If $l = 1$, then a lower $(1-\alpha)$ prediction limit H on the minimum Y_1 of a set of m future ordered observations $Y_1 \leq \ldots \leq Y_m$ is given by

$$
H = \begin{cases}
\left(X_1 + \dfrac{S}{m}\left[\left(\dfrac{n}{(1-\alpha)(n+m)} \right)^{\frac{1}{r-1}} - 1 \right] \right), & \text{if } \alpha \geq \dfrac{m}{n+m}, \\[2.5em]
\left(X_1 - \dfrac{S}{n}\left[\left(\dfrac{m}{\alpha(n+m)} \right)^{\frac{1}{r-1}} - 1 \right] \right), & \text{if } \alpha < \dfrac{m}{n+m}.
\end{cases}
$$

(45)

Remark 2. Let us assume that the parent distributions are the two-parameter exponential

$$
F_\theta(x) = 1 - \exp\left(-\frac{x-\theta_2}{\theta_1} \right), \quad x \geq \theta_2, \ \theta_1 > 0,
$$

(46)

where $\theta = (\theta_1, \theta_2)$, and the Pareto distribution

$$
F_\theta(x) = 1 - (\theta_2 / x)^{1/\theta_1}, \quad x \geq \theta_2 > 0, \ \theta_1 > 0.
$$

(47)

Let X be a random variable with the Pareto distribution (47), and define $Y = \ln X$. Then Y becomes a random variable with the exponential distribution (46), where θ_2 is replaced by $\ln\theta_2$. Therefore it is enough to consider only the exponential distribution, because the results for the Pareto distribution are easily obtained from those for the exponential distribution.

3   Numerical Example

An industrial firm has the policy to replace a certain device, used at several locations in its plant, at the end of 24-month intervals. It doesn't want too many of these items to fail before being replaced. Shipments of a lot of devices are made to each of three firms. Each firm selects a random sample of 5 items and accepts his shipment if no failures occur before a specified lifetime has accumulated. The manufacturer wishes to take a random sample and to calculate the lower prediction limit so that all shipments will be accepted with a probability of 0.95. The resulting lifetimes (rounded off to the nearest month) of an initial sample of size 15 from a population of such devices are given in Table 1.

Table 1. The resulting lifetimes

Observations														
x_1	x_2	x_3	x_4	x_5	x_6	x_7	x_8	x_9	x_{10}	x_{11}	x_{12}	x_{13}	x_{14}	x_{15}
8	9	10	12	14	17	20	25	29	30	35	40	47	54	62
Lifetime (in number of month intervals)														

Goodness-of-fit testing. It is assumed that

$$X_i \sim f_\theta(x) = \frac{\delta}{\sigma} x^{\delta-1} \exp[-(x^\delta - \mu)/\sigma], \quad (x \geq \mu, \sigma, \delta > 0), \quad i = 1(1)15, \tag{48}$$

where the parameters μ and σ are unknown; (δ=0.87). Thus, for this example, $r = n = 15$, $k = 3$, $m = 5$, $1-\alpha = 0.95$, $X_1^\delta = 6.1$, and $S = 170.8$. It can be shown that the

$$U_j = 1 - \left(\frac{\sum_{i=2}^{j+1}(n-i+1)(X_i^\delta - X_{i-1}^\delta)}{\sum_{i=2}^{j+2}(n-i+1)(X_i^\delta - X_{i-1}^\delta)} \right)^j, \quad j = 1(1)n-2, \tag{49}$$

are i.i.d. $U(0,1)$ rv's (Nechval *et al.* [13]). We assess the statistical significance of departures from the left-truncated Weibull model by performing the Kolmogorov-Smirnov goodness-of-fit test. We use the K statistic (Muller *et al.* [14]). The rejection region for the α level of significance is $\{K \geq K_{n;\alpha}\}$. The percentage points for $K_{n;\alpha}$ were given by Muller *et al.* [14]. For this example,

$$K = 0.220 < K_{n=13;\alpha=0.05} = 0.361. \tag{50}$$

Thus, there is not evidence to rule out the left-truncated Weibull model. It follows from (8) and (40), for

$$\alpha = 0.05 < \frac{km}{n+km} = 0.5, \tag{51}$$

that

$$H = \left(x_1^\delta - \frac{s}{n} \left[\left(\frac{km}{\alpha(n+km)} \right)^{\frac{1}{n-1}} - 1 \right] \right)^{1/\delta} = \left(6.1 - \frac{170.8}{15} \left[\left(\frac{15}{0.05(15+15)} \right)^{\frac{1}{14}} - 1 \right] \right)^{1/0.87} = 5.$$

(52)

Thus, the manufacturer has 95% assurance that no failures will occur in each shipment before $H = 5$ month intervals.

4   Conclusion and Future Work

In this paper we propose the technique of constructing unbiased simultaneous prediction limits on observations or functions of observations in all of k future samples under parametric uncertainty of the underlying distribution. These unbiased simultaneous prediction limits are based on a previously available complete or type II censored sample from the same distribution. We present an equation for this type of unbiased simultaneous prediction limits which holds for any distribution and any statistic from the previous sample when a prediction limit for a single future sample is available. The exact prediction limits are found and illustrated with a numerical example. The methodology described here can be extended in several different directions to handle various problems that arise in practice. We have illustrated the proposed methodology for the two-parameter exponential and Weibull distributions. Application to other distributions could follow directly.

Acknowledgments. This research was supported in part by Grant No. 09.1544 from the Latvian Council of Science and the National Institute of Mathematics and Informatics of Latvia.

References

1. Hahn, G.J., Meeker, W.Q.: Statistical Intervals – A Guide for Practitioners. John Wiley, New York (1991)
2. Patel, J.K.: Prediction Intervals – a Review. Communications in Statistics - Theory and Methods 18, 2393–2465 (1989)
3. Hahn, G.J.: Simultaneous Prediction Intervals to Contain the Standard Deviations or Ranges of Future Samples From a Normal Distribution. Journal of the American Statistical Association 67, 938–942 (1972)
4. Hahn, G.J.: A Prediction Interval on the Means of Future Samples from an Exponential Distribution. Technometrics 17, 341–345 (1975)
5. Hahn, G.J., Nelson, W.: A Survey of Prediction Intervals and Their Applications. Journal of Quality Technology 5, 178–188 (1973)
6. Mann, N.R., Schafer, R.E., Singpurwalla, J.D.: Methods for Statistical Analysis of Reliability and Life Data. John Wiley, New York (1974)

7. Fertig, K.W., Mann, N.R.: One-sided Prediction Intervals for at Least p out of m Future Observations from a Normal Population. Technometrics 19, 167–177 (1977)
8. Nechval, N.A., Berzins, G., Purgailis, M., Nechval, K.N.: Improved Estimation of State of Stochastic Systems via Invariant Embedding Technique. WSEAS Transactions on Mathematics 7, 141–159 (2008)
9. Nechval, N.A., Purgailis, M., Berzins, G., Cikste, K., Krasts, J., Nechval, K.N.: Invariant Embedding Technique and Its Applications for Improvement or Optimization of Statistical Decisions. In: Al-Begain, K., Fiems, D., Knottenbelt, W. (eds.) ASMTA 2010. LNCS, vol. 6148, pp. 306–320. Springer, Heidelberg (2010)
10. Nechval, N.A., Purgailis, M., Cikste, K., Berzins, G., Nechval, K.N.: Optimization of Statistical Decisions via an Invariant Embedding Technique. In: Proceedings of the World Congress on Engineering 2010, WCE 2010, London, June 30- July 2. Lecture Notes in Engineering and Computer Science, pp. 1776–1782 (2010)
11. Nechval, N.A., Purgailis, M., Nechval, K.N., Strelchonok, V.F.: Optimal Predictive Inferences for Future Order Statistics via a Specific Loss Function. IAENG International Journal of Applied Mathematics 42, 40–51 (2012)
12. Fisher, R.A.: Two New Properties of Mathematical Likelihood. Proceedings of the Royal Society A 144, 285–307 (1934)
13. Nechval, N.A., Nechval, K.N.: Characterization Theorems for Selecting the Type of Underlying Distribution. In: Proceedings of the 7th Vilnius Conference on Probability Theory and 22nd European Meeting of Statisticians, pp. 352–353. TEV, Vilnius (1998)
14. Muller, P.H., Neumann, P., Storm, R.: Tables of Mathematical Statistics. VEB Fachbuchverlag, Leipzig (1979)

Adaptive Stochastic Airline Seat Inventory Control under Parametric Uncertainty

Nicholas Nechval[1], Maris Purgailis[1], Uldis Rozevskis[1], and Konstantin Nechval[2]

[1] University of Latvia, EVF Research Institute, Statistics Department,
Raina Blvd 19, LV-1050 Riga, Latvia
Nicholas Nechval, Maris Purgailis, Uldis Rozevskis
nechval@junik.lv
[2] Transport and Telecommunication Institute, Applied Mathematics Department,
Lomonosov Street 1, LV-1019 Riga, Latvia
konstan@tsi.lv

Abstract. Airline seat inventory control is a very profitable tool in the airline industry. The problem of adaptive stochastic airline seat inventory control lies at the heart of airline revenue management. This problem concerns the allocation of the finite seat inventory to the stochastic customer demand that occurs over time before the flight is scheduled to depart. The objective is to find the right combination of customers of various fare classes on the flight such that revenue is maximized. In this paper, the static and dynamic policies of stochastic airline seat inventory control (airline booking) are developed under parametric uncertainty of underlying models, which are not necessarily alternative. For the sake of simplicity, but without loss of generality, we consider (for illustration) the case of nonstop flights with two fare classes. The system developed is able to recognize a situation characterized by the number of reservations made by customers of the above fare classes at certain moment of time before departure. The proposed policies of the airline seat inventory control are based on the use of order statistics of cumulative customer demand, which have such properties as bivariate dependence and conditional predictability. Dynamic adaptation of the system to airline customer demand is carried out via the bivariate dependence of order statistics of cumulative customer demand. Dynamic optimization of the airline seat allocation is carried out via the conditional predictability of order statistics. The system makes on-line decisions as to whether to accept or reject any customer request using established decision rules based on order statistics of the current cumulative customer demand. The computer simulation results are promising.

Keywords: Airlines, stochastic demand, airline booking, adaptive control.

1 Introduction

Passenger reservations systems have evolved from low level inventory control processes to major strategic information systems. Today, airlines and other transportation companies view revenue management systems and related information

A. Dudin and K. De Turck (Eds.): ASMTA 2013, LNCS 7984, pp. 308–323, 2013.

technologies as critical determinants of future success. Indeed, expectations of revenue gains that are possible with expanded revenue management capabilities are now driving the acquisition of new information technology. Each advance in information technology creates an opportunity for more comprehensive reservations control and greater integration with other important transportation planning functions.

It is common practice for airlines to sell a pool of identical seats at different prices according to different booking classes to improve revenues in a very competitive market. In other words, airlines sell the same seat at different prices according to different types of travelers (first class, business and economy) and other conditions. The question then arises whether to offer seats at a relatively low price at a given time with a given number of seats remaining or to wait for the possible arrival of a higher paying customer. Assigning seats in the same compartment to different fare classes of customers in order to improve revenues is a major problem of airline seat inventory allocation. This problem has been considered in numerous papers. For details, the reader is referred to a review of yield management, as well as perishable asset revenue management, by Weatherford et al. [1], and a review of relevant mathematical models by Belobaba [2].

This paper deals with the airline seat allocation problem when customers for different fare levels are booked into a common seating pool in the aircraft. The following assumptions are made: (1) single-leg flight: bookings are made on the basis of a single departure and landing; no allowance is made for the possibility that bookings may be part of larger trip itineraries, (2) independent demands: the demands for different fare classes are stochastically independent, (3) low before high demands: the lowest fare reservations requests arrive first, followed by the next lowest, etc., (4) no cancellations: cancellations, no-shows and overbooking are not considered, (5) nested classes: any fare class can be booked into seats not taken by bookings in lower fare classes, (6) fare classes: the business and economy fare classes are considered.

The first purpose of this paper is to improve the static and dynamic policies of the airline seat inventory allocation on the basis of the 'unbiasedness performance index'. The static and dynamic policies (unbiased) are more efficient (from the point of view of airline revenue management) as compared with the policies, where the unknown parameters of the airline customer demand models are estimated and then treated as if they were the true values. At the initial stage of airline booking it may be used the static policy of seat inventory allocation, and at the fundamental stage may be used the dynamic policy.

The second purpose of this paper is to introduce the idea of prediction of a future cumulative customer demand for the seats on a flight via the order statistics from the underlying distribution, where only the functional form of the distribution is specified, but some or all of its parameters are unspecified. This idea allows one to use the technique of invariant embedding of sample statistics in a performance index in order to eliminate the unknown parameters from the problem [3-7]. The technique represents a simple and computationally attractive statistical method based on the constructive use of the invariance principle in mathematical statistics. Unlike the Bayesian approach, an invariant embedding technique is independent of the choice of priors, i.e., subjectivity of investigator is eliminated from the problem. It allows one to find the improved invariant statistical decision rules, which have smaller risk than any of the well-known traditional statistical decision rules, and to use the previous and current sample data as completely as possible.

2   State-of-the-Art and Progress Beyond

A major problem of airline seat allocation is to sell the same seat at different prices according to different types of travelers (first class, business and economy) and other conditions in order to improve revenues. This problem has been considered in numerous papers. Littlewood [8] was the first to propose a solution method of the airline seat allocation problem for a single-leg flight with two fare classes. The idea of his scheme is to equate the marginal revenues in each of the two fare classes. He suggests closing down the low fare class when the certain revenue from selling low fare seat is exceeded by the expected revenue of selling the same seat at the higher fare. That is, low fare booking requests should be accepted as long as

$$c_2 \geq c_1 \Pr\{Y_1 > u_1\}, \tag{1}$$

where c_1 and c_2 are the high and low fare levels respectively, Y_1 denotes the demand for the high fare (or business) class, u_1 is the number of seats to protect for the high fare class and $\Pr\{Y_1 > u_1\}$ is the probability of selling more than u_1 protected seats to high fare class customers. The smallest value of u_1 that satisfies the above condition is the number of seats to protect for the high fare class, and is known as the protection level of the high fare class customers. The concept of determining a protection level for the high fare class can also be seen as setting a booking limit, a maximum number of bookings, for the low fare class. Both concepts restrict the number of bookings for the low fare class in order to accept bookings for the high fare class.

It should be remarked that there is no protection level for the low fare (or economy) class; u_2 is the booking limit, or number of seats available, for the low fare class; the low fare class is open as long as the number of bookings in this class remains less than this limit. Thus, (u_1+u_2) is the booking limit or number of seats available for the high fare class at time. The high fare class is open as long as the number of bookings in this and low classes remain less than this limit.

Richter [9] gave a marginal analysis, which proved that (1) gives an optimal allocation (assuming certain continuity conditions). Optimal policies for more than two classes have been presented independently by Curry [10], Wollmer [11], Brumelle and McGill [12], and Nechval et al. [13-15]. For a comprehensive and up-to-date overview of the area we refer to McGill and van Ryzin [16] containing a bibliography of over 190 references.

It will be noted that in practice of airline booking usually the different performance indexes are used to maximize the total expected revenue from non-nested and nested fare classes.

The performance index, which is based on a partitioning of the seats in the planes and can be used to determine the optimal allocation of seats between two independent (i.e., non-nested) fare classes, subject to the total capacity constraint, is given as follows.

Maximize the total expected revenue for a single leg-flight with two fare classes,

$$I_1(u_1,u_2 \mid \theta_1,\theta_2) = E_{\theta_2}\{c_2 \min(u_2,Y_2)\} + E_{\theta_1}\{c_1 \min(u_1,Y_1)\}, \tag{2}$$

subject to

$$u_1 + u_2 \le u, \quad u_j \ge 0 \quad \text{for } j = 1,2, \tag{3}$$

where

$$E_{\theta_j}\{c_j \min(u_j,Y_j)\} = c_j[E_{\theta_j}\{Y_j \mid Y_j \le u_j\}\Pr\{Y_j \le u_j\} + E_{\theta_j}\{Y_j \mid Y_j > u_j\}\Pr\{Y_j > u_j\}]$$

$$= c_j\left[\int_0^{u_j} y_j f_{\theta_j}(y_j)dy_j + \int_{u_j}^{\infty} u_j f_{\theta_j}(y_j)dy_j\right], \quad j = 1,2, \tag{4}$$

Y_2 denotes the demand for the low fare (or economy) class, $f_{\theta_j}(y_j)$ represents the underlying probability density function of the demand Y_j for the jth fare class ($j=1, 2$) with the parameter θ_j (in general, vector), u is the total capacity of the cabin to be shared among the two fare classes. A simple application of the Lagrange multipliers technique leads to the optimal solution satisfying

$$\frac{c_2}{c_1}\Pr\{Y_2 > u - u_1\} = \Pr\{Y_1 > u_1\}, \tag{5}$$

which is the well-known optimality condition that the marginal value of adding a single extra seat is equal for both classes. Note that this condition only holds when high-fare demand is unlikely to exceed capacity, but otherwise the discount rate wouldn't have been introduced in the first place.

Now we describe how booking limits can be determined by nesting seat allocations. The performance index which can be used to determine the optimal allocation of seats between two dependent (i.e., nested) fare classes, subject to the total capacity constraint, is given as follows.

Maximize the total expected revenue for a single leg-flight with two fare classes,

$$I_2(u_1,u_2 \mid \theta_1,\theta_2) = E_{\theta_2}\{c_2 \min(u_2,Y_2)\} + E_{\theta_2}\{E_{\theta_1}\{c_1 \min(u_1 + u_2 - Y_2,Y_1)\}\}, \tag{6}$$

subject to (3), where

$$E_{\theta_2}\{E_{\theta_1}\{c_1 \min(u_1 + u_2 - Y_2,Y_1)\}\} =$$

$$= c_1[E_{\theta_2}\{E_{\theta_1}\{c_1 \min(u_1 + u_2 - Y_2,Y_1)\}\} \mid Y_2 \le u_2\}\Pr\{Y_2 \le u_2\}$$

$$+ E_{\theta_2}\{E_{\theta_1}\{c_1 \min(u_1,Y_1)\}\} \mid Y_2 > u_2\}\Pr\{Y_2 > u_2\}]$$

$$= \int_0^{u_2} E_{\theta_1}\{c_1 \min(u_1 + u_2 - Y_2, Y_1)\} f_{\theta_2}(y_2) dy_2 + \int_{u_2}^{\infty} E_{\theta_1}\{c_1 \min(u_1, Y_1)\} f_{\theta_2}(y_2) dy_2$$

$$= c_1 \left[\int_0^{u_2} \left(\int_0^{u_1 + u_2 - y_2} y_1 f_{\theta_1}(y_1) dy_1 + \int_{u_1 + u_2 - y_2}^{\infty} (u_1 + u_2 - y_2) f_{\theta_1}(y_1) dy_1 \right) f_{\theta_2}(y_2) dy_2 \right.$$

$$+ \left. \int_{u_2}^{\infty} \left(\int_0^{u_1} y_1 f_{\theta_1}(y_1) dy_1 + \int_{u_1}^{\infty} u_1 f_{\theta_1}(y_1) dy_1 \right) f_{\theta_2}(y_2) dy_2 \right]. \tag{7}$$

A simple application of the Lagrange multipliers technique leads to the optimal solution satisfying

$$\frac{c_2}{c_1} = \Pr\{Y_1 > u_1\}. \tag{8}$$

Thus, using the Lagrange multipliers technique, we proved that the rule (1) indeed is optimal.

Comparing (5) and (8) learns that the solution (5) always protects more seats for the high-fare class than the optimal protection level from (8). In order to carry larger numbers of high-fare passengers, it is not necessary to reserve an accordingly large number of seats. Nesting can often accommodate the remaining part of high-fare demand. Lower protection levels lead to higher load factors.

3   Airline Booking Policies Which Are Used in Practice

3.1   Static Airline Booking Policy under Complete Information

It will be noted that (1) represents the static policy of airline seat allocation (or airline booking) under complete information. If F_θ, the probability distribution function of Y_1 with the parameter θ (in general, vector), is continuous and strictly increasing, the definition (1) of u_1 is equivalent to

$$u_1 = \arg(\overline{F}_\theta(u_1) = \gamma) \tag{9}$$

where

$$\gamma = c_2 / c_1, \tag{10}$$

$$\overline{F}_\theta(\tau_j) = 1 - F_\theta(\tau_j). \tag{11}$$

3.2   Static Airline Booking Policy under Parametric Uncertainty

In practice, under parametric uncertainty, i.e. when the parameter θ is unknown, the performance index,

$$\overline{F}_{\hat{\theta}}(u_1) = \gamma, \tag{12}$$

is usually used to construct the static policy given by

$$u_1 = \arg\left(\overline{F}_{\hat{\theta}}(u_1) = \gamma\right), \tag{13}$$

where $\hat{\theta}$ represents the maximum likelihood estimator of θ. The performance index (12) is named as 'maximum likelihood performance index'. The static policy (13) based on (12) is named as '*static maximum likelihood airline booking policy*'.

3.3   Dynamic Airline Booking Policy under Parametric Uncertainty

The static policy of airline booking is optimal as long as no change in the probability distributions of the customer demand is foreseen. However, information on the actual customer demand process can reduce the uncertainty associated with the estimates of demand. Hence, repetitive use of a static policy over the booking period, based on the most recent demand and capacity information, is the general way to proceed.

4   Airline Booking Policies Proposed in the Paper

4.1   Static Unbiased Airline Booking Policy under Parametric Uncertainty

This policy is based on the performance index,

$$E_\theta\{\overline{F}_\theta(u_1)\} = \gamma, \tag{14}$$

which takes into account (9) and the previous data of cumulative customer demand Y_1 for the seats on a flight. It allows one to construct the static airline booking policy given by

$$u_1^{(\text{unb})} = \arg(E_\theta\{\overline{F}_\theta(u_1)\} = \gamma), \tag{15}$$

where $u_1 \equiv u_1(\hat{\theta})$, $\hat{\theta}$ represents either the maximum likelihood estimator of θ or sufficient statistic S for θ, i.e., $u_1 \equiv u_1(S)$. The performance index (14) is named as 'unbiasedness performance index'. The static policy (15), which is based on (14), is named as '*static unbiased airline booking policy*'.

The relative bias of the static airline booking policy is given by

$$r(u_1) = \frac{|E_\theta\{\overline{F}_\theta(u_1)\} - \gamma|}{\gamma}100\%. \tag{16}$$

4.2   Dynamic Airline Booking Policy under Complete Information

In this section, we consider a flight for a single departure date with m predefined reading dates at which the dynamic policy is to be updated, i.e., the booking period before departure is divided into m readings periods: $(0, \tau_1], (\tau_1, \tau_2], ..., (\tau_{m-1}, \tau_m]$ determined by the m reading dates: $\tau_1, \tau_2, ..., \tau_m$. These reading dates are indexed in increasing order: $0<\tau_1<\tau_2< \cdots <\tau_m$, where $(\tau_{m-1}, \tau_m]$ denotes the reading period immediately preceding departure, and τ_m is at departure. Typically, the reading periods that are closer to departure cover much shorter periods of time than those further from departure. For example, the reading period immediately preceding departure may cover 1 day whereas the reading period 1-month from departure may cover 1 week.

Let us suppose that the cumulative passenger demand for the high fare class at the kth reading date (time τ_k, $1 \leq k \leq m$) is Y_{1k} representing the kth order statistic from the underlying distribution with the probability distribution function $G_\theta(y_{1k})$, where θ is a parameter (in general, vector). In other words, Y_{1k} represents the number of seats sold for the customers of the high fare class at the kth reading date. We assume that the cumulative passenger demands for the high and low fare classes are stochastically independent. Each booking of a seat of the high fare class generates average revenue of c_1. Each booking of a seat of the low fare class generates average revenue of c_2, where $c_2<c_1$. Let u_{1k} be an individual protection level for the high fare class at time τ_k (the kth reading date). This many seats are protected for the high fare class from the low fare class. There is no protection level for the low fare class; u_{2k} is the booking limit for the low fare class at time τ_k; the low fare class is open as long as the number of bookings in this class remains less than this limit. Thus, $(u_{1k}+u_{2k})$ is the booking limit for the high fare class at time τ_k. The high fare class is open as long as the number of bookings in this and low classes remain less than this limit. The maximum number of seats that may be booked by fare classes in the next at time τ_k prior to flight departure is the number of unsold seats u_k°.

Under the complete information, the dynamic airline booking policy is given by

$$u_{1k} = \arg(\overline{G}_\theta(u_{1k} \mid y_{1k}) = \gamma), \quad k = 1, 2, ..., m-1, \tag{17}$$

where

$$\overline{G}_\theta(u_{1k} \mid y_{1k}) = 1 - G_\theta(u_{1k} \mid y_{1k}), \tag{18}$$

$G_\theta(u_{1k} \mid y_{1k})$ represents the conditional probability distribution function of the mth order statistic Y_{1m}. The number of unsold seats protected for the high fare class from the low fare class in the next at time τ_k prior to flight departure is the number of unsold seats, u_{1k}°, which is given by

$$u_{1k}^\circ = \min(u_k^\circ, u_{1k} - y_{1k}). \tag{19}$$

4.3   Dynamic Unbiased Airline Booking Policy under Parametric Uncertainty

Under the parametric uncertainty, the dynamic unbiased airline booking policy is given by

$$u_{1k}^{(\text{unb})} = \arg(E_\theta\{\overline{G}_\theta(u_{1k} \mid y_{1k})\} = \gamma), \quad k = 1, 2, ..., m-1, \tag{20}$$

where $u_{1k} \equiv u_{1k}(\hat{\theta})$, $\hat{\theta}$ represents either the maximum likelihood estimator of θ or sufficient statistic S for θ, i.e., $u_{1k} \equiv u_{1k}(S)$. The number of unsold seats protected for the high fare class from the low fare class in the next at time τ_k prior to flight departure is the number of unsold seats u_{1k}°, which is given by

$$u_{1k}^{\circ(\text{unb})} = \min(u_k^{\circ}, u_{1k}^{(\text{unb})} - y_{1k}). \tag{21}$$

5   Mathematical Preliminaries

Theorem 1. Let $X_1 \leq ... \leq X_k$ be the first k ordered observations (order statistics) in a sample of size m from a continuous distribution with some probability density function $f_\theta(x)$ and distribution function $F_\theta(x)$, where θ is a parameter (in general, vector). Then the joint probability density function of $X_1 \leq ... \leq X_k$ and the lth order statistics X_l $(1 \leq k < l \leq m)$ is given by

$$g_\theta(x_1, ..., x_k, x_l) = g_\theta(x_1, ..., x_k)g_\theta(x_l \mid x_k), \tag{22}$$

where

$$g_\theta(x_1, ..., x_k) = \frac{m!}{(m-k)!}\prod_{i=1}^{k}f_\theta(x_i)[1-F_\theta(x_k)]^{m-k}, \tag{23}$$

$$g_\theta(x_l \mid x_k) = \frac{(m-k)!}{(l-k-1)!(m-l)!}\left[\frac{F_\theta(x_l)-F_\theta(x_k)}{1-F_\theta(x_k)}\right]^{l-k-1}\left[1-\frac{F_\theta(x_l)-F_\theta(x_k)}{1-F_\theta(x_k)}\right]^{m-l}\frac{f_\theta(x_l)}{1-F_\theta(x_k)}$$

$$= \frac{(m-k)!}{(l-k-1)!(m-l)!}\sum_{j=0}^{l-k-1}\binom{l-k-1}{j}(-1)^j\left[\frac{1-F_\theta(x_l)}{1-F_\theta(x_k)}\right]^{m-l+j}\frac{f_\theta(x_l)}{1-F_\theta(x_k)}$$

$$= \frac{(m-k)!}{(l-k-1)!(m-l)!}\sum_{j=0}^{m-l}\binom{m-l}{j}(-1)^j\left[\frac{F_\theta(x_l)-F_\theta(x_k)}{1-F_\theta(x_k)}\right]^{l-k-1+j}\frac{f_\theta(x_l)}{1-F_\theta(x_k)} \tag{24}$$

represents the conditional probability density function of X_l given $X_k = x_k$.

Proof. The joint density of $X_1 \leq ... \leq X_k$ and X_l is given by

$$g_\theta(x_1,...,x_k,x_l) = \frac{m!}{(l-k-1)!(m-l)!}\prod_{i=1}^{k}f_\theta(x_i)\,[F_\theta(x_l)-F_\theta(x_k)]^{l-k-1}f_\theta(x_l)[1-F_\theta(x_l)]^{m-l}$$

$$= g_\theta(x_1,...,x_k)g_\theta(x_l\,|\,x_k). \qquad (25)$$

It follows from (23) and (25) that

$$g_\theta(x_l\,|\,x_1,...,x_k) = \frac{g_\theta(x_1,...,x_k,x_l)}{g_\theta(x_1,...,x_k)} = g_\theta(x_l\,|\,x_k), \qquad (26)$$

i.e., the conditional distribution of X_l, given $X_i = x_i$ for all $i = 1,...,k$, is the same as the conditional distribution of X_l, given only $X_k = x_k$, which is given by (24). This ends the proof. $\qquad\qquad\square$

Corollary 1.1. The conditional probability distribution function of X_l given $X_k = x_k$ is

$$P_\theta\{X_l \le x_l\,|\,X_k = x_k\} = 1 - \frac{(m-k)!}{(l-k-1)!(m-l)!}\sum_{j=0}^{l-k-1}\binom{l-k-1}{j}\frac{(-1)^j}{m-l+1+j}\left[\frac{1-F_\theta(x_l)}{1-F_\theta(x_k)}\right]^{m-l+1+j}$$

$$= \frac{(m-k)!}{(l-k-1)!(m-l)!}\sum_{j=0}^{m-l}\binom{m-l}{j}\frac{(-1)^j}{l-k+j}\left[\frac{F_\theta(x_l)-F_\theta(x_k)}{1-F_\theta(x_k)}\right]^{l-k+j}. \qquad (27)$$

Corollary 1.2. Let $X_1 \le ... \le X_k$ be the first k order statistics in a sample of size m from the two-parameter Weibull distribution with the probability density function

$$f_\theta(x) = \frac{\delta}{\beta}\left(\frac{x}{\beta}\right)^{\delta-1}\exp\left[-\left(\frac{x}{\beta}\right)^\delta\right] \quad (x>0), \qquad (28)$$

where $\theta = (\beta,\delta)$, $\beta>0$ and $\delta>0$ are the scale and shape parameters, respectively. Then the conditional probability distribution function of X_l given $X_k = x_k$ is

$$P_\theta\{X_l \le x_l\,|\,X_k = x_k\} = 1 - \frac{(m-k)!}{(l-k-1)!(m-l)!}\sum_{j=0}^{l-k-1}\binom{l-k-1}{j}\frac{(-1)^j}{m-l+1+j}\left[\exp\left(-\frac{x_l^\delta - x_k^\delta}{\beta^\delta}\right)\right]^{m-l+1+j}. \qquad (29)$$

Theorem 2. If in (29) the scale parameter β is unknown, then the predictive probability distribution function of X_l based on (x_k,δ) is given by

$$P_\delta\left\{\left(\frac{X_l}{X_k}\right)^\delta \le \left(\frac{x_l}{x_k}\right)^\delta\right\} = 1 - \frac{m!}{(l-k-1)!(m-l)!}$$

$$\times \sum_{j=0}^{l-k-1} \binom{l-k-1}{j} \frac{(-1)^j}{m-l+1+j} \prod_{s=0}^{k-1} \left[\left[\left(\frac{x_l}{x_k} \right)^\delta - 1 \right] (m-l+1+j) + (m-k+1+s) \right]^{-1}. \quad (30)$$

Proof. We reduce (29) to

$$P_\theta \left\{ \left(\frac{X_l}{X_k} \right)^\delta \leq \left(\frac{x_l}{x_k} \right)^\delta \left| \left(\frac{X_k}{\beta} \right)^\delta = \left(\frac{x_k}{\beta} \right)^\delta \right. \right\} = 1 - \frac{(m-k)!}{(l-k-1)!(m-l)!} \sum_{j=0}^{l-k-1} \binom{l-k-1}{j}$$

$$\times \frac{(-1)^j}{m-l+1+j} \left[\exp(-w[v^\delta - 1]) \right]^{m-l+1+j} = P_\delta \left\{ V^\delta \leq v^\delta \mid W = w \right\}, \quad (31)$$

where $V = X_l / X_k$ is the ancillary statistic whose distribution does not depend on the parameter β. Since X_k does not depend on V, $W = (X_k / \beta)^\delta$ is the pivotal quantity, whose distribution is known and does not depend on the parameters β and δ, we eliminate the parameter β from the problem as

$$P_\delta \{X_l \leq x_l\} = \int_0^\infty P_\theta \{X_l \leq x_l \mid X_k = x_k\} g_\theta(x_k) dx_k, \quad (32)$$

where

$$g_\theta(x_k) = \frac{m!}{(k-1)!(m-k)!} F_\theta^{k-1}(x_k)[1 - F_\theta(x_k)]^{m-k} f_\theta(x_k), \quad x_k \in (0, \infty), \quad (33)$$

represents the probability density function of the kth order statistic X_k. Indeed, it follows from (33) that

$$g_\theta(x_k) dx_k = \frac{m!}{(k-1)!(m-k)!} \left[1 - \exp\left(-\left(\frac{x_k}{\beta} \right)^\delta \right) \right]^{k-1} \exp\left(-\left(\frac{x_k}{\beta} \right)^{\delta(m-k)} \right) \exp\left(-\left(\frac{x}{\beta} \right)^\delta \right) d\left(\frac{x}{\beta} \right)^\delta$$

$$= \frac{m!}{(k-1)!(m-k)!} [1 - e^{-w}]^{k-1} e^{-w(m-k+1)} dw = g(w) dw. \quad (34)$$

It follows from (31) and (34) that

$$P_\delta \{V^\delta \leq v^\delta\} = \int_0^\infty P_\delta \{V^\delta \leq v^\delta \mid W = w\} g(w) dw$$

$$= 1 - \frac{m!}{(l-k-1)!(m-l)!} \sum_{j=0}^{l-k-1} \binom{l-k-1}{j} \frac{(-1)^j}{m-l+1+j} \left(\prod_{s=0}^{k-1} \left[(v^\delta - 1)(m-l+1+j) + (m-k+1+s) \right] \right)^{-1}.$$

(35)

Now (30) follows from (35). This ends the proof.   □

Corollary 2.1. If the parameter $\delta = 1$, i.e., we deal with the exponential distribution, then the predictive probability distribution function of X_l based on x_k is given by

$$P\left\{ \left(\frac{X_l}{X_k} \right) \leq \left(\frac{x_l}{x_k} \right) \right\} = 1 - \frac{m!}{(l-k-1)!(m-l)!}$$

$$\times \sum_{j=0}^{l-k-1} \binom{l-k-1}{j} \frac{(-1)^j}{m-l+1+j} \left(\prod_{s=0}^{k-1} \left[\left(\frac{x_l}{x_k} - 1 \right)(m-l+1+j) + (m-k+1+s) \right] \right)^{-1}. \quad (36)$$

Theorem 3. Let $X_1 \leq \ldots \leq X_k$ be the first k ordered observations from a sample of size m from the two-parameter Weibull distribution (28). Then the joint probability density function of the pivotal quantities

$$W_2 = \frac{\hat{\delta}}{\delta}, \quad W_3 = \left(\frac{\hat{\beta}}{\beta} \right)^{\hat{\delta}}, \quad (37)$$

conditional on fixed $z^{(k)} = (z_i, \ldots, z_k)$, where $Z_i = (X_i / \hat{\beta})^{\hat{\delta}}$, $i = 1, \ldots, k$, are ancillary statistics, any $k-2$ of which form a functionally independent set, $\hat{\beta}$ and $\hat{\delta}$ are the estimators of β and δ, based on the first k ordered observations ($X_1 \leq \ldots \leq X_k$) from a sample of size m from the two-parameter Weibull distribution (28), such that W_2 and W_3 are the pivotal quantities (in particular, the maximum likelihood estimators of β and δ,

$$\hat{\beta} = \left(\left[\sum_{i=1}^{k} x_i^{\hat{\delta}} + (m-k)x_k^{\hat{\delta}} \right] / k \right)^{1/\hat{\delta}} \quad (38)$$

and

$$\hat{\delta} = \left[\left(\sum_{i=1}^{k} x_i^{\hat{\delta}} \ln x_i + (m-k)x_k^{\hat{\delta}} \ln x_k \right) \left(\sum_{i=1}^{k} x_i^{\hat{\delta}} + (m-k)x_k^{\hat{\delta}} \right)^{-1} - \frac{1}{k} \sum_{i=1}^{k} \ln x_i \right]^{-1} \quad (39)$$

respectively, lead to the pivotal quantities W_2 and W_3) is given by

$$f(w_2, w_3 \mid z^{(k)}) = \vartheta^\bullet(z^{(k)}) w_2^{k-1} \prod_{i=1}^{k} z_i^{w_2} w_3^{kw_2-1} \exp\left(-w_3^{w_2} \left[\sum_{i=1}^{k} z_i^{w_2} + (m-k)z_k^{w_2} \right] \right)$$

$$= \vartheta^{\bullet}(\mathbf{z}^{(k)}) w_2^{k-2} \prod_{i=1}^{k} z_i^{w_2} w_3^{w_2 (k-1)} \exp\left(-w_3^{w_2} \left[\sum_{i=1}^{k} z_i^{w_2} + (m-k) z_k^{w_2} \right] \right) w_2 w_3^{w_2-1}$$

$$= f(w_2 \mid \mathbf{z}^{(k)}) f(w_3 \mid w_2, \mathbf{z}^{(k)}), \quad w_2 \in (0, \infty), \quad w_3 \in (0, \infty), \tag{40}$$

where

$$\vartheta^{\bullet}(\mathbf{z}^{(k)}) = \left[\int_0^{\infty} \Gamma(k) w_2^{k-2} \prod_{i=1}^{k} z_i^{w_2} \left(\sum_{i=1}^{k} z_i^{w_2} + (m-k) z_k^{w_2} \right)^{-k} dw_2 \right]^{-1} \tag{41}$$

is the normalizing constant,

$$f(w_2 \mid \mathbf{z}^{(k)}) = \vartheta(\mathbf{z}^{(k)}) w_2^{k-2} \prod_{i=1}^{k} z_i^{w_2} \left(\sum_{i=1}^{k} z_i^{w_2} + (m-k) z_k^{w_2} \right)^{-k}, \quad w_2 \in (0, \infty), \tag{42}$$

$$\vartheta(\mathbf{z}^{(k)}) = \left[\int_0^{\infty} w_2^{k-2} \prod_{i=1}^{k} z_i^{w_2} \left(\sum_{i=1}^{k} z_i^{w_2} + (m-k) z_k^{w_2} \right)^{-k} dw_2 \right]^{-1}, \tag{43}$$

$$f(w_3 \mid w_2, \mathbf{z}^{(k)}) = \frac{\left[\sum_{i=1}^{k} z_i^{w_2} + (m-k) z_k^{w_2} \right]^{k}}{\Gamma(k)} w_3^{w_2 (k-1)}$$

$$\times \exp\left(-w_3^{w_2} \left[\sum_{i=1}^{k} z_i^{w_2} + (m-k) z_k^{w_2} \right] \right) w_2 w_3^{w_2-1}, \quad w_3 \in (0, \infty). \tag{44}$$

Proof. The joint density of $X_1 \le \ldots \le X_k$ is given by

$$f_{\theta}(x_1, \ldots, x_k) = \frac{m!}{(m-k)!} \prod_{i=1}^{k} \frac{\delta}{\beta} \left(\frac{x_i}{\beta} \right)^{\delta-1} \exp\left(-\left(\frac{x_i}{\beta} \right)^{\delta} \right) \exp\left(-(m-k) \left(\frac{x_k}{\beta} \right)^{\delta} \right). \tag{45}$$

Using $\hat{\beta}$ and $\hat{\delta}$ (the maximum likelihood estimators of β and δ obtained from solution of (38) and (39)) and the invariant embedding technique [3-7], we transform (45) as follows:

$$f_{\theta}(x_1, \ldots, x_k) \, d\hat{\beta} \, d\hat{\delta} = \frac{m!}{(m-k)!} \prod_{i=1}^{k} x_i^{-1} \delta^k$$

$$\times \prod_{i=1}^{k}\left(\frac{x_i}{\beta}\right)^{\delta} \exp\left(-\sum_{i=1}^{k}\left(\frac{x_i}{\beta}\right)^{\delta} - (m-k)\left(\frac{x_k}{\beta}\right)^{\delta}\right)d\beta\,d\delta$$

$$= -\frac{m!}{(m-k)!}\hat{\beta}\hat{\delta}^{k}\prod_{i=1}^{k}x_i^{-1}w_2^{k-2}\prod_{i=1}^{k}z_i^{w_2}w_3^{w_2(k-1)}$$

$$\times \exp\left(-w_3^{w_2}\left[\sum_{i=1}^{k}z_i^{w_2} + (m-k)z_k^{w_2}\right]\right)w_2 w_3^{w_2-1}dw_2\,dw_3. \tag{46}$$

Normalizing (46), we obtain (40). This ends the proof. □

Theorem 4. If in (29) both parameters β and δ are unknown, then the predictive probability distribution function of X_l based on $(x_k, \hat{\delta})$ and conditional on fixed $\mathbf{z}^{(k)}$ is given by

$$P\left\{\left(\frac{X_l}{X_k}\right)^{\hat{\delta}} \le \left(\frac{x_l}{x_k}\right)^{\hat{\delta}}\middle| \mathbf{z}^{(k)}\right\} = 1 - \frac{m!}{(l-k-1)!(m-l)!}\int_0^{\infty}\sum_{j=0}^{l-k-1}\binom{l-k-1}{j}\frac{(-1)^j}{m-l+1+j}$$

$$\times \left(\prod_{s=0}^{k-1}\left[\left(\left[\frac{x_l}{x_k}\right]^{\hat{\delta}}\right)^{w_2} - 1\right](m-l+1+j) + (m-k+1+s)\right]\right)^{-1}f(w_2|\mathbf{z}^{(k)})dw_2. \tag{47}$$

Proof. We reduce (30) to

$$P_{\delta}\left\{(X_l/X_k)^{\hat{\delta}\left(\frac{\delta}{\hat{\delta}}\right)} \le (x_l/x_k)^{\hat{\delta}\left(\frac{\delta}{\hat{\delta}}\right)}\right\} = P\left\{V_2^{W_2} \le v_2^{w_2}\right\}$$

$$= 1 - \frac{m!}{(l-k-1)!(m-l)!}\sum_{j=0}^{l-k-1}\binom{l-k-1}{j}\frac{(-1)^j}{m-l+1+j}$$

$$\times \left(\prod_{s=0}^{k-1}[(v_2^{w_2}-1)(m-l+1+j) + (m-k+1+s)]\right)^{-1}, \tag{48}$$

where $V_2 = (X_l/X_k)^{\hat{\delta}}$ is the ancillary statistic whose distribution does not depend on the parameters β and δ. Since the pivotal quantity W_2, whose distribution is given by (42), does not depend on V_2, it follows from (48) and (42) that

$$P\left\{V_2 \le v_2 \mid \mathbf{z}^{(k)}\right\} = \int_0^{\infty} P\left\{V_2^{W_2} \le v_2^{w_2}\right\} f(w_2 \mid \mathbf{z}^{(k)}) dw_2, \tag{49}$$

where the unknown parameters β and δ are eliminated from the problem. Now (47) follows from (49). This ends the proof. $\qquad\qquad\square$

6   Illustrative Example of Airline Booking Policies

Let $X_1, ..., X_n$ be the random sample of the previous independent observations of the cumulative customer demand for the high fare class, which follow the exponential distribution with the probability density function (28) ($\delta = 1$), where the parameter β is unknown. Then the static policies of airline booking under parametric uncertainty are given as follows.

The static maximum likelihood airline booking policy follows from (13):

$$u_1^{(ml)} = \ln \gamma^{-S/n}, \tag{50}$$

where $S = \sum_{i=1}^n X_i$ is the sufficient statistic for β, with

$$V = S / \beta \sim f(v) = \frac{1}{\Gamma(n)} v^{n-1} \exp(-v), \quad v \ge 0, \tag{51}$$

and the relative bias,

$$r(u_1^{(ml)}) = \frac{|E_\theta\{\overline{F}_\theta(u_1^{(ml)})\} - \gamma|}{\gamma} 100\% = \frac{|(1 + \ln \gamma^{-1/n})^{-1} - \gamma|}{\gamma} 100\%. \tag{52}$$

If, say, $n=1$ and $\gamma=0.4$, then $r_{rb}(u_1^{(ml)}) = 30\%$. Thus, in this example the static maximum likelihood airline booking policy has the relative bias equal to 30%. It follows that the protection level for customers of the high fare class will be determined incorrectly. This may lead to serious loss.

The static unbiased airline booking policy follows from (15):

$$u_1^{(unb)} = [\gamma^{-1/n} - 1] S, \tag{53}$$

where the relative bias $r(u_1^{(unb)}) = 0$.

The dynamic unbiased airline booking policy follows from (20) and (36):

$$u_{1k}^{(unb)} = \arg\left(\frac{m!}{(m-k-1)!} \sum_{j=0}^{m-k-1} \binom{m-k-1}{j} \frac{(-1)^j}{1+j} \prod_{s=0}^{k-1}\left[\left(\frac{u_{1k}}{y_{1k}} - 1\right)(1+j) + (m-k+1+s)\right]^{-1} = \gamma\right),$$

$$k = 1, 2, ..., m-1, \tag{54}$$

$$u_{1k}^{\circ(\text{unb})} = \min(u_k^\circ, u_{1k}^{(\text{unb})} - y_{1k}).$$ (55)

7   Conclusion and Directions for Future Research

In this paper, we develop a new frequentist approach to improve predictive statistical decisions for airline seat allocation problems under parametric uncertainty of the underlying distributions of the cumulative customer demand. Frequentist probability interpretations of the methods considered are clear. Bayesian methods are not considered here. We note, however, that, although subjective Bayesian prediction has a clear personal probability interpretation, it is not generally clear how this should be applied to non-personal prediction or decisions. Objective Bayesian methods, on the other hand, do not have clear probability interpretations in finite samples. For constructing the improved statistical decisions, a new technique of invariant embedding of sample statistics in a performance index is proposed. This technique represents a simple and computationally attractive statistical method based on the constructive use of the invariance principle in mathematical statistics.

The methodology, which is developed in this paper for the use in the airline industry under parametric uncertainty of airline customer demand models, may be found to be useful in other industries such as hotels, car rental companies, shipping companies, etc. While the details of problems considered in the paper can change significantly from one industry to the next, the focus is always on making better demand decisions – and not manually with guess work and intuition – but rather scientifically with models and technology, all implemented with disciplined processes and systems.

The methodology described here can be extended in several different directions to handle various problems that arise in practice. We have illustrated the proposed methodology for scale distributions (such as the exponential distribution). Application to other distributions could follow directly.

Acknowledgments. This research was supported in part by Grant No. 06.1936, Grant No. 07.2036, and Grant No. 09.1014 from the Latvian Council of Science and the National Institute of Mathematics and Informatics of Latvia.

References

1. Weatherford, L.R., Bodily, S.E., Pfeifer, P.E.: Modeling the Customer Arrival Process and Comparing Decision Rules in Perishable Asset Revenue Management Situations. Transportation Science 27, 239–251 (1993)
2. Belobaba, P.P.: Airline Yield Management: an Overview of Seat Inventory Control. Transportation Science 21, 66–73 (1987)
3. Nechval, N.A., Berzins, G., Purgailis, M., Nechval, K.N.: Improved Estimation of State of Stochastic Systems via Invariant Embedding Technique. WSEAS Transactions on Mathematics 7, 141–159 (2008)

4. Nechval, N.A., Purgailis, M., Berzins, G., Cikste, K., Krasts, J., Nechval, K.N.: Invariant Embedding Technique and Its Applications for Improvement or Optimization of Statistical Decisions. In: Al-Begain, K., Fiems, D., Knottenbelt, W. (eds.) ASMTA 2010. LNCS, vol. 6148, pp. 306–320. Springer, Heidelberg (2010)
5. Nechval, N.A., Purgailis, M., Cikste, K., Berzins, G., Nechval, K.N.: Optimization of Statistical Decisions via an Invariant Embedding Technique. In: Proceedings of the World Congress on Engineering, WCE 2010, London, UK, June 30-July 2. Lecture Notes in Engineering and Computer Science, pp. 1776–1782 (2010)
6. Nechval, N., Nechval, K., Purgailis, M., Rozevskis, U.: Improvement of Inventory Control under Parametric Uncertainty and Constraints. In: Dobnikar, A., Lotrič, U., Šter, B. (eds.) ICANNGA 2011, Part II. LNCS, vol. 6594, pp. 136–146. Springer, Heidelberg (2011)
7. Nechval, N.A., Purgailis, M., Nechval, K.N., Strelchonok, V.F.: Optimal Predictive Inferences for Future Order Statistics via a Specific Loss Function. IAENG International Journal of Applied Mathematics 42, 40–51 (2012)
8. Littlewood, K.: Forecasting and Control of Passenger Bookings. In: Proceedings of the 12th AGIFORS Symposium, American Airlines, New York, pp. 95–117 (1972)
9. Richter, H.: The Differential Revenue Method to Determine Optimal Seat Allotments by Fare Type. In: Proceedings of the XXII AGIFORS Symposium, American Airlines, New York, pp. 339–362 (1982)
10. Curry, R.E.: Optimal Airline Seat Allocation with Fare Classes Nested by Origins and Destinations. Transportation Science 24, 193–203 (1990)
11. Wollmer, R.D.: An Airline Seat Management Model for a Single Leg Route when Lower Fare Classes Book First. Operations Research 40, 26–37 (1992)
12. Brumelle, S.L., McGill, J.I.: Airline Seat Allocation with Multiple Nested Fare Classes. Operations Research 41, 127–137 (1993)
13. Nechval, N.A., Nechval, K.N., Vasermanis, E.K.: Dynamic Adaptive Control of Airline Seat Inventory with Multiple Nested Fare Classes. In: Proceedings of the Third International Conference on System Identification and Control Problems (SICPRO 2004), Moscow, Russia, January 28-30, pp. 55–69 (2004); CD-ROM Paper #23007
14. Nechval, N.A., Nechval, K.N., Rozite, K., Vasermanis, E.K.: Optimal Airline Seat Inventory Control for Multi-Leg Flights. In: Proceedings of the 16th IFAC World Congress, Prague, Czech Republic, July 4-July 8, 6 pages, (2005), Paper 4876
    http://www.nt.ntnu.no/users/skoge/prost/proceedings/
    ifac2005/Papers/Paper4876.html
15. Nechval, N.A., Rozite, K., Strelchonok, V.F.: Optimal Airline Multi-Leg Flight Seat Inventory Control. In: Dubois, D.M. (ed.) Computing Anticipatory Systems. AIP (American Institute of Physics) Proceedings, Melville, New York, vol. 839, pp. 591–600 (2006)
16. McGill, J.I., van Ryzin, G.J.: Revenue Management: Research Overview and Prospects. Transportation Science 33, 233–256 (1999)

A Parametric Copula Approach for Modelling Shortest-Path Trees in Telecommunication Networks

David Neuhäuser[1], Christian Hirsch[1], Catherine Gloaguen[2], and Volker Schmidt[1]

[1] Ulm University, Institute of Stochastics,
Helmholtzstr. 18, 89069 Ulm, Germany
{david.neuhaeuser,christian.hirsch,volker.schmidt}@uni-ulm.de
http://www.uni-ulm.de/mawi/mawi-stochastik
[2] Orange Labs,
38-40 rue du General Leclerc, 92794 Issy-Moulineaux Cedex9, France
catherine.gloaguen@orange.com

Abstract. We extend the *Stochastic Subscriber Line Model* by the introduction of shortest-path trees which are obtained by splitting up the segment system of the typical serving zone at its crossings and endings. Due to reasons in the complex field of cost and capacity estimation in telecommunication networks, it is desirable to gain knowledge about distributional properties of the branches of these trees. The present paper shows how to obtain parametric approximation formulas for the univariate density functions of the lengths of the two main branches in shortest-path trees. Besides, we derive a joint bivariate distribution for the lengths of these branches by means of copula functions, i.e., we give a parametric composition formula of the marginals. These approximative parametric representation formulas can be used in order to prevent time consuming computer experiments.

Keywords: Parametric copula, parametric marginal distribution, stochastic geometry, network planning, Palm calculus, shortest–path tree, telecommunication network.

1 Introduction

In [4], [8] and [13], the *Stochastic Subscriber Line Model* (SSLM) has been developed and extended, especially in order to model access networks in urban areas. So far, the research focus has been put on so-called typical shortest-path lengths where engineers are mainly interested in minimising the total length of the telecommunication network in a city, see [4]. Besides this, also other cost functionals have to be considered for reasonable optimising. Physical links, e.g., optical fibres emanating from several nodes of lower order in the network are merged into thicker fibres at nodes of higher order. For cost estimation of the network, knowledge of capacities in the network is an important factor and so far, only preliminary results are available, see [12]. The idea in the present paper is to extract the so-called shortest-path tree from the typical segment system of a typical serving zone. Having information about this tree, research departments of telecommunication companies such as *Orange Labs* can draw conclusions about capacity problems and cost estimation of communication networks. As the geometry of

A. Dudin and K. De Turck (Eds.): ASMTA 2013, LNCS 7984, pp. 324–336, 2013.
© Springer-Verlag Berlin Heidelberg 2013

the shortest-path tree can become extremely complex, it is not clear how to describe its structure in a simple way which can also be useful for engineers. A general roadmap for attacking this problem is to build models of increasing complexity which are able to describe the tree in greater and greater details. In the present paper, we therefore provide the first step towards this goal by deriving an analytical approximation formula for the joint distribution of the two main branches of the shortest-path tree.

The paper is organised as follows. Section 2 briefly discusses the tools of stochastic geometry and their properties which are used in the SSLM. Especially Palm calculus, allowing to deal with typical cells of random tessellations as well as Cox processes are described. Besides this, we explain how to extract the shortest-path tree from the typical segment system in order to investigate capacity problems in telecommunication networks. Then, in Section 3, a simulation algorithm for the main branches of shortest-path trees is described. In particular, we provide approximation formulas for the density functions of their lengths. Besides, as the lengths of these branches cannot be assumed to be independent, a copula approach is used in order to model the correlation structure between them. Finally, Section 4 concludes the paper and gives an outlook to possible future research topics.

2 A Stochastic Model Representing Access Networks

2.1 The SSLM for Urban Areas

The SSLM has been introduced as a spatial network model using tools of stochastic geometry in each of its three parts, see [4]. These three parts are the geometrical support, the network component part and the topological part. For the convenience of the reader, we briefly discuss the model existing so far before we proceed to expand the SSLM by shortest-path trees.

Geometrical Support. Cables and fibres in telecommunication networks are installed along the road system of a city or town in order to reach as many customers as possible. The whole network of main roads, side streets, dead ends, etc. is modelled in the SSLM by stationary random geometric graphs. In the present paper, these graphs are restricted to stationary random tessellations in the Euclidean space $\mathbb{R}^2$. We call a subdivision of $\mathbb{R}^2$ into a sequence $\Xi_1, \Xi_2, \ldots$ of random convex and compact polygons a planar random tessellation T if the following three conditions are fulfilled.

1. $\bigcup_{i=1}^{\infty} \Xi_i = \mathbb{R}^2$,
2. $\text{int } \Xi_i \cap \text{int } \Xi_j = \emptyset$ for $i \neq j$,
3. $\#\{i : \Xi_i \cap B \neq \emptyset\} < \infty$ for each bounded $B \subset \mathbb{R}^2$.

Important examples of random tessellations considered in this paper are Poisson-Voronoi tessellations, Poisson line tessellations and Poisson-Delaunay tessellations. For further information on these random graphs, the reader can consult [10] and [11].

Placement of Components. Network nodes such as antennas, wire centre stations, service area interfaces as well as the subscribers themselves are also assumed to be located along the road system. We restrict our considerations to two-hierarchy-level networks, i.e., we investigate path properties between higher level components (HLC) and lower level components (LLC). Therefore, we model these nodes as point processes located on the edge system of the underlying tessellation T. To be more precise, we use two stationary Cox point processes (for further information, see [11]), say $X_H = \{X_{H,n}\}$ and $X_L = \{X_{L,n}\}$. Their random intensity measures are concentrated on the edge set $T^{(1)} = \bigcup_{n=1}^{\infty} \partial \Xi_n$ of the underlying tessellation T and are proportional to the 1-dimensional Hausdorff measure ν_1 on T, i.e., $\mathbb{E}X_H(B) = \lambda_\ell \, \mathbb{E}\nu_1(B \cap T^{(1)})$ and $\mathbb{E}X_L(B) = \lambda_\ell' \, \mathbb{E}\nu_1(B \cap T^{(1)})$ for each Borel set $B \subset \mathbb{R}^2$ and linear intensities $\lambda_\ell, \lambda_\ell' > 0$. The planar intensities are thus given by $\lambda_H = \lambda_\ell \gamma$ for HLC and $\lambda_L = \lambda_\ell' \gamma$ for LLC, where $\gamma = \mathbb{E}\nu_1([0,1]^2 \cap T^{(1)})$.

Topology Model and Palm Calculus. To define connection rules in the network, an LLC is assumed to be linked with its nearest HLC in the Euclidean sense. In other words, we consider the Voronoi tessellation $T_H = \{\Xi_{H,n}\}_{n \geq 1}$ which is generated by the Cox point process $X_H = \{X_{H,n}\}$ of high level components, i.e., the cell $\Xi_{H,n}$ around its nucleus $X_{H,n}$ is given by

$$\Xi_{H,n} = \{x \in \mathbb{R}^2 : \|x - X_{H,n}\| \leq \|x - X_{H,m}\| \text{ for all } m \neq n\},$$

where $\| \cdot \|$ denotes the Euclidean norm. Each LLC $X_{L,i}$ which is located within a so-called *serving zone* $\Xi_{H,n}$ is connected to the corresponding HLC $X_{H,n}$ along the edges of the underlying tessellation T, i.e., via the cable system along the roads of the city. The connection is arranged in a way such that the distance between LLC and corresponding HLC measured along $T^{(1)}$ is the smallest, i.e., we consider shortest paths, see [8]. In the present paper, we consider the Palm version X_H^* of X_H whose distribution can be interpreted as conditional distribution of X_H given that there is a HLC located at the origin $o = (0,0)^\top \in \mathbb{R}^2$. To be more precise, the distribution of X_H^* is given by the representation formula

$$\mathbb{E}g(X_H^*) = \frac{1}{\lambda_H} \, \mathbb{E} \sum_{i:X_{H,i} \in [0,1]^2} g(\{X_{H,n}\} - X_{H,i}),$$

where $g : \mathbb{L} \to [0,\infty)$ is an arbitrary measurable function and $\mathbb{L}$ denotes the family of all locally finite sets of $\mathbb{R}^2$. Note that in particular, we have by definition that $\mathbb{P}(o \in X_H^*) = 1$. In addition, the *typical Voronoi cell* Ξ_H^* of X_H is defined as the Voronoi cell associated with the cell centre o in the Voronoi tessellation constructed from X_H^*, i.e.,

$$\Xi_H^* = \{x \in \mathbb{R}^2 : \|x\| \leq \|x - X_{H,j}^*\| \text{ for all } j \geq 1\}.$$

Besides, let $S_{H,i} = \Xi_{H,i} \cap T^{(1)}$ denote the segment system of the serving zone $\Xi_{H,i}$ which belongs to the corresponding HLC $X_{H,i}$. Then, the *typical segment system* S_H^* is defined as the typical mark of the Cox point process of the HLC $X_{H,i}$ marked with the corresponding segment systems $S_{H,i}$, see Figure 1 for an illustration. For further details on marked point processes and Palm mark distribution, the reader is referred to [3] and [11].

2.2   Extracting Shortest-Path Trees

In this part of the present paper, we want to extend the SSLM by a further, new component, the so-called *shortest-path tree G*. Following any leaf, say v, in this tree towards
its root o corresponds to tracing the shortest path from v to o along $T^{(1)}$ in the typical
serving zone Ξ_H^*. An illustration of this situation can be found in Figure 2.

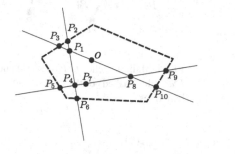
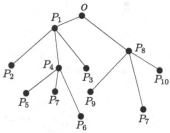

Fig. 1. Typical serving zone Ξ_H^*
(dashed) and corresponding segment
system S_H^* (solid)

Fig. 2. Extracted shortest-path tree G
with origin o as root

For cost estimation and capacity calculations in telecommunication networks, it is
useful to have knowledge of this tree with the lengths of its subparts, the number of links
etc. Focus is put on two main branches building the backbone and skeletal structure of
G. The first main branch is defined as the longest branch of the whole tree. It is called
the longest shortest path and denoted by LSP in the following. Its length, C_{LSP}, is a well-
defined random variable as the supremum over all path lengths is actually a maximum
due to the fact that we have a random but finite number of endpoints in the tree. Observe
that with probability 1, the origin o has two emanating edges (see Figure 1). Thus, we
can subdivide the shortest-path tree G into two subtrees, a half-tree G_1^h and a half-tree
G_2^h as shown in Figure 3. The graph G_1^h is defined as the half-tree which contains LSP.
The second main branch, denoted by LSP', is now defined as the longest branch in the
second half-tree G_2^h, see also Figure 4. Its length will be denoted by $C_{LSP'}$. Note that
$C_{LSP} \geq C_{LSP'}$ holds, but LSP' does not necessarily have to be the second longest branch
of G.

3   Modelling Shortest-Path Trees via Copulas

The goal in this section is to find suitable two-dimensional (joint) distribution functions
for the lengths C_{LSP} and $C_{LSP'}$ of the two main branches LSP and LSP'. In general, it
cannot be assumed that these lengths are independent random variables. Indeed, both
lengths certainly depend on the size of the typical serving zone in a sense that the
lengths of both LSP and LSP' are positively correlated with the size of Ξ_H^*. As a consequence, the bivariate joint distribution function cannot be written as product of the two
univariate marginal distribution functions, i.e.,

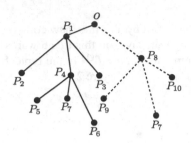

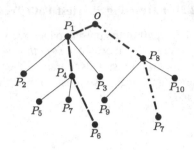

Fig. 3. Half-trees G_1^h (solid) and G_2^h (dashed) of G emanating from the root

Fig. 4. Main branches LSP (dashed) and LSP' (dot-dashed) of the corresponding half-trees

$$F_{(C_{LSP},C_{LSP'})}(x,y) \neq F_{C_{LSP}}(x) \cdot F_{C_{LSP'}}(y).$$

A first attempt could be to consider a non-parametric approach for the fitting of a multivariate density function. Although a good fit can be achieved for a specific choice of parameters λ_ℓ and γ, it is not suitable for our goals, as we want to obtain a parsimonious parametric model which applies for any values of λ_ℓ and γ. A second attempt could be to fit a well-known parametric multivariate distribution function to the data, such as e.g. a multivariate normal or t-distribution. However, as we will see in Section 3.1, the univariate distribution of $C_{LSP'}$ can be closely approximated by a mixed gamma-distribution and we are not aware of any commonly used multivariate distribution with mixed-gamma marginals. Both types of problems mentioned above can be avoided by using a family of parametric copulas.

A copula function, roughly speaking, combines marginal distributions to a joint distribution by adding some correlation structure in a way which has to be precised. In general, a function $K : [0,1]^2 \to [0,1]$ is called a 2-dimensional copula if there exists a probability space $(\Omega, \mathscr{F}, \mathbb{P})$ supporting a random vector $\mathbf{U} = (U_1, U_2)^\top$ such that

$$K(u_1, u_2) = \mathbb{P}(U_1 \leq u_1, U_2 \leq u_2), \quad u_1, u_2 \in [0,1],$$

and $U_i \sim U[0,1]$ for $i \in \{1,2\}$. Note that Sklar's theorem (see [7]) guarantees the existence of a (not necessarily parametric) copula function $K_C : [0,1]^2 \to [0,1]$ such that the bivariate joint distribution function of the random vector $\mathbf{C} = (C_{LSP}, C_{LSP'})^\top$ can be written as

$$F_\mathbf{C}(\mathbf{c}) = K_\mathbf{C}(F_{C_{LSP}}(c_1), F_{C_{LSP'}}(c_2)),$$

where $\mathbf{c} = (c_1, c_2)^\top$, $c_1, c_2 > 0$. Note that the density of $\mathbf{C}$ is given by

$$f_\mathbf{C}(\mathbf{c}) = f_{C_{LSP}}(c_1)\, f_{C_{LSP'}}(c_2) \cdot k_\mathbf{C}(F_{C_{LSP}}(c_1), F_{C_{LSP'}}(c_2)), \tag{1}$$

where $k_\mathbf{C}(u_1, u_2) = \frac{\partial^2}{\partial u_1 \partial u_2} K_\mathbf{C}(u_1, u_2)$ denotes the density function of the copula $K_\mathbf{C}$ and $f_{C_{LSP}}$ and $f_{C_{LSP'}}$ are the density functions of $F_{C_{LSP}}$ and $F_{C_{LSP'}}$, respectively.

A well-known tool in order to fit parametric models to data is the maximum-likelihood method. Suppose that we have parametric models $F_{C_{LSP}}(\cdot \mid \eta_1)$, $F_{C_{LSP'}}(\cdot \mid \eta_2)$

and $K_{\mathbf{C}}(\cdot \mid \eta)$ with parameter vectors η_1, η_2 and η for the marginals as well as for the copula, respectively, and assume that we have an i.i.d. sample $\mathbf{C}_i = (C_{LSP,i}, C_{LSP',i})^{\top}$, $i = 1, \ldots, n$ where n denotes the sample size. Considering (1), one obtains for the log-likelihood function $\log L$ the following representation

$$
\begin{aligned}
\log L(\eta_1 \eta_2, \eta) \\
= \sum_{i=1}^{n} \bigl(\log f_{C_{LSP}}(C_{LSP,i} \mid \eta_1) + \log f_{C_{LSP'}}(C_{LSP',i} \mid \eta_2) \\
+ \log \bigl[k_{\mathbf{C}}(F_{C_{LSP}}(C_{LSP,i} \mid \eta_1), F_{C_{LSP'}}(C_{LSP',i} \mid \eta_2) \mid \eta) \bigr] \bigr).
\end{aligned}
\tag{2}
$$

A quite annoying handicap of the maximum-likelihood method is the fact that maximising the log-likelihood function is a challenging numerical problem if we have several parameters in the model. One way out of this unpleasant situation is the usage of the so-called *parametric pseudo-maximum-likelihood method* in order to fit a suitable model, see [9]. More precisely, we follow a similar way as we do in maximum-likelihood estimation but this time, we have an optimisation process in two steps. First, the marginal distributions are estimated and represented by parametric families using the common maximum-likelihood method, each on its own. Second, we have to determine the best copula in a way which still has to be defined. Note that by estimating parameters of the marginal distributions and the copula separately, the pseudo-maximum-likelihood approach avoids a higher-dimensional optimisation.

In the following, we precisely describe the two optimisation steps for the marginals and copula cases and provide the numerical results.

3.1   Fitting Parametric Marginal Distributions

Parametric Density Function for C_{LSP}. The distribution of C_{LSP} depends both on the linear intensity λ_ℓ of the HLC and the length intensity γ of T. For Poisson-Voronoi tessellations (PVT), Poisson line tessellations (PLT) and Poisson-Delaunay tessellations (PDT) it was shown in [13] that we have a certain scaling invariance in our model and therefore it suffices to investigate the dependence of the distribution of C_{LSP} on the ratio $\kappa = \frac{\gamma}{\lambda_\ell}$ called scaling parameter. In contrast to the situation observed for the limit cases where $\kappa \to 0$ or $\kappa \to \infty$ (see [5]), for general values of κ it is hardly possible to derive an explicit formula for the distribution of C_{LSP}. In this section, we therefore aim at finding a suitable parametric representation formula for the density function $f_{C_{LSP}}$ of C_{LSP} which approximates $f_{C_{LSP}}$ sufficiently well. For three different types of random tessellations (PVT, PDT, PLT) and for various values of the scaling parameter κ, we run a sufficiently large number n of realisations of the typical segment system S_H^* via Monte Carlo simulation and extract the shortest-path tree G. According to the obtained histograms, we fit a suitable family of parametric distributions which can be characterised by just a few parameters. This can be achieved by manual choice of an eligible class of distributions and via subsequent maximum-likelihood estimation. It turns out that a suitable class of parametric distributions for C_{LSP} is the family of scaled Gamma distributions. To be more precise, it approximately holds that $C_{LSP} \sim \Gamma(k, \lambda)$ with density function

$$f_{\Gamma(k,\lambda)}(x;k,\lambda) = \frac{1}{\lambda^k \Gamma(k)} x^{k-1} \exp\left(-\frac{x}{\lambda}\right) \mathbb{1}_{[0,\infty)}(x), \tag{3}$$

for some shape parameter $k > 0$ and scale parameter $\lambda > 0$. This class of distributions is applicable to all three types (PVT, PDT, PLT) of underlying tessellations representing the infrastructure of the city, see Figure 5. Note that the parameters k and λ depend on the type of the underlying tessellation and on κ.

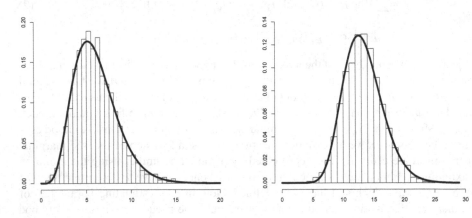

Fig. 5. Parametric densities of the scaled Gamma distribution for C_{LSP} where the underlying tessellation is a *PLT* with $\kappa = 20$ (left) and *PDT* with $\kappa = 120$ (right)

Parametric Density Function for $C_{LSP'}$. Next, we want to proceed in an analogous way as above and derive a parametric approximation formula for the density of $C_{LSP'}$. Looking at the histogram in Figure 6, the mindful reader may observe that, in contrast to the distribution of C_{LSP} (see Figure 5), we now have a bimodal type of distribution. In order to cope with this situation, the usage of a mixture of two scaled Gamma distributions is reasonable (note that this type of distribution indeed fulfills the necessary bimodality). To be more precise, we get that approximately $C_{LSP'} \sim \Gamma_{mix}^{\alpha,k,\lambda,\ell,\theta}$ with density function

$$f_{\Gamma_{mix}^{\alpha,k,\lambda,\ell,\theta}}(x;\alpha,k,\lambda,\ell,\theta)$$

$$= \left[\alpha \cdot \left(\frac{1}{\lambda^k \Gamma(k)} x^{k-1} \exp\left(-\frac{x}{\lambda}\right)\right)\right.$$

$$\left. + (1-\alpha) \cdot \left(\frac{1}{\theta^\ell \Gamma(\ell)} x^{\ell-1} \exp\left(-\frac{x}{\theta}\right)\right)\right] \cdot \mathbb{1}_{[0,\infty)}(x),$$

for some mixing parameter $\alpha \in [0,1]$, shape parameters $k, \ell > 0$ and scale parameters $\lambda, \theta > 0$. It turns out that this class of distributions is applicable to all three types of underlying tessellations which have been considered so far, i.e., PVT, PDT, PLT. Note that the parameters α, k, λ, ℓ and θ depend on the type of the underlying tessellation and on κ. Formally, we can also consider the distribution of C_{LSP} as a mixture of gamma distributions (indeed, with mixing parameter $\alpha \in \{0,1\}$).

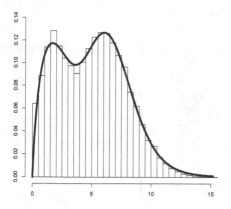 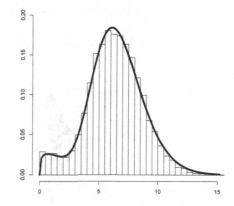

Fig. 6. Parametric densities of the mixture of two scaled Gamma distributions for $C_{LSP'}$ where the underlying tessellation is a *PDT* with $\kappa = 25$ (left) and *PVT* with $\kappa = 50$ (right)

3.2 Choosing a Suitable Copula Type

Now having information about the (marginal) distributions of the lengths of the two main branches each considered on its own, the remaining task is to find a suitable approximative model for the correlation structure between them. Similar as in the univariate case, some asymptotic results on the joint distribution of $\mathbf{C}$ can be obtained for the limit cases where $\kappa \to 0$ or $\kappa \to \infty$ (see [6]). However, for general values of κ it is hardly possible to derive an explicit formula for the joint distribution of $\mathbf{C}$. In order to achieve a good approximation formula for this distribution, we choose a copula approach, i.e., we consider several common parametric copulas and investigate which of them represents the correlation structure best, in a way which still has to be defined. Choosing a suitable type of copula is essential as the several types differ notably from each other, e.g., some have tail dependance while others have not, some allow negative correlation and others not, etc.

Preprocessing of Data. Before we can start to find an appropriate copula $K_{\mathbf{C}}$ adding information about the correlation of C_{LSP} and $C_{LSP'}$ when combining its marginal distributions in a joint distribution, we have to manage some difficulties resulting from special properties of our data. More precisely, our data is completely asymmetric in terms of the fact that for each shortest-path tree G, we have $C_{LSP} \geq C_{LSP'}$. This means that the scatterplot of a sample $\mathbf{C}_i$, $1 \leq i \leq n$, where n denotes the sample size, is completely located beneath the first angle bisector, see the left-hand side of Figure 7.

Scatterplots of data which was sampled from the vast majority of commonly used bivariate copula types are however more or less some kind of symmetric, i.e., the corresponding point pairs are located beneath as well as above the first angle bisector. This is due to the fact that for many copulas $K : [0,1]^2 \to [0,1]$ it holds that $K(u_1, u_2) = K(u_2, u_1)$ for any $u_1, u_2 \in [0,1]$. In order to manage this problem, each point pair $\mathbf{C}_i = (C_{LSP,i}, C_{LSP',i})^\top$ will be associated with an independent $U[0,1]$ distributed mark U_i. If $U_i < 0.5$, we put $\tilde{\mathbf{C}}_i = (C_{LSP,i}, C_{LSP',i})^\top$, and if $U_i \geq 0.5$, then we put $\tilde{\mathbf{C}}_i = (C_{LSP',i}, C_{LSP,i})^\top$.

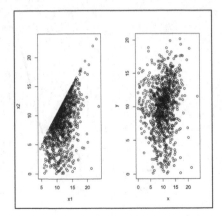

Fig. 7. Scatterplots of original and symmetrized data

The effect of the transformation $\mathbf{C}_i \mapsto \tilde{\mathbf{C}}_i$ can be seen on the right-hand side of Figure 7. In order to obtain the bivariate distribution of $\mathbf{C}_i$, we proceed now in the following way. First, we determine the distribution of $\tilde{\mathbf{C}}_i$ by using copulas. Then, we put

$$\mathbf{C}_i^{(1)} = \max\{\tilde{\mathbf{C}}_i^{(1)}, \tilde{\mathbf{C}}_i^{(2)}\}$$

and

$$\mathbf{C}_i^{(2)} = \min\{\tilde{\mathbf{C}}_i^{(1)}, \tilde{\mathbf{C}}_i^{(2)}\}.$$

Here, $\mathbf{C}_i^{(j)}$ and $\tilde{\mathbf{C}}_i^{(j)}$ denote the j-th component of $\mathbf{C}_i$ and $\tilde{\mathbf{C}}_i$, respectively, where $j \in \{1,2\}$.

Another possibility to bypass this problem would be to avoid the explicit ordering in $\mathbf{C}$, so that both branch lengths are identically distributed. Note that this would lead to a representation of the marginals as a mixture of C_{LSP} and $C_{LSP'}$. However, for certain values of κ, the branch lengths C_{LSP} and $C_{LSP'}$ can be dramatically different which makes a separate consideration of C_{LSP} and $C_{LSP'}$ necessary for capacity analysis of real-world telecommunication networks.

Note that the endpoints of LSP and LSP' are either located on the boundary of the typical cell Ξ_H^* or they form a so-called distance-peak, i.e., there is a point in the interior of the typical segment system S_H^* for which the shortest path to the origin is not unique (see [8] for further explanations). The point P_7 in Figure 1 illustrates that distance-peaks can occur with positive probability. Note however that in this example the point P_6 (and not the distance-peak P_7) is the endpoint of LSP, see also Figure 4. Another problem occurs if LSP and LSP' share the same distance-peak as endpoint (i.e., $C_{LSP,i} = C_{LSP',i}$). Then the distribution of $\mathbf{C}$ cannot be modelled by commonly used parametric copulas (since these are absolutely continuous with respect to 2-dimensional Lebesgue measure). In order to prevent this unsatisfying situation, we (temporarily) remove distance-peaks and treat them separately.

Gumbel Copula. Recall that we already determined parameters for the marginal distributions by using a pseudo-maximum-likelihood approach, see Section 3.1. Therefore, only the parameters of the copula model remain to be estimated. Copula types which are considered in the present paper are the Archimedean copulas of Clayton, Gumbel, and Frank type as well as the Gaussian and the t-copula. In order to find a suitable copula type and to prevent overfitting due to too many parameters, a reasonable decision tool in this context is Akaike's information criterion (AIC) which is defined as

$$AIC = 2\left(p - \log L(\hat{\eta})\right),$$

where p denotes the number of parameters in the model and $\log L(\hat{\eta})$ is the maximised log-likelihood where the log-likelihood function is given by

$$\log L(\eta) = \sum_{i=1}^{n} \log\left[k_{\mathbf{C}}\left(\hat{F}_{C_{LSP}}(C_{LSP,i}), \hat{F}_{C_{LSP'}}(C_{LSP',i}) \mid \eta\right)\right]. \tag{4}$$

Note that (4) is obtained from (2) by replacing the marginal distribution functions by their estimators (which are in fact nothing else than the empirical distribution functions $\hat{F}_{C_{LSP}}$ and $\hat{F}_{C_{LSP'}}$) and omitting summands which do not depend on the parameter vector η of the copula. The type of copula which has the smallest AIC value is now chosen to work with. This seems quite reasonable as maximising $\log L(\eta)$ and minimising the number of parameters p is the goal we aim at. It turned out that for all three underlying tessellation models (PVT, PLT, PDT) we can use the same family of copulas since the Gumbel copula minimises AIC among the five considered copula models for PVT, PDT and PLT. To be more precise, we use the copula

$$K_{\mathbf{C}}(u,v) = \exp\left(-((-\log u)^{\eta} + (-\log v)^{\eta})^{1/\eta}\right),$$

where we obtain $\eta \approx 1.21$ for each κ and for each type of the considered tessellations. This means that neither the type of the copula nor the value of its parameter η depend on the type of the underlying tessellation and the scaling parameter κ which is a quite surprising result.

Distance Peaks. In this section, we describe how to handle the distance peaks mentioned before, i.e., we suggest a model for those point pairs $\mathbf{C}_i = (C_{LSP,i}, C_{LSP',i})^{\top}$, where $C_{LSP,i} = C_{LSP',i}$. Due to the fact that those points are located along the first angle bisector $f(x) = x$, it suffices to consider the (conditional) univariate distribution of C_{LSP} given that LSP and LSP' have the same length. It turns out that we can model the distance peaks $\mathbf{C}_i = (C_{LSP,i}, C_{LSP',i})^{\top}$ by again using a scaled gamma distribution. Note that the parameters are different from those obtained in Section 3.1, but nevertheless also depend on the underlying tessellation T as well as the scaling parameter κ. In particular, we sample from a density function with representation formula

$$f(x;k,\lambda) = \frac{1}{\lambda^k \Gamma(k)} x^{k-1} \exp\left(-\frac{x}{\lambda}\right) \mathbb{1}_{[0,\infty)}(x).$$

Putting a realisation x of a random variable $X \sim \Gamma(k,\lambda)$ into both components of $\mathbf{C}_i$, i.e., $\mathbf{C}_i = (X,X)^\top$, yields the desired quantity. The probability ρ of distance-peaks can be easily estimated by the ratio

$$\widehat{\rho} = \frac{\#\,\text{distance-peaks in empirical data}}{n},$$

where n denotes the sample size.

Combining Simulated Data. Finally, we join data sampled from the Gumbel copula and from the scaled Gamma distribution of the distance peaks. To be more precise, for some large integer $N \geq 1$, we generate $\lfloor(1-\widehat{\rho})N\rfloor$ points from the copula model and $\lfloor\widehat{\rho}N\rfloor$ points from the distance-peak model and consider the union of these data. In Figures 8, 9 and 10, the reader can compare symmetrised empirical data which has already been extracted from the shortest-path tree G, to data which has been directly sampled by the copula approach presented above. For each type of tessellation and a wide range of κ, we obtain quite good results. For example, see Figure 8 for $\kappa = 375$ in the PVT case, Figure 9 for $\kappa = 20$ in the PDT case and Figure 10 for $\kappa = 120$ in the PLT case.

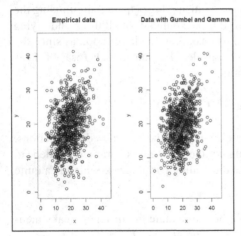 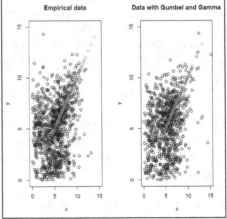

Fig. 8. Scatterplots of empirical data (left) and directly simulated data (right) where $\kappa = 375$ and T is a PVT

Fig. 9. Scatterplots of empirical data (left) and directly simulated data (right) where $\kappa = 20$ and T is a PDT

Note that with increasing values of κ, the number of distance peaks (grey points) compared to non-distance-peaks (black points) decreases for each of the three types of underlying tessellations.

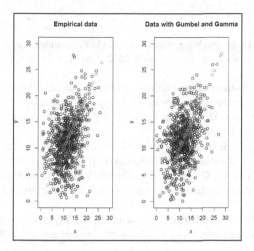

Fig. 10. Scatterplots of empirical data (left) and directly simulated data (right) where $\kappa = 120$ and T is a PLT

4   Conclusions and Outlook

We extended the SSLM by a further component, the shortest-path tree G which can be extracted out of the typical segment system S_H^*. In order to simulate structural characteristics of G directly instead of simulating Ξ_H^* respectively S_H^* and extracting the tree, we developed a method how to obtain the skeletal backbone – the main branches – of G. We used a pseudo-maximum-likelihood approach to achieve this. Parametric approximation formulas for the (marginal) density functions of the longest shortest-path-lengths C_{LSP} and $C_{LSP'}$ were derived as well as a parametric copula type representing the (joint) distribution of $\mathbf{C} = (C_{LSP}, C_{LSP'})$. In our future research, we will investigate further structural characteristics of the shortest-path tree G using the copula approach developed in the present paper. Moreover, other possible topics for future research could be the investigation of shortest-path trees with other types of underlying tessellations, e.g., iterated tessellations, or even completely different types of stationary random geometric graphs as β-skeletons (see [1]). Besides, further types of copulas used for the joint bivariate distribution for $\mathbf{C}$ can be investigated in order to examine if they yield even better results for directly simulated characteristics of shortest-path trees compared to those ones obtained in the present paper.

Acknowledgement. This work was supported by Orange Labs through Research grant No. 46146063-9241. Christian Hirsch was supported by a research grant from DFG Research Training Group 1100 at Ulm University.

References

1. Bose, P., Devroye, L., Evans, W., Kirkpatrick, D.: On the spanning ratio of Gabriel graphs and β-skeletons. In: Rajsbaum, S. (ed.) LATIN 2002. LNCS, vol. 2286, pp. 479–493. Springer, Heidelberg (2002)
2. Calka, P.: The distributions of the smallest disks containing the Poisson-Voronoi typical cell and the Crofton cell in the plane. Advances in Applied Probability 34, 702–717 (2002)
3. Daley, D.J., Vere-Jones, D.: An Introduction to the Theory of Point Processes, vol. I/II. Springer, New York (August 2005)
4. Gloaguen, C., Voss, F., Schmidt, V.: Parametric distributions of connection lengths for the efficient analysis of fixed access network. Annals of Telecommunications 66, 103–118 (2011)
5. Hirsch, C., Neuhäuser, D., Gloaguen, C., Schmidt, V.: First-passage percolation on random geometric graphs and its applications to shortest-path trees. Advances in Applied Probability (submitted)
6. Hirsch, C., Neuhäuser, D., Gloaguen, C., Schmidt, V.: Asymptotic properties of Euclidean shortest-path trees and applications in telecommunication networks. Working paper (under preparation)
7. Mai, J.-F., Scherer, M.: Simulating Copulas: Stochastic Models, Sampling Algorithms, and Applications. Imperial College Press, London (2012)
8. Neuhäuser, D., Hirsch, C., Gloaguen, C., Schmidt, V.: On the distribution of typical shortest-path lengths in connected random geometric graphs. Queueing Systems 71, 199–220 (2012)
9. Ruppert, D.: Statistics and Data Analysis for Financial Engineering. Springer Texts in Statistics, New York (2010)
10. Schneider, R., Weil, W.: Stochastic and Integral Geometry. Springer, Berlin (2008)
11. Stoyan, D., Kendall, W.S., Mecke, J.: Stochastic Geometry and its Applications. J. Wiley & Sons, Chichester (1995)
12. Voss, F., Gloaguen, C., Schmidt, V.: Capacity distributions in spatial stochastic models for telecommunication networks. Image Analysis and Stereology 28, 155–163 (2009)
13. Voss, F., Gloaguen, C., Schmidt, V.: Scaling limits for shortest path lengths along the edges of stationary tessellations. Advances in Applied Probability 42, 936–952 (2010)

Performance Modelling of Concurrency Control Schemes for Relational Databases

Rasha Osman, David Coulden, and William J. Knottenbelt

Department of Computing, Imperial College London
London SW7 2AZ, UK
{rosman,drc09,wjk}@doc.ic.ac.uk

Abstract. The performance of relational database systems is influenced by complex interdependent factors, which makes developing accurate models to evaluate their performance a challenging task. This paper presents a novel case study in which we develop a simple queueing Petri net model of a relational database system. The performance of the database system is evaluated for three different concurrency control schemes and compared to the results predicted by a queueing Petri net model. The results demonstrate the potential of our modelling approach in modelling database systems using relatively simple models that require minimal parameterization. Our models gave accurate approximations of the mean response times for shared and exclusive transactions with average prediction errors of 10% for high contention scenarios.

1 Introduction

It is now commonplace for organizations to each manage tens of petabytes of data. A large proportion of this data is stored in databases and is managed by database management systems (DBMSs). Despite the increasing use of NoSQL databases, relational databases are still the industry's main data model with forecasted steady growth to 2016 [6]. As users' expectations of performance and availability increases, the performance of these large DBMSs becomes a critical issue for organizations and vendors alike.

The performance engineering community has contributed many performance studies of database system components and several methodologies have been proposed for database system performance evaluation [9]. However, the impact of these studies on industry has been limited. One of the reasons may be that database system performance is affected by complex and interdependent interactions of physical and logical resources, which are difficult to represent using traditional modelling formalisms.

In previous work [2], we have demonstrated the suitability of queueing Petri nets (QPNs) as a modelling formalism for relational database contention. Queueing Petri nets [1] extend coloured stochastic Petri nets by incorporating queues and scheduling strategies into places forming *queueing places*, thus producing a powerful modelling formalism that has the synchronization capabilities of Petri nets while also being capable of modelling queueing behaviours. The queueing

A. Dudin and K. De Turck (Eds.): ASMTA 2013, LNCS 7984, pp. 337–351, 2013.
© Springer-Verlag Berlin Heidelberg 2013

places in QPNs allow for the accurate representation of lock scheduling as implemented in DBMSs, while the places and transitions naturally represent the flow of execution of a transaction in the system. Moreover, unlike previous studies of database concurrency control (see [9] for a survey), our models are able to reflect lock conflicts between read and update transactions.

DBMSs implement concurrency control through locking protocols. The most widely used protocol is Strict Two Phase Locking (Strict 2PL) [10]. Strict 2PL forces transactions to hold *exclusive locks* to modify data and *shared locks* to read data. For a transaction to acquire an exclusive lock on a data object, no other transaction should hold a shared or exclusive lock on that object. Transactions can acquire shared locks on data objects only if no transaction has an exclusive lock on the objects.

In this paper, we expand on our previous work and use QPNs to model a database system under three modes of Strict 2PL, specifically table-level, row-level and multi-version 2PL, each of which is implemented in commercial DBMSs. Our results demonstrate the capability of our modelling approach to reflect the dynamic operation of relational database systems using simple models that require minimal parameterization.

The rest of this paper is organized as follows. Section 2 describes the QPN model of the measured database system. In Sections 3, 4 and 5 we present the QPN models and results for table-level, row-level and multi-version 2PL, respectively. Section 6 concludes the paper and provides directions for future work.

2   General QPN Model of a Database System

Queueing places in QPN models consist of two components: the *queue*, and the *depository* where serviced tokens (customers) are placed. Tokens enter the queueing place through the firing of input transitions, as in other Petri nets; however, as the entry place is a queue they are placed in the queue according to the scheduling strategy of the queue's server. Once a token has been serviced it is deposited in the depository where it can be used in further transitions. Queueing places can have variable scheduling strategies and service distributions; these are known as *timed queueing* places. *Immediate queueing* places impose a scheduling discipline on arriving tokens without a delay. Due to space limitations, we refer the reader to [1] for a more detailed description of QPNs.

Figure 1 shows the general components of the QPN model for the modelled database system used in this work. The database system is composed of three tables, A, B and C. The system has two types of transactions: shared (read) and exclusive (update), with each client submitting one transaction type to the database server. The details of tables A and B are presented in their respective sections, as their components depend on the locking mechanism. There is no contention between transactions for table C and therefore no locking is modelled for table C. The *sleep* place is a timed queueing place representing the artificial delay for the exclusive transactions only.

The clients are represented by a timed queueing place with an infinite-server queue with an exponentially distributed think time with mean 200ms. The tokens in the *client* place have two colours; of which represent a client of one transaction type. The transactions enter the database through the *initial-processing* timed queueing place, which represents the delay for setting up the transaction when executing the BEGIN statement. A transaction will leave the database through the *final-processing* timed queueing place, which represents the time to commit the transaction and release its locks.

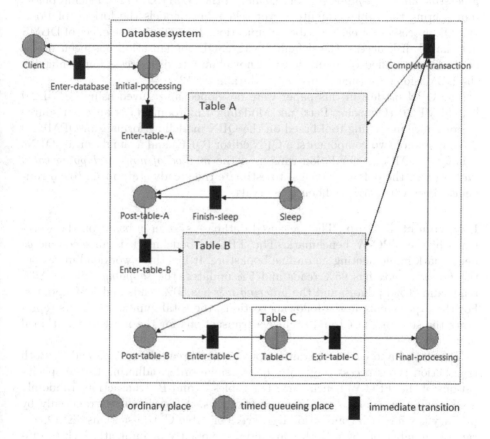

Fig. 1. General QPN model of the measured database system

A table is represented by a timed queuing place with an infinite server queue (*table-C* place in Fig. 1) that models transaction execution. Each type of transaction has its individual exponentially distributed service time that represents the execution time of the transaction when accessing the specific table. Here, we are assuming that the database table represents the main service centre for the transactions. This concept is inspired by previous work in modelling database systems using queuing networks [8] and is similar to other work that abstracts transaction CPU and disk execution by one service centre with an exponential service distribution [9]. When a transaction has been serviced at a table, it will continue to the next table or if it has finished executing it will continue to the *final-processing* place and then be passed back to the *client* place to repeat the process.

The QPN model reflects the logical execution of transactions on the database system. Therefore, the model does not directly model disk and CPU performance. However, the effects of processing are partially reflected in the *initial-processing* and *final-processing* places and in the $G/M/\infty - IS$ queueing places representing the tables. Infinite server scheduling models the forking of PostgreSQL processes for each database connection. To minimize the effect of DBMS automated disk access, the default PostgreSQL configuration has been modified[1]. This modified configuration will not eliminate disk access but configures the DBMS for performance instead of durability [12].

The QPN models in this paper were developed and solved using QPME2.0 [3]. QPME2.0 (Queueing Petri net Modeling Environment) is an open source performance modelling tool based on the QPN modelling formalism. QPME2.0 is composed of two components, a QPN editor (QPE) and a simulator for QPNs (SimQPN). All our simulation runs used the *method of non-overlapping batch means* (with the default settings) to estimate the steady state mean token residence times with 95% confidence intervals.

Experimental Setup. The measured database system is based on the workloads of the TPC-W benchmark. The TPC-W benchmark is an e-commerce benchmark implementing an on-line bookstore. It has three workload mixes [4]: the *browsing mix* has 95% reads and 5% updates, the *shopping mix* has 80% reads and 20% updates, and the *ordering mix* has 50% reads and 50% updates. For the experiments in the following sections, the total number of clients represents the sum of shared and exclusive transactions for the measured workload mix.

The shared and exclusive transactions access three tables, A, B and C. Both transaction types access the tables in the same order, adhering to the specifications of the TPC-W benchmark. For tables A and B, transactions randomly access one row out of a maximum of five rows. A row is chosen randomly by primary key for each transaction instance. For table C, transactions SELECT a random number of rows. Both transactions explicitly or implicitly lock the tables/rows in the appropriate lock mode (shared or exclusive) and read or modify one row for tables A and B and read a set of rows from table C. All tables have a primary key index, which is utilized by the transactions in the measured systems.

In order to simulate a TPC-W like workload in which update transactions are longer than read transactions [13], the execution time of the exclusive transaction

---

[1] The modified server configuration parameters are: fsync=off and synchronous_commit=off. The reader is referred to [12] for a definition of these parameters.

is artificially lengthened by 100ms, this is represented by the *sleep* place. To parameterize the QPN models, each transaction type is executed on the measured system in isolation (i.e. without any locking contention) and the mean service time for each QPN place was extracted from the DBMS logs. These are shown in Table 1.

The measured system was run on a virtual machine with four virtual processors@2.6GHz running Ubuntu Linux 10.04 64-bit and PostgreSQL 9.0 DBMS [12]. Each table is an average of 5MB in total size and has approximately 25 000 randomly generated rows. The mean response time of each transaction type is measured and compared to that emerging from the QPN model for the different transaction mixes.

Table 1. Mean service times (in milliseconds) for QPN model transaction types (colours). For table-level 2PL, the service times for the *final-processing* place include commit and lock statement mean execution times. All places have exponentially distributed service times, except for the row-level 2PL model, in which the *initial-processing* and *final-processing* places have a deterministic distribution with zero service time for the shared transaction. This is because the shared transaction in the row-level scenario is not contained within a BEGIN/END block.

QPN places	table-level		row-level		multi-version	
	shared	exclusive	shared	exclusive	shared	exclusive
initial-processing	0.06	0.03	0	0.04	0.03	0.03
table A	0.18	0.60	0.26	0.28	0.19	0.23
sleep	-	100.24	-	100.27	-	100.24
table B	0.12	0.18	0.13	0.23	0.11	0.19
table C	6.52	7.24	13.85	14.18	6.48	6.86
final-processing	0.14	0.74	0	0.05	0.03	0.04

3   QPN Model of Table-Level 2PL

Measured System. For this scenario, we model table-level Strict 2PL, which resembles the default locking method implemented in the MySQL default storage engine [7]. In order to implement table-level 2PL in PostgreSQL, each transaction type explicitly locks each table in the appropriate lock mode (shared or exclusive). The structure of the transactions is shown in Fig. 2. The shared transaction is composed of a set of smaller transactions; each represents an access to a table. This allows the shared transaction to release the shared lock on the table immediately after accessing the table. Therefore, exclusive transactions only wait for the shared transaction to leave a table, while shared transactions must wait for an exclusive transaction to commit before gaining access to any of the tables.

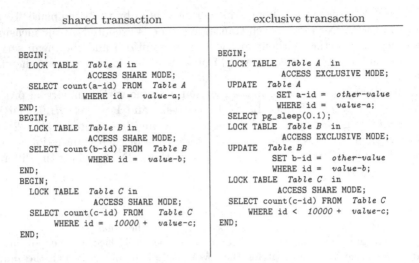

shared transaction	exclusive transaction

```
BEGIN;
  LOCK TABLE  Table A in
            ACCESS SHARE MODE;
  SELECT count(a-id) FROM  Table A
            WHERE id =  value-a;
END;
BEGIN;
  LOCK TABLE  Table B in
            ACCESS SHARE MODE;
  SELECT count(b-id) FROM  Table B
            WHERE id =  value-b;
END;
BEGIN;
  LOCK TABLE  Table C in
            ACCESS SHARE MODE;
  SELECT count(c-id) FROM   Table C
            WHERE id =  10000 +  value-c;
END;
```

```
BEGIN;
  LOCK TABLE  Table A  in
            ACCESS EXCLUSIVE MODE;
  UPDATE  Table A
            SET a-id =  other-value
            WHERE id =  value-a;
  SELECT pg_sleep(0.1);
  LOCK TABLE  Table B  in
            ACCESS EXCLUSIVE MODE;
  UPDATE  Table B
            SET b-id =  other-value
            WHERE id =  value-b;
  LOCK TABLE  Table C  in
            ACCESS SHARE MODE;
  SELECT count(c-id) FROM   Table C
            WHERE id <  10000 +  value-c;
END;
```

Fig. 2. Structure of shared and exclusive transactions for table-level 2PL

QPN Model. The general components of the QPN model of the measured system were discussed in Section 2. Here, we present the aspects of the QPN model that are related to the modelling of table-level 2PL. Figure 3 depicts the QPN model of table A, the model for table B (not shown) is identical. Transactions wishing to enter the *table-A* (or *table-B*) place must acquire a suitable lock on the table by entering the *lock-wait-A* place, in which they will wait for the lock on the table to be free. The *lock-wait-A* place is an immediate queueing place (shown with a bold outline in Fig. 3) with FIFO departure discipline. The table-level locking mechanism is represented using the *lock-store-A* place, which is an ordinary place containing *lock* tokens. A shared transaction will require one lock token and an exclusive transaction will require the maximum number of tokens defined for the *lock-store-A* place. By setting the number of lock tokens within the *lock-store-A* place to be equal to the maximum number of shared transactions, all shared transactions will be able to run simultaneously. In contrast, an exclusive transaction will be forced to wait if there is a least one shared transaction accessing the table. This process is modelled within the *acquire/release-lock-A* transition.

The shared transactions release table locks on table A (and B) immediately after leaving the *table-A* place by depositing a token in the *prepare-lock-release-A* place, enabling the *acquire/release-lock-A* immediate transition, which returns one lock token to the *lock-store-A* place. Exclusive transactions release locks after leaving the *final-processing* place (Fig. 1) by depositing a token in the *prepare-lock-release-A/B* place of table A and B, enabling the *acquire/release-lock-A/B* transition, which returns the maximum number of tokens back to the *lock-store-A/B* place.

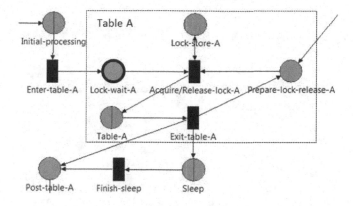

Fig. 3. QPN model of table A for table-level 2PL

Results. The results for the browsing, shopping and ordering workloads are presented in Fig. 4. The QPN prediction error percentages are in Table 2. For the browsing workload, the model underestimates the performance of the shared transaction for 30 clients and less, with an average underestimation of 27%, while accurately predicting the mean response times for exclusive transactions. This can be attributed to the model not representing the processing which may favour the concurrently executing shared transactions. For client numbers of 50 and above, we see the opposite effect; the model overestimates performance for both transactions, with average error of 19% for both transaction types. Here, the system is empty (four exclusive transactions at 80 clients) therefore no contention is present, allowing other variables, such as processing and disk access to dominate transaction execution time. These factors are not explicitly represented in the model. Nonetheless, the model follows the trend of degradation of performance with increasing number of clients for both transaction types.

For the shopping and ordering workloads, when locking contention starts to dominate transaction execution (i.e. the number of exclusive transactions is greater than four) the model gives an excellent prediction of mean response times for both transaction types. The mean error for both transaction types is less than 6% for the shopping workload, and less than 2% for the ordering workload. We note that the modelling error for 10 clients for the shopping workload is similar to that of 10 and 20 clients for the browsing workload, as both scenarios have the same level of contention.

4   QPN Model of Row-Level 2PL

Measured System. Row-level Strict 2PL is the default locking mechanism implemented in Microsoft SQL Server [5]. In row-level 2PL, a shared transaction acquires a shared row-lock for the duration of the read statement only. Exclusive transactions hold exclusive row-locks on rows until transaction commit, thus

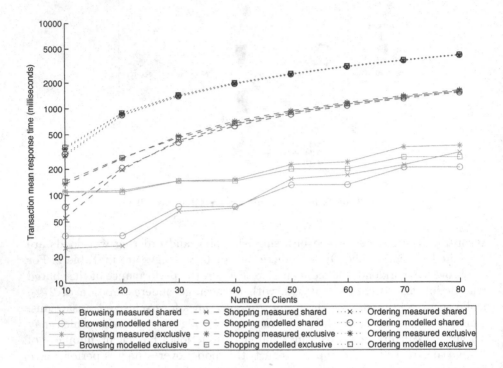

Fig. 4. Mean response time for TPC-W workload for table-level 2PL

Table 2. Error percentages for table-level 2PL

| | browsing error (%) | | shopping error (%) | | ordering error (%) | |
# of clients	shared	exclusive	shared	exclusive	shared	exclusive
10	37.79	-2.24	33.52	6.94	4.43	3.72
20	31.15	-3.93	4.73	2.21	1.29	0.88
30	13.37	-0.61	-4.57	-4.02	1.51	1.34
40	4.08	-3.52	-4.72	-4.67	1.29	1.15
50	-14.87	-11.34	-3.84	-3.99	1.23	1.16
60	-23.09	-16.41	-3.52	-3.68	1.25	1.17
70	-6.23	-23.35	-3.54	-3.69	1.38	1.35
80	-32.79	-26.08	-3.17	-3.23	1.35	1.34

preventing other shared and exclusive transactions from accessing these rows. The structure of the transactions is shown in Fig. 5.

To implement row-level Strict 2PL in PostgreSQL we utilize the default locking behavior of PostgreSQL statements. The shared transaction's SELECT statement includes the FOR SHARE clause, which implicitly acquires a shared lock on the retrieved rows thus preventing exclusive transactions from locking these rows [12]. A side effect of this is the increased processing needed to mark the row as locked. In addition, the shared transaction is not enclosed in a BEGIN/END block so that the shared lock is released upon statement completion. The exclusive transaction is similar to the exclusive transaction of table-level 2PL just without explicit table locks. Here the UPDATE statement acquires an implicit exclusive lock on the modified row, which is held until transaction commit.

shared transaction	exclusive transaction

SELECT count(a-id) FROM *Table A*
     WHERE id = *value-a*
     FOR SHARE;
SELECT count(b-id) FROM *Table B*
     WHERE id = *value-b*
     FOR SHARE;
SELECT count(c-id) FROM *Table C*
   WHERE id = *10000 + value-c;*
	BEGIN;
	  UPDATE *Table A*
	      SET a-id = *other-value*
	      WHERE id = *value-a;*
	  SELECT pg_sleep(0.1);
	  UPDATE *Table B*
	      SET b-id = *other-value*
	      WHERE id = *value-b;*
	  SELECT count(c-id) FROM *Table C*
	      WHERE id < *10000 + value-c;*
	END;

Fig. 5. Structure of shared and exclusive transactions for row-level 2PL

QPN Model. Figure 6 shows the QPN model of table A for row-level 2PL. This model differs from the QPN model for table-level locking in the representation of the locking mechanism. Here, transactions are blocked waiting on row-locks to be released. The waiting queue for each row of the five rows is represented by the *lock-wait-A-row* place, which is an immediate queueing place. The *used-lock-store-A* and *unused-lock-store-A* are complementary ordinary places that hold the *lock* tokens, one token (lock) colour for each row. The maximum number of each *lock* token colour defined for the *unused-lock-store-A* place is greater than the maximum number of shared transactions. Hence, the total number of tokens, irrespective of colour, in the *unused-lock-store-A* place is more than five times the maximum number of shared transactions. The *used-lock-store-A* place holds identical token colours as that of the *unused-lock-store-A* place but their number is set to zero.

When the *enter-table-A* (or B) transition is enabled, a transaction entering table A randomly chooses a row and enters the corresponding *lock-wait-A-row* place to request the appropriate lock for the row. A shared transaction in a specific *lock-wait-A-row* place will require one of the corresponding lock tokens from the *unused-lock-store-A* place then deposit it into the *used-lock-store-A* place and enter the table for service. An exclusive transaction will require the maximum defined number of lock tokens for the corresponding row and deposit them into the *used-lock-store-A* place and enter the table for service. Therefore, all shared transactions will be able to access a row simultaneously if no exclusive transaction is currently updating that row. An exclusive transaction will be forced to wait if there is at least one shared transaction or an exclusive transaction accessing the row. This process is modelled within the *acquire-lock-A* transition.

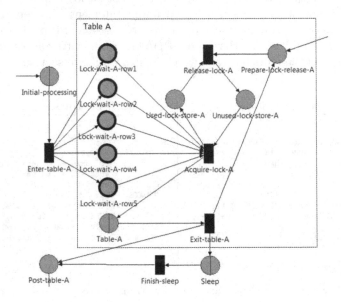

Fig. 6. QPN model of table A for row-level 2PL

The shared transactions release row-locks on table A (or B) immediately after leaving the *table-A* place by depositing a token in the *prepare-lock-release-A* place, enabling the *release-lock-A* immediate transition. The *release-lock-A* transition randomly chooses a token colour that corresponds to a row that has been locked by a shared transaction, i.e. token colours that have a nonzero amount in the *used-lock-store-A* and the *unused-lock-store-A* places. Then one token is removed from each of the *used-lock-store-A* and the *unused-lock-store-A* places and two tokens are deposited into the *unused-lock-store-A* place[2].

Exclusive transactions release locks after leaving the *final-processing* place (Fig. 1) by depositing a token in the *prepare-lock-release-A* place of table A (and B), enabling the *release-lock-A* transition. The *release-lock-A* transition randomly chooses a token colour that corresponds to a row that has been locked by an exclusive transaction, i.e. token colours that have the maximum number of tokens in the *used-lock-store-A* place. In the same fashion as shared transactions,

---

[2] This functionality should have been modelled with an inhibitor arc; however, QPME2.0 does not support inhibitor arcs [3].

this number of tokens is removed from the *used-lock-store-A* place and deposited into the *unused-lock-store-A* place.

Results. Figure 7 shows that transaction performance is now better than that of table-level locking, as contention is at the row-level, with transactions blocked only if they want to acquire conflicting locks on a row. The error percentages of the QPN model are in Table 3.

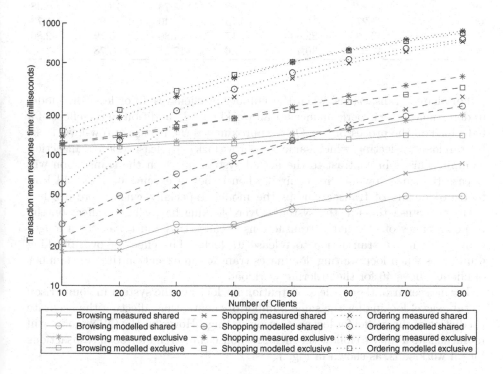

Fig. 7. Mean response time for TPC-W workload for row-level 2PL

For the exclusive transaction, the model's accuracy for the browsing workload is similar to that of the table-level locking QPN. For the shopping workload, as the number of clients increase over 60, the model overestimates the performance of the exclusive transaction. This is likely due to the effect of increased processing time for the exclusive transaction due to increased SELECT FOR SHARE statements in the system. The effect of low contention is clear in the ordering workload; the model underestimates the performance of the exclusive transaction for number of clients 30 and below, then its accuracy increases as the number of exclusive transactions increase. The model's level of accuracy increases when the lock contention is similar to that of the shopping and ordering workloads of table-level locking. This is most likely at 40 clients for the ordering workload.

Table 3. Error percentages for row-level 2PL

	browsing error (%)		shopping error (%)		ordering error (%)	
# of clients	shared	exclusive	shared	exclusive	shared	exclusive
10	17.55	-1.84	27.73	1.08	43.97	9.86
20	14.96	-3.62	31.93	3.34	37.21	13.70
30	13.92	-3.75	24.72	2.92	23.51	10.45
40	4.39	-6.20	11.90	-0.42	14.59	5.09
50	-5.74	-9.36	1.98	-5.78	10.60	0.38
60	-21.27	-15.60	-5.73	-10.40	6.92	-1.73
70	-32.80	-22.96	-11.17	-14.80	5.79	-2.80
80	-43.60	-30.41	-15.06	-17.55	3.78	-3.54

For shared transactions, when the number of clients is 40 or less, the model underestimates their performance for all workloads. In addition to the effect of a light load on the processor, the QPN approximates the release of row-level locks by randomly selecting which same type row-lock to release upon transaction commit. This is in contrast to the actual system in which the transaction will release the same row-lock it acquired when it began execution. This will lead to over-blocking of transactions in the model, especially when there are less exclusive transactions in the system. Over-blocking has a higher effect on the response times of the shared transactions because they must wait for a longer running exclusive transaction to release its locks. The effect of over-blocking diminishes when lock waiting dominates transaction execution time for number of clients above 40 for the ordering workload.

In this scenario, there is less contention for locks in the system in comparison to table-level locking for the same number of clients, thus affecting the accurate prediction of the model for low contention settings. However, the QPN model still maintains its ability to follow the mean response time trend for both transactions for all workloads, as shown in Fig. 7.

5   QPN Model of Multi-version 2PL

Measured System. Multi-version Strict 2PL is similar to the default locking mechanism for PostgreSQL [12]. In this case, shared transactions read snapshots of the data representing the most recent consistent state regardless of the current state of the database [11]. Therefore, shared transactions do not block exclusive transactions and exclusive transactions do not block shared transactions. Alternatively, an exclusive transaction modifying a row blocks other exclusive transactions from accessing that row. This is because exclusive transactions modify the current consistent state of the database and hold exclusive locks on rows until transaction commit. The structure of both transactions is similar to the transactions of table-level 2PL, without explicit LOCK TABLE statements, and the shared transaction is enclosed within one BEGIN/END block.

QPN Model. The QPN model for table A for multi-version 2PL is identical to the QPN model for row-level 2PL in Fig. 6, with the addition of an edge between the *enter-table-A* transition and the *table-A* place. This edge represents the shared transactions bypassing lock checking and accessing the rows of table A. There is no edge between the *exit-table-A* transition and the *prepare-lock-release-A* place, as no locks are held by shared transactions. Each token colour in the *unused-lock-store-A* place is initialized to one, i.e. only one exclusive transaction can hold a lock on a given row at any time. Lock release for exclusive transactions is identical to that of row-level locking except that only one token is transferred for each exclusive transaction when releasing its locks.

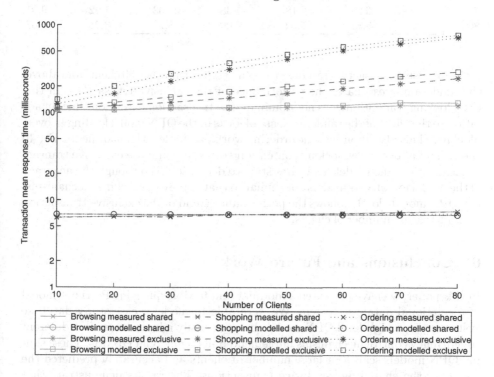

Fig. 8. Mean response time for TPC-W workload for multi-version 2PL

Results. In this scenario, the shared transactions run through the system unaffected by the exclusive transactions. This lowers the contention for locks in the system in comparison to table-level and row-level locking for the same number of clients. From Fig. 8 and Table 4, the QPN model gives an excellent prediction of mean response times for the shared transaction for all workloads. As the model does not represent the processing effects on the shared transaction, it overestimates their performance for small number of clients and underestimates their performance for larger number of clients.

For the exclusive transaction, the model gives an excellent prediction for the browsing workload. For the shopping workload, the model gives accurate estimates

Table 4. Error percentages for multi-version 2PL

| | browsing error (%) | | shopping error (%) | | ordering error (%) | |
# of clients	shared	exclusive	shared	exclusive	shared	exclusive
10	8.48	0.92	9.07	3.62	8.18	12.24
20	5.56	0.81	5.40	9.43	5.46	21.83
30	2.30	3.22	6.82	14.80	5.98	21.97
40	0.60	2.93	0.29	18.51	-1.11	17.83
50	-1.84	5.87	-1.51	20.23	0.32	13.67
60	-3.70	5.37	-2.98	21.01	-0.59	10.74
70	-6.99	8.18	-5.38	21.03	-1.92	9.31
80	-8.81	7.89	-9.00	18.79	-3.21	6.87

of the exclusive transaction response times up to 20 clients. For 30 clients and above, the model underestimates the performance of the exclusive transaction. In this case, locking contention does not dominate the transaction execution time, in addition to the effect of the random release of locks in the QPN model leading to over-blocking. This also applies to the ordering workload for less than 30 clients. For 40 clients and above, when locking contention starts to dominate exclusive transaction execution, the model error rates start to decrease. Even though the accuracy of the model for all workloads is not similar to that of previous locking mechanisms, the QPN model closely follows the performance trend of the exclusive transaction with increasing number of clients.

6   Conclusions and Future Work

In this paper, we have presented a modular and flexible queueing Petri net model of a relational database system. We were able to model a case study database system using a high-level QPN model that was easily adapted to reflect different concurrency control mechanisms implemented in commercial DBMSs. These simple QPN models scaled for large number of clients and accurately predicted the trends of the shared and exclusive transactions. The results demonstrate that the QPN models were able to give accurate predictions of the mean response times of transactions for moderate and high contention workloads.

The QPN models presented were simple models of the database logical level that do not require detailed modelling of the underlying hardware architecture. The models have the potential to be applied to logical layer database modelling to evaluate lock contention for systems in which hardware provisioning is sufficient. For such scenarios, these models are able to give indications of the ability of the DBMS, and therefore the database design, to cope with different workloads under different concurrency control schemes.

For low contention scenarios or when the bottleneck is at the physical hardware layer, a layered model in which logical and physical resources are represented would perhaps be more suitable. More realistic row access scenarios representing skew in data access and data distribution will need to be investigated. Moreover, as the QPN models have an intuitive structure that directly reflects the database design, they are a suitable candidate for an automated mapping tool between database system specifications and QPN models. These are issues for future work.

References

1. Bause, F.: Queueing Petri Nets–A Formalism for the Combined Qualitative and Quantitative Analysis of Systems. In: Fifth Intl. Workshop Petri Nets and Performance Models (1993)
2. Coulden, D., Osman, R., Knottenbelt, W.J.: Performance Modelling of Database Contention using Queueing Petri Nets. In: 4th ACM/SPEC International Conference on Performance Engineeering (2013)
3. Kounev, S., Spinner, S.: QPME 2.0 User's Guide. Karlsruhe Institute of Technology, Am Fasanengarten 5, 76131 Karlsruhe, Germany (2011),
   http://descartes.ipd.kit.edu/fileadmin/
   user_upload/descartes/QPME/QPME-UsersGuide.pdf
4. Menasce, D.A.: TPC-W: a benchmark for e-commerce. IEEE Internet Computing 6(3), 83–87 (2002)
5. Microsoft Corporation: Set Transaction Isolution Level, Transact-SQL (2013),
   http://msdn.microsoft.com/en-us/library/ms173763%28v=sql.110%29.aspx
6. Olofson, C.W.: Worldwide Relational Database Management Systems 2012–2016 Forecast. International Data Corporation, Doc # 236273 (2012),
   http://www.idc.com
7. Oracle Corporation: MySQL 5.6 Reference Manual. Internal Locking Methods (2013), http://dev.mysql.com/doc/refman/5.6/en/internal-locking.html
8. Osman, R., Awan, I., Woodward, M.E.: QuePED: Revisiting Queueing Networks for the Performance Evaluation of Database Designs. Simulation Modelling Practice and Theory 19(1), 251–270 (2011)
9. Osman, R., Knottenbelt, W.J.: Database System Performance Evaluation Models: A Survey. Performance Evaluation 69(10), 471–493 (2012)
10. Ramakrishnan, R., Gehrke, J.: Database management systems. McGraw-Hill, Boston (2003)
11. Silberschatz, A., Korth, H.F., Sudarshan, S.: Database System Concepts. McGraw-Hill (2011)
12. The PostgreSQL Global Development Group: PostgreSQL 9.0.12 Documentation (2012), http://www.postgresql.org/docs/9.0/static/index.html
13. The Transaction Processing Performance Council: TPC-W BENCHMARK ver-sion 2 (2003), http://www.tpc.org/tpcw/

Modeling and Simulation of Mobility of Crowds

Sushma Patil[1], Eitan Altman[1],
Manjesh Kumar Hanawal[1], and Julio Rojas-Mora[2]

[1] Team MAESTRO, INRIA, Sophia Antipolis, France
[2] Institute of Statistics, Universidad Austral de Chile, Chile

Abstract. Mobility models studied in the networking community usually assume independence between the movement of individuals. While this may well model sparse networks, there are many scenarios that might not follow this assumption. In contrast, within other communities, such as road traffic engineering, biology and computer graphics, models of mobility usually take into account the dependence of the mobility pattern of an individual with respect to that of its neighbors. Our goal in this paper is to study how this dependence impacts the performance measure from the networking point of view. In particular, we implement a bio-inspired model for mobility of crowds and, by simulation, we study how mobility influences the performance measures of a distributed network. We perform statistical analysis on the samples obtained through simulations. In particular, we study the distribution of the message delivery time and show that it is light tailed, with exponential tail distribution.

Keywords: Mobility, flocking, statistical analysis.

1 Introduction

Mobility of nodes in a wireless network have a direct influence on the performance of various protocols used for message delivery; performance metrics like delay in message delivery, energy consumed, probability of successful message transmission, etc., are functions of mobility pattern of nodes. It is desirable to know the performance of a protocol by testing it on real world traces of the node mobility patterns collected through experiments. However, often many a times traces of the desired scenario are not available as the wireless network may not yet be deployed. In such scenario, one relies on the data generated from synthetic models to study network performance. Synthetic models that describe a mobility scenario under consideration, in a realistic fashion, can be complex. The models available in literature tradeoff between ease of implementation/analysis and being more realistic. Depending on which synthetic model best captures node mobility, the choice of the synthetic model is made for evaluation of the performance of a protocol.

Quite often, simple mobility models like Random Waypoint or its variants [6] are assumed as they are relatively easy to implement and analyze [5]. These models assume that movement of each mobile node (MN) is independent of others, and also their movement does not depend on their past locations or speeds.

A. Dudin and K. De Turck (Eds.): ASMTA 2013, LNCS 7984, pp. 352–363, 2013.

These assumptions are justifiable in a very sparse network where MNs are spread far from each other and hence their movement can be assumed to be uncorrelated. These models, though easy to analyze, lead to unrealistic scenarios, like sharp turns or sudden stops. To avoid such cases models are proposed that correlate a node's mobility on its past movements to varied degrees of generality, but with increasing complexity of tractability. In Gauss-Markov models [7], current location of a node is made to depend on its past location through a controlled parameter. For very thorough surveys on mobility models see [4] and [14].

In many scenarios MNs may not move independent of each other, but in a group, or following a particular reference node in the group. For example, a group of people visiting a museum or touring a city following an instructor. To cater to this kind of dependent mobility, several models are proposed by the network community. In the Reference Point Group Mobility (RPGM) model all of the nodes follow a path traveled by a logical center [8]. The logical center may be made to follow a predefined path or can follow random waypoint mobility. The other MNs move according to some random waypoint in the vicinity of the logical center. Several other useful group mobility models can be derived as variants of RPGM: column group mobility, nomadic group mobility, pursue mobility [4], etc.

We shall be interested in crowd movements where MNs are humans carrying wireless terminals. Here the MNs may not follow any particular MN, nor their movement is independent of each other. In such crowd movement, they avoid colliding with each other and maintain a safe distance from their neighbors. For example, people moving in a market or a busy commercial area. In this paper, we aim to study such crowd mobility where mobility of a MN depends on its neighbors. In biology, such kind of mobility is extensively studied to understand movement of flock of birds, school of fish or heard of animals. Rules governing such mobility are extremely interesting if one wishes to recreate patterns of such flock movements through artificial life. Indeed, scientists in the field of computer graphics synthesize mobility of such crowds to create beautiful patterns [1]. To synthesize such patterns they implement rules that govern how mobility of an entity in the group depends on its neighbors.

Craig Reynolds [9] proposed a systematic way to synthesize aggregate motion of crowds using computer graphics. He referred to each entity in the crowd as 'boid'. The synthesized movement of the boids is based on his finding that the complex auto-organization of the group into complex macroscopic patterns is determined by three simple microscopic rules that each individual in the group follows: (i) Boids try to fly towards the center of mass of neighboring boids. (ii) Boids try to keep a small distance away from other objects (including other boids). (iii) Boids try to match velocity with near boids. If each boid is implemented as an independent actor, which navigates according to its local perception of the dynamic environment obeying the above rules, the interaction of the boids results in the aggregate action that looks very synchronized as if centrally controlled.

The primary aim of [9] is to recreate beautiful patterns observed in nature. Variants of the rules proposed by Reynolds are used to study local interaction and evolutionary patterns in biology [1]. In one of such studies, Ariel Dolan [2] implemented creatures that guard their territory against intruders, and study evolutions of populations. He referred to such life-like creature as 'Floy', which interact with their neighbors as follows: (i) stay close to your fellows but not too close. (ii) if you see an intruder, move towards it and attack. The first rule of Dolan is similar to the first two rules of Reynolds, but they differ in how the Floys track its neighbors: In Dolan's Floy model, each Floy tracks two of its randomly chosen neighbors and tries to be close to them, whereas in Reynolds' model, each boid tries to be in the center of its neighbors that are within a certain radius. Also, in Dolan's model there is no matching of velocities among the Floys.

In Reynolds' boids model the three rules make the mobility of boids highly dependent on each other; the boids stay close to each other, and result in patterns that give a feeling that all the boids are guided centrally. However, relaxed rules of Dolan's Floys model makes Floys to avoid too much togetherness, which lead them to move in an ensemble that spreads over a wide region. Each Floy may stray away from its group and wander somewhat randomly in a given territory around its group members. But it will not go too far away, as soon it will tend to move towards its group due to the first rule. This behavior is similar to that of a crowd movement where the nodes move from one location to other. Obviously, MNs avoid colliding with each other and do not stray far from its neighboring fellows while it moves in a given territory. Thus, we can use rules similar to that in the Floy model to simulate and study crowd movements in ad hoc networks.

Our aim in this paper is to study performance in our crowd mobility model, from the networking point of view. We consider the delay tolerant network scenario. Each MN in the crowd can have a message that another MN in the crowd is interested. The MN which have this message (source MN) can spread the message in the crowd by giving it to any other MN that comes close to it. The message is thus relayed among the crowd and can finally reach the MN that is interested in this message (destination MN). We shall study the time taken for a message sent by a source MN, using the other MNs as relays, to reach the destination MN. In particular, we shall be interested in identifying the tail distribution of the message transfer delay and its statistics, i.e., the probability that the message takes larger than a threshold time to reach destination. Such analysis is of primary importance as a message may become irrelevant if it does not reach the destination within a stipulated time. We also study how the transmission distance influences the message delivery time. This is equivalent to analyzing energy-delay trade off in wireless networks [15], as larger transmission range means more power transmission with lesser delay and vice versa.

The paper is organized as follows: In Section 2, we discuss the simulator we built to study crowd mobility based on Craig Reynolds' boid model. In Section 3 we discuss how we adopt the boid model to study network crowd movement and explain the simulation settings. In Section 4 we discuss the statistical results of the message delivery time. In Section 5 we discuss the effect of transmission range on the message delivery time. Finally, in Section 6 we discuss the possible extensions and summary of our work.

2   Crowd Mobility Simulator

In this section we discuss the simulator we built to study crowd mobility based on Reynolds' boid model. We refer to each boid as a node, henceforth.

Below we discuss how each node updates its location and velocity according to Reynolds' three rules discussed in the introduction. In the 2-D version of the simulator each node begins with a fixed velocity v_0 at an arbitrary location in a rectangular region. In applying rule 1 (cohesion), each node looks for other nodes within a radius, say R_1, and finds the position that is the center of mass with respect to the location of these nodes. In applying rule 2 (separation), each nodes looks for other nodes within a radius, say R_2, and finds the position that is the center of mass with respect to the the locations of these nodes. In applying rule 3 (alignment), each node looks for relative velocity of the other nodes within a distance, say R_3, with respect to itself, and finds the average velocity. The new velocity is obtained by combining the values obtained from the three rules with weights γ_1, γ_2, and γ_3, respectively, and adding to the current velocity. For example, let $A_1 = (a_1, b_1)$, $A_2 = (a_2, b_2)$, and $A_3 = (a_3, b_3)$ be the values obtained from rule 1,2, and 3, respectively, for a node which has current velocity $V_n = (Vx_n, Vy_n)$ and is at position $X_n = (x_n, y_n)$. Here the first component in each vector corresponds to the x-axis, and the second to the y-axis. Then, the node moves to the new position $X_{n+1} = (x_{n+1}, y_{n+1})$ with velocity $V_{n+1} = (Vx_{n+1}, Vy_{n+1})$, given by

$$V_{n+1} = V_n + \gamma_1 A_1 - \gamma_2 A_2 + \gamma_3 A_3$$

and

$$X_{n+1} = X_n + \Delta * V_{n+1},$$

where Δ is a constant that governs how fast the nodes update their position.

The weights γ_1, γ_2, and γ_3 are positive constants that govern the relative importance assigned to each rule. When a node comes close to another node, it should decrease its velocity in the current direction to avoid collision. This is accounted by taking minus sign next to γ_2 in the velocity recursion. By adjusting these parameters one gets varied degree of dependency among the nodes' movement. For example, if γ_1 is large while γ_2 and γ_3 are small, then nodes stay very close to each other, but seem to change directions and come close to each other often. If γ_2 is large while γ_1 and γ_3 are small, then nodes are equally spaced from each other but move more randomly and change direction often. If γ_3 is large while γ_1 and γ_2 are small, then all the nodes align themselves in a particular direction and continue to move in the same direction till they hit the boundaries. By varying these weights one gets varied degree of randomness in the movements of the nodes. We put an upper limit on the velocity of each node. If the velocity of a node exceeds this limit, denoted by v_{max}, then its velocity is set to v_{max}.

Fig. 1. Snap shot of Crowd Mobility simulator: Blue node has message

Fig. 2. Snap shot of Crowd Mobility Simulator: Blue nodes have the message and green ones are yet to receive it

We implemented Reynolds' three rules in Java with an interface to control the weights. A snapshot of our implementation is shown in figures 1 and 2. In the simulator, the number of nodes (N), and the parameters R_1, R_2, and R_3 can be set. If the values of R_1, R_2, and R_3 are high, i.e., nodes take into account movement of other nodes from a larger area in deciding their new velocity, and the movement becomes more dependent. When the values of these parameters are decreased, the movement tends to become more random. Thus, by changing various parameters in the simulator, the movement can be changed from being completely random to perfectly synchronized.

In building the aforementioned simulator, we used the Dolan's Java templates which he developed to implement his Floy model. We implemented Reynolds' boid with the help of these templates and built in various functionalities to control movements of boids that help us study crowd mobility in ad hoc networks as discussed in the following sections. The simulator and its source code are available online at [17].

3   An Adaptation of Boids Model to Networking

In crowd movement, people try to avoid colliding with each other by maintaining a safe distance, while also not straying far away from its neighbors. This scenario can be simulated by appropriately setting the weights γ_1, γ_2, and γ_3 and the parameters R_1, R_2, and R_3 in the simulator. Note that in crowd mobility, velocity matching is not an important criterion. However, avoiding collisions and not straying away far from neighbors are important conditions, with the former having more priority. These criteria correspond to setting $\gamma_3 = 0$, and $\gamma_2 > \gamma_1$ in the simulator.

We adopt the boid model to study crowd movement in wireless ad hoc networks. In particular, we will be interested in scenarios like delay tolerant networks (DTNs) where end-to-end connectivity between source and destination is not available. Everybody in the crowd is assumed to have a transmitter and receiver, and can transmit, receive and relay messages. We henceforth refer to each boid-like entity or each person in the crowd as a "mobile node" (MN). The MNs are battery powered and can not transmit beyond a certain transmission range. Let T denote the transmission range of each MN. They can transmit or receive messages when within distance T from other MNs. If the other MNs already have this message, then they simply ignore the message.

In this preliminary study, we consider epidemic routing scenario in a delay tolerant network [11], i.e., a MN having the message transmits it to all the MNs that are within the transmission range T. We leave other protocols like, two hop routing, direct contact, spray and wait [13] for future study. Also, we assume that the transmissions are error free and instantaneous. Further, all the nodes are identical and participate equally in relaying the message. In the simulator it is possible to set a certain fraction of MNs not to participate in message spreading. However, in this paper we restrict to the case where all the nodes participate in message spreading.

In epidemic routing the copy of the message is available with a large number of MNs and it is likely that it will reach the destination with high probability. But the time required for the message to reach the destination, or the delay in delivery of the message, can be still very high. For example, small MN density, small transmission range, can result in longer time for two MNs to be within transmission distance of each other hence leading to large time delays. In DTNs, the message delivery is of primary interest than the time taken it to be delivered at the destination, but the sooner the message gets delivered the better: in some case the message may lose its relevance if it fails to reach within a stipulated time. Thus, it is of interest to know probability of message taking too long to be delivered at the destination. Or, more generally, to know the tail distribution of the delay.

4   Experimental Setup and Statistical Analysis

4.1   Experimental Setup

In this section we discuss simulation setup and data collection for empirical study of the delay distribution. In the simulator we set the parameters as follows:

$R_1 = R_3 = 100, R_2 = 25, \gamma_1 = 0.001, \gamma_2 = 1.0, \gamma_3 = 0, v_0 = 2, \Delta = 0.095, T = 25, v_{\max} = 20$. The choice of these parameters is based on visual monitoring, such that the simulated mobility looks close to that of a crowd mobility. The above parameters are set as default values in the simulator. The simulation is run with 50 MNs that start at random locations with MN_1 as the source and MN_n as the destination. In each run, we set a warmup period of 30 seconds before collecting any data. Once this warmup period lapses, we start a timer at the destination and the note the time when it receives the message for the first time. We made 500 runs to collect samples of the random variable D.

In this paper, we present statistical analysis with the parameter set to default values. However, we note that the same observations continues to hold for other set of parameters as well. In the following subsections, we explain the statistical analysis for one set of samples collected from the simulator. These samples are available online at [17].

4.2   Statistical Analysis

In this section we analyze the delay distribution using statistical tools available in MATLAB. We obtain the empirical distribution using the samples generated through the simulator. It is observed that the exponent of the tail distribution is linear, and the slope is obtained using linear regression. Finally, we derive the theoretical distribution that best represents the tail of the delay distribution.

Let $F_D(\cdot)$ denote the cumulative distribution function (CDF) of D. We will characterize the delay distribution by evaluating the empirical distribution (ECDF) of D which we denote as $\hat{F}_D(\cdot)$. We also calculated upper and lower bounds (see Figure 3) that corresponds to 95% confidence intervals of the ECDF [12].

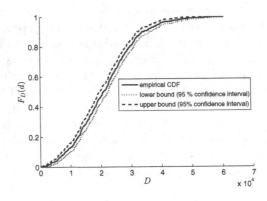

Fig. 3. empirical CDF of delay

We next evaluate the exponent of the tail distribution, i.e., rate of change or slope of:

$$y(d) = -\log(1 - \hat{F}_D(d)).$$

This function is plotted in Figure 4, as the ECDF, with its upper and lower confidence interval bounds. As seen from Figure 4, $y(d)$ is an almost linear function for large values of D.

4.3 Linear Regression of $y(d)$

In this subsection we evaluate the slope of $y(d)$ using linear regression. Let $d = (d_1, d_2, \cdots, d_m)$ and $f = (f_1, f_2, \cdots, f_m)$ denote the time sequence and the corresponding values of the cumulative probability. Define $y_i = -\log(1 - f_i)$ for each $i = 1, 2, \cdots m$ and denote $y = (y_1, y_2, \cdots, y_m)$. As our interest is in the behavior of the tail distribution, we take into account only the last 40% of sample of the function (d, y) to evaluate the slope. Let M denote the size of this truncated vector (d, y), and re-index its samples from 1 to M. For the samples we have, the sample size after truncation is $M = 201$.

Before we proceed to apply regression analysis on the samples, let us first verify that there is significant linear relationship between d and y using regression the t-test. The *correlation coefficient* for the pair (d, y) is $R = 0.9960$, and the corresponding *test statistics* for the data is:

$$t = R\sqrt{\frac{M - 2}{1 - R^2}} = 156.7324,$$

whereas the critical value is 1.972 at *significance level* 0.05. Clearly, at significance level 0.05, there is a significant linear relation between d and y as $t > 1.9720$.

We next evaluate the slope and intercept of the linear function that best represents the linear relation between d and y. These values are given, respectively, as follows:

$$m = \frac{M \sum d_i y_i - (\sum d_i)(\sum y_i)}{M \sum d_i^2 - (\sum d_i)^2} = 0.0001443$$

and:

$$b = \frac{\sum y_i - m \sum d_i}{M} = -2.602.$$

The best fit linear function in shown in figure 4 in a continuous line. Note that, in the region of interest, its tail, the linear function lies within the upper and lower confidence interval bounds. Thus the logarithm of the tail of the ECDF can be approximated by the linear line with slope and intercept as computed above, and the confidence interval for this fit is 95 %. Also, the adjusted R^2 value as a measure of the *coefficient of determination* for this fit is 0.9919. Hence, with a high confidence value, we can approximate $y(d)$ by a linear function, which in turn implies that tail of the empirical distribution can be approximated by an exponential function, i.e.:

$$\Pr\{D \geq d\} \approx \exp\{-md\} \quad for \ \ d > d^*,$$

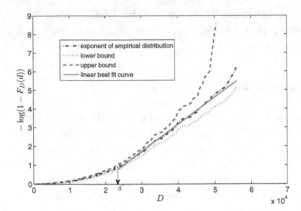

Fig. 4. Exponent of tail distribution

where $d^* = d_{N-M}$ is the threshold beyond which we apply the regression to fit $y(d)$ to the data, with m as the slope of the best fit linear function. Note that in arriving at the exponential approximation for the tail distribution we only took into account the slope of the linear regression and not the intercept. In the next section we study the goodness of fit of the exponential distribution to the tail of the empirical distribution by studying *quantile-quantile* plots and performing a Kolmogorov-Smirnov test.

4.4   Quantile-Quantile Plot

The exponential approximation of the empirical distribution is good when the samples are larger than the threshold d^*. Let $\bar{D}$ denote the distribution of the delay conditioned that it is larger than d^*. Then the CDF of $\bar{D}$ is given by

$$F_{\bar{D}}(d) = \begin{cases} 1 - \exp\{-m(d - d^*)\}, & \text{if } d \geq d^* \\ 0, & \text{otherwise} \end{cases} \tag{1}$$

The q-quantile of this CDF for a given $p \in [0 \ 1]$ can be evaluated as

$$F_{\bar{D}}^{-1}(p) = d^* - \frac{\log(1 - p)}{m},$$

where $F_{\bar{D}}^{-1}(p)$ is the inverse of the CDF of the tail. We compare this theoretical quantile function with the quantile of the ECDF obtained from the samples collected. For a fair comparison, we discard all the samples that are smaller that d^* in computing the ECDF. The q-q plot is shown in figure 5. Note that the quantile functions closely follows the 45 degree line, with deviations only at the edges. This is an expected behavior as we are approximating an exponential function with a linear fit. Thus, we overstimate the quantiles at the lower end of the curve, while overstimating them at the upper end. Nevertheless, the correlation coefficient between the two quantile functions is 0.9938. Hence, the exponential approximation of the tail distribution is a good fit.

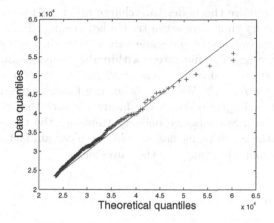

Fig. 5. q-q plot

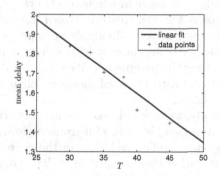

Fig. 6. Transmission range vs mean delay

We next perform the Kolomogrov-Smirnov (KS) test on the data samples. The test statistic for this test is

$$D = \max_d |F_{\bar{D}}(d) - \hat{F}_{\bar{D}}(d)| = 0.0978,$$

where $\hat{F}_{\bar{D}}$ denotes the ECDF of the samples that takes value larger than d^* in the data. At a significance level of 5%, the hypothesis that the samples are drawn from the exponential distribution with $\lambda = m$ is not rejected by the KS test. Thus, it can be believed that the tail is exponentially distributed.

5   Delay-Energy Tradeoff

In this section we study the effect of transmission range T on the mean message delivery time. If nodes can transmit at higher power, then they can transmit the message successfully over a larger distance. However, since the nodes are battery powered, this will require the nodes to recharge often. In Figure 6 we plot the mean message delivery time versus the transmission range T. In generating the plots we set the parameters to the same values given in Section 3. Note that if $T < 25$, then the nodes come rarely within the transmission range of each other as they will try to maintain a separation of 25 units ($R_2 = 25$). Thus, we generate the plot for $T \geq 25$. We also obtain the linear regression of the data points and plot it in Figure 6. From the figure we note that there is a linear relation between the mean message delivery time and the transmission range. Thus, to decrease the mean transmission delay by a certain factor, the MS need to increase the transmission range by the same factor.

6 Conclusion

Mobility modeling is crucial to analyze the performance of wireless networks. To understand the performance of a protocol, simplifying assumptions, like independence of node mobility, are made in wireless networks, which is not realistic. In this paper, we proposed a model to study crowd mobility that captures dependency among the movements of individuals in a group. We implemented a simulator in which the degree of dependency of movements among the individuals can be varied by controlling the parameters. We established through systematic statistical analysis that the distribution of message delivery time is light tailed with exponential distribution.

In this preliminary study, we have not take into account errors in transmissions and minimum contact duration for successful transmission. In future, we like to bring in these aspects into simulations. Also, in the current model, when two nodes come close to each other they sometimes make a complete U-turn, which is not realistic. We would like to make such turns smooth, and study how this impacts the results and validate our observations thorough sensitivity analysis.

Acknowledgement. The authors would like to thank the Indo French Centre for Applied Mathematics (IFCAM) and the GANESH project for the financial support.

References

1. http://www.red3d.com/cwr/boids/
2. http://www.aridolan.com/ofiles/eFloys.html
3. http://www.aridolan.com/ofiles/Floys2.aspx
4. Camp, T., Boleng, J., Davies, V.: A Survey of Mobility Models for Ad Hoc Network Research. Wireless Communication and Mobile Computing (WCMC): Special Issue on Mobile Ad Hoc Networking: Research. Trends and Applications 2(5), 483–502 (2002)
5. Bettstetter, C., Resta, G., Santi, P.: The Node Distribution of the Random Waypoint Mobility Model for Wireless Ad Hoc Networks. IEEE Trans. Mobile Computing 2(3), 257–269 (2003)

6. Hyytia, E., Lassila, P., Virtamo, J.: Spatial Node Distribution of the Random Waypoint Mobility Model with Applications. IEEE Trans. Mobile Computing 5(6), 680–694 (2006)
7. Liang, B., Haas, Z.: Predictive distance-based mobility management for PCS networks. In: Proceedings of the Joint Conference of the IEEE Computer and Communications Societies, INFOCOM (March 1999)
8. Hong, X., Gerla, M., Pei, G., Chiang, C.: A group mobility model for ad hoc wireless networks. In: Proceedings of the ACM International Workshop on Modeling and Simulation of Wireless and Mobile Systems (MSWiM) (August 1999)
9. Reynolds, C.: Flocks, Herd, and Schools: A Distributed Behavioral Model. Computer Graphics 21(4), 25–34 (1987); SIGGRAPH 1987
10. Groenevelt, R., Nain, P.: Message delay in MANETs. In: Proc. of ACM SIGMETRICS, Banff, Canada, pp. 412-423 (June 6, 2005); see also Groenevelt, R.: Stochastic Models for Mobile Ad Hoc Networks. PhD Thesis, University of Nice-Sophia Antipolis (April 2005)
11. Altman, E., Basar, T., De Pellegrini, F.: Optimal monotone forwarding policies in delay tolerant mobile ad-hoc networks. In: Proc. of ACM/ICST Inter-Perf. ACM, Athens (October 24, 2008)
12. Greenwood, M.: The natural duration of cancer. In: Reports on Public Health and Medical Subjects, vol. 33, pp. 1–26. Her Majestys Stationery Office, London
13. Spyropoulos, T., Psounis, K., Raghavendra, C.S.: Spray and wait: an efficient routing scheme for intermittently connected mobile networks. In: Proceedings of the 2005 ACM SIGCOMM Workshop on Delay-tolerant Networking, WDTN 2005, pp. 252–259 (2005)
14. Roy, R.R.: Handbook of Mobile Ad Hoc Networks for Mobility Models. Springer, Heidelberg
15. Small, T., Haas, Z.J.: Resource and Performance Tradeoffs in Delay-Tolerant Wireless Networks. In: SIGCOMM 2005 Workshops, Philadelphia, PA, USA, August 22-26 (2005)
16. Agarwal, R., Banerjee, A., Gauthier, V., Becker, M., Yeo, C.K., Lee, B.S.: Achieving Small-World Properties using Bio-Inspired Techniques in Wireless Networks. The Computer Journal 55(8), 909–931 (2012)
17. http://www-sop.inria.fr/members/Manjesh_Kumar.Hanawal/sushma/mobility.html
18. http://www-sop.inria.fr/members/Manjesh_Kumar.Hanawal/sushma/Draft.pdf

The Age of Information in Gossip Networks

Jori Selen[1], Yoni Nazarathy[2,3], Lachlan L.H. Andrew[3], and Hai L. Vu[3]

[1] Department of Mechanical Engineering, Eindhoven University of Technology,
Eindhoven, Netherlands
j.selen@student.tue.nl
[2] School of Mathematics and Physics, The University of Queensland,
Brisbane, Australia
y.nazarathy@uq.edu.au
[3] Faculty of ICT, Swinburne University of Technology,
Melbourne, Australia
{landrew,hvu}@swin.edu.au

Abstract. We introduce models of gossip based communication networks in which each node is simultaneously a sensor, a relay and a user of information. We model the status of ages of information between nodes as a discrete time Markov chain. In this setting a gossip transmission policy is a decision made at each node regarding what type of information to relay at any given time (if any). When transmission policies are based on random decisions, we are able to analyze the age of information in certain illustrative structured examples either by means of an explicit analysis, an algorithm or asymptotic approximations. Our key contribution is presenting this class of models.

Keywords: Gossip Networks, Discrete Time Markov Chains, Approximations, Minima of Random Variables.

1 Introduction

We consider gossip networks in which the nodes wish to maintain an updated situation awareness view of the information sensed by all other nodes in the network. Using the *gossip* paradigm [8,18], this is done by having nodes transmit both their own sensed information and information that they have received from others. Thus nodes act as sensors, relays and receivers. Bandwidth is limited and communication channels are imperfect, thus the decision of what and when to transmit may often greatly affect performance. A natural application for gossip networks is intelligent transport systems (ITS) in which vehicles wirelessly share information relating to traffic congestion, road conditions and route alternatives, in order to improve safety and reduce congestion [9,20]. In this setting, gossiping is a suitable way to overcome the frequent changes in network topology.

The decision at each node of whether to transmit and what to transmit, are typically taken so as to minimize some measure of cost. Natural measures include the *ages of information* between the various node pairs, where the age of information at node i of information sensed at node j is defined as the difference

A. Dudin and K. De Turck (Eds.): ASMTA 2013, LNCS 7984, pp. 364–379, 2013.

between the current time and the time-stamp found on the most recent sensor measurement from j received (perhaps through relays) at i.

Our aim is to introduce simple Markovian age of information models together with preliminary performance analysis results. Such models may influence network planning, protocol design and synthesis of efficient control methods. For the specific examples in this paper, it is easy to generate efficient deterministic transmission policies, but the analysis we carry out here is a first step toward studying more complex networks in which randomized policies are beneficial.

A fundamental question in the design of gossip networks is the following: *In order to help the greater good, how should a node balance relaying with transmitting its own information?* This paper sets the tone for treatment of this question by means of performance analysis and optimal policy design. For the specific case of ring networks, we give an answer based on asymptotics.

There has been much work focusing on either information aggregation [6,17,19] or the age of information in gossip networks [2,3,7,10]. The former dealt with the problem of computing aggregates based on some functions, such as sum, average or quantile of a set of data distributed over the nodes of a gossip network, and studied the performance of protocols in terms of convergence and the optimization of neighbour selection (i.e. strategy). The latter looked at the age of information via either analyzing the evolution of processes that gossip one message or content [3,10] or characterizing the distribution of latency (i.e. age) over the network of many nodes [2,7].

In particular, both models in [2,7] are based on a mean field analysis with the networks size tending to infinity. The model in [7] yields a set of partial differential equations that uniquely describe a system that allowed opportunistic content updates as in our work but without interference or a lossy wireless channel. The model in [2], on the other hand, is based on a discrete-time Markov chain which could possibly be extended to account for a lossy channel but without a content update. Finally, [11] considers a lossy channel, and uses model checking and Monte Carlo simulation to investigate the performance of a probabilistic broadcast gossip protocol.

Asymptotic results for a problem related to the age of information have been studied under the name of first-passage percolation [14]. Results in that field typically consider a single piece of information spreading on an infinite two dimensional lattice, and consider properties such as the shape of the region which has obtained information by a given time [22], or the variance of the time until the information reaches a given location [4]. Much less work has considered irregular networks, although there has been some study of the Dirichlet triangulation of a two-dimensional Poisson process [24], geometric graph networks $\mathbb{R}^d$ [12], Erdosh-Renyi networks [15] and scale-free networks [5].

Our models and flavour of results are different in that we propose a simpler Markovian framework that can provide explicit formulae for the stationary distribution of the age of information in some specific cases. Using this framework the mean age at each node is also obtained for arbitrary tree networks, while the same is achieved via asymptotic analysis for ring networks. A further distinctive feature is that our models are suited to real-time data that is continuously up-dated. This differs from models where one big file is being transferred, or sensor network models where the key aim is to conserve energy, as in [13].

The rest of this paper is organized as follows. Section 2 introduces the age of information models. These are specialized to linear, tree and ring networks in Sect. 3, where we also present some basic results for the mean and variance of the age of information and motivate the understanding of rings. Section 4 presents some non-trivial explicit and algorithmic solutions for specific structured exam-ples. Section 5 presents asymptotic approximations for structured ring networks with a simple policy where we also answer the question of the balance between relaying and transmitting one's own information.

2   Age of Information Modeling

We consider networks of a finite number of nodes, in which sensing, transmission and reception occurs at discrete (slotted) time instances. The age of information process, $\{A_{i,j}(n),\ n = 0, 1, 2, \ldots\}$ is such that $A_{i,j}(n)$ is the age of the informa-tion that node i has about node j at time n. Thus for example if $A_{1,3}(n) = 15$, we know that at time n, node 1's most updated view regarding the sensed infor-mation at node 3 is from time $n - 15$.

We denote the sequence of information transmissions indicators by $\{I_{i,j}(n)\}$, where $I_{i,j}(n) = 1$ if and only if at time n node i has broadcast its information regarding node j, otherwise $I_{i,j}(n) = 0$. Note that $I_{i,i}(n)$ indicates if a node broadcasts its own sensed information.

We assume some sort of channel model in which the received packets at a given time n at every node depend on the transmitted packets in the whole network at time n and some other possible random effects that are independent for different n, yet follow the same probabilistic law. This may describe essentially any form of time-independent communication channel without memory. At time n the resulting receptions of packets are a random function of $I_{i,j}(n)$ for all i, j and are denoted by $R_{i,j}(n)$ where $R_{i,j}(n) = 1$ if and only if j received a packet sent by node i (containing any form of sensor information, original or relayed). Using $\wedge$ to denote the minimum, the dynamics of the age process are

$$A_{i,j}(n+1) = \begin{cases} \left(A_{i,j}(n)\ \wedge\ \bigwedge_{\{k:R_{k,i}(n)I_{k,j}(n)=1\}} A_{k,j}(n)\right) + 1, i \neq j, \\ 0, \hspace{5.5cm} i = j. \end{cases} \tag{1}$$

As (1) illustrates, age increases by 1 at each time slot, unless "fresh information" is received. Each node i is only interested in the "freshest" information about j and therefore compares the minimum age of information that was received (on all receptions k) with the current age of information stored in node i. The channel plays a role here in determining how $I(n)$ is mapped to $R(n)$: $I(n)$ determines all transmissions made on the network and this in turn (perhaps taking interference into account) determines all receptions.

Randomness enters (1) through both the channel and possibly through the transmission decisions $I(n)$ in case they are random. In this paper we shall take $\{I(n)\}$ to be a (multi-dimensional) i.i.d. sequence. We refer to this as having *Bernoulli policies*, i.e., the decision of what to transmit at any time instant is based on the time-invariant probability distribution of $I(n)$. In this case it is clear that (1) together with some initial distribution, defines a discrete time Markov chain.

For a network of N nodes where each node is assumed to have a sensor, the state space of the Markov chain is $\mathbb{Z}_+^{N^2-N}$. Transitions on this space are either of the form (a) incrementing a coordinate by 1 (no new reception) or (b) shifting a coordinate to equal the value of another coordinate plus 1 (new reception of fresh information). Showing that the Markov chain has a single irreducible countably infinite class (nicely represented as a subset of $\mathbb{Z}_+^{N^2-N}$), is non-periodic and is positive recurrent, is straight-forward under quite general assumptions on the channels and the transmission policy. We shall skip these details as they are non-instructive. (Positive recurrence can be established by means of a linear Lyapunov function.)

Finding explicit performance measures, most importantly finding the stationary distribution, marginals of the stationary distribution or their mean, poses a much greater challenge. In the remainder of the paper we focus on introductory special structured examples on which the behaviour can by analyzed.

3   Structured Models

In order to get some insight into the behaviour of age of information models of the form (1), we look at some structured examples. To do so, we assume that the channel is represented by a directed graph, indicating which nodes can directly communicate. The graph determines the possible paths in which information may flow from sensor to user. The minimal attainable age of information, $A_{i,j}(n)$, is then the shortest path on the graph from j to i. In case there is no such path, the $A_{i,j}(\cdot)$ component of the Markov chain is ignored.

Linear and Tree Networks

As a first structured example, consider a *directed linear network* with infinitely many nodes. See Fig. 1a. In this situation we assume the channel is such that information from node k can be directly transmitted only to node $k+1$. While channel interference may be taken into account, the model is insightful enough even in the case of perfect channel conditions. The choice that each node faces at any time instant is what information to transmit: its own or that of some node to the left of it. A Bernoulli policy is then determined by a probability distribution, $\{p_i, \ i = 0, 1, 2, \ldots\}$ such that each node k transmits or relays information about node $k-i$ with probability p_i.

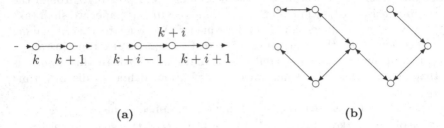

Fig. 1. (a) A directed linear network. (b) A tree.

For this class of networks, finding the marginal distribution of age is a simple task. We assume stationarity and thus suppress the dependence on the time n. Denote by $A_{k+i,k}$ the age of information at some arbitrary node $k+i$ with respect to the information from node k. Then, for infinitely long networks, the random variables $A_{k+i,k}$ have the same distribution for every k, thus for shorthand we write A_i. Now the time it takes information to propagate from node k to node $k+i$ is distributed as the sum of i independent geometric random variables (each with support $\{1,2,\dots\}$) having parameters $p_0, p_1, \dots, p_{i-1}$. Hence we have,

$$\mathbb{E}[A_i] = \sum_{j=0}^{i-1} \frac{1}{p_j}, \qquad \text{Var}(A_i) = \sum_{j=0}^{i-1} \frac{1-p_j}{p_j^2}.$$

A similar line of argumentation can be applied to infinite or finite trees as in Fig. 1b. Since there is only one path[1] that information can take between any two nodes we again have that the set $\{k : R_{k,i}(n)I_{k,j}(n) = 1\}$ appearing in (1) contains at most one element. Thus the distribution of the age of information can be represented as a sum of independent geometric random variables (whose parameters depend generally both on the Bernoulli policy and on possible channel interference, in a straight-forward way). Further details are in [21].

Ring Networks

For modeling of situations in which information may travel on more than one route, a natural first step is to consider ring networks as in Fig. 2. For brevity we consider networks with an even number of nodes, say $2M$ and assume ideal channels (a channel in which every transmitted packet is received). Each node transmits packets of information to its two closest neighbours. Assuming rotational symmetry, it is sufficient to study the distribution of the age of information with respect to a single source, say node 1. The age of information at node i is then given by $A_{i,1}$, $i = 1, \dots, 2M$, for shorthand we write A_i.

Let us introduce a global coordinate variable θ, defined for $i = 1, \dots, 2M$, by

$$\theta := \frac{i-1-M}{M} \in \{-M/M, (-M+1)/M, \dots, 0, \dots, (M-1)/M\}.$$

---

[1] Throughout, we ignore redundant receptions in which a node receives information it has already relayed.

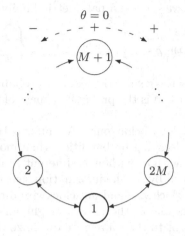

Fig. 2. Ring network of $2M$ nodes with node 1 the source of information

It is now convenient to use the value $Z_\theta := A_{M(\theta+1)+1}$. Fig. 2 illustrates both the node numbering i and the coordinate variable θ.

At every time slot, each node decides which sensor information it should relay (its own sensor information is also an option). Using Bernoulli policies, node i transmits information it knows about node j with a probability depending on the "angle" of j relative to i, namely θ. Denote this probability $q(\theta)$.

Our aim is to study the marginal distribution Z_θ. For each θ, Z_θ is the minimum of the age of information coming from the clockwise and anticlockwise directions. Information flowing back to the source is redundant, so these are equivalent to the age of information processes in a ring network with clockwise or anticlockwise *directed* transmission, respectively. Using the same reasoning as in the linear network, the age of information in one direction is distributed as a sum of independent geometric random variables in a ring network with directed transmission. We denote the directed age of information in the clockwise direction by $X_\theta^{(+)}$ and in the anticlockwise direction by $X_\theta^{(-)}$. To exemplify, $X_\theta^{(+)}$ is the age of information of the node corresponding to θ with respect to the source, node 1, when ignoring information coming from the anticlockwise direction.

Using the same reasoning as in the directed linear network, we note that $X_\theta^{(+)}$ is a sum of independent geometric random variables, with

$$\mathbb{E}[X_\theta^{(+)}] = \sum_{d=-M}^{\theta M-1} \frac{1}{q(d/M)}, \qquad \mathrm{Var}(X_\theta^{(+)}) = \sum_{d=-M}^{\theta M-1} \frac{1-q(d/M)}{q(d/M)^2}.$$

In the anticlockwise directed transmission $(X_\theta^{(-)})$ and with $q(\cdot)$ symmetric with respect to the distance from the source, the mean and variance are expressed in the same way except for the interchange of θ by $(1-\theta)$ in the summation.

As a Bernoulli policy, we suggest a parametric family of distributions:

$$q(\theta) := \begin{cases} \beta, & \theta = -1, \\ C\alpha^{|M(|\theta|-1)|}, & \theta > -1, \end{cases} \quad \text{where} \quad C := \begin{cases} \frac{1-\beta}{2M-1}, & \alpha = 1, \\ \frac{(1-\beta)(1-\alpha)}{2\alpha - \alpha^M(\alpha+1)}, & \alpha < 1, \end{cases}$$

where $\alpha \in (0, 1]$ describes the geometric decay in probability when moving away from the source and $\beta \in (0, 1)$ is the probability mass of the source transmitting its own information.

This family allows various behaviours: A uniform transmission probability ($\alpha = 1$) or alternatively decaying probabilities when moving further away from the source ($\alpha < 1$), both with or without a different probability of transmitting at the source as determined by β. The information sent by the source is usefully transmitted in both the clockwise and anticlockwise direction, whereas relayed information only benefits one of the relay's neighbours. This suggests that β should give a higher weight to the source; we optimize β in Sect. 5.

4   Explicit and Algorithmic Solutions

Finding the stationary distributions, their marginals or the means of our models is in general not straightforward. Nevertheless in this section we report some successful results. In doing so we illustrate a recurring pattern in these types of models: Using marginal distributions to find joint distributions.

The most basic model is a sensor node transmitting to a receiver, where there is a chance of $\lambda \in (0, 1)$ for successful reception. In this case the age of information at the receiver follows a specific GI/M/1 type Markov chain (c.f. [1], Sect. XI.3) in which transitions increment the state by one with probability $(1 - \lambda)$ or reset the state to 0 with probability λ. As with all GI/M/1 (scalar) Markov chains, the stationary distribution is geometric, in this case with parameter $(1 - \lambda)$ and support $\{0, 1, \ldots\}$. We shift the support to $\{1, 2, \ldots\}$ to accommodate the minimal possible age, 1. In general the value of λ may be influenced by both the channel properties and the transmission policy. For example we may have $\lambda = pq$ where p is the chance of receiving a packet conditional on it being transmitted and q is the chance of transmitting.

This GI/M/1 type stationary distribution can be used as a "building block" for finding the (multi-dimensional) stationary distributions of more complicated models. We illustrate this now for two types of models: star networks and a small ring, further examples and details are in [21].

Star Networks

Consider star networks as illustrated in Fig. 3. Transmissions take place from the source node to N receivers. We denote a version of the steady state age of information at node i with respect to the source node by A_i. What is then the joint distribution of $A_1, \ldots, A_N$?

To illustrate the solution approach we first consider the case of $N = 2$. Let $\lambda_\emptyset$, $\lambda_{\{1\}}$, $\lambda_{\{2\}}$ and $\lambda_{\{1,2\}}$ denote the respective probabilities that reception occurs at

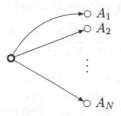

Fig. 3. A star topology

neither node, node A_1 only, node A_2 only, or both nodes. The transition diagram of this model is shown in Figure 4.

Let $\pi_{i,j} := \mathbb{P}(A_1 = i, A_2 = j)$. Then,

$$\pi_{1,1} = \lambda_{\{1,2\}} \sum_{i=1}^{\infty} \sum_{j=1}^{\infty} \pi_{i,j} , \tag{2a}$$

$$\pi_{i,1} = \lambda_{\{2\}} \sum_{j=1}^{\infty} \pi_{(i-1),j}, \quad i \geq 2 , \tag{2b}$$

$$\pi_{1,j} = \lambda_{\{1\}} \sum_{i=1}^{\infty} \pi_{i,(j-1)}, \quad j \geq 2 , \tag{2c}$$

$$\pi_{i,j} = \lambda_{\emptyset} \cdot \pi_{(i-1),(j-1)}, \quad i,j \geq 2 . \tag{2d}$$

Now a key observation is that in (2b)-(2c) there is summation over one entire coordinate, therefore we can use the marginal distributions. For nodes $k = 1, 2$, let $c_k = 1 - (\lambda_{\{3-k\}} + \lambda_{\{1,2\}})$ denote the probability of no reception on the other node, $3 - k$. Then as in the GI/M/1 type Markov chain described above, the marginal distributions are given by

$$\pi_i^{(A_k)} := \mathbb{P}(A_k = i) = (1 - c_{3-k})c_{3-k}^{i-1}, \quad k = 1, 2, \ i = 1, 2, \ldots .$$

Since $\pi_{1,1} = \lambda_{\{1,2\}}$, these marginal distributions imply the equilibrium equations simplify to

$$\pi_{1,1} = \lambda_{\{1,2\}} ,$$
$$\pi_{i,1} = \lambda_{\{2\}} \pi_{i-1}^{(A_1)} = \lambda_{\{2\}} c_2^{i-2}(1 - c_2), \quad i \geq 2 ,$$
$$\pi_{1,j} = \lambda_{\{1\}} \pi_{j-1}^{(A_2)} = \lambda_{\{1\}} c_1^{j-2}(1 - c_1), \quad j \geq 2 ,$$
$$\pi_{i,j} = \lambda_{\emptyset} \cdot \pi_{(i-1),(j-1)}, \quad i,j \geq 2 .$$

These then yield the stationary distribution

$$\pi_{i,j} = \begin{cases} \lambda_{\emptyset}^{i-1}\lambda_{\{1,2\}}, & i = j , \\ \lambda_{\emptyset}^{j-1}\lambda_{\{2\}}c_2^{i-j-1}(1 - c_2), & i > j , \\ \lambda_{\emptyset}^{i-1}\lambda_{\{1\}}c_1^{j-i-1}(1 - c_1), & i < j . \end{cases}$$

After some straightforward calculations this yields

$$\mathrm{Cov}(A_1, A_2) = \frac{\lambda_\emptyset \lambda_{\{1,2\}} - \lambda_{\{1\}} \lambda_{\{2\}}}{(\lambda_{\{1\}} + \lambda_{\{1,2\}})(\lambda_{\{2\}} + \lambda_{\{1,2\}})(1 - \lambda_\emptyset)}.$$

It can now be verified that if there is no interaction between the communication links, i.e., $(\lambda_{\{1\}} + \lambda_{\{1,2\}})(\lambda_{\{2\}} + \lambda_{\{1,2\}}) = \lambda_{\{1,2\}}$, then there is a product form solution to $\pi_{i,j}$ and the covariance is 0. Otherwise, the covariance is non-zero and can be used to get LMMSE (linear minimum mean squared error estimates) of A_k based on A_{3-k}. We do not discuss this further here.

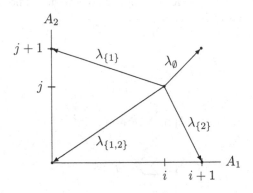

Fig. 4. Markov chain transition diagram for a star with $N = 2$

The idea of a network with $N = 2$ can now be generalized to arbitrary N by recursive usage of marginal distributions of some lower order. We describe this in brief and present an algorithm for calculating the *exact* stationary distribution.

The Bernoulli policies and i.i.d. channel conditions imply that we may essentially have λ_B for any receiving subset of nodes B. We let D denote some proper subset of the set of all nodes in order to consider smaller networks in the recursive specification that follows. When computing the joint distribution of a subset D of all the nodes, we need to know $\lambda_{B;D}$, which are the probabilities of successful reception on the nodes in the set B, given that we only consider receptions on the nodes in the subset D of the full network. That is, we ignore transmissions to the nodes in the complement of D.

Let $D = \{i_1, i_2, \ldots, i_{|D|}\}$. To find

$$\pi^{(D)}_{a_{i_1}, \ldots, a_{i_{|D|}}} := \mathbb{P}(A_{i_1} = a_{i_1}, \ldots, A_{i_{|D|}} = a_{i_{|D|}}),$$

let $j_1, j_2, \ldots, j_{|D|}$ be a permutation of D such that $0 \le a_{j_1} \le a_{j_2} \le \ldots \le a_{j_{|D|}}$, and let $\tilde{\pi}^{(D)}_{a_{j_1}, a_{j_2}, \ldots, a_{j_{|D|}}} = \pi^{(D)}_{a_{i_1}, a_{i_2}, \ldots, a_{i_{|D|}}}$. Then $\tilde{\pi}$ can be calculated recursively by Alg. 1.

Algorithm 1. Joint distribution of $|D|$ nodes in a network of N nodes

1: **if** $a_{j_1} = \ldots = a_{j_{|D|}}$ **then**
2:     $m = |D|$
3: **else**
4:     $m = \min\{k : a_{j_k} < a_{j_{k+1}}\}$
5: **end if**
6: **if** $a_{j_{|D|}} = 1$ **then**
7:     $\tilde{\pi}^{(D)}_{1,\ldots,a_{j_{|D|}}} = \lambda_{D;D}$
8: **else**
9:     Let $F = D \setminus \{j_1, \ldots, j_m\}$ and its complement is F^c.
10:

$$
\tilde{\pi}^{(D)}_{a_{j_1}, a_{j_2}, \ldots, a_{j_{|D|}}} = \begin{cases} \lambda^{a_{j_1}-1}_{\emptyset;D} \cdot \tilde{\pi}^{(D)}_{1,\ldots,1,a_{j_{m+1}}-a_{j_1}+1,\ldots,a_{j_{|D|}}-a_{j_1}+1}, & a_{j_1} > 1 \\ \lambda_{F^c;C} \cdot \tilde{\pi}^{(F)}_{a_{j_{m+1}}-1,\ldots,a_{j_{|D|}}-1}, & a_{j_1} = 1 \end{cases}
$$

11: **end if**

Similarly to the $N = 2$ case, the probability at the point $(1, 1, \ldots, 1)$ equals the probability of reception on all nodes in the network; see line 7. For any state in the interior of the state space, i.e., the smallest age satisfies $a_{i_1} > 1$, we can compute the probability by moving back along the diagonal to the nearest (hyper) plane or edge and using the knowledge that there is a geometric decay along the diagonals. This is shown in the first part of the equation on line 10. If we are already on a (hyper) plane or edge, we can use the marginal distribution of all the other nodes that have a strictly positive age (see second part of line 10). See report [21] for an illustration in the case of $N = 3$.

A Small Ring

Let us now consider the smallest non-trivial ring: a ring with $2M = 4$ nodes. We exploit now the fact that $A_2 = A_4$ and denote them by $A_{2;4}$. We then allow this "virtual" node $A_{2;4}$ to transmit over two separate channels to the node diametrically opposite the source. Thus we can represent the steady state age of information by $A_3 \in \{2, 3, \ldots\}$ and $A_{2;4} \in \{1, 2, \ldots, A_3\}$. See Fig. 5.

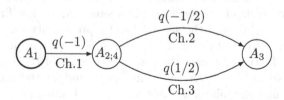

Fig. 5. Alternate representation of a network with four nodes where node 1 is the source. There are three channels.

Denote $\pi_{i,j} := \mathbb{P}(A_{2;4} = i, A_3 = j)$. Observe that the marginal distribution of $A_{2;4}$ is geometric with parameter $q(-1) = \beta$ and support $\{1, 2, \ldots\}$, as we found earlier in this section. Let $\pi_i^{(A_{2;4})} = \mathbb{P}(A_{2;4} = i)$. Similarly to the star, this value appears in the balance equations of $\pi_{i,j}$. These equations are based on reception probabilities on subsets of the channels denoted by λ_B, where B is a set of channels. For example $\lambda_{\{2,3\}} = (1 - q(-1))q(-1/2)q(1/2)$.

$$\pi_{1,2} = \left(\lambda_{\{1,2\}} + \lambda_{\{1,3\}} + \lambda_{\{1,2,3\}}\right) \pi_1^{(A_{2;4})}, \tag{3a}$$

$$\pi_{2,2} = \left(\lambda_{\{2\}} + \lambda_{\{3\}} + \lambda_{\{2,3\}}\right) \pi_1^{(A_{2;4})}, \tag{3b}$$

$$\pi_{1,j} = \lambda_{\{1\}} \sum_{i=1}^{j-1} \pi_{i,j-1} + \left(\lambda_{\{1,2\}} + \lambda_{\{1,3\}} + \lambda_{\{1,2,3\}}\right) \pi_{j-1}^{(A_{2;4})}, \quad j \geq 3, \tag{3c}$$

$$\pi_{i,j} = \left(\lambda_{\{2\}} + \lambda_{\{3\}} + \lambda_{\{2,3\}}\right) \pi_{i-1}^{(A_{2;4})} + \lambda_\emptyset \pi_{i-1,j-1}, \quad i = j, \ i \geq 3, \tag{3d}$$

$$\pi_{i,j} = \lambda_\emptyset \pi_{i-1,j-1}, \quad i \neq j, \ i \geq 2, \ j \geq 3. \tag{3e}$$

Algorithm 2 uses these equations to calculate $\{\pi_{i,j}, \ i, j \leq K\}$ *exactly* for any K.

Algorithm 2. Joint distribution of $A_{2;4}$ and A_3

Use the known $\pi_{i,j}^{(A_{2;4})}$ and set $\pi_{1,2}$ and $\pi_{2,2}$.
for $j = 3 : K$ **do**   ▷ Iterate until a bounding box of size K is reached.
    for $i = 1 : j$ **do**
        if $i = 1$ **then**
            Calculate $\pi_{i,j}$ using (3c), based on $\pi_{j-1}^{(A_{2;4})}$ and $\pi_{i,j-1}$.
        else if $i = j$ **then**
            Calculate $\pi_{i,j}$ using (3d), based on $\pi_{i-1}^{(A_{2;4})}$ and $\pi_{i-1,j-1}$.
        else
            Calculate $\pi_{i,j}$ using (3e), based on $\pi_{i-1,j-1}$.
        end if
    end for
end for

We now present a numerical example. We compute the joint distribution of $A_{2;4}$ and A_3 for two sets of transmission parameters (α, β). The first is a uniform policy ($\alpha = 1, \beta = \frac{1}{2M}$), and the second has its probability mass concentrated around the source, ($\alpha = 0.1, \beta = \frac{2}{2M}$). Figure 6 shows the joint distribution found by Alg. 2. In the first case the probability mass of the joint distribution is more widely spread out over the state space and in the latter it is more concentrated around the minimum ages, i.e. $a_{2;4} = 1$ and $a_3 = 2$.

A similar approach to that of Alg. 2 can essentially be applied to networks with more nodes. However, this is analytically demanding and becomes impractical. Even for a network with 5 nodes there are 5 possible transmissions and thus 2^5 different subsets of B in λ_B and many more equations in comparison to (3a)-(3e). We therefore shift our attention to approximations.

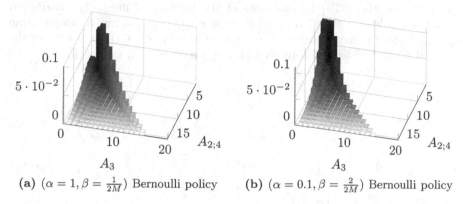

(a) $(\alpha = 1, \beta = \frac{1}{2M})$ Bernoulli policy   (b) $(\alpha = 0.1, \beta = \frac{2}{2M})$ Bernoulli policy

Fig. 6. Joint distribution of $A_{2;4}$ and A_3 for two different policies, using Alg. 2

5 Asymptotic Approximations in Rings

In this section we present an asymptotic evaluation of ring networks with $\alpha = 1$ and some β. We revisit the question presented in the introduction: How should a node balance transmitting its own information against relaying? Alternatively, what is a good value for β? Our analysis is based on the representation

$$Z_\theta = \tilde{X}_{\frac{M-1}{M}} + \left(\tilde{X}_\theta^{(+)} \wedge \tilde{X}_\theta^{(-)}\right),$$

where $\tilde{X}_{\frac{M-1}{M}}$ represents the age at the neighbouring nodes of the source (both have the same age) and $\tilde{X}_\theta^{(+)}, \tilde{X}_\theta^{(-)}$ represent the age difference between the node in question and the neighbouring nodes of the source, in the clockwise and anticlockwise directions respectively, based on directed transmission.

For large M, we are guided by the central limit theorem to use a Gaussian approximation for each of the directed transmissions, i.e., the Negative Binomially distributed $\tilde{X}_\theta^{(+)}$ and $\tilde{X}_\theta^{(-)}$ are approximately normally distributed with

$$\mu_\theta^{(+)} := \mathbb{E}[\tilde{X}_\theta^{(+)}] = ((1+\theta)M-1)C^{-1}, \qquad \mu_\theta^{(-)} := \mathbb{E}[\tilde{X}_\theta^{(-)}] = ((1-\theta)M-1)C^{-1},$$

and standard deviations, $\sigma_\theta^{(+)} := \sqrt{\mu_\theta^{(+)}(C^{-1} - 1)}$, $\sigma_\theta^{(-)} := \sqrt{\mu_\theta^{(-)}(C^{-1} - 1)}$ respectively. We now have

$$Z_\theta \approx^d \hat{Z}_\theta := \tilde{X}_{\frac{M-1}{M}} + \left(\mathcal{N}_{\tilde{X}_\theta^{(+)}} \wedge \mathcal{N}_{\tilde{X}_\theta^{(-)}}\right),$$

where $\approx^d$ informally denotes approximate equality in distribution and the $\mathcal{N}$ variables are independent versions of normal random variables with the aforementioned parameters. In this paper we do not formalize this as a weak-convergence result (as $M \to \infty$). This technical hurdle is left for future research.

In [16] (see also [23]) the moments of the minima of normally distributed random variables are given. We exploit these results here to find approximating expressions for the mean and variance of Z_θ. Denoting the CDF and PDF of the standard normal distribution by $\Phi(\cdot)$ and $\phi(\cdot)$ respectively, we obtain

$$\mathbb{E}[\hat{Z}_\theta] = \frac{1}{\beta} + \mu_\theta^{(+)}\Phi\left(\frac{-\bar{\mu}}{\Delta}\right) + \mu_\theta^{(-)}\Phi\left(\frac{\bar{\mu}}{\Delta}\right) - \Delta\phi\left(\frac{-\bar{\mu}}{\Delta}\right), \tag{4}$$

$$\mathbb{E}[\hat{Z}_\theta^2] = \frac{-1}{\beta} + \omega^{(+)}\Phi\left(\frac{-\bar{\mu}}{\Delta}\right) + \omega^{(-)}\Phi\left(\frac{\bar{\mu}}{\Delta}\right) - \bar{\mu}\Delta\phi\left(\frac{-\bar{\mu}}{\Delta}\right), \tag{5}$$

where $\bar{\mu} := \mu_\theta^{(+)} - \mu_\theta^{(-)}$, $\Delta := \sqrt{\left(\sigma_\theta^{(+)}\right)^2 + \left(\sigma_\theta^{(-)}\right)^2}$, $\omega^{(+)} := \left(\mu_\theta^{(+)}\right)^2 + \left(\sigma_\theta^{(+)}\right)^2$ and $\omega^{(-)} := \left(\mu_\theta^{(-)}\right)^2 + \left(\sigma_\theta^{(-)}\right)^2$.

We conjecture that for any $\theta \in [-1, 1]$, $\lim_{M\to\infty} \mathbb{E}[Z_{[\theta]}]/\mathbb{E}[\hat{Z}_\theta] = 1$ and the same for the variance (here $[\theta]$ denotes the nearest value that θ may attain over the grid). We have verified this conjecture numerically by means of extensive Monte-Carlo simulations. As an illustration we compare the curves for $2M = 30$ nodes in Fig. 7.

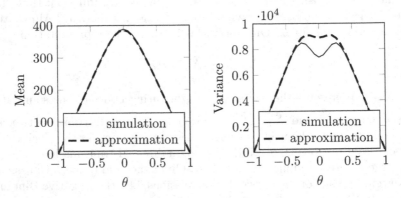

Fig. 7. Comparison of the mean and variance obtained from simulation with the approximation values $\mathbb{E}[\hat{Z}_\theta]$ and $\mathrm{Var}[\hat{Z}_\theta]$. This is for $2M = 30$.

To observe convergence of the variance to the "volcano curve", we also ran long simulations for $2M = 100$. The results are not displayed here. Our numerical experiments have also clearly indicated that $\mathbb{E}[Z_\theta] \geq \mathbb{E}[\hat{Z}_\theta]$ and $\mathrm{Var}(Z_\theta) \leq \mathrm{Var}(\hat{Z}_\theta)$. We leave proofs of these inequalities for future work.

The Asymptotically Best β

We now optimize the transmission policy with respect to minimizing the mean age of information at the node corresponding to θ. For $\theta = 0$ we can simplify (4). We know that the mean and variance of $\tilde{X}_0^{(+)}$ and $\tilde{X}_0^{(-)}$ are equal and we omit the superscripts $(+), (-)$. This leads to the following expression:

$$\mathbb{E}[\hat{Z}_0] = \frac{1}{\beta} + \mu_0 - \frac{\sigma_0}{\sqrt{\pi}}.$$

The mean μ_0 is $O(M^2)$, whereas σ_0 is $O(\sqrt{M^3})$ and both scale with $1/(1 - \beta)$. Hence the mean dominates the standard deviation for large M, and thus

$$\mathbb{E}[\hat{Z}_0] \approx \frac{1}{\beta} + \frac{2M^2}{1 - \beta}$$

for large M. This is minimized for $\hat{\beta}^* = \sqrt{2}/(2M)$ for large M and $\theta = 0$. For $\theta \neq 0$, again $\sigma_\theta = o(\mu_\theta)$ whence $|\bar{\mu}/\Delta| \to \infty$ and $\hat{\beta}^* = \sqrt{\frac{2}{1-|\theta|}}/(2M)$. We numerically compute the β^* values for various fractions θ using (4), summarized in Fig. 8. Observe the converge of β^* to $\hat{\beta}^*$ as $M \to \infty$.

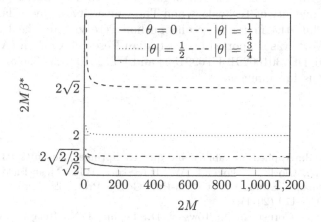

Fig. 8. Optimal values for β for various angles θ for increasing network size $2M$

We have thus found that in large rings, if the overall goal is to maintain timely information at the farthest node from each sensor, then each node should transmit its own information about 40% more frequently than the information of some other node. This finding is of course based on a series of assumptions and stylized modeling assumptions. Yet it can perhaps serve as a rule of thumb for gossip networks, if there is no better alternative.

6 Conclusion

We have developed a simple Markovian framework for the design and analysis of a gossip protocol in tree or ring topology networks where information is probabilistically updated by each individual node and sent over bandwidth-limited lossy wireless channels. Using the framework, we presented some basic results for the mean, the variance, and distribution of age in the studied star networks and a small ring, including non-trivial explicit and algorithmic solutions to obtain the age of information distribution. For large ring networks, we obtained asymptotic forms for the age of information using normal approximations and explored the optimal strategy to forward information in such a network.

Future work will deal with the extension of the framework beyond the linear (or tree) and ring network topologies where new asymptotic approximations could be developed. For most applications, including ITS, information about nearby nodes is more important than information about distant nodes. Hence it will be useful to consider both optimizing weighted means of ages of information, and also coarsely aggregating information as it emanates further from its source. We also wish to settle some of the conjectures laid out regarding the ring asymptotics.

Acknowledgment. This research began while the first author was visiting Swinburne University of Technology and The University of Queensland and also took place while the first and second author were visiting the University of Haifa. This work was supported by Australian Research Council (ARC) grants DP130100156, DE130100291, FT0991594 and FT120100723. The authors thank Ivo Adan for useful comments.

References

1. Asmussen, S.: Applied probability and queues. Springer, Berlin (2003)
2. Bakhshi, R., Cloth, L., Fokkink, W., Haverkort, B.R.: Mean-Field Framework for Performance Evaluation of Push-Pull Gossip Protocols. Performance Evaluation 68, 157–179 (2011)
3. Banerjee, N., Corner, M.D., Towsley, D., Levine, B.N.: Relays, Base Stations, and Meshes: Enhancing Mobile Networks with Infrastructure. In: MobiCom 2008, pp. 81–91. ACM, New York (2008)
4. Benjamini, I., Kalai, G., Schramm, O.: First Passage Percolation has Sublinear Distance Variance. The Annals of Probability 31(4), 1970–1978 (2003)
5. Bhamidi, S., van der Hofstad, R., Hooghiemstra, G.: Extreme Value Theory, Poisson-Dirichlet Distributions, and First Passage Percolation on Random Networks. Advances in Applied Probability 42(3), 706–738 (2010)
6. Boyd, S., Ghosh, A., Prabhakar, B., Shah, D.: Randomized Gossip Algorithms. IEEE/ACM Trans. Networking (TON), 2508–2530 (2006)
7. Chaintreau, A., Le Boudec, J.-Y., Ristanovic, N.: The Age of Gossip: Spatial Mean Field Regime. In: SIGMETRICS 2009, pp. 109–120. ACM, New York (2009)
8. Demers, A., Greene, D., Hauser, C., Irish, W., Larson, J., Shenker, S., Sturgis, H., Swinhart, D., Terry, D.: Epidemic Algorithms for Replicated Database Maintenance. ACM SIGOPS Operating Systems Review 22, 8–32 (1988)
9. Dimitrakopoulos, G.A., Demestichas, P.: Intelligent Transportation Systems. IEEE Vehicular Technology Magazine 5(1), 77–84 (2010)
10. Eugster, P., Guerraoui, R., Handurukande, S., Kermarrec, A.-M., Kouznetsov, P.: Lightweight Probabilistic Broadcast. ACM Trans. Computer Systems, 341–374 (2003)

11. Fehnker, A., Gao, P.: Formal Verification and Simulation for Performance Analysis for Probabilistic Broadcast Protocols. In: Kunz, T., Ravi, S.S. (eds.) ADHOC-NOW 2006. LNCS, vol. 4104, pp. 128–141. Springer, Heidelberg (2006)
12. Friedrich, T., Sauerwald, T., Stauffer, A.: Diameter and Broadcast Time of Random Geometric Graphs in Arbitrary Dimensions. In: Asano, T., Nakano, S.-i., Okamoto, Y., Watanabe, O. (eds.) ISAAC 2011. LNCS, vol. 7074, pp. 190–199. Springer, Heidelberg (2011)
13. Goseling, J., Boucherie, R.J., van Ommeren, J.-K.: Energy-Delay Tradeoff in Wireless Network Coding. Accepted for Publication in Performance Evaluation.
14. Hammersley, J.M., Welsh, D.J.A.: First-Passage Percolation, Subadditive Processes, Stochastic Networks, and Generalized Renewal Theory. Bernoulli 1713 Bayes 1763 Laplace 1813, 61–110 (1965)
15. van der Hofstad, R., Hooghiemstra, G., van Mieghem, P.: First-Passage Percolation on the Random Graph. Probability in the Engineering and Informational Sciences 15(2), 225–237 (2001)
16. Hunter, J.T.: Renewal Theory in Two Dimensions: Asymptotic Results. Advances in Applied Probability 6, 546–562 (1974)
17. Jelasity, M., Montresor, A., Babaoglu, O.: Gossip-based aggregation in large dynamic networks. ACM Trans. Computer Systems 23(3), 219–252 (2005)
18. Karp, R., Schindelhauer, C., Shenker, S., Vocking, B.: Randomized Rumor Spreading. In: FOCS 41, pp. 565–574. IEEE Computer Society (2000)
19. Kempe, D., Dobra, A., Gehrke, J.: Gossip-Based Computation of Aggregate Information. In: FOCS, pp. 482–491. IEEE Computer Society (2003)
20. Papadimitratos, P., La Fortelle, A., Evenssen, K., Brignolo, R., Cosenza, S.: Vehicular Communication Systems: Enabling Technologies, Applications, and Future Outlook on Intelligent Transportation. IEEE Commun. Mag. 47(11), 84–95 (2009)
21. Selen, J.: The Age of Information of Situation Awareness Networks. Master project report, Eindhoven University of Technology (2012)
22. Smythe, R.T., Wierman, J.C.: First-Passage Percolation on the Square Lattice. Advances in Applied Probability 9(1), 38–54 (1977)
23. Tong, Y.L.: The Multivariate Normal Distribution. Springer-Verlag Inc., New York (1990)
24. Vahidi-Asl, M.Q., Wierman, J.C.: First-Passage Percolation on the Voronoi Tessellation and Delaunay Triangulation. Random Graphs 87, 341–359 (1990)

Efficient Steady State Analysis
of Multimodal Markov Chains

David Spieler and Verena Wolf

Saarland University, Germany

Abstract. We consider the problem of computing the steady state distribution of Markov chains describing cellular processes. Our main contribution is a numerical algorithm that approximates the steady state distribution in the presence of multiple modes. This method tackles two problems that occur during the analysis of systems with multimodal distributions: stiffness preventing fast convergence of iterative methods and largeness of the state space leading to excessive memory requirements and prohibiting direct solutions. We use drift arguments to locate the relevant parts of the state space, that is, parts containing $1 - \epsilon$ of the steady state probability. In order to separate the widely varying time scales of the model we apply stochastic complementation techniques. The memory requirements of our method are low because we exploit accurate approximations based on inexact matrix vector multiplications. We test the performance of our method on two challenging examples from biology.

1 Introduction

The randomness inherent to the interactions between molecular species plays an important role when studying cellular processes. In particular if species are present in low copy numbers, stochastic models have proven to be an adequate description [1, 6, 11, 13, 18]. For instance there may be only few copies of a gene or of some important regulatory molecule. The most widely used approach to model such systems is based on the chemical master equation (CME). This equation describes the dynamics of a Markov jump process whose state variables encode the number of molecules of the different species, i.e., if the model includes n different chemical species then a state is a vector $(x_1, \ldots, x_n)$ and the transitions of the process correspond to chemical reactions between different molecules.

Approaches for the analysis of such Markov chains are either based on the generation of trajectories of the process (often referred to as Gillspie simulation [9]) or they aim at a numerical solution of the CME [5, 14, 17]. Numerical approaches mostly focus on transient solutions, however, for many gene regulatory networks the equilibrium distribution reveals important qualitative and quantitative properties such as multimodality and oscillatory behavior [7, 8]. Thus, the computation of the steady state distribution is of great importance. Unfortunately, in most cases an efficient solution is hindered by both the very large or even infinite state space of the underlying Markov chain and its stiff dynamics. In particular if the distribution is multimodal the main part of the steady state probability mass concentrates on several attracting regions. Typically, the locations of such attractors are not known a-priori, but even if their locations are known, the

A. Dudin and K. De Turck (Eds.): ASMTA 2013, LNCS 7984, pp. 380–395, 2013.

problem is that they are usually separated by low-probability regions which renders the application of standard numerical and simulative methods very difficult. The reason is that switching from one attractor to the other is a rare event and a large number of trajectories is needed to estimate the amount of probability mass of one attractor relative to the mass of the other attractors. Similarly, standard numerical approaches, which are often iterative methods, show slow convergence properties.

For some systems it might be possible to obtain a truncation of the state space that contains all attractors and is of numerically tractable size [3]. In this case, however, the problems related to stiffness are still present. For example, the case study that was considered by Milias-Argeitis and Lygeros [16] could not be solved using the tool PRISM which provides a large number of different numerical methods to obtain the steady state distribution of a Markov chain. Even after a reasonable amount of one million iterations and a relative precision of 1e–15 none of these methods converged. However, since for this example the state space is small (19600 states) it can be solved exactly with direct methods implemented in Matlab using a sparse matrix representation. A three-dimensional extension of that system, the tri-stable toggle switch [12], containing around 1.5 million states[1] can not be handled by Matlab within a maximum computation time of one day.

The purpose of this paper is to develop and analyze a numerical method that approximates the steady state distribution of a large (or infinite) ergodic Markov chain with one or several attractors. We propose an algorithm that overcomes both the problems of largeness of the state space and the problems related to stiffness. Our method provides answers to the following questions:

a) How can we find the attractors of the system, i.e., finite regions containing most of the steady state probability?
b) If there is more than one attractor, how can we efficiently compute the steady state probability of each attracting region?

Note that once the total probabilities of the attractors are known, the probabilities of the individual states can be computed with stochastic complementation [15].

To solve problem a) we use the geometric bounding techniques developed in [3] to first find a region containing all attractors, i.e, a single region containing at least 1-ϵ of the total steady state probability mass. In order to separate multiple attractor regions in b), we numerically explore the most likely paths leading from one attractor to the other. Our method does *not* explore the whole state space but allows to direct the exploration behavior. Once we have found the attractor regions, we compute their relative strength using a novel approach combining stochastic complementation and inexact matrix-vector multiplication.

Concerning related work we remark that steady state analysis has been widely considered in the area of queuing and computer systems [19] but to the best of our knowledge, specific solutions for systems with multi-modal distributions as they occur in biological applications have not been developed. A dynamic state space exploration has been proposed by de Souza e Silva and Ochoa [4] for performance and dependability

---

[1] 1.5 million is the number of states after an appropriate truncation of the infinite state space which is necessary because Matlab (and also PRISM) can only analyze finite systems.

analysis based on finite Markov models. Their method can only be used for the approximation of the steady state probability conditioned on a certain subset. The same applies to [3] where also only a single attractor is considered. In contrast, here we propose a technique to overcome the numerical problems, that arise due to the stiffness inherent to multimodal systems, by considering the most likely paths leading from one attractor to any other. In particular, here we construct an attractor Markov chain to obtain the steady state probability of each attractor relative to the others. The closest work to ours is that of Milias-Argeitis and Lygeros [16] where the authors consider the same problem but combine stochastic complementation with a simulative approach. Here, we focus on numerical solutions and show that they are superior to the simulative approach in [16] both in terms of computation time and accuracy. In addition, we describe how to locate the attractors of the system which was not considered by Milias-Argeitis and Lygeros.

The outline of the paper is as follows: We start with a short description of necessary mathematical background in Section 2. In Section 3 we concentrate on locating attractors and present a novel approximation algorithm for computing the steady state probabilities in the case of several attractors. In addition we analyze the approximation error of the method. An evaluation of the performance of the algorithm for two case studies follows in Section 4 and we finally conclude the paper in Section 5.

2   Mathematical Background

We consider an ergodic continuous-time Markov chain $\{X(t) \mid t \in \mathbb{R}_{\geq 0}\}$ with infinitesimal generator matrix $\mathbf{Q}$ that has entries q_{ij}. In the sequel, we always use bold face for matrices and vectors and normal face for the corresponding scalar entries. Note that for $i \neq j$ the entry q_{ij} of $\mathbf{Q}$ is nonnegative. It specifies the transition rate between states i and j and the negative sum of the off-diagonal entries $-\sum_{j \neq i} q_{ij}$ is equal to the diagonal entry q_{ii}. The matrix $\mathbf{Q}$ may be implicitly given by some high level specification language such as guarded commands, Petri nets or process algebra terms. Often the matrix $\mathbf{Q}$ is very large or even infinite but rather sparse and a high level specification provides a concise description assuming that $\mathbf{Q}$ has a finite number of non-zero entries per row. Such a matrix can be specified by a (finite) number of guarded commands of the form guard |- rate -> update; where guard is a Boolean predicate over the variables (with domain $\mathbb{N}$ and declared in the variables header) which determines whether corresponding transition is enabled, rate is an expression evaluating to a nonnegative transition rate, and update is a rule describing the change of the system variables if the transition is performed. Syntactically, update is a list of statements, each assigning to a variable an expression over variables. For simplicity, we assume that all variable assignments are reachable. Otherwise, the set of reachable states must be explicitly defined. Note that transition rates q_{ij} are calculated by summing up the rates of all commands whose guard is true in i and whose update corresponds to a transition from i to j.

Example 1. We consider the *genetic toggle switch* in E. coli which is a synthetic gene regulatory network that consists of two genes each of which encodes a protein that represses the other [8]. The corresponding commands are stated below.

```
variables x,y;
true          |- c1*(1+y^n1)^(-1) -> x:=x+1;
(x>0)         |- a1*x             -> x:=x-1;
true          |- c2*(1+x^n2)^(-1) -> y:=y+1;
(y>0)         |- a2*y             -> y:=y-1;
```

Variables x and y represent the number of proteins encoded by the two genes. The protein production of one gene is repressed by the other gene because the rate $\lambda_1(x,y) = c_1 \cdot (1 + y^{n_1})^{-1}$ (rate $\lambda_2(x,y) = c_2 \cdot (1 + x^{n_2})^{-1}$) decreases as y (as x) increases where the parameters n_1 and n_2 regulate the repression strength. For $n_1, n_2 \geq 1$, two attractors emerge that represent two possible modes of the network, i.e, x is high and y is low, and vice versa. Parameters c_1, c_2 and a_1, a_2 influence the positions of the two attractors and the amount of probability that each attractor contains. The row in $\mathbf{Q}$ that corresponds to a state $i = (x,y)$ with $x > 0$, $y > 0$ has five non-zero entries (all four guards evaluate to true and the diagonal entry), e.g. for $j = (x+1,y)$ the entry is $q_{ij} = c_1 \cdot (1 + y^{n_1})^{-1}$.

The steady state distribution π of an ergodic Markov chain is given by the solution of the equation system

$$\pi \cdot \mathbf{Q} = 0, \tag{1}$$

which has a unique solution up to the normalization condition $\pi \cdot \mathbf{e} = 1$ where $\mathbf{e}$ is a column vector with all entries one. In the sequel we also use the *embedded matrix* $\mathbf{E}$, which is defined as

$$\mathbf{E} = \mathbf{I} + |\mathbf{diag}(\mathbf{Q})|^{-1} \cdot \mathbf{Q},$$

where $\mathbf{diag}(\mathbf{Q})$ denotes the diagonal matrix obtained from $\mathbf{Q}$ by setting all off-diagonal entries to 0 and the matrix operator $|\cdot|$ denotes the matrix obtained by taking the component-wise absolute values. The stochastic matrix $\mathbf{E}$ contains the time-independent transition probabilities of X, that is, entry e_{ij} is the probability to enter state j at the next jump of X when the current state is i. For some finite set $\mathcal{A}$ we define the *stochastic complement* of $\mathbf{Q}$ as

$$\overline{\mathbf{Q}} = \mathbf{Q}_{\mathcal{AA}} + \mathbf{Q}_{\mathcal{A\bar{A}}} \cdot \sum_{i=0}^{\infty} \left(\mathbf{E}_{\bar{A}\bar{A}}\right)^i \cdot \mathbf{E}_{\bar{A}\mathcal{A}}, \tag{2}$$

where $\bar{\mathcal{A}}$ is the set of all states not in $\mathcal{A}$ and for $\mathbf{Q}$ we use the block structure

$$\mathbf{Q} = \begin{bmatrix} \mathbf{Q}_{\mathcal{AA}} & \mathbf{Q}_{\mathcal{A\bar{A}}} \\ \mathbf{Q}_{\bar{A}\mathcal{A}} & \mathbf{Q}_{\bar{A}\bar{A}} \end{bmatrix}.$$

The matrices $\mathbf{E}_{\mathcal{A\bar{A}}}$ and $\mathbf{E}_{\bar{A}\bar{A}}$ denote the corresponding blocks of $\mathbf{E}$. Intuitively, $\overline{\mathbf{Q}}$ represents the behavior of the Markov chain when the transitions outside of $\mathcal{A}$ are immediate, that is, it considers the rates of transitions within $\mathcal{A}$ (matrix $\mathbf{Q}_{\mathcal{AA}}$) or for leaving $\mathcal{A}$ (matrix $\mathbf{Q}_{\mathcal{A\bar{A}}}$) but time does not evolve until the set $\mathcal{A}$ is re-entered after i transitions in $\bar{\mathcal{A}}$ (matrix $\left(\mathbf{E}_{\bar{A}\bar{A}}\right)^i \cdot \mathbf{E}_{\bar{A}\mathcal{A}}$). The matrix $\overline{\mathbf{Q}}$ is the infinitesimal generator[2] of an ergodic

---

[2] A real-valued quadratic matrix is an infinitesimal generator matrix if all off-diagonal entries are non-negative and if the diagonal entries are equal to the negative sum of the off-diagonal entries.

Markov chain and the corresponding steady state distribution $\overline{\pi}$ coincides with that of $\mathbf{Q}$ conditioned on the set $\mathcal{A}$ [3, 15], that is, for a state $i \in \mathcal{A}$ we have

$$\overline{\pi}_i = \pi_i / \sum_{j \in \mathcal{A}} \pi_j. \tag{3}$$

3   Approximation Algorithm

In this section we propose an approximation algorithm for the efficient computation of the steady state distribution. Since we consider Markov chains with a population structure, certain regions of the state space are attracting regions, i.e., regions corresponding to local maxima of the steady state distribution. Consider for instance the steady state distribution of the genetic toggle switch in Fig. 1b. The regions where $x \in [0, 2]$ and $y \in [5, 60]$ and where $x \in [14, 105]$ and $y \in [0, 2]$ for $(x, y) \in \mathbb{N}^2$ are attracting regions. In Section 3.3 we devise an algorithm to locate attractors and determine their extend.

First, we assume that the locations of all attractors of the Markov chain are known and concentrate on the problem of approximating the probability of each attractor, i.e, on the amount p_i of probability that attractor A_i has where

$$p_i = \sum_{\ell \in A_i} \pi_\ell, \quad \sum_{i=1}^n p_i > 1 - \epsilon$$

and n is the number of attractors. Later, we will explain how a finite set containing more than $1 - \epsilon$ of the probability mass can be determined for a given $\epsilon > 0$. Moreover, we will propose a heuristic to locate the attractors of the system.

3.1   Approximation of Attractor Probabilities

We partition the state space of the Markov chain into attractors $\mathcal{A}_i$ with $i \in \{1, \ldots, n\}$ and the set $\mathcal{I}$ which contains all states that are not part of an attractor, i.e., if S is the set of all states then $S = \mathcal{A}_1 \,\dot{\cup}\, \ldots \,\dot{\cup}\, \mathcal{A}_n \,\dot{\cup}\, \mathcal{I}$. Now, we proceed in two steps:

(a) We approximate for each $\mathcal{A}_i$ the conditional steady state distribution $\overline{\pi}^{(i)}$ (cf. Eq. 3).
(b) We approximate the steady state distribution $\hat{\pi}$ of the aggregated system, i.e., the Markov chain that consists of the macro states $\mathcal{A}_1, \ldots, \mathcal{A}_n$, which yields an approximation of $p_1, \ldots, p_n$.

Note that in step (b) we use the fact that ϵ is small and that the probability mass in $\mathcal{I}$ is negligible. Given approximations of p_i and $\overline{\pi}^{(i)}$, the steady state probability of a state $j \in \mathcal{A}_i$ can be approximated as well since

$$\pi_j = p_i \cdot \overline{\pi}_j^{(i)}.$$

For step (a) we remark that the exact conditional distributions $\overline{\pi}^{(i)}$ are the (unique) solutions of

$$\overline{\pi}^{(i)} \cdot \overline{\mathbf{Q}}^{(i)} = 0 \text{ and } \overline{\pi}^{(i)} \cdot \mathbf{e} = 1$$

where $\overline{\mathbf{Q}}^{(i)}$ is the stochastic complement of $\mathbf{Q}$ w.r.t. the set $\mathcal{A}_i$. The problem is that in order to obtain the exact entries of $\overline{\mathbf{Q}}^{(i)}$, the whole state space has to be considered (see Eq. 2) which is infeasible for large Markov chains. Therefore, the behavior outside of $\mathcal{A}_i$ has to be approximated in an appropriate way. By abuse of notation, for a vector $\mathbf{v}$ let $\mathbf{diag}(\mathbf{v})$ denote the diagonal matrix $\mathbf{X}$ with entries $x_{ij} = v_i$ if $i = j$ and 0 otherwise. An approximation $\tilde{\pi}^{(i)} \approx \overline{\pi}^{(i)}$ is given by

$$\tilde{\pi}^{(i)} \cdot (\mathbf{Q}_{ii} - \mathbf{diag}(\mathbf{Q}_{ii} \cdot \mathbf{e})) = \mathbf{0}, \quad \tilde{\pi}^{(i)} \cdot \mathbf{e} = 1 \tag{4}$$

where $\mathbf{Q}_{ii}$ is the block matrix induced by the partitioning w.r.t. the attractor sets, i.e.,

$$\mathbf{Q} = \begin{bmatrix} \mathbf{Q}_{11} & \cdots & \mathbf{Q}_{1n} & \mathbf{Q}_{1\mathcal{I}} \\ \vdots & \ddots & \vdots & \vdots \\ \mathbf{Q}_{n1} & \cdots & \mathbf{Q}_{nn} & \mathbf{Q}_{n\mathcal{I}} \\ \mathbf{Q}_{\mathcal{I}1} & \cdots & \mathbf{Q}_{\mathcal{I}n} & \mathbf{Q}_{\mathcal{I}\mathcal{I}} \end{bmatrix}.$$

Note that the block $\mathbf{Q}_{i\ell}$ contains all rates of transitions from $\mathcal{A}_i$ to $\mathcal{A}_\ell$ and that $\mathbf{Q}_{ii} - \mathbf{diag}(\mathbf{Q}_{ii} \cdot \mathbf{e})$ is a generator matrix. Thus, in the corresponding Markov chain the transitions for leaving the set $\mathcal{A}_i$ are redirected to the state from which the set is left.

We remark that for solving Eq. 4, direct methods such as Gaussian elimination can be applied since typically the attractors are of tractable size. Moreover, we can improve the approximation of the local steady state distribution by considering an enlarged set $\mathcal{A}_i \cup \mathcal{B}_i$ where $\mathcal{B}_i$ contains the states reachable from $\mathcal{A}_i$ within a certain number of transitions, e.g., in the experimental results in Section 4 we chose $\mathcal{B}_i$ as the set of states reachable from $\mathcal{A}_i$ within 15 steps. Let $\tilde{\pi}^{(i)}$ denote the obtained approximate distribution on $\mathcal{A}_i \cup \mathcal{B}_i$. In this case we re-scale $\tilde{\pi}^{(i)}$ such that it yields a distribution on $\mathcal{A}_i$, i.e., for $j \in \mathcal{A}_i$ we use the approximation

$$\overline{\pi}_j^{(i)} \approx \tilde{\pi}_j^{(i)} \cdot (\textstyle\sum_{k \in \mathcal{A}_i} \tilde{\pi}_k^{(i)})^{-1}.$$

For step (b) we consider the embedded matrix w.r.t. the partitioning induced by the attractor regions, that is,

$$\mathbf{E} = \begin{bmatrix} \mathbf{E}_{11} & \cdots & \mathbf{E}_{1n} & \mathbf{E}_{1\mathcal{I}} \\ \vdots & \ddots & \vdots & \vdots \\ \mathbf{E}_{n1} & \cdots & \mathbf{E}_{nn} & \mathbf{E}_{n\mathcal{I}} \\ \mathbf{E}_{\mathcal{I}1} & \cdots & \mathbf{E}_{\mathcal{I}n} & \mathbf{E}_{\mathcal{I}\mathcal{I}} \end{bmatrix} = \mathbf{I} + |\mathbf{diag}(\mathbf{Q})|^{-1} \cdot \mathbf{Q}.$$

Extending the stochastic complement to several sets, we define

$$\overline{\mathbf{Q}} = \begin{bmatrix} \mathbf{Q}_{11} & \cdots & \mathbf{Q}_{1n} \\ \vdots & \ddots & \vdots \\ \mathbf{Q}_{n1} & \cdots & \mathbf{Q}_{nn} \end{bmatrix} + \begin{bmatrix} \mathbf{Q}_{1\mathcal{I}} \\ \vdots \\ \mathbf{Q}_{n\mathcal{I}} \end{bmatrix} \cdot (\textstyle\sum_{k=0}^{\infty} \mathbf{E}_{\mathcal{I}\mathcal{I}}^k) \cdot \begin{bmatrix} \mathbf{E}_{\mathcal{I}1} & \cdots & \mathbf{E}_{\mathcal{I}n} \end{bmatrix}$$

$$= \begin{bmatrix} \mathbf{Q}_{11} + \mathbf{Q}_{1\mathcal{I}} \cdot (\sum_{k=0}^{\infty} \mathbf{E}_{\mathcal{I}\mathcal{I}}^k) \cdot \mathbf{E}_{\mathcal{I}1} & \cdots & \mathbf{Q}_{1n} + \mathbf{Q}_{1\mathcal{I}} \cdot (\sum_{k=0}^{\infty} \mathbf{E}_{\mathcal{I}\mathcal{I}}^k) \cdot \mathbf{E}_{\mathcal{I}n} \\ \vdots & \ddots & \vdots \\ \mathbf{Q}_{n1} + \mathbf{Q}_{n\mathcal{I}} \cdot (\sum_{k=0}^{\infty} \mathbf{E}_{\mathcal{I}\mathcal{I}}^k) \cdot \mathbf{E}_{\mathcal{I}1} & \cdots & \mathbf{Q}_{nn} + \mathbf{Q}_{n\mathcal{I}} \cdot (\sum_{k=0}^{\infty} \mathbf{E}_{\mathcal{I}\mathcal{I}}^k) \cdot \mathbf{E}_{\mathcal{I}n} \end{bmatrix}$$

$$= [\overline{\mathbf{Q}}_{ij}]_{i,j \in \{1,\ldots,n\}}.$$

The block $\overline{\mathbf{Q}}_{ii}$ of the complement w.r.t. all attractor sets should not be confused with the generator matrix $\overline{\mathbf{Q}}^{(i)}$ which is the stochastic complement of $\mathbf{Q}$ w.r.t. attractor set $\mathcal{A}_i$ only. Now we are able to define the *Attractor Markov Chain* with generator matrix

$$\widehat{\mathbf{Q}} = \begin{bmatrix} \widehat{q}_{11} & \cdots & \widehat{q}_{1n} \\ \vdots & \ddots & \vdots \\ \widehat{q}_{n1} & \cdots & \widehat{q}_{nn} \end{bmatrix} = \left[\widehat{q}_{ij} \right]_{i,j \in \{1,\dots,n\}},$$

where

$$\widehat{q}_{ij} = \overline{\pi}^{(i)} \, \overline{\mathbf{Q}}_{ij} \, \mathbf{e} = \overline{\pi}^{(i)} \cdot \left[\mathbf{Q}_{i\mathcal{I}} \cdot \left(\sum_{k=0}^{\infty} \mathbf{E}_{\mathcal{I}\mathcal{I}}^k \right) \cdot \mathbf{E}_{\mathcal{I}j} + \mathbf{Q}_{ij} \right] \cdot \mathbf{e}. \tag{5}$$

Rate $\widehat{q}_{ij}$ is the aggregated rate of the transition from attractor $\mathcal{A}_i$ to $\mathcal{A}_j$ ignoring the time spent outside the attractor regions. To approximate $\widehat{\mathbf{Q}}$ we first replace $\overline{\pi}_j^{(i)}$ by its approximation from step a). Then, the idea is to use inexact matrix-vector multiplications for computing $\widehat{\mathbf{Q}}$ row-wise as illustrated in Algorithm 1 and 2.

Thus, the output of Algorithm 1 is an approximation of the i-th row of $\widehat{\mathbf{Q}}$. Based on a small significance threshold δ, we consider only the relevant summands, similar to approximation algorithms for the transient distribution of a Markov chain [5, 10]. In other words, we use inexact matrix-vector multiplications while propagating the vector $\overline{\pi}^{(i)} \cdot \mathbf{Q}_{i\mathcal{I}}$ through the infinite sum. We first define W as the set of all states reachable from $\mathcal{A}_i$ within one step (see line 1 of Algorithm 1) given that the system is in steady state, where the total rate mass at which $\mathcal{A}_i$ is left is defined as θ (line 2). In the while-loop in line 4 we consider states that are reachable from $\mathcal{A}_i$ within K transitions where K is the number of times the while loop is executed. This corresponds to a truncation of the infinite sum in Eq. (5) after K summands. In Algorithm 2 we expand the current set W by adding all states reachable within a single transition. We remark that in line 4 of Algorithm 2 we only consider transitions of states in $\mathcal{I}$. This ensures that the while loop will terminate after a finite number of steps because eventually the set $\mathcal{I}$ is left and the sets $\mathcal{A}_i$ are absorbing. In line 5 we compute the corresponding partial sum according to Eq. 5. Note that $q_{rs} \cdot q_{rr}^{-1}$ corresponds to the entries in $\mathbf{E}_{\mathcal{I}\mathcal{I}}$ and $\mathbf{E}_{\mathcal{I}j}$. Thus we propagate

---

Algorithm 1. attract($\mathbf{Q}, \{\mathcal{A}_j\}_{1 \leq j \leq n}, i, \delta$)

---

1 $W \leftarrow \{r \mid [\overline{\pi}^{(i)} \mathbf{Q}_{i\mathcal{I}}]_r > 0\}$;

2 $\theta \leftarrow \overline{\pi}^{(i)} \cdot \mathbf{Q}_{i\mathcal{I}} \cdot \mathbf{e}$;

3 $\mathbf{m} \leftarrow$ new HashMap(S,[0, 1]) such that $m_r = [\overline{\pi}^{(i)} \cdot \mathbf{Q}_{i\mathcal{I}}]_r$ for all $r \in W$;

4 **while** $W \cap \mathcal{I} \neq \emptyset$ **do**

5 $\qquad [W', \mathbf{m}] \leftarrow$ advance($\mathbf{Q}, \mathcal{A}_1 \cup \dots \cup \mathcal{A}_n, W, \mathbf{m}$);

6 $\qquad W \leftarrow \{r \mid r \in W' \wedge m_r \geq \delta \cdot \theta\}$;

7 $\qquad$ delete all entries with keys $r \in W \setminus W'$ in $\mathbf{m}$;

8 **end**

9 **for** $j \in \{1, \dots, n\}$ **do**

10 $\qquad \widehat{q}_{ij} \leftarrow \sum_{r \in \mathcal{A}_j} m_r + \overline{\pi}^{(i)} \cdot \mathbf{Q}_{ij} \cdot \mathbf{e}$;

11 **end**

12 **return** $[\widehat{q}_{i1} \ \cdots \ \widehat{q}_{in}]$

Algorithm 2. advance($\mathbf{Q}, \mathcal{A}, W, \mathbf{v}$)

1   $W' \leftarrow \emptyset$;
2   $\mathbf{v}' \leftarrow$ new HashMap($S, [0, 1]$);
3   **for** $r \in W \setminus \mathcal{A}$ **do**
4   **for** s *with* $q_{rs} > 0$ **do**
5       $v'_s \leftarrow v'_s + q_{rs} \cdot q_{rr}^{-1} \cdot v_r$;
6       $W' \leftarrow W' \cup \{s\}$;
7   **end**
8     $v'_r \leftarrow v'_r - q_{rr} \cdot q_{rr}^{-1} \cdot v_r$;
9   **end**
10   **return** $[W \cup W', \mathbf{v}']$;

---

the vector $\overline{\pi}^{(i)} \cdot \mathbf{Q}_{i\mathcal{I}}$ step-by-step in the embedded Markov chain represented by $\mathbf{E}_{\mathcal{I}\mathcal{I}}$ and $\mathbf{E}_{\mathcal{I}j}$ which yields an approximation of the vector $\overline{\pi}^{(i)} \cdot \mathbf{Q}_{i\mathcal{I}} \cdot (\sum_{\ell=0}^{k-1} \mathbf{E}_{\mathcal{I}\mathcal{I}}^{\ell})$ where k is the number of times the while-loop has been executed so far. In line 6 and 7 we find all relevant states, i.e, states that contribute at least δ of the total rate mass θ. The states that contribute less than δ are dropped. Note that this ensures that among the states reachable from $\mathcal{A}_i$ within K transitions, we consider only those that are relevant. The for-loop in lines 9 and 10 finally adds the last summand $\overline{\pi}_i \cdot \mathbf{Q}_{ij} \cdot \mathbf{e}$ (cf. Eq. 5).

The approximation proposed in Algorithm 1 and 2 corresponds to a reachability analysis during which the main part of the rate mass drifts back towards the attractor. The remaining part drifts to the other attractors where the truncation based on δ ensures that the most likely paths between the attractors are explored (they correspond to the subset $W \cap \mathcal{I}$), i.e., we do not explore regions holding an negligible portion of the total rate mass θ. More precisely, regions that are rarely visited when switching between attractors will not be included in W.

Once an approximation of $\widehat{\mathbf{Q}}$ is obtained we can compute the steady state distribution $\hat{\pi}$ of the attractor Markov chain using, for instance, a direct solution method such as Gaussian elimination. Note that the attractor Markov chain has only n states. Thus, even if the transition rates in $\widehat{\mathbf{Q}}$ differ widely an efficient solution is possible.

3.2 Approximation Error Analysis

We first consider the error made in step (a), i.e., when approximating $\overline{\pi}^{(i)}$. For each state j in $\mathcal{A}_j$ that has a transition to a state outside of $\mathcal{A}_i$ we construct a Markov chain with generator $\tilde{\mathbf{Q}}^{(j)}$ where $\tilde{\mathbf{Q}}^{(j)}$ is equal to $\mathbf{Q}_{ii}$ except that from the j-th column we subtract the vector $\mathbf{Q}_{ii} \cdot \mathbf{e}$, i.e., we redirect all transitions leaving $\mathcal{A}_i$ to state j. Solving the steady state distributions for all $\tilde{\mathbf{Q}}^{(j)}$ and taking the component-wise minima and maxima gives bounds on $\overline{\pi}_i$ [2]. It is of course computationally quite expensive to bound the error in this way except if the attractor is very small. The approximation obtained in step (a) yields a solution which lies within these bounds since any redirection induces a distribution that lies in the polyhedral hull spanned by the steady state distributions of $\tilde{\mathbf{Q}}^{(j)}$ [2]. The bounds correspond to the worst case and in practice our approximation is much better than the bounds obtained in the way that we describe above.

For step (b), we note that some parts of the total rate mass θ is dropped because of the threshold δ in line 6. Consequently, when we approximate the rows of $\widehat{\mathbf{Q}}$ using

Algorithm 1, i.e.,

$$\widehat{\mathbf{Q}} \approx \widetilde{\mathbf{Q}} = \begin{bmatrix} \text{attract}(\mathbf{Q}, \{\mathcal{A}_i\}_{1 \le i \le n}, 1, K, \delta) \\ \vdots \\ \text{attract}(\mathbf{Q}, \{\mathcal{A}_i\}_{1 \le i \le n}, n, K, \delta) \end{bmatrix},$$

the resulting matrix $\widetilde{\mathbf{Q}}$ will not be an infinitesimal generator matrix since the row sum will be smaller than zero. The reason is that for the off-diagonal entries we obtain underapproximations as

$$\widetilde{q}_{ij} = \overline{\pi}^{(i)} \cdot \overline{\mathbf{Q}}_{ij} \cdot \mathbf{e} = \overline{\pi}^{(i)} \cdot \mathbf{Q}_{ij} \cdot \mathbf{e} + \underbrace{\overline{\pi}^{(i)} \cdot \mathbf{Q}_{i\mathcal{I}} \cdot \left(\sum_{k=0}^{\infty} \mathbf{E}_{\mathcal{II}}^k \right) \cdot \mathbf{E}_{\mathcal{I}j} \cdot \mathbf{e}}_{\ge \sum_{r \in \mathcal{A}_j} m_r} \ge \widetilde{q}_{ij}.$$

The diagonal entries $\widetilde{q}_{ii}$, however, contain the total rate at which attractor $\mathcal{A}_i$ is left. Given that we have the exact local steady states $\overline{\pi}^{(i)}$, we bound the approximation error made during the computation of $\widehat{\pi}$ (the steady state of $\widehat{\mathbf{Q}}$) as follows. The idea is to consider the slack rate mass vector $\mathbf{s} = -\widetilde{\mathbf{Q}}\mathbf{e}$ whose i-th entry is the difference between the total rate mass at which macro state $\mathcal{A}_i$ is left and the approximated rate mass that accumulated in the attractors (including $\mathcal{A}_i$). In order to transform $\widetilde{\mathbf{Q}}$ into a generator matrix, the amount s_i has to be distributed among the elements in the i-th row (including the diagonal entry since it is possible to enter $\mathcal{I}$ from $\mathcal{A}_i$ and return to $\mathcal{A}_i$). Bounds for the approximation error can be derived by considering all "extreme cases" where vector $\mathbf{s}$ is added to the j-th column. Formally, we define the matrices $\widetilde{\mathbf{Q}}^{(j)}$ with entries

$$\widetilde{q}_{ik}^{(j)} = \begin{cases} \widetilde{q}_{ik} & \text{if } k \ne j, \text{ and} \\ \widetilde{q}_{ik} + s_i & \text{otherwise} \end{cases}$$

for $1 \le j \le n$. Now, if $\widetilde{\pi}^{(j)}$ is the unique steady state of $\widetilde{\mathbf{Q}}^{(j)}$, for the steady state $\widehat{\pi}_i$ of $\widehat{\mathbf{Q}}$ we have [2, 3]

$$\widehat{\pi}_i \in [\min_j \widetilde{\pi}_i^{(j)}, \max_j \widetilde{\pi}_i^{(j)}].$$

Since the number of attractor regions n is small, we can compute the bounds on $\widehat{\pi}_i$ efficiently using direct methods. Multiplication of the above bounds on $\widehat{\pi}_i$ with the conditional steady state probability $\overline{\pi}_j^{(i)}$ yields a bounded approximation of the steady state probability π_j for state $j \in A_i$. Note that the probability of all states in $\mathcal{I}$ is approximated as zero. It is possible to combine the bounds from steps (a) and (b) by executing Algorithm 1 twice where $\overline{\pi}^{(i)}$ is replaced by its upper and lower bound, respectively.

Choosing threshold δ: The accuracy of the approximation depends on the parameter δ. It is possible to dynamically control the amount of rate mass that gets "lost" due to the truncation by repeating a step of the reachability analysis with a lower value for δ. In this case, one has to determine a-priori an upper bound on the rate mass lost per step. Moreover, if δ is chosen too high it might even happen that $\widehat{\mathbf{Q}}$ is not ergodic (for $\delta = 0$ the attractor Markov chain is always ergodic [3, 15]). In such a case the computation has to be repeated using a smaller value for δ. In our experiments we chose $\delta \in [1e - 20, 1e - 8]$ and we always obtained an ergodic $\widehat{Q}$. Moreover, we found that the approximation of $\hat{\pi}$ is accurate even for $\delta = 1e - 8$. In fact, choosing $\delta \in [1e - 20, 1e - 10]$ was also suggested in [5].

3.3 Locating Attracting Regions

Let $\epsilon \in (0, 1)$ be a small probability bound. In order to determine a finite set containing $1 - \epsilon$ of the probability mass, we consider the expected drift defined as $d_i(t) = \frac{d}{dt} E[g(X(t)) \mid X(t) = i]$ w.r.t. a Lyapunov function g that assigns to state i the non-negative Lyapunov value $g(i)$. We assume that g has the property that the set of all states i with $g(i) \leq l$ is finite for any $l \in \mathbb{N}$ and drop the argument t of d since X is time-homogeneous. In this case the drift vector $\mathbf{d}$ that contains the entries d_i does not depend on t and is given by $\mathbf{d} = \mathbf{Q} \cdot \mathbf{g}$ where $\mathbf{g}$ is the column vector of Lyapunov values. Intuitively the entries of $\mathbf{d}$ describe how the current g-value $g(X(t))$ will change on average within the next infinitesimal time. The following theorem provides a way of truncating infinite systems based on drift arguments.

Theorem 1 ([3]). *If the expected drift is finite in all states and non-negative only in finitely many states then the Markov chain is ergodic. Moreover, for a given $\epsilon \in (0, 1)$, if the expected drift is bounded by some constant $c \in \mathbb{R}$ and if the set $C = \{i \mid d_i > c \cdot (1 - \epsilon^{-1})\}$ is finite then $\sum_{i \in C} \pi_i > 1 - \epsilon$.*

It can be shown that for every ergodic Markov chain there exists a Lyapunov function such that the expected drift is bounded and C is finite. We found that for the type of systems that we consider the squared Euclidean distance meets these conditions in nearly all cases.

Example 2. Consider the toggle switch of Example 1. We choose the Lyapunov function $g(x, y) = x^2 + y^2$. It is easy to see that $\{(x, y) \in \mathbb{N}^2 \mid x^2 + y^2 \leq l\}$ is finite for any l. Furthermore, we use the parameter set $c_1 = 60$, $c_2 = 30$, $n_1 = 3$, $n_2 = 2$, and $a_1 = a_2 = 1$ from [16] and the corresponding drift is given by

$$d(x, y) = -2\left(x^2 + y^2\right) + \left(\frac{120}{1+y^3} + 1\right) x + \left(\frac{60}{1+x^2} + 1\right) y + \frac{120}{1+y^3} + \frac{60}{1+x^2}$$

and it is obviously finite for all states. The drift is bounded from above by $c = 1891$. In Figure 1a we illustrate the set $C = \{i \mid d_i > c \cdot (1 - \epsilon^{-1})\}$ for $\epsilon = 0.01$. Note that according to Theorem 1, C contains 99% of the steady state probability.

The theorem above provides a way to truncate a large system in an appropriate way. The problem of locating attractors is, however, not fully solved since the set C contains all attractors of the system but it might also contain low probability regions. The drift d does not provide enough information to locate attractors. Regions where the drift is close to zero are, of course, good candidates for attractors but since d is an expectation, there might be states with $d_i \approx 0$ where the system does only reside for a short while. Similarly, there might be states with high absolute drift and high steady state probability.

A simple way of locating attractors is to generate a long trajectory of the system e.g. by employing Gillespie simulation [9]. Then, attractors are those regions where the process resides most of the time, i.e., regions that have a small probability of exiting.

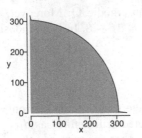
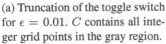
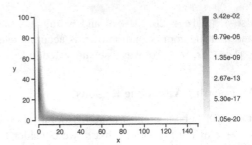

(a) Truncation of the toggle switch for $\epsilon = 0.01$. C contains all integer grid points in the gray region.

(b) steady state distribution of the toggle switch model on a logarithmic scale).

Fig. 1. Geometric bounds and steady state distribution of the toggle switch model

The problem is that once an attractor is found it is unlikely that the system switches to another attractor rendering this strategy inefficient. If many trajectories are generated that start from different initial states, it is more likely that all attractors are found. It is however not clear how to determine the extents of an attractor and when a simulation run can be finished.

We suggest an alternative strategy where a transient analysis based on a dynamically truncated state space is performed from several starting points. The starting points can be chosen at random, based on prior knowledge about the system or based on approximations such as the mean field of the Markov chain. The transient analysis is a modification of the method suggested by Didier et al. [5]. It will follow the main part of the probability mass and converge when an attractor is found. Assume that W is the set of all starting points and that the vector $\mathbf{v}$ assigns equal probability to all states in W and zero to all other states. Then we perform an approximate reachability analysis similar to the analysis in the while-loop in Algorithm 1. We use Algorithm 2 by iteratively calling $[W, \mathbf{v}] \leftarrow$advance$(\mathbf{Q}, \emptyset, W, \mathbf{v})$ to move the probability mass in the embedded Markov chain but after each call we remove all states in W that have probability less than δ. Again $\delta > 0$ is a small threshold which is either fixed during the analysis or chosen in an adaptive way to ensure that the total number of states in W remains small. We iterate until the set W becomes stable, i.e., does not grow or shrink. Note that in order to guarantee that all attractors are found an exploration of the whole set C, which contains $1 - \epsilon$ of the steady state probability mass, is necessary. Our experimental results, however, show that even with the high choice $\delta = 1e - 7$ we can locate all attractor regions of the examples that we considered.

4   Experimental Results

In the following, we apply our approach to two case studies from biology. The first case study was also considered in [16] and since experimental results were reported, we can compare our results to those in [16]. All computations were done on a 2.66 GHz machine with 4 GB of RAM using a single core, i.e., a machine with the same specifications as in [16]. The two models that we consider are structurally similar (both

Table 1. Results of the steady state analysis of the genetic toggle switch case study

Method	δ	$\widehat{\pi}_1$	$\widehat{\pi}_2$	time (seconds)
Exact	–	0.4727196615367557	0.5272803384632443	463
[16]	–	0.4759	0.5241	290
Ours	1e–5	0.5	0.5	60.1 + 0.170 + 0.304
Ours	1e–7	0.47	0.53	60.1 + 0.170 + 0.307
Ours	1e–8	0.473	0.527	60.1 + 0.170 + 0.312
Ours	1e–9	0.4727	0.5273	60.1 + 0.170 + 0.313
Ours	1e–20	0.47272	0.52728	60.1 + 0.170 + 0.367

are genetic switches) but differ in the number of attractors (two and three). Moreover, the attractors of the second model have very different probabilities while for the first model the probability mass distributes nearly equally. We remark that these two models are exemplary for gene regulatory networks with multi-modal steady states.

4.1 Genetic Toggle Switch

In the Table 1 we list the experimental results of the approximation described in the previous sections applied to the genetic toggle switch (see Examples 1 and 2). We first truncated the infinite system as described in Example 2. In order to locate the attractor regions, we determined starting points by considering the mean field of the model. The expected molecule numbers $(x(t), y(t))$ can be approximated as

$$\tfrac{d}{dt}(x,y) = \left[c_1 \cdot (1 + y^{n_1})^{-1} - d_1 \cdot x \quad c_2 \cdot (1 + x^{n_2})^{-1} - d_2 \cdot y \right].$$

We solved $\tfrac{dx}{dt} = 0$ using the HOM4PS2 package (which took less than one second). This yielded the three real valued (rounded) solutions $(0, 30)$, $(60, 0)$, and $(3, 3)$ which lie inside the geometric bounds given by the set C. We performed a truncation-based transient analysis starting in the two points with truncation threshold of $\delta = 1e - 7$. The method converged after 421 steps, i.e., when the set of significant states remained unchanged. This took 60.1 seconds of computation time. The resulting significant sets correspond to our approximation of the attractor regions $\mathcal{A}_1$ and $\mathcal{A}_2$. Note that the probability mass that started in the third point $(3, 3)$ distributed among the sets $\mathcal{A}_1$ and $\mathcal{A}_2$ implying that it corresponds to an unstable fixed point of the mean field. The sets that we derived for $\delta = 1e - 8$ are slightly smaller than the regions

$$\mathcal{A}_1 = \{(x,y) \in \mathbb{N}^2 \mid 0 \le x \le 2 \text{ and } 5 \le y \le 60\}, \text{ and}$$
$$\mathcal{A}_2 = \{(x,y) \in \mathbb{N}^2 \mid 14 \le x \le 105 \text{ and } 0 \le y \le 2\}.$$

defined in [16]. Using a threshold of $\delta = 1e - 8$ for our method resulted in slightly larger sets. In order to allow a comparison with [16], we will use the regions defined above for the approximation of the attractor probabilities.

To obtain a more accurate approximation of $\overline{\pi}^{(1)}$ and $\overline{\pi}^{(2)}$, we used in step (a) of the method enlarged sets $\mathcal{A}_1$ and $\mathcal{A}_2$ by including those states reachable within 15 steps (see Section 3.1). Then, we constructed the attractor Markov chain using Algorithm 1 and 2. Note that the chain has only the two states $\mathcal{A}_1$ and $\mathcal{A}_2$. In the second column in Table 1 we list the final result of our approximation for different values of the truncation threshold δ (see Algorithm 1). Moreover, since the truncated model (containing those states inside the geometric bounds) was of tractable size, we used the direct solution methods implemented in Matlab to compute the "exact solution" circumventing problems related to stiffness. We also list the results given in [16] for comparison where the number of digits corresponds to the computational precision. The number of listed digits corresponds to the accuracy of the approximation. In our method, the computation time consists of the time needed to locate the attractor regions (left), to compute the local steady states (middle) and the time needed to compute the attractor probabilities $\widehat{\pi}_i$ using Algorithm 1 (right). The time needed to compute the geometric bounds and to solve the steady state of the mean-field was negligible (below 0.03 seconds each). The final steady state distribution of this model is depicted in Figure 1b.

Our method is magnitudes faster than the exact solution using Matlab and the simulation based approach in [16]. Note that in contrast to [16], our method allows for an identification of the attractor regions. Also, the precision of our algorithm is much higher than the results presented in [16]. From the table, it can also be seen that the accuracy of the approximation is very good for thresholds $\delta \leq 1e - 8$. The computation time is short even if $\delta = 1e - 20$.

4.2   Tri-Stable Genetic Switch

In order to show the applicability of our approach in the presence of several attractors, we also analyzed a model with three attractors. The *tri-stable toggle switch* [12] consists of three chemical species x, y, and z, where each species represses the production of the other two species. The guarded commands are given below.

```
variables x,y,z;
true   |- c12*(1+y^n12)^(-1)+c13*(1+z^n13)^(-1)  -> x:=x+1;
(x>0)  |- d1*x                                    -> x:=x-1;
true   |- c21*(1+x^n21)^(-1)+c23*(1+z^n23)^(-1)  -> y:=y+1;
(y>0)  |- d2*y                                    -> y:=y-1;
true   |- c31*(1+x^n31)^(-1)+c32*(1+z^n32)^(-1)  -> z:=z+1;
(z>0)  |- d3*z                                    -> z:=z-1;
```

The variables x, y, and z represent the number of proteins encoded by the three genes. The parameters n_i, $i \in \{1, 2, 3\}$ regulate the repression strength among the proteins and c_i and d_i control protein production and degradation. For the analysis, we chose the parameter set $c_{12} = c_{13} = 60.0$, $c_{21} = c_{23} = 30.0$, $c_{31} = c_{32} = 50.0$, $\mathbf{d} = [d_1 \ d_2 \ d_3] = \mathbf{e}^T$ and $n_{12} = n_{13} = n_{21} = n_{23} = n_{31} = n_{32} = 3$. First, we computed geometric bounds for $\epsilon = 0.01$ as depicted in Figure 2a which took 2.641 seconds. Next, we computed the possible steady states of the mean field by solving $\frac{d\mathbf{x}}{dt} = \mathbf{0}$ for

$$\frac{d\mathbf{x}}{dt} = \left[\frac{c_{12}}{1+x_2^{n12}} + \frac{c_{13}}{1+x_3^{n13}} \quad \frac{c_{21}}{1+x_1^{n21}} + \frac{c_{23}}{1+x_3^{n23}} \quad \frac{c_{31}}{1+x_1^{n31}} + \frac{c_{31}}{1+x_2^{n31}} \right] - \mathbf{x} \cdot \mathbf{d}^T,$$

where $\mathbf{x} = (x_1, x_2, x_3)^T$ represents the three protein populations x, y, z using the HOM4PS2 package (which took less than 0.030 seconds). We got the five (rounded)

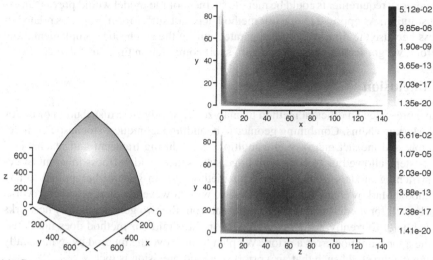

(a) Geometric bounds of the tri-stable toggle switch for $\epsilon = 0.01$.

(b) steady state distribution of the tri-stable toggle switch model on a logarithmic scale.

Fig. 2. Geometric bounds and steady state distribution of the tri-stable toggle switch

solutions points $[60\ 0\ 50]$, $[0\ 30\ 50]$, $[60\ 30\ 0]$, $[2\ 3\ 7]$, and $[8\ 3\ 2]$ located within the geometric bounds. Truncation-based transient analysis with threshold $\delta = 1e - 7$ revealed that only the first three points correspond to attracting regions. The sets stabilized after 493 iterations and the total computation time was 136.4 seconds. We will refer to these three regions as $\mathcal{A}_1$, $\mathcal{A}_2$, and $\mathcal{A}_3$ where

$$\mathcal{A}_1 \approx \{\mathbf{x} \in \mathbb{N}^3 \mid 10 \le x_1 \le 110, 0 \le x_2 \le 3, 10 \le x_3 \le 95\},$$
$$\mathcal{A}_2 \approx \{\mathbf{x} \in \mathbb{N}^3 \mid 0 \le x_1 \le 3, 5 \le x_2 \le 70, 5 \le x_3 \le 95\},\text{ and}$$
$$\mathcal{A}_3 \approx \{\mathbf{x} \in \mathbb{N}^3 \mid 5 \le x_1 \le 110, 5 \le x_2 \le 70, 0 \le x_3 \le 3\}.$$

For approximating the local steady states, we enlarged each set $\mathcal{A}_i$ by those states reachable within 15 steps. We ran Algorithm 1 and list the results in Table 2. A plot of the full steady state distribution is given in Figure 2b. The time needed to compute the local

Table 2. Results of the steady state analysis of the tri-stable toggle switch case study

δ	$\widehat{\pi}_1$	$\widehat{\pi}_2$	$\widehat{\pi}_3$	time (seconds)
1e–9	0.995	0.001	0.004	2.641 + 136.4 + 246.011 + 1.194
1e–10	0.9952	0.0013	0.0035	2.641 + 136.4 + 246.011 + 1.338
1e–11	0.99516	0.00132	0.00352	2.641 + 136.4 + 246.011 + 1.472
1e–20	0.995165	0.001315	0.003520	2.641 + 136.4 + 246.011 + 4.536

steady state was 246 seconds and to compute $\widehat{\pi}$ was below five seconds for any precision. Note that for this model, we could not solve the linear equation using Matlab since the corresponding linear equation system has more than 1.5 million equations. But even if the memory requirements could be met the stiffness of the model would prevent an efficient solution. As opposed to that our method does not suffer from problems related to stiffness because the time scales are separated during the stochastic complementation. Therefore, we got accurate results and very short computation times in Table 2.

5   Conclusion

We have presented an efficient method to analyze the steady state distribution of multimodal Markov chains. Combining geometric bounding techniques, stochastic complementation, and inexact matrix vector multiplication during transient and reachability analysis steps allowed us to cope with huge state spaces and stiffness inherent to genetic switching models from biology. Our method can in general be applied to any multi-modal Markov chain. For future work we will, however, explore whether it works similarly well for models that are very different from the two gene regulatory networks analyzed here. Currently, for a given truncation threshold our method does not determine the approximation error a priori. We plan to improve the method by dynamically choosing the threshold such that an a priori specified precision is met.

References

[1] Blake, W.J., Kaern, M., Cantor, C.R., Collins, J.J.: Noise in eukaryotic gene expression. Nature 422, 633–637 (2003)
[2] Courtois, P.-J., Semal, P.: Bounds for the positive eigenvectors of nonnegative matrices and for their approximations by decomposition. Journal of the ACM 31(4), 804–825 (1984)
[3] Dayar, T., Hermanns, H., Spieler, D., Wolf, V.: Bounding the equilibrium distribution of markov population models. NLAA 18(6), 931–946 (2011)
[4] de Souza e Silva, E., Ochoa, P.M.: State space exploration in Markov models. SIGMETRICS Perform. Eval. Rev. 20(1), 152–166 (1992)
[5] Didier, F., Henzinger, T.A., Mateescu, M., Wolf, V.: Fast adaptive uniformization of the chemical master equation. In: Proc. of HIBI, pp. 118–127. IEEE Computer Society, Washington, DC (2009)
[6] Elowitz, M.B., Levine, M.J., Siggia, E.D., Swain, P.S.: Stochastic gene expression in a single cell. Science 297, 1183–1186 (2002)
[7] Forger, D.B., Peskin, C.S.: Stochastic simulation of the mammalian circadian clock. PNAS 102(2), 321–324 (2005)
[8] Gardner, T.S., Cantor, C.R., Collins, J.J.: Construction of a genetic toggle switch in Escherichia coli. Nature 403(6767), 339–342 (2000)
[9] Gillespie, D.T.: A general method for numerically simulating the time evolution of coupled chemical reactions. J. Comput. Phys. 22, 403–434 (1976)
[10] Henzinger, T.A., Mateescu, M., Wolf, V.: Sliding window abstraction for infinite Markov chains. In: Bouajjani, A., Maler, O. (eds.) CAV 2009. LNCS, vol. 5643, pp. 337–352. Springer, Heidelberg (2009)
[11] Larson, D.R., Singer, R.H., Zenklusen, D.: A single molecule view of gene expression. Trends in Cell Biology 19(11), 630–637 (2009)
[12] Lohmueller, J., Neretti, N., Hickey, B., Kaka, A., Gao, A., Lemon, J., Lattanzi, V., Goldstein, P., Tam, L.K., Schmidt, M., Brodsky, A.S., Haberstroh, K., Morgan, J., Palmore, T., Wessel, G., Jaklenec, A., Urabe, H., Gagnon, J., Cumbers, J.: Progress toward construction and modelling of a tri-stable toggle switch in e. coli. IET, Synthetic Biology 1(1.2), 25–28 (2007)

[13] McAdams, H.H., Arkin, A.: Stochastic mechanisms in gene expression. PNAS 94, 814–819 (1997)

[14] Menz, S., Latorre, J., Schütte, C., Huisinga, W.: Hybrid stochastic–deterministic solution of the chemical master equation. Multiscale Modeling & Simulation 10(4), 1232–1262 (2012)

[15] Meyer, C.D.: Stochastic complementation, uncoupling markov chains, and the theory of nearly reducible systems. SIAM Review 31, 240–272 (1989)

[16] Milias-Argeitis, A., Lygeros, J.: Efficient stochastic simulation of metastable Markov chains. In: CDC-ECE, pp. 2239–2244. IEEE (2011)

[17] Munsky, B.: The Finite State Projection Approach for the Solution of the Master Equation and its Applications to Stochastic Gene Regulatory Networks. PhD thesis, University of California (June 2008)

[18] Paulsson, J.: Summing up the noise in gene networks. Nature 427(6973), 415–418 (2004)

[19] Stewart, W.J.: Introduction to the Numerical Solution of Markov Chains. Princeton University Press (1995)

A Tight Bound on the Throughput of Queueing Networks with Blocking

Jean-Sébastien Tancrez[1], Philippe Chevalier[1,2], and Pierre Semal[1]

[1] Louvain School of Management, Université catholique de Louvain (UCL), Belgium
[2] Center for Operations Research and Econometrics (CORE), UCL, Belgium
{js.tancrez,philippe.chevalier,pierre.semal}@uclouvain.be

Abstract. In this paper, we present a bounding methodology that allows to compute a tight bound on the throughput of fork-join queueing networks with blocking and with general service time distributions. No exact models exist for queueing networks with general service time distributions and, consequently, bounds are the only certain information available. The methodology relies on two ideas. First, probability mass fitting (PMF) discretizes the service time distributions so that the evolution of the modified system can be modelled by a discrete Markov chain. Second, we show that the critical path can be computed with the discretized distributions and that the same sequence of jobs offers a bound on the original throughput. The tightness of the bound is shown on computational experiments (error on the order of one percent). Finally, we discuss the extension to split-and-merge networks and the approximate estimations of the throughput.

Keywords: Queueing Networks, General Distributions, Blocking, Performance Evaluation, Bounds.

1 Introduction

In this work, we are interested in the modelling of open queueing networks with finite capacity and with general service time distributions. Queueing networks have shown to be of broad practical interest. They arise in many application fields (production systems, supply chains, communication and computer networks, road traffic flow, healthcare, etc.) and have proven to be very useful tools to evaluate the performance of complex systems. Queueing networks with finite capacity are common in practice, when buffer space is limited (e.g. finite storage space in production systems). The limited capacity gives rise to two phenomena: the starving and the blocking. First, a station is said to be starved when it cannot begin to serve a new unit because a previous buffer is empty. Second, a station is said to be blocked when it cannot get rid of an item because a next buffer is full. Accordingly, queueing networks with finite capacity are also called queueing networks with blocking. Obviously, increasing the buffer sizes allows to limit these phenomena, and to reduce the efficiency losses.

In this paper, we firstly focus on the analysis of fork-join queueing networks (FJQN) with blocking. Split and merge configurations are then discussed in

A. Dudin and K. De Turck (Eds.): ASMTA 2013, LNCS 7984, pp. 396–415, 2013.

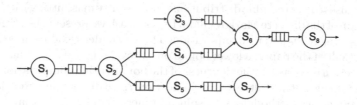

Fig. 1. A fork-join queueing network (FJQN)

Section 7. A FJQN is a feedforward open queueing network (not closed) in which the nodes are linked arbitrarily without forming loops. The stations can have several input or output stations, but each buffer has exactly one upstream station and one downstream station [1]. When a service ends in a station, one job is taken from each upstream buffer and one job is sent to each downstream buffer. An example of FJQN is given in Figure 1. Each station is given an index $i = 1, 2, \ldots, m$, with m being the number of stations, and the finite size of the buffer between stations i and j is denoted $b(i, j)$. In a fork station (e.g. S_2 in Figure 1), one item is taken from the previous buffer (except if it is empty, the station is starved in this case), disassembled into several pieces, and put in the next buffers (except if they are full, the station is blocked). In a join station (e.g. S_6 in Figure 1), items are taken from the previous buffers (except if they are empty, the station is starved), assembled into one unit, and the latter is put in the next buffer (except if it is full, the station is blocked). Such queueing systems arise in many applications areas such as manufacturing (assembly/disassembly systems) or parallel processing. Note that a tandem queue, used to model a production line for example, is a particular case of FJQN.

In this work, our goal is to investigate queueing networks with general service time distributions. In practice, the service time may follow a large range of distributions, depending on the particular application and on the collected data, and histograms in particular. However, no exact models exist for queueing networks with general service time distributions and, consequently, bounds are the only certain information available. They offer some certainty in the performance evaluation of complex systems, about the quality of the estimations. It can be used to test approximations or simulation models, or to develop approximations with known accuracy. However, very few such bounds have been proposed in the literature (see Section 2). In this paper, we propose a bounding methodology for queueing networks with finite capacity and with general service time distributions (with finite support). More specifically, we find lower bounds on the expected cycle time of fork-join queueing networks and split networks. The cycle time is defined as the time between two units leaving the system (in steady state). In other words, it is simply the reciprocal of the throughput. In manufacturing, the expected cycle time corresponds to the expected time to produce one unit (reciprocal of the expected number of items produced per time unit).

In the next section, we review the literature on queueing networks modelling and on existing bounding methodologies. In Section 3, we introduce the discretization method we use to get tractable distributions (probability mass fitting) and then the construction of the discrete Markov chain. In Section 4, we present the critical path, its relevant properties and its computation. These two ideas lead us in Section 5 to the main result of the paper: a lower bound on the cycle time of queueing networks. Next, in Section 6, we assess the tightness of the bound, using computational experiments on various network structures. In Section 7, we discuss an extension of the proposed bounding methodology to split and merge configurations and we show that an accurate approximation of the cycle time can also be computed. Finally, we conclude in Section 8.

2   Literature Review

Queueing networks with general service time distributions cannot be analyzed exactly, i.e. their expected performance measures cannot be computed exactly. Either they are approximately analyzed, or bounds are computed. In order to analytically model such networks, and measure their performance, the distributions have to be transformed, as a preliminary step. They have to be transformed to tractable distributions, so that the modified queueing networks can be analyzed. In most cases, this means using phase-type distributions, so that the Markov theory can be applied. This can be considered as the first step of the global modelling process. It is important to note that this first step, the transformation of the original distributions, unavoidably introduces an error in the subsequent performance analysis. In other words, any modelling of a queueing network with general distributions is approximate. There is no exact model in the general case. In order to fit the original distributions to tractable distributions, moments fitting is most commonly used, i.e. phase-type distributions sharing the same first moments as the original distributions are built [2–4]. The maximum likelihood estimation [5, 6] and the minimum distance estimation [7] are other options [8]. Lang and Arthur [9] conducted a vast experimental study of several fitting methods, based on moments fitting as well as on likelihood maximization. They concluded that a "moderate order (order 4 or order 8)" is satisfactory for the "phase-type behaved" distributions while a higher order is necessary for other distributions (low variance, steep increase or decrease, multiple modes, finite support). In a previous article, we proposed an alternative method to build tractable discrete distributions: probability mass fitting [10]. This method is used in the present paper (see Section 3.1).

In this context, many analytical models of queueing networks have been proposed supposing phase-type distributions (or the particular case of exponential distributions) (see [8, 11–13] for reference books). They are of two types: exact or approximate. For non-trivial configurations, the main exact methods are called state models, and consist in the exact modelling of the evolution of the system by a Markov Chain. Their limitation comes from the explosion of the state space size. Consequently, approximate methods have been proposed. The most popular models are based on the idea of decomposing the system into smaller subsystems, and then including back the interdependencies between the subsystems (see, for example, the survey [14] for the decomposition method and [15] for the expansion method). These analytical models can be used as the second step of the global modelling process of queueing networks with general distributions, after transforming the distributions to tractable phase-type distributions. Again, note that even if exact models are used in this second step, the global modelling process is not exact, as an error is unavoidably introduced by the distribution fitting in the first step.

When analyzing the performance of queueing networks with general distributions, bounds are the only options in order to get some certainty, to get guaranteed quality of the estimation. This is our goal in this paper. We aim at proposing a lower bound on the expected cycle time (i.e. an upper bound on the expected throughput). The literature proposing bounds on the performance measures of queueing networks can be divided in two types: assuming phase-type service time distributions or not. Most papers assume phase-type distributions, and often exponential distributions (i.e. the results are not valid with general distributions). In this case, exact solutions can be computed, theoretically. Researchers thus look for approximations and bounds which are quicker to compute. In the following, we cite the main bounding methodologies proposed in the literature, supposing phase–type distributions. A first approach relies on simplifying assumptions on the original system, or on modifications of it. It is then proved that it leads to bounds on the performance of the system (see [16] for unified view proposition). An early contribution was made by van Dijk and Lamond [17]. They considered a two station tandem queue with exponential service time distributions and a Poisson arrival process and, modifying the capacity limitations, got product form lower and upper bounds on the throughput. Shanthikumar and Jafari [18] then used a similar method and linked it to a decomposition technique. Tcha et al. [19] extended the idea to more complex systems. Liu and Buzacott [20] proposed other throughput bounds related to a decomposition technique. To get a lower bound on the cycle time (upper bound on the throughput), they increase the capacity of all buffers to infinity, except for the buffer of the first subsystem. Onvural and Perros [21] proposed another modification of the network, using an equivalent closed network. An alternative bounding methodology is known as the bounded aggregation [12, 22]. It decomposes the underlying Markov process, i.e. its state space, derives bounds on the stationary probabilities, with various levels of accuracy and complexity, and, from them, gets bounds on the performance measures of the system. Kumar and coauthors proposed to use constraints on the behavior of the system to obtain a linear program which leads to upper and lower bounds on the throughput [23, 24]. Another approach, called Performance Bound Hierarchy, has been proposed by Eager and Sevcik [25]. Their recursive algorithm computes bounds whose tightness improves with the number of recursive steps.

Concerning queueing networks with general distributions, as in our case, few papers have proposed bounds. The first approach relies on the concavity and monotonicity properties of queueing systems [12, 26, 27]. Increasing the buffer capacities leads to lower bounds on the cycle time (the throughput is improved) while decreasing them leads to upper bounds. However, these properties are useful only if the modified network is tractable, which can be the case with zero or infinite buffer capacities. Accordingly, it is impossible to evaluate the impact of the size of the buffers, and the achieved accuracy is poor. A second approach is the bounding methodology developed by Baccelli and Makowski [28] for systems with synchronization constraints and general distributions. It can be related to the theoretical framework developed by Baccelli et al. [29]. Baccelli and Liu [30] applied the methodology to stochastic decision free Petri nets, which can model fork-join queueing networks with finite buffers. Similarly to our work, this methodology relies on the structural property of the system, in the form of a recursion equation. The authors show that stochastic ordering of the service times leads to stochastic ordering of the cycle times and then bounds on the asymptotic expected cycle time. However, direct relations between the service times are required (additional terms are not allowed). Furthermore, the tightness of the proposed bounds has not been investigated. In previous works [10, 31], using a different methodology, we proposed bounds on the throughput of tandem queues with general service time distributions and finite capacity. We showed that they cannot be found as corollaries of the results by Baccelli and Liu [30], except for two particular cases. Using computational experiments, we showed in [10] that this (different) bounding methodology leads to errors on the order of 10%. Finally, Nakade [32] studied tandem queues with general service time distributions and zero buffer sizes. He found upper bounds on the cycle time using an equivalent synchronous tandem queue and lower bounds using recursive equations with respect to the waiting times. On some numerical examples, "the bounds give at most 8% relative errors", and compare favorably to the bounds shown in [17, 18].

In this paper, we present a new bounding methodology for queueing networks with general distributions. As in classical analytical models, the method consists of two steps, distribution fitting and analysis of the modified system, but, using the notion of critical path, we are able to bound the approximation error made in the first step, which applies probability mass fitting. The methodology offers tight lower bounds on the expected cycle time of complex queueing networks (fork-join and split) with general service time distributions and with finite buffers of any size. In this sense, our contribution comes from the generality of the modelled configurations as well as from the tightness of the bound (on the order of 1%).

3   Modelling Method

In this section, we briefly present the modelling method we use. Its main origicality comes in the first step, i.e. in the building of tractable distributions. We propose an alternative method called probability mass fitting (PMF) to build discrete phase–type distributions. After PMF, the evolution of the queueing network can be modelled by a discrete Markov chain, from which the performances can be evaluated. This modelling method has been introduced and discussed in [10] for tandem queues. In the following, we first briefly present probability mass fitting and, then, the global modelling method. Later, the modelling method will be used as a basis to compute the bound on the cycle time.

3.1   Probability Mass Fitting

In order to build tractable discrete phase-type distributions, we propose a method called probability mass fitting (PMF). PMF is quite intuitive: the probability masses on regular intervals are computed and aggregated on a single value in the corresponding interval. PMF transforms a given finite distribution into a discrete one by aggregating the probability mass distributed in the interval $((j-1)\tau+\alpha, j\tau+\alpha]$ on the point $j\tau$, for $j = 2, \ldots, a$, and the mass in $[0, \tau+\alpha]$ on τ. The first interval is treated differently in order to avoid a discrete value on zero. The parameter τ gives the size of the interval between two discrete values (which has to be constant), as well as the size of the interval on which the probability mass is computed. The parameter a gives the number of discrete values, and α gives the shift of the probability mass intervals compared to the corresponding discrete values.

To formally formulate PMF, we need to introduce some notations. Each job is given an index $k = 1, 2, \ldots$, where the first job to leave the system is given index 1. In the following, the random variable representing the service time of $w_{i,k}$, the job k on station i, is denoted $l(w_{i,k})$. A sample run in the sample space $\Omega = \{\omega\}$ is denoted $r(\omega)$ (see Figure 3 for an illustration). The realization of $l(w_{i,k})$ in a sample run r, i.e. the time the job $w_{i,k}$ takes in this particular run r, is denoted $l^r(w_{i,k})(\omega)$. For the sake of readability, (ω) will be omitted in the rest of the text. Moreover, we fix, without loss of generality, that a run starts at time zero. We denote $l_\alpha(w_{i,k})$ the random variable giving the discretized service time corresponding to the original service time of job $w_{i,k}$, i.e. $l(w_{i,k})$. Its realization in the sample run is denoted $r\ l_\alpha^r(w_{i,k})$. With these notations, the PMF can be formulated as follows:

$$l_\alpha^r(w_{i,k}) \stackrel{\Delta}{=} \tau, \qquad \text{if } l^r(w_{i,k}) \le \tau + \alpha, \forall r, i, k,$$
$$\stackrel{\Delta}{=} \left\lceil \frac{l^r(w_{i,k})-\alpha}{\tau} \right\rceil \tau, \quad \text{if } l^r(w_{i,k}) > \tau + \alpha, \forall r, i, k.$$

Note that, by default, we will use $\alpha/\tau = 1/2$, i.e. aggregating the probability mass in the middle of the interval, as it is the most natural option. Probability mass fitting has some interesting properties. By definition, it is simple, it preserves the shape of the original distribution and it is refinable, i.e. the fitting can be improved by increasing the number a of discretization steps. It is easily seen that $l_\alpha^r(w_{i,k}) - \tau \le l^r(w_{i,k}) \le l_\alpha^r(w_{i,k}) + \alpha$. In [10], we show that these bounds on the service time can be extended to bounds on the cycle time, $c_\alpha - \tau \le c \le c_\alpha + \alpha$, where c and c_α are the cycle time in the original and in the discretized time, respectively. We also show that this result can be seen as a corollary of the results presented in [30], only if $\alpha = \tau$ for the lower bound and if $\alpha = 0$ for the upper bound, and not in other cases. These bounds have an accuracy in the order of 10%. Finally, we note that the application of PMF requires the support of the ditribution to be finite (or to assume so). The main weaknesses of PMF lies in the fact that it is not well suited for service time distributions including rare events, or a long tail (i.e. with a non-negligible probability for a service time to be much larger than the mean).

3.2   Markov Model

The global method, modelling queueing networks with general service time distributions, can be described as follows. First, the original, general, service time distributions are transformed to discrete phase-type distributions by probability mass fitting. Note that the step size τ has to be the same for each station's distribution. Then, the modified system can be analyzed, using the Markov theory. Various state-of-the art models could be applied, exact or approximative (see Section 1). In this paper, aiming to show a bound on the cycle time, we use a classical exact model, namely a state model. The evolution of the system is simply described by a discrete Markov chain whose states are the possible combinations of the stages of the various stations and the number of units in the buffers. The performance of the queueing system can then easily be evaluated from the Markov chain, and its state's stationary probabilities.

The method is best illustrated on a simple example. Let us consider the simplest possible network: a two station tandem queue (see Figure 2.a). The buffer size equals one. The original service time distributions are generally distributed and shown in Figure 2.b. Probability mass fitting is applied with $a = 2$ (to keep it drawable) and $\alpha/\tau = 1/2$. It leads to the discrete distributions given in Figure 2.c. They can be represented as discrete phase-type distributions (Figure 2.d). Then, the second step of the method comes into play: the system is analytically modelled. The Markov chain given in Figure 2.e lists all the possible recurrent states of the system and the transitions between these states. The first symbol of a state refers to the first station, the second to the buffer and the third to the second station. Each station can be starved (S), blocked (B) or in some stage of service (for example, 1 means that the station already spent one time step serving the current unit). Each buffer is described by the number of units waiting in it (0 or 1 in this case). For example, state B12 means that the first station is blocked, that the buffer is full and that the second station already worked during two time steps on the current job. From a state, the transition to a new state depends on a station ending its current job or not. From state 112 for example, in the next time step, the second station will finish its current job, and begin to serve the unit which is waiting in the buffer. So, if the first station ends its job, he will put the unit in the buffer and begin a new job, leading to state 111. On the other hand, if the first station continues to serve the same unit, the new state will be 201. Finally, the Markov chain can be built applying this logic to every state, and the performances of the system can be computed from the stationary probabilities. Note that we suppose the network to be saturated, i.e. with infinite supply and demand (sources, like S_1 and S_3 in Figure 1, are never starved, and sinks, like S_7 and S_8, are never blocked). The blocking policy is supposed to be "blocking after service", i.e. if the next buffer is full when a station ends its job, the job waits in the station. These two last assumptions are not restrictive as

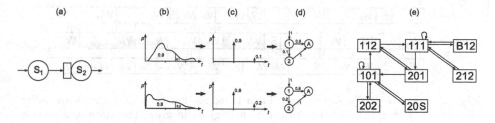

Fig. 2. Steps of the modelling method, applied to a two stations tandem queue

virtual stations could be used to model arrivals or demand, and as the method could be fitted to other blocking policies.

4   Critical Path

From the method presented in previous section, the evolution of a queueing network with general service time distributions can be modelled. In this section, we present the second key idea allowing to prove and compute the proposed bound: the critical path.

4.1   Synchronization Constraints

The critical path is a sequence of jobs that covers the running time, the length of the run. Its existence relies on the synchronization constraints of the queueing network, also called evolution equation, or state recursion [28]. This property can be found in [1] for fork-join queueing networks with blocking (with "blocking before service" policy). We present the synchronization constraints for FJQN with "blocking after service" in the next property. They are the basis which allows to define, build and compute the critical path. They rely on the fact that a station i can start a job k only if three conditions are satisfied. First, each previous station should have finished job k (1). Second, the previous job on station i, job $k - 1$, has to be finished (2). Third, the previous job $k - 1$ should not be blocked in station i, and there should thus be some room left for this job $k - 1$ in each following buffer (3). Moreover, once all these conditions are satisfied, there is no reason to wait and, so, $w_{i,k}$ starts exactly when the last of these three conditions becomes satisfied.

Property 1. (Synchronization constraints of a FJQN) Given a fork-join queueing network, the moment a job $w_{i,k}$ starts, in a sample run r, is given by the following equation:

$$t^r_{start}(w_{i,k}) = \max \big[\ \max_{j \in E(i)} [\, t^r_{end}(w_{j,k}) \,], \tag{1}$$

$$t^r_{end}(w_{i,k-1}), \tag{2}$$

$$\max_{j \in F(i)} [\, t^r_{start}(w_{j,k-b(i,j)-1}) \,] \ \big]. \tag{3}$$

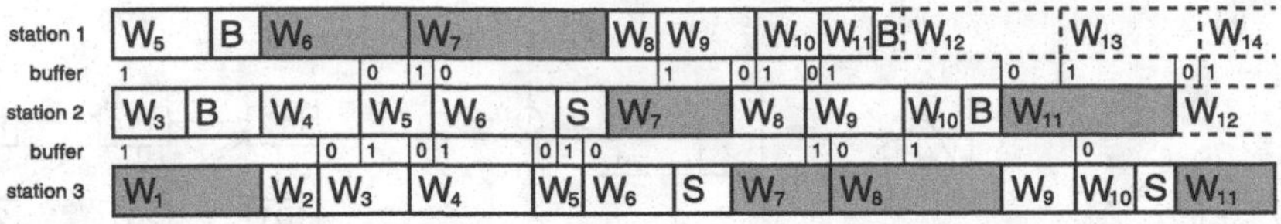

station 1 | W_5 | B | W_6 | W_7 | W_8 W_9 | W_{10} W_{11} B W_{12} | W_{13} | W_{14}

buffer 1 | 0 | 1 0 | 1 | 0 1 | 0 1 | 0 | 1 | 0 1

station 2 | W_3 | B | W_4 | W_5 W_6 | S W_7 | W_8 W_9 | W_{10} B W_{11} | W_{12}

buffer 1 | 0 1 | 0 1 | 0 1 0 | 1 0 | 1 | 0

station 3 | W_1 | W_2 W_3 | W_4 | W_5 W_6 S W_7 | W_8 | W_9 W_{10} S W_{11}

Fig. 3. Gantt chart of a sample run, for a three station tandem queue, with buffer sizes $b(1,2) = b(2,3) = 1$. The critical path is given in gray. The state of a station is represented either by a letter (B for blocked, S for starved) or by the job currently served. The state of a buffer is represented by the number of units carried.

Where $E(i)$ ($F(i)$) denotes the set of stations which directly precede (follow) a station i, and $t^r_{start}(w_{i,k})$ and $t^r_{end}(w_{i,k})$ denote the moments at which job $w_{i,k}$ starts and ends in the run r. Moreover, the starting time of a job is always equal to the ending time of another job, as we have: $t^r_{start}(w_{j,k-b(i,j)-1}) = \max\{t^r_{end}(w_{j,k-b(i,j)-2}), \max_{h\in F(j)}[t^r_{end}(w_{h,k-b(i,j)-b(j,h)-3})], \ldots\}$.

4.2 Definition and Construction

Here, we give the formal definition of the critical path, show its existence and explain how it can be built. From this point on, we will assume the sample runs to be finite. During a sample run r, n^r units are served. (A sink station is chosen to be the "reference" station, and given index m. The end of the run corresponds to the end of the n^rth job on this station.) The critical path of the run r, denoted $cp(r)$, is defined as the sequence of jobs that covers the run r, without gap and without overlap. By definition, the length of a run r can thus be written as a sum of job lengths:

$$l(r) = \sum_{w_{i,k} \in cp(r)} l^r(w_{i,k}).$$

It can be seen from this equation that the critical path offers the way to relate the individual service times $l^r(w_{i,k})$ and the length $l(r)$ of the run, i.e. the time to serve the n^r units or, equivalently, $t^r_{end}(w_{m,n^r})$. The notion of critical path is illustrated in Figure 3, where we see that the critical path allows to cover the sample run by a sequence of jobs.

The critical path can be built quite easily. Starting with the last job that leaves the network (at the end of run r), we look which job end, in this precise run, has triggered its start (in other words which term among (1-3) equals $t^r_{start}(w_{m,n^r})$). This new job will be part of the critical path and, from it, the next job can be found in the same way. Repeating this process, we can proceed backwards in time until the start of the run. From Property 1, and as every job start is triggered by the end of another job, every run r has at least one critical path. Furthermore, one may also observe on Figure 3 that the course of the critical path among the stations can be related to the station state before each job of the path. Indeed, the predecessor of a job in the critical path can be deduced from the state of the same station just before this job. If the station is previously working, the predecessor is obviously a job on the same station (and this corresponds to the term (2) in Property 1). If the station is previously starved, the predecessor is a job on the previous station for which the current station was waiting (term (1), see for example $w_{3,11}$ to $w_{2,11}$ in Figure 3). If the station is previously blocked, the predecessor is a job on the following station which was blocking the current station (term (3), see for example $w_{1,6}$ to $w_{3,1}$ in Figure 3). This characteristic is the key idea which allows us to compute the critical path, as explained in the next section.

4.3   Computation

In this section, we show how the critical path can be computed in the discretized run, in other words with the discretized service time distributions. This will ultimately allow us to compute our bound. By the computation of the critical path, we mean the computation of the steady-state conditional probabilities $p_\alpha^{cp}(i, s)$ that, if the system is in state s, the critical path lies in station i. For example, in Figure 2, $p_\alpha^{cp}(1, \text{B12}) = 0$ and $p_\alpha^{cp}(2, \text{B12}) = 1$, as the critical path cannot lie on a blocked station. Note that we use the subscript α because the critical path is, and can only be, computed in the discretized run.

The question is thus how the probabilities $p_\alpha^{cp}(i, s)$ can be computed. They can be computed thanks to the fact that the predecessor of a job in the critical path can be deduced from the state of the same station just before this job (see previous subsection). Let us consider a transition from a system state s to another system state s' and analyze the behavior of the critical path while this transition is encountered (see Figure 4). We can infer in which station the critical path will lie in the system state s if we know in which station cp lies in s' and if we know this station status (W, B or S) in state s. The critical path will lie in the (working) station i of s in three cases. First, going backward in time, cp will stay in station i in state s if it was already in i in state s' (see Figure 4.a). Second, cp will jump, going backward in time, to station i in s if it was on a following station i_f in state s', with $i_f \in F(i)$, and if the station was starved by station i previously, i.e. $s_{i_f} = S^i$ (see Figure 4.b). Third, cp will jump, going backward in time, to station i in s if it was on preceding station i_e in state s', with $i_e \in E(i)$, and if the station was blocked by station i previously, i.e. $s_{i_e} = B^i$ (see Figure 4.c). To get the probability $p_\alpha^{cp}(i, s)$ that the critical path lies on station i if the system is in state s, we thus simply have to consider these three cases for each possible transition from state s to one of its successors and weight each transition by its probability. We finally get the following equation, $\forall i, s$:

$$p_\alpha^{cp}(i, s) = 1_{\{s_i \neq S, B\}} \sum_{s' \in \text{Suc}(s)} p[s \to s'] \left(p_\alpha^{cp}(i, s') + p_{\text{fol}} + p_{\text{prec1}} + p_{\text{prec2}} + \dots \right),$$

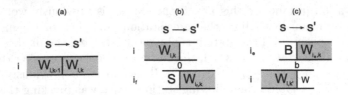

Fig. 4. The predecessor of a job in the critical path can be deduced from the state of the same station just before this job: (a) working, (b) starved, or (c) blocked

with

$$p_{\text{fol}} = \mathbf{1}_{\{s_{i_f}=S_i,i\in E(i_f)\}}\, p_\alpha^{cp}(i_f,s'),$$

$$p_{\text{prec1}} = \mathbf{1}_{\{s_{i_e}=B_i,i\in F(i_e)\}}\, p_\alpha^{cp}(i_e,s'),$$

$$p_{\text{prec2}} = \mathbf{1}_{\{s_{i_e}=B_i,s_{i_d}=B_{i_e},i\in F(i_e),i_e\in F(i_d)\}}\, p_\alpha^{cp}(i_d,s'),$$

Where $\mathbf{1}_{\{condition\}}$ is the indicator function, i.e. it equals one if the *condition* is satisfied and zero otherwise, $p[s \to s']$ is the transition probability between the system states s and s', and $\text{Suc}(s)$ is the set of successors of s, i.e. $s' \in \text{Suc}(s)$ iff $p[s \to s'] > 0$.

Regarding the complexity, to compute the probabilities $p_\alpha^{cp}(i,s)$, a linear system of equations has thus to be solved. The number of unknowns of this system equals the number of stations m (index i) times the number of system states (index s). In first approximation, the number of systems states is proportional to the number of individual station and buffer states combinations (see Figure 2.e), i.e. $\prod_{i=1}^{m} a_i \cdot \prod_{i=2}^{m} b_i$ with a_i the number of discrete values in the discretized service time distribution of station i, and b_i the size of the buffer preceding station i. Note however that the matrix is sparse and structured, what makes the solving of the system quicker. The cost of the critical path computation (using the Gaussian elimination implemented in MATLAB® on a 2.16 GHz usual PC, 2 GB RAM) is illustrated on Table 1 for tandem queues. The case of fork-join and split-and-merge queueing networks are similar. This system of equations has to be solved to compute the bound proposed in this paper. The complexity is thus the main limitation of the proposed bounding methodology. It increases when the number of discrete values a increases, in other words when the PMF discretization refines. There is thus a clear trade-off between complexity and accuracy, which can be directly controlled by the parameter a.

4.4   Properties

In this section, we give two properties of the critical path which will directly be used in order to prove the bound in the next section. First, let us consider the critical path of a sample run r. It is defined as the sequence of jobs, $cp(r) = \{w_{i,k}\}$, that covers r without gap and without overlap. We would like to know the behavior of this sequence of jobs $\{w_{i,k}\}$ in another run. The equation of Property 1 is valid for any run: a job $w_{i,k}$ cannot be started before all the

Table 1. Transition matrix size, sparsity (under parenthesis) and computational time (in italic), when computing the critical path, for tandem queues with a total storage space equal to two. The number m of stations and the number a of discretization steps vary.

a	$m = 2$	$m = 3$	$m = 4$	$m = 5$
4	84 (3.4%)	558 (0.79%)	2656 (0.23%)	11190 (0.07%)
	0.1 sec.	*0.2 sec.*	*1 sec.*	*13 sec.*
6	176 (1.8%)	1530 (0.34%)	8884 (0.08%)	43815 (0.02%)
	0.1 sec.	*0.5 sec.*	*7.7 sec.*	*250 sec.*
8	300 (1.1%)	3186 (0.17%)	21276 (0.04%)	117375 (0.01%)
	0.2 sec.	*1.7 sec.*	*53 sec.*	*2000 sec.*

jobs on the right hand side of (1-3) are finished. This is just a static structural property of the system, independent of the run. Consequently, as two consecutive jobs in the sequence $cp(r) = \{w_{i,k}\}$ satisfy this structural property, in another run, the same sequence of jobs will not show any overlap neither. The absence of overlap is independent of the run considered. However, which precise job end will trigger the start of job $w_{i,k}$, i.e. which term of equation (1-3) will be satisfied at equality, depends on the service times and thus on the particular run we consider. In another run, gaps could thus appear between the jobs of the sequence $\{w_{i,k}\}$. In conclusion, while it forms the critical path in the run r, the sequence of jobs $\{w_{i,k}\}$ will just be a non-overlapping path, maybe with gaps, in another run. Obviously, in a given run, the sum of the lengths of the jobs composing a non-overlapping path is always shorter than the length of the critical path, as the first may include gaps. In particular, when focusing on an original run and the one discretized with our modelling method, we get the following property.

Property 2. The sequence of jobs making the critical path in the discretized run is a non-overlapping path (possibly with gaps) in the original run, and vice versa.

The next property relates the cardinality $|cp(r_\alpha)|$ of the critical path of the discretized run r_α, i.e. the number of jobs in it, to the number n^r of units served during the run r_α. We denote c_α the expected cycle time computed in the discretized run, and $l_\alpha(w_{cp})$ the expected length of a job in the critical path in the discretized run.

Property 3. The ratio between the number of jobs in the critical path and the number of units served can be computed as follows:

$$\lim_{n^r \to \infty} \frac{|cp(r_\alpha)|}{n^r} = \frac{c_\alpha}{l_\alpha(w_{cp})}.$$

Proof. The expected cycle time c_α is the expected time between two units leaving the system, i.e. the length of the run divided by the number of units served. As the critical path covers the run, the expected length of a critical job is given by the length of the

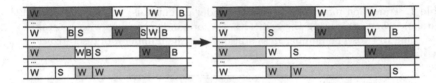

Fig. 5. Examples of critical paths and corresponding non-overlapping paths, in an original run (left) and in the discretized run (right), for a tandem queue made of four stations

run divided by the number of critical jobs. The property then simply follows from the following equation:

$$\frac{c_\alpha}{l_\alpha(w_{cp})} = \lim_{n^r \to \infty} \frac{l(r_\alpha)/n^r}{l(r_\alpha)/|cp(r_\alpha)|} = \lim_{n^r \to \infty} \frac{|cp(r_\alpha)|}{n^r}. \qquad \square$$

5 Tight Lower Bound

At this stage, we are able to show the main result of the paper: a computable and tight lower bound on the cycle time of general fork-join queueing networks. The core idea comes from Property 2 given in the previous subsection. The sequence of jobs which constitutes the critical path in the discretized run forms a non-overlapping path in the original sample run, thus shorter than the length of the original run. The critical path in the discretized run thus leads to a lower bound in the original running time, and, importantly, this bound can be computed (see Section 4.3).

Before proving this result, let us illustrate it in Figure 5. The left-hand side shows an original run r and the right-hand side depicts the corresponding discretized run r_α. On the left hand-side, the critical path is colored in light gray. Its length gives the real time to serve three units (but cannot be computed exactly). On the right-hand side, the critical path is colored in dark gray (and can be computed). The corresponding non-overlapping path (Property 2) is given in dark gray on the left-hand side. As this path may include gaps (at the end of the run in Figure 5), it is smaller than the light critical path in the original run, and offers a lower bound on the real time to serve three units. In a sense, discretizing may appear as a trick to get accessible information about the original run. As the running time cannot be computed in the original run (left in Figure 5), we rely on the discretized run (right), which can be modelled and where the critical path can be computed. We then go back to the original run to find the same sequence of jobs, whose length can be inferred (see the next proposition), leading to a lower bound. Moreover, it is reasonable to think that the critical path in the discretized run leads to a non-overlapping path with few gaps, and thus to a tight lower bound.

In the next proposition, we show that the length of the non-overlapping path in the original run (dark sequence in the left-hand side of Figure 5) can be computed, and thus, that a good bound can be computed on the expected throughput (infinite run). The proof relies on two ideas. First, the probability for a job of this non-overlapping path to be on a given station and to have its length in a given interval can be deduced from the computation of the critical path in the discretized run. Second, the expected length of such a job (knowing its discretized length) is independent of the fact that it belongs to the non-overlapping path and can thus be computed.

Proposition 1. (Lower bound on the expected cycle time of a FJQN)
The modelling method, and the critical path computation, allows to compute the following lower bound on the expected cycle time c of a fork-join queueing network with general distributions:

$$c \geq \frac{c_\alpha}{l_\alpha(w_{cp})} \sum_{i=1}^{m} \sum_{j=1}^{a} P[l_\alpha(w_{i,k}) = j\tau \mid w_{i,k} \in cp(r_\alpha)] \cdot E[l(w_{i,k}) \mid l_\alpha(w_{i,k}) = j\tau]. \quad (4)$$

where r_α is a discretized run of infinite length ($n^r \to \infty$), c_α is the expected cycle time in r_α, $l_\alpha(w_{cp})$ is the expected length of a critical job in r_α, $l(w_{i,k})$ is the original service time of job $w_{i,k}$ and $l_\alpha(w_{i,k})$ is the discretized service time of job $w_{i,k}$.

Proof. Let us suppose that we take a given $w_{i,k}$ in the discretized run r_α (right-hand side of Figure 5) and that its length equals $j\tau$. We know that the length of $w_{i,k}$ in the original run r (left-hand side of Figure 5) is between $(j-1)\tau + \alpha$ (0 if $j = 1$) and $j\tau + \alpha$, i.e. in the interval for which the probability mass is aggregated on $j\tau$. As $w_{i,k}$ is chosen in the discretized run, independently of its length in the original run, we have no clue about its exact position in the interval. The original length in r (known to be in a given interval $[(j-1)\tau + \alpha, j\tau + \alpha]$) of a job which is chosen because it belongs to the critical path in the discretized run r_α is independent of the fact that it belongs to this critical path $cp(r_\alpha)$. We have: $E[l(w_{i,k}) \mid l_\alpha(w_{i,k}) = j\tau \ \& \ w_{i,k} \in cp(r_\alpha)] = E[l(w_{i,k}) \mid l_\alpha(w_{i,k}) = j\tau]$. Moreover using Property 3, the right-hand side term of inequality (4) can be rewritten as follows:

$$\left(\lim_{n^r \to \infty} \frac{|cp(r_\alpha)|}{n^r} \right) \sum_{i=1}^{m} \sum_{j=1}^{a} P[l_\alpha(w_{i,k}) = j\tau \mid w_{i,k} \in cp(r_\alpha)].$$

$$E[l(w_{i,k}) | l_\alpha(w_{i,k}) = j\tau \ \& \ w_{i,k} \in cp(r_\alpha)] = \left(\lim_{n^r \to \infty} \frac{|cp(r_\alpha)|}{n^r} \right) \cdot E[l(w_{i,k}) \mid w_{i,k} \in cp(r_\alpha)].$$

The expression $|cp(r_\alpha)| \cdot E[l(w_{i,k}) \mid w_{i,k} \in cp(r_\alpha)]$ is the number of critical jobs in the discretized run r_α multiplied by their average length in the original run r. It thus gives the expected length, in the original run r, of the sequence of jobs making the critical path in the discretized run r_α (i.e. the dark sequence on the left-hand side of Figure 5). As it is only a non-overlapping path (with gaps) in the original run r, it is shorter than the global original time to serve n^r units (the light gay sequence on the left-hand side of Figure 5). Formally, this can be written as follows, and it completes the proof.

$$\left(\lim_{n^r \to \infty} \frac{|cp(r_\alpha)|}{n^r} \right) E[l(w_{i,k}) \mid w_{i,k} \in cp(r_\alpha)] \leq \left(\lim_{n^r \to \infty} \frac{|cp(r)|}{n^r} \right) E[l(w_{i,k}) \mid w_{i,k} \in cp(r)] = c. \quad \square$$

It is essential to see that this lower bound is computable. Each term can be computed. The two terms implying the critical path $cp(r_\alpha)$ (i.e. $l_\alpha(w_{cp})$ and $P[l_\alpha(w_{i,k}) = j\tau \mid w_{i,k} \in cp(r_\alpha)]$) are easily inferred from the critical path computation, i.e. from the probabilities $p_\alpha^{cp}(i, s)$ (see Section 4.3). The expected cycle time c_α in the discretized run is easily deduced from the steady-state probabilities of the Markov chain modelling the queueing network (see Section 3.2). Finally, $E[l(w_{i,k}) \,|\, l_\alpha(w_{i,k}) = j\tau]$ can be computed from the original service time distributions, which are known.

It can be argued that the proposed lower bound is tight because the critical path does not differ much from the discretized to the original run. Furthermore, when the number of discretization steps a increases, the length of a discretized job becomes closer to the length of the original job. Consequently, the lower bound becomes tighter when the PMF discretization is refined. At the limit, when a goes to infinity, the discretized run tends to be equivalent to the original run, and the bound thus converges to the exact cycle time. Finally, note that the proposed bound relies on both points presented previously in the paper. First, the critical path, and its computation, is obviously essential here. Second, the lower bound also relies on the idea of probability mass fitting as it uses the fact that the probability mass in an interval is aggregated in one value $j\tau$.

6   Computational Experiments

In order to assess the tightness of the proposed bound, and to study its behavior, we compare it to simulation results. Various network configurations have been tested: tandem, fork and join, with two, three, or four stations (various possibilities in this case). The total buffer space of a network goes from zero to four and is supposed to be balanced among the buffers. Each configuration has been tested with various service time distributions, arbitrary chosen among the ten following distributions: uniform(0,1), beta(1.3,1), beta(2,2), beta(4,4), beta(5.5,6), beta(8,8), beta(10,9), triangular(0,1,0.5), triangular(0.2,1,0.3) and triangular(0.1,0.9,0.6). Their support is in the interval $[0, 1]$ (sometimes smaller). These distributions have been chosen to be fairly realistic (e.g. in manufacturing) but different from each other. The means had to be chosen similar (from 0.48 to 0.57) to be realistic, and to keep the system balanced. The coefficients of variation ranges from 0.212 to 0.577. In total, 900 FJQN have been analyzed. Moreover, regarding the modelling method parameters, we varied the number a of discretization steps (4, 6 and 8 steps) to understand its impact. In total, we thus report the results of 2700 experiments. We believe they are representative, knowing that the complexity limits the size of the analyzed queueing networks. Also note that the PMF shift parameter α was also tested ($\alpha/\tau = 0, 0.25, 0.5, 0.75, 1$), but showed to have little influence, so that we focus here on $\alpha/\tau = 1/2$, i.e. PMF aggregating the probability mass in the middle of the interval.

Tables 2 and 3 illustrate the levels of accuracy reached by the lower bound on the cycle time. They give the average relative error, in percent, between the bound and the result of the simulation (i.e. $(sim - bound)/sim$). Table 2 aims to illustrate the influence of the configuration of the network on the tightness of the bound. First of all, we see that the bound is tight, and that its accuracy improves when the PMF discretization is refined. On average, the level of accuracy reached is 0.4% with eight discretization steps, 0.7% with six and 1.4% with only four

Table 2. Average bound tightness reached for various FJQN network topologies, with various numbers of stations m. The number a of discretization steps varies, and $\alpha/\tau = 0.5$.

	Tandem			Fork		Join	
a	$m = 2$	$m = 3$	$m = 4$	$m = 3$	$m = 4$	$m = 3$	$m = 4$
4	0.83%	1.30%	1.64%	1.27%	1.55%	1.30%	1.62%
6	0.38%	0.61%	0.80%	0.61%	0.75%	0.64%	0.79%
8	0.21%	0.36%	0.51%	0.36%	0.48%	0.38%	0.51%

Table 3. Average bound tightness reached for various storage spaces (B_Σ stands for $\sum_{i,j} b(i,j)$). The number a of discretization steps varies, and $\alpha/\tau = 0.5$.

a	$B_\Sigma = 0$	$B_\Sigma = 1$	$B_\Sigma = 2$	$B_\Sigma = 3$	$B_\Sigma = 4$
4	1.98%	1.78%	1.38%	1.03%	0.87%
6	0.98%	0.87%	0.66%	0.49%	0.40%
8	0.58%	0.51%	0.39%	0.29%	0.17%

steps. We remind that this accuracy, as well as the bound characteristic, is valid for general service time distributions. Furthermore, it can be seen on Table 2 that the accuracy reached for tandem, fork or join networks is very similar. The number of stations composing the network has however a significant influence, even if the bound stays tight for larger networks. Table 3 shows that the bound tends to tighten when the storage space of the network is increased. In summary, concerning the configuration of the network, while the topology does not seem to matter much, the tightness of the bound seems to deteriorate when the number of stations increases, and it seems to improve when the storage space increases.

Ideally, these results should be compared to those of concurrent methods in the literature. However, as explained in Section 2, few bounding methods exist for networks with general service time distributions. The effect of the distribution fitting on the global modelling process is barely studied in general, and bounds exist essentially for queueing networks with phase-type distributions (often exponential). To the best of our knowledge, the accuracy of the bounds proposed in [28, 30] has not been investigated. Nakade [32] proposed upper and lower bounds on the expected cycle time of tandem queues with general service time distributions and zero buffer sizes. On some numerical examples, he shows that "the bounds give at most 8% relative errors", and compare favorably to the bounds shown in [17, 18] (for systems with exponentially distributed service times). On systems with blocking-after-service and uniform distributions, Nakade [32] gets a 5.6% average relative error for tandem queues with zero buffer sizes. The proposed bounding methodology thus compares favorably in terms of accuracy (less than 1%) and in terms of the scope of application (more configurations and buffer sizes). In previous work, we proposed another bounding methodology, based on the result of the modelling methodology, without taking advantage of the critical path computation (see Section 3.1 or [10] for more details). The tightness of the upper and lower bounds is directly related to the step size τ, and is clearly not as good. For example, for three station tandem queues (with various buffer sizes), with $a = 8$ and $\alpha/\tau = 0.5$, the average relative error equals 9.9%. This can be compared to the 0.36% tightness reached with the present bound (see Table 2). However, note that the upper bound on the expected cycle time found by the previous methodology can be computed without additional cost, and that, combining both methodologies, our approach thus allows to bound the cycle time from below and from above.

The proposed approach offers a tight bound on the expected throughput or cycle time of queueing networks with general service time distributions. It provides some certainty about the computed estimation. On the contrary, the approximate methods commonly used in practice offer little certainty about the quality of their estimations. To illustrate this, we applied the classical cycle time estimation presented in [33] for two station tandem queues. We compute it for 2500 tandem queues using the same test parameters as previously: buffer sizes ranging from 0 to 4 and service time distributions chosen randomly from the ten distributions listed previously. We get a 5.7% average relative absolute error (which can be compared with the accuracy of the bound and of the approximation given on the first column of Table 2 and Table 4, respectively). The cycle time is sometimes overevaluated and sometimes underevaluated, and a 60.3% maximum relative error has been observed (1.41% with the proposed bound on the reported tests, with $a = 6$).

7   Extensions

In this section, we discuss the extension of our bounding methodology to split-and-merge networks and the approximate estimations of the throughput. In split and merge open queueing networks with blocking [34], the stations can have several input or output stations, but, unlike in FJQNs, one buffer is associated to each station and a station exclusively takes items from this single buffer. Items are not assembled or disassembled, one item entering the system leads to exactly one unit leaving the system. In a split configuration, the item is routed to one of the following stations, with some routing probabilities. In a merge configuration, the merge station takes the items from the buffer, which gathers the items from every preceding station.

Our bounding methodology can be straightforwardly extended to split configurations. The critical path in the discretized run leads to a non-overlapping path in the original run (which can be computed) and thus to a lower bound. Indeed, the methodology mainly relies on two ideas: the probability mass fitting and the critical path. Obviously, the former is independent of the network configuration. The latter can be extended to split networks as its existence basically follows from the fact that a job start is always triggered by another job end. Experiments on 300 split configurations, with $a = 8$, led to average bound tightness of 0.38% for networks with 3 stations and 0.57% with 4 stations (0.70% and 0.98%

Table 4. Average accuracy of the approximation for various fork-join and split-and-merge networks, with various numbers of stations m. The number a of discretization steps varies, and $\alpha/\tau = 0.5$.

	Tandem			Fork		Join		Split		Merge	
a	$m=2$	$m=3$	$m=4$	$m=3$	$m=4$	$m=3$	$m=4$	$m=3$	$m=4$	$m=3$	$m=4$
4	0.57%	0.47%	0.46%	0.51%	0.46%	0.50%	0.46%	1.55%	1.20%	3.50%	3.01%
6	0.22%	0.20%	0.21%	0.22%	0.21%	0.22%	0.21%	0.56%	0.47%	1.22%	1.29%
8	0.12%	0.12%	0.13%	0.13%	0.13%	0.13%	0.13%	0.28%	0.24%	0.56%	0.66%

with $a = 6$, see [35] for details). Concerning merge configurations, the extension of the bounding methodology turns out to be trickier, if not infeasible. Indeed, with merge stations, the sequence of jobs which makes the critical path in the discrete run is not necessarily non-overlapping in the original run: the station feeding a starved station could be different from the discrete to the original run, and, in such a case, an overlap could appear in the sequence of job.

In the core of the paper, we focus on showing a bound on the cycle time, which offers some certainty on the quality of the estimation. However, the cycle time can also easily be evaluated from the modelling method described in Section 3 (without computing the critical path), using the cycle time of the system in the discretized time (with $\alpha/\tau = 0.5$). To illustrate the accuracy of the cycle time approximation, we present in Table 4 the results of a test on 1500 queueing networks. The configurations, the service time distributions and the storage spaces are the same as for the bound tests (see Section 6, and [35]). Overall, we see that, besides offering some certainty thanks to the bounds, our modelling method leads to accurate approximate performance evaluation for open queuing networks with general distributions.

8   Conclusion

In this paper, we proposed a computable tight lower bound on the expected cycle time of queueing networks with blocking and with general service time distributions. The methodology is presented for fork-join queueing networks and then extended to split configurations. As the service time distributions are general, an exact analysis is impossible, and a bound is thus the only way to get certainty about the quality of the performance evaluation. In few words, the bounding methodology works as follows. The distributions are discretized by probability mass fitting, i.e. the probability masses on regular intervals are aggregated on a particular value of the corresponding interval. With the discretized distributions, the critical path, a sequence of jobs which covers the run, can be computed. In the original run, this sequence of jobs is non-overlapping, shorter than the running time, and thus leads to a lower bound on the cycle time. Moreover, this sequence is close to the critical path in the original run, and this becomes more and more true when the discretization is refined, i.e. the bound becomes tighter and tighter.

In order to assess the tightness of the bound, we compared the results of the method to simulation results. It showed the good accuracy of the methodology: the average relative error equals 1.4%, 0.7% and 0.4% with 4, 6 and 8 discretization steps, for fork-join queueing networks. Note that there is a trade-off between the accuracy and the complexity. We also argued that the methodology can be straightforwardly extended to split configurations and showed that it leads to similar tightness. Finally, we showed that the approximation of the cycle time, given by the cycle time of the queueing network (fork-join or split-and-merge) with the discretized distributions, is accurate (0.5%, 0.2% and 0.13% with 4, 6 and 8 discretization steps, for FJQN). Future research could aim to extend the bounding methodology to even more network configurations, or to couple it with decomposition methods in order to lower the complexity and analyze larger networks.

References

1. Dallery, Y., Liu, Z., Towsley, D.: Equivalence, reversibility, symmetry and concavity properties in fork/join queueing networks with blocking. Journal of the Association for Computing Machinery 41, 903–942 (1994)
2. Whitt, W.: Approximating a point process by a renewal process, i: Two basic methods. Operations Research 30(1), 125–147 (1982)
3. Johnson, M., Taaffe, M.: Matching moments to phase distributions: Mixtures of erlang distributions of common order. Stochastic Models 5(4), 711–743 (1989)
4. Bobbio, A., Horváth, A., Telek, M.: Matching three moments with minimal acyclic phase type distributions. Stochastic Models 21, 303–326 (2005)
5. Asmussen, S., Nerman, A., Olsson, M.: Fitting phase-type distributions via the em algorithm. Scandinavian J. of Statistics 23, 419–441 (1996)
6. Bobbio, A., Horváth, A., Scarpa, M., Telek, M.: Acyclic discrete phase type distributions: Properties and a parameter estimation algorithm. Performance Evaluation 54, 1–32 (2003)
7. Varah, J.M.: On fitting exponentials by nonlinear least squares. SIAM J. on Sci. and Statist. Comput. 6(1), 30–44 (1985)
8. Perros, H.: Queueing Networks with Blocking: Exact and Approximate Solutions. Oxford University Press, New York (1994)
9. Lang, A., Arthur, J.: Parameter approximation for phase-type distributions. In: Chakravarthy, S., Alfa, A. (eds.) Matrix Analytical Methods in Stochastic Models, pp. 151–206. Marcel Dekker, New York (1997)
10. Tancrez, J.S., Chevalier, P., Semal, P.: Probability masses fitting in the analysis of manufacturing flow lines. Annals of Operations Research 182, 163–191 (2011)
11. Altiok, T.: Performance Analysis of Manufacturing Systems. Springer, New York (1996)
12. Balsamo, S., de Nitto Personé, V., Onvural, R.: Analysis of Queueing Networks with Blocking. Kluwer Academic Publishers, Dordrecht (2001)
13. Buzacott, J., Shanthikumar, J.: Stochastic Models of Manufacturing Systems. Prentice-Hall, Englewood Cliffs (1993)
14. Dallery, Y., Frein, Y.: On decomposition methods for tandem queueing networks with blocking. Operations Research 41(2), 386–399 (1993)

15. Kerbache, L., Smith, J.M.: The generalized expansion method for open finite queueing networks. European Journal of Operational Research 32, 448–461 (1987)
16. van Dijk, N.: Bounds and error bounds for queueing networks. Annals of Operations Research 79, 295–319 (1998)
17. van Dijk, N., Lamond, B.: Simple bounds for finite single-server exponential tandem queues. Operations Research 36(3), 470–477 (1988)
18. Shanthikumar, J., Jafari, M.: Bounding the performance of tandem queues with finite buffer spaces. Annals of Operations Research 48, 185–195 (1994)
19. Tcha, D.W., Paik, C.H., Lee, W.T.: Throughput upper bounds for open markovian queueing networks with blocking. Computers & Industrial Engineering 28(2), 351–365 (1995)
20. Liu, X.G., Buzacott, J.: A decomposition-related throughput property of tandem queueing networks with blocking. Queueing Systems 13, 361–383 (1993)
21. Onvural, R.O., Perros, H.G.: Some equivalencies between closed queueing networks with blocking. Performance Evaluation 9(2), 111–118 (1989)
22. Courtois, P.J., Semal, P.: Computable bounds for conditional steady-state probabilities in large markov chains and queueing models. IEEE Journal on Selected Areas in Communications 4(6), 926–937 (1986)
23. Kumar, S., Kumar, P.: Performance bounds for queueing networks and scheduling policies. IEEE Trans. on Automatic Control 39(8), 1600–1611 (1994)
24. Kumar, S., Srikant, R., Kumar, P.: Bounding blocking probabilities and throughput in queueing networks with buffer capacity constraints. Queueing Systems 28, 55–77 (1998)
25. Eager, D., Sevcik, K.: Bound hierarchies for multiple-class queuing networks. Journal of the Association for Computing Machinery 33(1), 179–206 (1986)
26. Tsoucas, P., Walrand, J.: Monotonicity of throughput in non-markovian networks. Journal of Applied Probability 26, 134–141 (1989)
27. Dallery, Y., Gershwin, S.: Manufacturing flow line systems: a review of models and analytical results. Queueing Systems 12, 3–94 (1992)
28. Baccelli, F., Makowski, A.: Queueing models for systems with synchronization constraints. Proceedings of the IEEE 77, 138–161 (1989)
29. Baccelli, F., Cohen, G., Olsder, G.J., Quadrat, J.P.: Synchronization and linearity. Wiley, New York (1992)
30. Baccelli, F., Liu, Z.: Comparison properties of stochastic decision free petri nets. IEEE Trans. on Automatic Control 37, 1905–1920 (1992)
31. Tancrez, J.S., Semal, P., Chevalier, P.: Histogram based bounds and approximations for production lines. European Journal of Operational Research 197, 1133–1141 (2009)
32. Nakade, K.: New bounds for expected cycle times in tandem queues with blocking. European Journal of Operational Research 125(1), 84–92 (2000)
33. Hopp, W., Spearman, M.: Factory Physics: Foundations of Manufacturing Management. Irwin, Burr Ridge (1996)
34. Altiok, T., Perros, H.: Open networks of queues with blocking: Split and merge configurations. IIE Transactions 18(3), 251–261 (1986)
35. Tancrez, J.S.: Modelling Queueing Networks with Blocking using Probability Mass Fitting. PhD thesis, Université catholique de Louvain (2009)

Semi-Product-Form Solution
for PEPA Models with Functional Rates

Nigel Thomas[1] and Peter G. Harrison[2]

[1] School of Computing Science, Newcastle University, UK
nigel.thomas@ncl.ac.uk
[2] Department of Computing, Imperial College London, UK
pgh@doc.ic.ac.uk

Abstract. We consider the problem of finding a separable solution for
the equilibrium state probabilities in a Markovian process algebra model,
in which the action rates may depend on the behaviour of other components. To do this we consider regular cycles in the underlying state space
and show that a semi-product form solution exists when the functions describing the action rates have specific forms. The approach is illustrated
with two examples, one a generalised version of a known state-dependent
queueing network and the other in the domain of security protocols.

1 Introduction

In stochastic process algebra (SPA) the state space explosion problem, where
each additional component causes a multiplicative increase in the size of the
global state space, is particularly significant when there are many instances of
the same type of component (so-called *massively parallel systems*). Such models
may be extremely concise to specify, but even when the state space is represented
in a very efficient way and folded or lumped [4], they may still far exceed the
computational capacity available for solution. There have been many attempts
to find efficient solutions to large SPA models. Recent results include fluid approximations [11, 15, 19, 22] and mean value analysis [23, 24]. However, these
results are either approximate or of limited applicability. It is clear that the
modeller needs a variety of tools at their disposal and there is no single solution
to this problem. The aim of this paper is to extend the existing practical methods by attempting to derive a separable solution for a class of model where the
transitions might depend on the system (rather than component) state.

Many of the approaches to efficiently solving SPA models have been based on
concepts of decomposition originally derived for queueing networks [14, 20, 21].
Applying such approaches to stochastic process algebra allows the concepts to be
understood in a more general modelling framework and applied to non-queueing
models. Work by Harrison on the *Reversed Compound Agent Theorem* (RCAT)
has exploited properties of the reversed process to derive product form solutions
for models expressed in stochastic process algebra [5–8]. This has resulted in a
body of work defining the identification of product forms at the syntactic level,

A. Dudin and K. De Turck (Eds.): ASMTA 2013, LNCS 7984, pp. 416–430, 2013.

based on corresponding active and passive actions in synchronising components. The method outlined in this paper can be used to consider classes of model not amenable to solution by the standard RCAT method. Furthermore, the resultant decomposition is not strictly a product form, as it includes terms relating to the global state space. Nevertheless, we can exploit the same properties of the reversed process to derive expressions based on the cycles which arise in the underlying CTMC and hence an efficient scalable decomposition.

Most product form results in queueing networks rely on regular structure within the underlying state space. Arrival rates at a given node are clearly dependent on the number of jobs at preceding nodes, however service rates are generally constant or depend only on the local state of a component (in a limited fashion). There are a number of existing results where particular transition rates may depend on the state of other components, e.g. [2, 3, 12]. In the work presented here we consider the case where any transition rate may depend on some function of the current global state, generalising our previous results in this area [6, 10]. Clearly this creates an apparently strong dependency between components and it is counter-intuitive that a product form solution could exist in such situations. In fact, as we show, the resulting decomposition will include terms relating to the rate functions and hence the global state. As such this is not strictly a product form solution, so we use the term *semi-product form*.

Bonald and Proutiere [1] consider a general queueing network model and show that it is insensitive to the service time distributions (and hence has a semi-product form solution) subject to a general balance condition. The first example we include here is subject to the same condition. However, our approach differs from that taken in [1] in that we use the reversed process and Kolmogorov's generalised criteria to find conditions (equivalent to those of Bonald and Proutiere [1]) on the rate functions. This gives a simpler and more intuitive means of deriving the semi-product form solution.

The remainder of this paper is organised as follows. In the next section a brief overview of the Markovian process algebra PEPA is presented, followed by an outline of the proposed method, including coverage of the underlying theory and an approach to finding reversed component specifications. Section 4 then presents two examples of the application of the methodology in practice: one a generalised version of a known state-dependent queueing network and the other in the domain of security protocols. The paper concludes in Section 5 with a summary and a discussion of further research directions.

2   PEPA

A formal presentation of PEPA is given in [13]; in this section a brief informal summary is presented. PEPA, being a Markovian Process Algebra, only supports actions that occur at times that are negative exponentially distributed. Specifications written in PEPA are Markovian and can be mapped to a continuous time Markov chain (CTMC). Systems are specified in PEPA in terms of *activities* and *components*. An activity (α, r) is described by the type of the activity, α, and the parameter, or rate, of the associated negative exponential distribution, r. This rate may be any positive real number, or given as unspecified using the symbol $\top$. It is important to note that in this paper the unspecified rate is not used.

The syntax for describing components is given as:

$$A \mid (\alpha, r).P \mid P + Q \mid P/L \mid P \underset{\mathcal{L}}{\bowtie} Q$$

$A \overset{def}{=} P$ gives the constant A the behaviour of the component P. The component $(\alpha, r).P$ performs the activity of type α at rate r and then behaves like P. The component $P + Q$ behaves either like P or like Q, the resultant behaviour being given by the first activity to complete.

Concurrent components can be synchronised, $P \underset{\mathcal{L}}{\bowtie} Q$, such that activities in the cooperation set $\mathcal{L}$ involve the participation of both components. In PEPA the shared activity occurs at the slowest of the rates of the participants and if a rate is unspecified in a component, the component is passive with respect to activities of that type. The shorthand notation $P||Q$ is used to mean $P \underset{\emptyset}{\bowtie} Q$ and $\prod_{i=1}^{N} P_i$ is used to mean the parallel composition of the components P_i where i takes the values 1 through to N, i.e. $P_1|| \ldots ||P_N$. In addition, we employ a further shorthand for synchronisation over many identical components, first introduced in [15], whereby $P[N]$ is taken to mean N copies of the component P, synchronised on the empty set, i.e. $P|| \ldots ||P$ where there are N instances of component P.

The component P/L behaves exactly like P except that the activities in the set L are concealed, their type is not visible and instead appears as the unknown type τ. In this paper we do not make use of hiding, although non-shared actions could be hidden in a model if that was desirable. The action set $\mathcal{A}(P)$ is defined as the set of actions which are currently enabled in the component P.

Typically the rate of actions is given as a constant, however PEPA also supports a *functional rate*, which is a rate whose value is a function of the system state (or more commonly, a subset of the system state). Clearly the existence of a functional rate breaks the compositional principle, since the evolution of a component may become dependent on the behaviour of other components even if the action is not synchronised. For this reason in this paper we only employ functional rates within shared actions as a means of specifying alternative forms of synchronisation.

We consider only models which are cyclic, that is, every derivative of components P and Q are reachable in the model description $P \underset{\mathcal{L}}{\bowtie} Q$. Necessary conditions for a cyclic model may be defined on the component and model definitions without recourse to examining the entire state space of the model.

3   Outline of the Method

For every stationary Markov process, there is a reversed process with the same state space and the same steady state probability distribution, i.e. $\pi_{\mathbf{x}} = \pi'_{\mathbf{x}}$, where $\pi_{\mathbf{x}}$ and $\pi'_{\mathbf{x}}$ are the steady state probabilities of being in state $\mathbf{x}$ in the forward and reversed process respectively. Furthermore, the forward and reversed processes are related by the transitions between states; there will be a non-zero transition rate between states $\mathbf{x}$ and $\mathbf{x}'$ in the reversed process, $q'_{\mathbf{x},\mathbf{x}'}$ iff there is a non-zero transition rate between states $\mathbf{x}'$ and $\mathbf{x}$ in the forward process, $q_{\mathbf{x}',\mathbf{x}}$. A special case is the *reversible* process, where the reversed process is stochastically identical to the forward process; an example is the M/M/1 queue. The reversed process is easily found if we already know the steady state probability distribution (see Kelly [16] for example).

The forward and reversed probability fluxes balance at equilibrium, i.e.

$$\pi'_{\mathbf{x}} q'_{\mathbf{x},\mathbf{x}'} = \pi_{\mathbf{x}'} q_{\mathbf{x}',\mathbf{x}} \tag{1}$$

and so, since $\pi_{\mathbf{x}} = \pi'_{\mathbf{x}}$, we find:

$$q'_{\mathbf{x},\mathbf{x}'} = \frac{\pi_{\mathbf{x}'} q_{\mathbf{x}',\mathbf{x}}}{\pi_{\mathbf{x}}}$$

Kolmogorov's criteria utilise these balance equations to relate the forward and reversed transitions directly, as follows:

$$\pi_{\mathbf{x}} q_{\mathbf{x},\mathbf{x}'} \pi_{\mathbf{x}'} q_{\mathbf{x}',\mathbf{x}} = \pi'_{\mathbf{x}'} q'_{\mathbf{x}',\mathbf{x}} \pi'_{\mathbf{x}} q'_{\mathbf{x},\mathbf{x}'}$$

Since $\pi_{\mathbf{x}} = \pi'_{\mathbf{x}}$, we find:

$$q_{\mathbf{x},\mathbf{x}'} q_{\mathbf{x}',\mathbf{x}} = q'_{\mathbf{x}',\mathbf{x}} q'_{\mathbf{x},\mathbf{x}'}$$

More generally, we observe that

$$q_{\mathbf{x}_1,\mathbf{x}_2} q_{\mathbf{x}_2,\mathbf{x}_3} \cdots q_{\mathbf{x}_n,\mathbf{x}_1} = q'_{\mathbf{x}_1,\mathbf{x}_n} \cdots q'_{\mathbf{x}_3,\mathbf{x}_2} q'_{\mathbf{x}_2,\mathbf{x}_1}$$

That is, for any cycle length n, the product of the transition rates in the forward process must equal the product of the transition rates in the reversed process of the same cycle taken in reverse order. This is referred to as the second of Kolmogorov's generalised criteria.

In order to find a possible (semi) product form solution, we first identify a notional reversed model of the system. That is, an equivalent model where time is reversed. For any given formalism there are a number of rules that can be followed to *structurally reverse* any given model. In the approach here we reverse each component of a model to derive a *notional* reversed model, i.e. a model with the same system equation, but each component is reversed. If this notional reversed model is the actual reversed CTMC, then Kolmogorov's criteria will be satisfied and a (semi) product form solution will exist. Hence we can use Kolmogorov's criteria as a test for our notional reversed model, to see if it is the actual reversed CTMC. If we cannot apply Kolmogorov's criteria, then a product form may still exist for the underlying CTMC, but not with the component structure given.

3.1   Deriving Reversed PEPA Components

In theory it is possible to compute a reversed model directly from the global state space. However, this has a number of significant drawbacks. Firstly, it would require explicitly deriving the entire state space, which we are trying to avoid since we are interested in efficiently solving very large models. Secondly, a reversed model derived from the global state space would not have any component structure, so deriving a decomposed solution would require additional effort (to provide that structure, if possible). Thirdly, deriving a large reversed model is computationally expensive as every reversed transition needs to be computed.

The method introduced in the previous section relies on the identification of a notional reversed model. This notional reversed model is formed by reversing each of the components in the model. This approach avoids the problems associated with direct computation of the true reversed process (as long as components are not too large). However, calculation of the reversed rates is still not trivial in general and it is clearly possible (indeed probable) that the notional reversed model will not be the actual reversed model, i.e. a product-form solution with this component structure does not exist.

Deriving the structure of a reversed PEPA component is a relatively simple process. For every forward action in the original model, there must be a corresponding reversed action in the reversed model. Thus, if there is a PEPA statement in the forward model of the form,

$$P \stackrel{def}{=} (\alpha, r).Q$$

Then in the reversed model there will be a statement of the form,

$$\overline{Q} \stackrel{def}{=} (\overline{\alpha}, \overline{r}).\overline{P}$$

where, $\overline{P}$ and $\overline{Q}$ are reversed derivatives, $\overline{\alpha}$ is a reversed action and $\overline{r}$ is the associated reversed rate. The naming of reversed actions is only important when considering shared actions. Clearly, if there is more than one derivative which is a precursor of Q in the forward model, then the definition of $\overline{Q}$ will contain a choice between reversed actions, i.e. suppose,

$$P_1 \stackrel{def}{=} (\alpha_1, r_1).Q$$
$$P_2 \stackrel{def}{=} (\alpha_2, r_2).Q$$

Then,

$$\overline{Q} \stackrel{def}{=} (\overline{\alpha_1}, \overline{r_1}).\overline{P_1} + (\overline{\alpha_2}, \overline{r_2}).\overline{P_2}$$

Similarly, if there is a choice in the specification of P in the forward process, then there will be multiple precursor derivatives in the reversed model for $\overline{P}$, i.e. suppose,

$$P \stackrel{def}{=} (\alpha_1, r_1).Q_1 + (\alpha_2, r_2).Q_2$$

Then,

$$\overline{Q_1} \stackrel{def}{=} (\overline{\alpha_1}, \overline{r_1}).\overline{P}$$
$$\overline{Q_2} \stackrel{def}{=} (\overline{\alpha_2}, \overline{r_2}).\overline{P}$$

In many situations some reversed rates may be found by solving equations arising from the conservation of rates from the forward to the reversed processes (from Kolmogorov) – essentially solving the marginal local balance equations of the forward and reversed components symbolically. That is, the sum of the rates of actions in derivative P must equal the sum of the rates of the actions in the reversed derivative $\overline{P}$. Hence, if we have a PEPA component of the form,

$$P_1 \overset{def}{=} (\alpha_1, r_1).P_2$$
$$P_2 \overset{def}{=} (\alpha_2, r_2).P_3$$
$$P_3 \overset{def}{=} (\alpha_3, r_3).P_1$$

then it is trivial to derive the reversed rates in the reversed model as follows,

$$\overline{P}_1 \overset{def}{=} (\overline{\alpha}_1, r_1).\overline{P}_3$$
$$\overline{P}_2 \overset{def}{=} (\overline{\alpha}_2, r_2).\overline{P}_1$$
$$\overline{P}_3 \overset{def}{=} (\overline{\alpha}_3, r_3).\overline{P}_2$$

In the presence of choice this calculation becomes a little more difficult, but still feasible if all the rates are constant, e.g.

$$P_1 \overset{def}{=} (\alpha_1, r_1).P_2 + (\beta, r_4).P_3$$
$$P_2 \overset{def}{=} (\alpha_2, r_2).P_3$$
$$P_3 \overset{def}{=} (\alpha_3, r_3).P_1$$

The reversed rates in $\overline{P}_3$ are found by symbolically solving the local balance equations as follows. In the forward process we know that,

$$(r_1 + r_4)\pi_1 = r_3\pi_3$$
$$r_2\pi_2 = r_1\pi_1$$
$$r_3\pi_3 = r_4\pi_1 + r_2\pi_2$$

where, π_i is the marginal steady state probability of being in behaviour P_i. In the reversed process we have,

$$a_1\pi_1 = a_2\pi_2 + a_3\pi_3$$
$$a_2\pi_2 = a_4\pi_3$$
$$(a_3 + a_4)\pi_3 = a_1\pi_1$$

Therefore, from Kolmogorov's criteria,

$$r_1 r_2 r_3 = a_1 a_4 a_2$$

and,

$$r_4 r_3 = a_1 a_3$$

Thus the reversed rates in the reversed component are given as follows,

$$\overline{P_1} \stackrel{def}{=} (\overline{\alpha}_1, r_1).\overline{P_3} + (\overline{\beta}, r_4).P3$$
$$\overline{P_2} \stackrel{def}{=} (\overline{\alpha}_2, r_2).\overline{P_1}$$
$$\overline{P_3} \stackrel{def}{=} (\overline{\alpha}_3, \frac{r_1 r_3}{r_1 + r_4}).\overline{P_2} + (\overline{\alpha}_3, \frac{r_3 r_4}{r_1 + r_4}).\overline{P_1}$$

4   Example 1: A Three Node Towsley Model with State Dependent Routing

In order to illustrate the approach, we consider a simple example based on the classic adaptive routing model proposed by Towsley [18]. The model is illustrated in Figure 1.

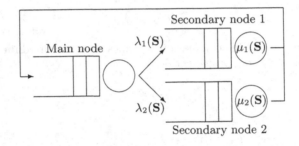

There is one main node and two secondary nodes ($k = 1, 2$) in a closed queueing network. Following service at the main node a job will proceed to either secondary node according to some (state dependent) probability. Following service at a secondary node a job will return to the main node with certainty. This model can be specified in PEPA as follows:

$$Main \stackrel{def}{=} (main_1, \lambda_1(\mathbf{S})).Main + (main_2, \lambda_2(\mathbf{S})).Main$$
$$Secondary_1 \stackrel{def}{=} (service_1, \mu_1(\mathbf{S})).Secondary_1$$
$$Secondary_2 \stackrel{def}{=} (service_2, \mu_2(\mathbf{S})).Secondary_2$$

$$Task_0 \stackrel{def}{=} (main_1, \lambda_1(\mathbf{S})).Task_1 + (main_2, \lambda_2(\mathbf{S})).Task_2$$
$$Task_1 \stackrel{def}{=} (service_1, \mu_1(\mathbf{S})).Task_0$$
$$Task_2 \stackrel{def}{=} (service_2, \mu_2(\mathbf{S})).Task_0$$

$$(Main||Secondary_1||Secondary_2) \underset{\mathcal{L}}{\bowtie} Task_0[N]$$

Where $\mathcal{L} = \{main_1, main_2, service_1, service_2\}$ and $\mathbf{S} = \{i_1, i_2\}$ is the global system state.

In Towsley's model the rates are fixed and only the routing probabilities are state dependent. Here we make all rates and probabilities state dependent. The service rates at the secondary nodes are $\mu_k(i_1, i_2)$, $k = 1, 2$; $\lambda_k(i_1, i_2)$ denotes the service rate at the main node of jobs proceeding to node k, $k = 1, 2$, where i_1 and i_2 are the number of jobs at secondary nodes 1 and 2 respectively. Obviously there will be $n - i_1 - i_2$ jobs at the main node when the total population is n .

The individual components in the model can be reversed to find a notional reversed model, specified in PEPA as follows.

$$Main \stackrel{def}{=} (main_1, \overline{\mu}_1(\mathbf{S})).Main + (main_2, \overline{\mu}_2(\mathbf{S})).Main$$
$$Secondary_1 \stackrel{def}{=} (service_1, \overline{\lambda}_1(\mathbf{S})).Secondary_1$$
$$Secondary_2 \stackrel{def}{=} (service_2, \overline{\lambda}_2(\mathbf{S})).Secondary_2$$

$$Task_0 \stackrel{def}{=} (main_1, \overline{\mu}_1(\mathbf{S})).Task_1 + (main_2, \overline{\mu}_2(\mathbf{S})).Task_2$$
$$Task_1 \stackrel{def}{=} (service_1, \overline{\lambda}_1(\mathbf{S})).Task_0$$
$$Task_2 \stackrel{def}{=} (service_2, \overline{\lambda}_2(\mathbf{S})).Task_0$$

$$(Main \| Secondary_1 \| Secondary_2) \underset{\mathcal{L}}{\bowtie} Task_0[N]$$

Note that we have not yet defined $\overline{\mu}_k$ and $\overline{\lambda}_k$, nor do we know that this notional reversed model is the true reversed, i.e. we do not yet know that the model has a product form.

The original (forward) model and the notional reversed model give rise to four *minimal cycles*, corresponding to

1. A $main_1$ action followed by a $service_1$ action.
2. A $main_2$ action followed by a $service_2$ action.
3. A $main_1$ action followed by a $main_2$ action followed by a $service_1$ action followed by a $service_2$ action.
4. A $main_2$ action followed by a $main_1$ action followed by a $service_2$ action followed by a $service_1$ action.

We do not need to consider more cycles as longer cycles do not give any additional conditions because they are always formed from combinations of these minimal cycles.

The Kolmogorov criteria for the minimal cycles are given as follows:

$$\lambda_1(i_1, i_2)\mu_1(i_1 + 1, i_2) = \overline{\mu}_1(i_1, i_2)\overline{\lambda}_1(i_1 + 1, i_2) \tag{2}$$

$$\lambda_2(i_1, i_2)\mu_2(i_1, i_2 + 1) = \overline{\mu}_2(i_1, i_2)\overline{\lambda}_2(i_1, i_2 + 1) \tag{3}$$

$$\lambda_1(i_1, i_2)\lambda_2(i_1 + 1, i_2)\mu_1(i_1 + 1, i_2 + 1)\mu_2(i_1, i_2 + 1) =$$
$$\overline{\mu}_2(i_1, i_2)\overline{\mu}_1(i_1, i_2 + 1)\overline{\lambda}_2(i_1 + 1, i_2 + 1)\overline{\lambda}_1(i_1 + 1, i_2) \tag{4}$$

$$\lambda_2(i_1, i_2)\lambda_1(i_1, i_2 + 1)\mu_2(i_1 + 1, i_2 + 1)\mu_1(i_1 + 1, i_2) =$$

$$\overline{\mu}_1(i_1, i_2)\overline{\mu}_2(i_1 + 1, i_2)\overline{\lambda}_1(i_1 + 1, i_2 + 1)\overline{\lambda}_2(i_1, i_2 + 1) \tag{5}$$

Hence, it follows that,

$$\overline{\mu}_1(i_1, i_2) = x_1\lambda_1(i_1, i_2)$$
$$\overline{\mu}_2(i_1, i_2) = x_2\lambda_2(i_1, i_2)$$
$$\overline{\lambda}_1(i_1, i_2) = \mu_1(i_1, i_2)/x_1$$
$$\overline{\lambda}_2(i_1, i_2) = \mu_2(i_1, i_2)/x_2$$

where x_1 and x_2 are constants.

Thus we find the necessary condition for satisfying the Kolmogorov criteria to be:

$$\frac{\lambda_1(i_1, i_2)}{\lambda_1(i_1, i_2 + 1)} \frac{\mu_1(i_1 + 1, i_2 + 1)}{\mu_1(i_1 + 1, i_2)} =$$

$$\frac{\lambda_2(i_1, i_2)}{\lambda_2(i_1 + 1, i_2)} \frac{\mu_2(i_1 + 1, i_2 + 1)}{\mu_2(i_1, i_2 + 1)} \tag{6}$$

Since the Kolmogorov criteria are satisfied, a product form solution exists when (6) is satisfied. This solution will have the form,

$$\pi_{ij} = \pi_{00} \prod_{m=0}^{i-1} \frac{\lambda_1(m, 0)}{\mu_1(m + 1, 0)} \prod_{n=0}^{j-1} \frac{\lambda_2(i, n)}{\mu_2(i, n + 1)} \quad , i + j \leq N$$

where π_{00} is the steady state probability that all the jobs are located at the main node.

4.1   Fixed Rates

In Towsley's paper [18] the rates are fixed, but the routing probabilities are state dependent: $\mu_k(i_1, i_2) = \mu_k$, $\lambda_k(i_1, i_2) = p_k(i_1, i_2)\lambda$, where $p_2(i_1, i_2) = 1 - p_1(i_1, i_2)$. Hence,

$$(1 - p_1(i_1, i_2))p_1(i_1, i_2 + 1) =$$
$$(1 - p_1(i_1 + 1, i_2)p_1(i_1, i_2) \tag{7}$$

This is clearly satisfied if

$$p_1(i_1, i_2) = \frac{i_1 + a}{b(i_1 + i_2) + c}$$

Recall that $p_2(i_1, i_2) = 1 - p_1(i_1, i_2)$, hence

$$p_k(i_1, i_2) = \frac{i_k + a_k}{i_1 + i_2 + c} \tag{8}$$

where $c = a_1 + a_2$. Equation (8) is equivalent to Towsley's result, although there are clearly also other p_k's satisfying (7); i.e. Towsley's result was a special case. Our result simply extends to more than two secondary nodes.

4.2 State Dependent Rates

Now recall equation (6):

$$\frac{\lambda_1(i_1, i_2)}{\lambda_1(i_1, i_2 + 1)} \frac{\mu_1(i_1 + 1, i_2 + 1)}{\mu_1(i_1 + 1, i_2)} =$$

$$\frac{\lambda_2(i_1, i_2)}{\lambda_2(i_1 + 1, i_2)} \frac{\mu_2(i_1 + 1, i_2 + 1)}{\mu_2(i_1, i_2 + 1)}$$

This is a generalisation of Towsley's result where the service rates are state dependent. Equation (6) is clearly satisfied if

- Rates are either dependent on the total number of jobs at all secondary nodes, or only on the number of jobs at the target secondary node:
$\lambda_k(i_1, i_2) = \lambda_k f_\lambda(\sum_{\forall j} i_j)$ or $\lambda_k(i_1, i_2) = \lambda_k(i_k)$
and
$\mu_k(i_1, i_2) = \mu_k f_k(\sum_{\forall j} i_j)$ or $\mu_k(i_1, i_2) = \mu_k(i_k)$
- Functional dependence is the same at all secondary nodes:

$$\lambda_k(i_1, i_2) = \lambda_k f(i_1, i_2)$$

and

$$\mu_k(i_1, i_2) = \mu_k f(i_1, i_2).$$

- Arrival and departure rates at each secondary node are dependent on the total number of jobs at all other secondary nodes:

$$\lambda_k(i_1, i_2) = \lambda_k f_k(\sum_{j \neq k} i_j)$$

and

$$\mu_k(i_1, i_2) = \mu_k f_k(\sum_{j \neq k} i_j).$$

Where $f(.)$, $f_k(.)$ and $f_\lambda(.)$ are arbitrary functions.

4.3 Minimal Cycles and Multiple Dimensions

In the derivation above we have employed the notion of *minimal cycles* without explanation. The reader will have observed that the longest minimal cycles used are simple squares, consisting of services at the main node of one type, then the other type, followed by services at the secondary nodes of the first type and then the second type. Why, one might ask, do we not consider multiple services of each type? The reason is simply that we do not need to, as such cycles would produce multiple redundant equations, even in the case of state dependent rates. The proof of this condition is not complex for this example, an outline of the general proof is given in [9].

It is a simple matter to consider more secondary nodes simply by modifying the model as follows.

$$Main \overset{def}{=} \sum_{i=1}^{K}(main_i, \overline{\mu}_i(\mathbf{S})).Main$$

$$Secondary_i \overset{def}{=} (service_i, \overline{\lambda}_i(\mathbf{S})).Secondary_i \ , \ i = 1, \ldots, K$$

$$Task_0 \overset{def}{=} \sum_{i=1}^{K}(main_i, \overline{\mu}_i(\mathbf{S})).Task_i$$

$$Task_i \overset{def}{=} (service_i, \overline{\lambda}_i(\mathbf{S})).Task_0 \ , \ i = 1, \ldots, K$$

$$(Main||Secondary_1|| \ldots ||Secondary_K) \underset{\mathcal{L}}{\bowtie} Task_0[N]$$

$$\mathcal{L} = \{main_i, service_i\} \ , \ i = 1, \ldots, K$$

Such a model clearly gives rise to more minimal cycles. Intuitively it might seem that the longest minimal cycles would be formed when there is a sequence of all possible $main_i$ actions followed by all possible $service_i$ actions (in a different order). However, a closer inspection reveals all cycles can in fact be constructed by considering minimal cycles of maximum length 4, i.e. consisting only of actions shared between $Task_i$, $Main$ and at most two $Secondary_i$'s.

5   Example 2: A Non-repudiation Protocol

Non-repudiation protocols are used to prevent participants in a communication from later falsely denying that they took part in that communication. There are many such protocols, with different properties. The one depicted here, first proposed by Zhou and Gollmann [27], utilises a secure server, known as a *Trusted Third Party*, or TTP. The protocol is defined as follows.

$$(request) \ 1.A \rightarrow TTP : f_{NRO}, TTP, B, M, NRO$$
$$(publish1\&$$
$$getByA1) \ 2.A \leftrightarrow TTP : f_{NRS}, A, B, T_s, L, NRS$$
$$(sendB) \ 3.TTP \rightarrow B : A, L, NRO$$
$$(sendTTP) \ 4.B \rightarrow TTP : f_{NRR}, L, NRR$$
$$(publish2\&$$
$$getByB) \ 5.B \leftrightarrow TTP : L, M$$
$$(publish2\&$$
$$getByA2) \ 6.A \leftrightarrow TTP : f_{NRD}, T_d, L, NRR, NRD$$

The action names preceding each item are included for convenience for the reader to observe the correlation between this definition and the following PEPA specification, taken from [25].

$$TTP \stackrel{def}{=} (publish_1, F_1(\mathbf{S})).TTP + (publish_2, F_2(\mathbf{S})).TTP$$
$$+(sendB, F_3(\mathbf{S})).TTP$$

$$AB_0 \stackrel{def}{=} (request, rt_1).AB_1$$

$$AB_1 \stackrel{def}{=} (publish_1, F_1(\mathbf{S})).AB_2$$

$$AB_2 \stackrel{def}{=} (getByA_0, rga_0).AB_3$$

$$AB_3 \stackrel{def}{=} (sendB, F_3(\mathbf{S})).AB_4$$

$$AB_4 \stackrel{def}{=} (sendTTP, rt_2).AB_5$$

$$AB_5 \stackrel{def}{=} (publish_2, F_2(\mathbf{S})).AB_6$$

$$AB_6 \stackrel{def}{=} (getByB, rgb).AB_7 + (getByA, rga).AB_8$$

$$AB_7 \stackrel{def}{=} (getByA, rga).AB_9$$

$$AB_8 \stackrel{def}{=} (getByB, rgb).AB_9$$

$$AB_9 \stackrel{def}{=} (work, rw).AB_0$$

$$System = TTP \underset{\{publish_1, publish_2, sendB\}}{\bowtie} AB_0[N]$$

The model depicts a single TTP with N client pairs (*Alice* and *Bob*, denoted *AB*). The protocol utilises publishing in a public space (e.g. a bulletin board) by the TTP, from where the clients each download. In the specification above we are only concerned with contention on the publication by the TTP, although it would also be possible to specify an additional component to act as a web server. It is clear that the PEPA model does not include the message contents given in the formal protocol definition. As such this PEPA specification could not be used to perform any kind of security analysis on this model, instead we are concerned solely with understanding the scalability of the TTP.

In this model actions performed by, or with, the TTP ($publish_1$, $publish_2$ and $sendB$) are given functional rates, where the function may be dependent on the global state space $\mathbf{S}$. If we assume that these functions are only dependent on the number of components behaving as AB_1, AB_3 and AB_5 then, for convenience, we can simplify the model to hide actions which are not shared and combine sequences of multiple hidden actions within a component without loss of generality, as follows.

$$TTP \stackrel{def}{=} (publish_1, F_1(\mathbf{S}^*)).TTP + (publish_2, F_2(\mathbf{S}^*)).TTP$$
$$+(sendB, F_3(\mathbf{S}^*)).TTP$$

$$AB_1 \stackrel{def}{=} (publish_1, F_1(\mathbf{S}^*)).AB_2$$

$$AB_2 \stackrel{def}{=} (\tau, rga_0).AB_3$$

$$AB_3 \stackrel{def}{=} (sendB, F_3(\mathbf{S}^*)).AB_4$$

$$AB_4 \stackrel{def}{=} (\tau, rt_2).AB_5$$

$$AB_5 \stackrel{def}{=} (publish_2, F_2(\mathbf{S}^*)).AB_6$$

$$AB_6 \stackrel{def}{=} (\tau, rX).AB_1$$

$$System = TTP \underset{\{publish_1, publish_2, sendB\}}{\bowtie} AB_6[N]$$

Where, $\mathbf{S}^* = \{s_1, s_3, s_5\}$.

It is now a simple matter to derive the notional reversed model.

$$\overline{TTP} \stackrel{def}{=} (\overline{publish_1}, F_1(\mathbf{S}^*)).\overline{TTP} + (\overline{publish_2}, F_2(\mathbf{S}^*)).\overline{TTP}$$
$$+ (\overline{sendB}, F_3(\mathbf{S}^*)).\overline{TTP}$$
$$\overline{AB}_1 \stackrel{def}{=} (\overline{publish_1}, F_1(\mathbf{S}^*)).\overline{AB}_6$$
$$\overline{AB}_2 \stackrel{def}{=} (\tau, rga_0).\overline{AB}_1$$
$$\overline{AB}_3 \stackrel{def}{=} (\overline{sendB}, F_3(\mathbf{S}^*)).\overline{AB}_2$$
$$\overline{AB}_4 \stackrel{def}{=} (\tau, rt_2).\overline{AB}_3$$
$$\overline{AB}_5 \stackrel{def}{=} (\overline{publish_2}, F_2(\mathbf{S}^*)).\overline{AB}_4$$
$$\overline{AB}_6 \stackrel{def}{=} (\tau, rX).\overline{AB}_5$$
$$System = TTP \underset{\{publish_1, publish_2, sendB\}}{\bowtie} \overline{AB}_1[N]$$

A semi-product form can now be derived (if it exists) of the form:

$$\pi_{\mathbf{S}} = \pi_{\mathbf{S}_0} \frac{rX^{s_1}}{\prod_{i=1}^{s_1} F_1(i,0,0)} \left(\frac{rx}{rga_0} \right)^{s_2} \frac{rX^{s_3}}{\prod_{j=1}^{s_3} F_3(s_1,j,0)} \left(\frac{rx}{rt_2} \right)^{s_4} \frac{rX^{s_5}}{\prod_{k=1}^{s_5} F_3(s_1,s_3,k)}$$

Where $\mathbf{S} = \{s_1, s_2, s_3, s_4, s_5, s_6\}$ and $\mathbf{S}_0 = \{0,0,0,0,0,N\}$.

The conditions for this result to exist are found by again considering the minimal cycles. In this case we are only concerned with cycles involving the shared actions, so we can ignore the τ actions (nb. if we included the τ actions then they would cancel out in the forward and reversed rate products).

1. A $publish_1$ followed by a $sendB$ followed by a $publish_2$.
2. A $publish_1$ followed by a $publish_2$ followed by a $sendB$.
3. A $sendB$ followed by a $publish_2$ followed by a $publish_1$.
4. A $sendB$ followed by a $publish_1$ followed by a $publish_2$.
5. A $publish_2$ followed by a $publish_1$ followed by a $sendB$.
6. A $publish_2$ followed by a $sendB$ followed by a $publish_1$.

Using Kolmogorov's criteria, cycle 2) gives,

$$F_1(s_1, s_3, s_5)F_2(s_1 - 1, s_3 + 1, s_5)F_3(s_1, s_3 + 1, s_5 - 1) =$$
$$F_2(s_1, s_3, s_5)F_1(s_1, s_3 + 1, s_5 - 1)F_3(s_1 - 1, s_3 + 1, s_5)$$

Cycle 4) gives,

$$F_3(s_1, s_3, s_5)F_1(s_1, s_3 - 1, s_5 + 1)F_2(s_1 - 1, s_3, s_5 + 1) =$$
$$F_1(s_1, s_3, s_5)F_3(s_1 - 1, s_3, s_5 + 1)F_2(s_1, s_3 - 1, s_5 + 1)$$

And cycle 6) gives,

$$F_2(s_1, s_3, s_5)F_3(s_1 + 1, s_3, s_5 - 1)F_1(s_1 + 1, s_3 - 1, s_5) =$$

$$F_3(s_1, s_3, s_5)F_2(s_1 + 1, s_3 - 1, s_5)F_2(s_1 + 1, s_3, s_5 - 1)$$

which hold if

$$F_i(\mathbf{S}^*) = F_i(s_1 + s_3 + s_5)$$

Or,

$$F_i(\mathbf{S}^*) = c_i F(\mathbf{S}^*)$$

Where the c_i's are constants and $F(\mathbf{S}^*)$ is an arbitrary function.

6   Conclusions and Further Work

This paper has demonstrated the use of a practical and intuitive method to derive separable steady state solutions for a class of PEPA models. The task of mechanising this method has not yet been tackled and is an obvious next step in facilitating its use. In this paper we have demonstrated how to find a notional reversed component symbolically. However, it is also possible to find a notional reversed component directly from the solution of the marginal steady state equations – sometimes simpler than dealing with non-linear equations that arise from Kolmogorov's criteria. In the case where the number of actions in a component is large there may be a significant saving in computational effort by taking this more pragmatic approach.

The task of finding a numerical solution does not stop with identifying a product form. Product form solutions of finite models, such as those considered here, generally include a normalising constant so that the resultant sum of all possible probabilities is equal to one. Finding this constant in general is non-trivial. There are a number of approaches that might be used to find the normalising constant, but perhaps the one which is most obvious in this context is mean value analysis [23]. The current characterisations of mean value analysis for PEPA do not cover the class of model considered in this paper. However, the arrival theorem, on which mean value analysis depends, holds for any product form queueing network [17] and it should therefore be feasible to extend the result in [23] to the class of model with functional rates considered here.

References

1. Bonald, T., Proutiere, A.: Insensitivity in processor-sharing networks. Performance Evaluation 49, 193–209 (2002)
2. Boucherie, R., van Dijk, N.: Product forms for queueing networks with state-dependent multiple job transitions. Advances in Applied Probability 23, 152–187 (1991)
3. Chao, X., Miyazawa, M., Pinedo, M.: Queueing Networks: Customers, Signals and Product Form Solutions. Wiley (1999)
4. Derisavi, S., Hermanns, H., Sanders, W.: Optimal state-space lumping in Markov chains. Information Processing Letters 87, 309–315 (2003)
5. Harrison, P.G.: Turning back time in Markovian process algebra. Theoretical Computer Science 290(3), 1947–1986 (2003)

6. Harrison, P.G.: Reversed processes, product-forms and a non-product-form. Linear Algebra and Its Applications 386, 359–381 (2004)
7. Harrison, P.G., Lee, T.T.: Separable equilibrium state probabilities via time reversal in Markovian process algebra. Theoretical Comp. Sci. 346(1), 161–182 (2005)
8. Harrison, P.G.: Product-forms and functional rates. Performance Evaluation 66, 660–663 (2009)
9. Harrison, P.G.: Process algebraic non-product-forms. Electronic Notes in Theoretical Computer Science 151(3) (2006)
10. Harrison, P.G., Thomas, N.: Product form solution in PEPA via the reversed process. In: Kouvatsos, D. (ed.) Next Generation Internet. LNCS, vol. 5233, pp. 343–356. Springer, Heidelberg (2011)
11. Hayden, R., Bradley, J.: A fluid analysis framework for a Markovian process algebra. Theoretical Computer Science 411(22), 2260–2297 (2010)
12. Henderson, W., Taylor, P.: State-dependent Coupling of Quasireversible Nodes. Queueing Systems: Theory and Applications 37(1/3), 163–197 (2001)
13. Hillston, J.: A Compositional Approach to Performance Modelling. Cambridge University Press (1996)
14. Hillston, J.: Exploiting Structure in Solution: Decomposing Compositional Models. In: Brinksma, E., Hermanns, H., Katoen, J.-P. (eds.) FMPA 2000. LNCS, vol. 2090, pp. 278–314. Springer, Heidelberg (2001)
15. Hillston, J.: Fluid flow approximation of PEPA models. In: Proceedings of QEST 2005, pp. 33–43. IEEE Computer Society (2005)
16. Kelly, F.P.: Reversibility and stochastic networks. Wiley (1979)
17. Sevcik, K., Mitrani, I.: The distribution of queueing network states at input and output instants. JACM 28(2), 358–371 (1981)
18. Towsley, D.: Queuing Network Models with State-Dependent Routing. Journal of the ACM 27(2), 323–337 (1980)
19. Thomas, N.: Using ODEs from PEPA models to derive asymptotic solutions for a class of closed queueing networks. In: 8th Workshop on Process Algebra and Stochastically Timed Activities, PASTA (2008)
20. Thomas, N., Gilmore, S.: Applying quasi-separability to Markovian process algebra. In: Proceedings 6th International Workshop on Process Algebra and Performance Modelling (1998)
21. Thomas, N., Bradley, J., Thornley, D.: Approximate solution of PEPA models using component substitution. In: IEE Proceedings - Computers and Digital Techniques, vol. 150(2), pp. 67–74 (2003)
22. Thomas, N., Zhao, Y.: Fluid flow analysis of a model of a secure key distribution centre. In: Proceedings of the 24th Annual UK Performance Engineering Workshop, Imperial College London (2008)
23. Thomas, N., Zhao, Y.: Mean value analysis for a class of PEPA models. The Computer Journal 54(5), 643–652 (2011)
24. Tribastone, M.: Approximate Mean Value Analysis of Process Algebra Models. In: Proceedings of MASCOTS, pp. 369–378 (2011)
25. Zhao, Y., Thomas, N.: Comparing Methods for the Efficient Analysis of PEPA Models of Non-repudiation Protocols. In: Proceedings of 15th International Conference on Parallel and Distributed Systems, pp. 821–827 (2009)
26. Zhao, Y., Thomas, N.: Efficient solutions of a PEPA model of a key distribution centre. Performance Evaluation 67(8), 740–756 (2010)
27. Zhou, J., Gollmann, D.: An efficient non-repudiation protocol. In: Proceedings of the 10th Computer Security Foundations Workshop (CSFW 1997). IEEE Computer Society (1997)

Trading Off Subtask Dispersion and Response Time in Split-Merge Systems

Iryna Tsimashenka and William J. Knottenbelt

Imperial College London, 180 Queen's Gate,
London SW7 2AZ, United Kingdom
{it09,wjk}@doc.ic.ac.uk

Abstract. In many real-world systems incoming tasks split into subtasks which are processed by a set of parallel servers. In such systems two metrics are of potential interest: response time and subtask dispersion. Previous research has been focused on the minimisation of one, but not both, of these metrics. In particular, in our previous work, we showed how the processing of selected subtasks can be delayed in order to minimise expected subtask dispersion and percentiles of subtask dispersion in the context of split-merge systems. However, the introduction of subtask delays obviously impacts adversely on task response time and maximum sustainable system throughput. In the present work, we describe a methodology for managing the trade off between subtask dispersion and task response time. The objective function of the minimisation is based on the product of expected subtask dispersion and expected task response time. Compared with our previous methodology, we show how our new technique can achieve comparable subtask dispersion with substantial improvements in expected task response time.

Keywords: Split-Merge System, Subtask Dispersion, Response Time, Trade-Off.

1 Introduction

In the modern world we are surrounded by systems in which incoming tasks can naturally be decomposed into subtasks that are processed by a set of parallel servers. Completed subtasks are held in an output buffer pending arrival of its sibling subtasks. When all child subtasks of a given task have completed service and are present in the output buffer, the task is deemed to have completed service. Two classes of performance metrics are of concern in such systems: those related to task response time and those related to subtask dispersion (i.e. the time between the arrival of the first and last subtasks originating from a given task in the output buffer). The former class has been the subject of extensive research [2, 3, 6, 7, 9–11, 13, 14, 16, 18, 21, 22], while the latter class has received less attention [19, 20], although reducing subtask dispersion can be of critical importance in certain real-life contexts.

Consider by way of example the processing of customer orders in an automated warehouse. Incoming orders (tasks) are made up of several items (subtasks), each of which must be retrieved from a different part of the warehouse. Partially completed orders must be held in an output buffer, and each order can only be released from the output buffer and dispatched to the customer when all items making up the order

A. Dudin and K. De Turck (Eds.): ASMTA 2013, LNCS 7984, pp. 431–442, 2013.
© Springer-Verlag Berlin Heidelberg 2013

have been retrieved. The output buffer space is often limited and difficult to manage on account of its high utilisation, so it is important to keep subtask dispersion low. At the same time keeping task response time low (i.e. increasing system throughput) is an important concern. Another example is a restaurant in which customer orders (tasks) consisting of different menu items (subtasks) must be concurrently prepared such that all dishes for a particular table of customers are ready at roughly the same time. In the mean time, partially completed orders for tables are held on a service counter (the output buffer), which should not be overburdened. Simultaneously a good standard of customer service dictates that customers should receive their orders in reasonable time.

Parallel queueing systems are natural modelling abstractions for these kinds of systems. Here we focus on an analytically tractable subclass of such systems, namely split-merge systems. In our previous research [19,20] we showed that reduction of subtask dispersion in such systems can be achieved through the application of judiciously chosen delays to subtask processing. This naturally has an adverse impact on task response time (as characterised in [12]), and therefore on maximum sustainable system throughput. This adverse impact increases with workload intensity, and can even result in previously stable systems becoming unstable. There is therefore an urgent need to define an analtyical framework which allows effective balancing of subtask dispersion and response time concerns. To this end, we formally characterise the subtask dispersion–response time trade-off in split-merge systems as an optimisation problem. The objective function is the product of the expected subtask dispersion and expected task response times. We were inspired in this research by the survey on multi-objective optimisation techniques by Marler and Arora [15], in particular the exposition of product methods, as well as the work of Gandhi, Harchol-Balter et al. [8] which explored energy–performance trade-offs in server farms by means of an objective function based on the energy–response time product (ERP).

The remainder of this paper organised as follows. Section 2 presents definitions of split-merge queueing systems, and important results related to the theory of homogeneous and heterogeneous order statistics. Section 3 characterises the subtask dispersion–response time trade-off in split-merge systems and presents a methodology for its optimisation based on a modified Newton's method. Section 4 presents a case study which illustrates the benefits of our proposed approach in comparison to approaches based on the unique minimisation of either subtask dispersion or task response time. Section 5 concludes and considers directions for future work.

2 Preliminaries

2.1 Split-Merge Systems

As shown in Fig. 1, a *split-merge system* consists of split and merge points, a queue before the split point (a split queue) and N heterogeneous parallel servers with queueing capability after service (merge buffers). Tasks enter the split queue according to a Poisson process with mean rate λ. Whenever all parallel servers are idle and the split queue is not empty, a task is taken from the head of the split queue and is injected into the system, splitting into N subtasks at the split point. Each subtask is sent to its allocated parallel server where it is served according to a general service time distribution with mean $1/\mu_i$, $i = 1, ..., N$. Completed subtasks enter a merge buffer. When all subtasks (originating from a particular task) are present in the merge buffers, the original task exits the system via the merge point.

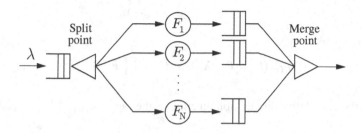

Fig. 1. Split–Merge queueing model

In the Fig. 1 the merge buffers have been shown as separate entities but in many real applications they share the same physical space, which we will term the *output buffer*.

2.2 Theory of Order Statistics [5]

The theory of order statistics studies the behaviour and properties of arranged random variables and the statistics derived from them [5].

Definition 1. *Let the increasing sequence* $X_{(1)}, X_{(2)}, \ldots, X_{(n)}$ *be a permutation of the real-valued random variables* $X_1, X_2, \ldots, X_n$, *i.e. the* X_i *arranged in increasing order* $X_{(1)} \leqslant X_{(2)} \leqslant \ldots \leqslant X_{(n)}$. *Then* $X_{(i)}$ *is the ith order statistic for* $i = 1, 2, \ldots, n$.

The random variables X_i are typically assumed to be identically and independently distributed (iid) with cumulative distribution function $F(t)$, but of course the $X_{(i)}$ are dependent because of the ordering. The extreme values $X_{(1)}$ and $X_{(n)}$ are the minimum and maximum order statistics respectively. $T = X_{(n)} - X_{(1)}$ is the *range*.

2.3 Theory of Heterogeneous Order Statistics

Relaxing the assumption that the X_i should be identically distributed leads to the theory of heterogeneous order statistics [4, 20].

Definition 2. *We consider n independent, real-valued random variables* $X_1, \ldots, X_n$ *where each* X_i *has an arbitrary probability distribution* $F_i(t)$ *and probability density function* $f_i(t) = F_i'(t)$. *In this case of "heterogeneous" (or independent, but not necessarily identically distributed) random variables, we call the corresponding order statistics heterogeneous order statistics.*

The cumulative distribution functions of the minimum and maximum order statistics are:

$$F_{(1)}(t) = Pr\{X_{(1)} \leqslant t\} = 1 - \prod_{i=1}^{n}[1 - F_i(t)],$$

and

$$F_{(n)}(t) = Pr\{X_{(n)} \leqslant t\} = \prod_{i=1}^{n} F_i(t).$$

In addition the cumulative distribution function of the range $X_{(n)} - X_{(1)}$ is [20]:

$$F_{range}(t) = \sum_{i=1}^{n} \int_{-\infty}^{\infty} f_i(x) \prod_{j=1, j\neq i}^{n} [F_j(x+t) - F_j(x)] \, dx \qquad (1)$$

3 Subtask Dispersion–Response Time Trade-Off

3.1 Metrics for Split-Merge Systems

Two important metrics of operational interest in split-merge systems are subtask dispersion, defined as the difference between the arrival times of the first and last subtasks originating from a given task in the output buffer, and task response time, defined as the difference between the arrival time of a task in the split queue and the time at which the task leaves the system via the merge point. There is a conflicting tension between these two metrics. Indeed, in previous research we have considered how to minimise mean subtask dispersion [12, 19] and percentiles of subtask dispersion [20] by delaying the processing of the subtasks in such a way as to cluster subtask completion times. Whilst we apply a constraint to ensure that no delay is added to the bottleneck server, the introduction of subtask delays does naturally have an adverse impact on mean task response time (as quantified in [12]), with a corresponding reduction in maximum sustainable system throughput.

3.2 Application of Heterogeneous Order Statistics to Split-Merge Systems

We consider a split-merge system with i parallel servers. Suppose $X_i \sim F_i(x)$ is a random variable that describes the (heterogenous) service time distribution of the ith parallel server. Then the heterogeneous order statistics $X_{(i)}$ correspond to ordered subtask completion times (since in a split-merge system all subtasks originating from a given task start service at the same time). The minimum heterogeneous order statistic $X_{(1)}$ corresponds to the time of first subtask completion and the maximum heterogeneous order statistic $X_{(n)}$ corresponds to the time of last subtask completion (equivalently, task response time). The range $X_{(n)} - X_{(1)}$ corresponds to subtask dispersion.

3.3 Introducing Deterministic Subtask Delays

As in our previous research, in order to provide a means to control the subtask dispersion–response time trade-off we introduce a vector of deterministic delays $\mathbf{d} = (d_1, d_2, ...d_n)$, where element d_i represents the delay applied to the processing times of the ith parallel server. The random variables that describe the parallel service times with applied delays are: $X_i^{\mathbf{d}} \sim F_i(x - d_i) \; \forall i$ with corresponding heterogeneous order statistics $X_{(1)}^{\mathbf{d}}, X_{(2)}^{\mathbf{d}}, \ldots, X_{(n)}^{\mathbf{d}}$.

3.4 An Objective Function for the Subtask Dispersion–Response Time Trade-Off

In order to express the subtask dispersion–response time trade-off for a split-merge system subject to subtask delay vector $\mathbf{d}$ we create an objective function formed from the product of mean task response time ($\mathbb{E}[R_{\lambda,\mathbf{d}}]$) and mean subtask dispersion ($\mathbb{E}[D_{\mathbf{d}}]$). The former metric is computed using the Pollaczek-Khinchine formula for mean task response time in an $M/G/1$ queue with service distribution $X_{(n)}^{\mathbf{d}}$. This is because a split-merge system is conceptually equivalent to an $M/G/1$ queue whose service time is the maximum of its set of parallel service times (giving mean service time $\mu^{-1} = \mathbb{E}[X_{(n)}^{\mathbf{d}}]$). The latter metric is computed as the expected difference between the maximum and minimum heterogeneous order statistics. We note dependence between order statistics is irrelevant when considering mean values due to the linearity property of expectation operator, i.e. $\mathbb{E}[D_{\mathbf{d}}] = \mathbb{E}[X_{(n)}^{\mathbf{d}} - X_{(1)}^{\mathbf{d}}] = \mathbb{E}[X_{(n)}^{\mathbf{d}}] - \mathbb{E}[X_{(1)}^{\mathbf{d}}]$.

We thus express the trade-off as a function of $\mathbf{d}$ and λ:

$$T(\mathbf{d}, \lambda) = \mathbb{E}[R_{\mathbf{d},\lambda}]\mathbb{E}[D_{\mathbf{d}}] = \mathbb{E}[R_{\mathbf{d},\lambda}](\mathbb{E}[X_{(n)}^{\mathbf{d}}] - \mathbb{E}[X_{(1)}^{\mathbf{d}}]) \tag{2}$$

$$= \left(\frac{\rho + \lambda\mu Var[X_{(n)}^{\mathbf{d}}]}{2(\mu - \lambda)} + \mu^{-1} \right)$$

$$\times \left(\int_0^\infty 1 - \prod_{i=1}^n F_i(x - d_i)dx - \int_0^\infty \left(1 - \left(1 - \prod_{i=1}^n (1 - F_i(x - d_i))\right)\right) dx \right)$$

The variance of $X_{(n)}^{\mathbf{d}}$ can be computed as:

$$Var[X_{(n)}^{\mathbf{d}}] = 2 \int_0^\infty x\left(1 - \prod_{i=1}^n F(x - d_i)\right) dx - \left(\int_0^\infty 1 - \prod_{i=1}^n F(x - d_i) dx \right)^2$$

In the above we have assumed that each major component of the objective function (i.e. mean subtask dispersion and mean task response time) should be given equal weighting. We note that, in line with the treatment of weighted product methods in [15], each component can be raised to a different exponent (> 1) in order to express a preference about the relative importance of the components.

3.5 Optimising the Objective Function

We seek the vector of subtask delays $\mathbf{d}_{\min}$ which minimises $T(\mathbf{d}, \lambda)$. That is,

$$\mathbf{d}_{\min} = \arg \min_{\mathbf{d}} T(\mathbf{d}, \lambda) \tag{3}$$

We can apply Newton's method to find d_{min} iteratively:

$$d_{k+1} = d_k - \gamma [H_{T(d_k,\lambda)}]^{-1} \nabla T(d_k, \lambda), \ k \geq 0 \qquad (4)$$

where $H_{T(d_k,\lambda)}$ is the Hessian matrix (matrix of second order partial derivatives) of the objective function $T(d_k, \lambda)$ and d_k is the kth iterate of the subtask delay vector. The initial subtask delay vector is chosen heuristically as:

$$d_0 = ((\max_i \mathbb{E}[X_i] - \mathbb{E}[X_1])(1 - \rho), \ldots, (\max_i \mathbb{E}[X_i] - \mathbb{E}[X_n])(1 - \rho)) \qquad (5)$$

where $\rho = \lambda/\mu$.

In Eq. (4) $\gamma < 1$ is a constant introduced to satisfy the Wolfe conditions [23, 24]. The step size γ needs to be chosen to be small enough to support convergence yet large enough to make rapid progress towards the minimum. There are two inequalities which must hold to ensure this. Firstly:

$$T\Big(d_k + \gamma\big(-H_{T(d_k,\lambda)}^{-1}\nabla T(d_k,\lambda)\big),\lambda\Big) \leq \\ T(d_k,\lambda) + c_1\gamma\nabla T^T(d_k,\lambda)\big(-H_{T(d_k,\lambda)}^{-1}\nabla T(d_k,\lambda)\big) \qquad (6)$$

And secondly:

$$\nabla T^T\Big(d_k + \gamma(-H_{T(d_k,\lambda)}^{-1}\nabla T(d_k,\lambda)),\lambda\Big)\big(-H_{T(d_k,\lambda)}^{-1}\nabla T(d_k,\lambda)\big) \geq \\ c_2\nabla T^T(d_k,\lambda)\big(-H_{T(d_k,\lambda)}^{-1}\nabla T(d_k,\lambda)\big) \qquad (7)$$

Here c_1 and c_2 are constants, which should be chosen such that $0 < c_1 < c_2 < 1$, $c_1 \ll c_2$. Practically, for the purposes of Newton and quasi-Newton methods it is recommended to set c_1 as 10^{-4} and for c_2 to take on 0.9 [17]. Inequality (6) above corresponds to the Armijo rule [1] which guarantees that the step size γ will decrease the objective function sufficiently, while Inequality (7) ensures the curvature condition.

3.6 Implementation

We have implemented the above optimisation technique in C++. Using Newton's method, we begin by initialising d_0 according to Eq. (5) and choose an appropriate γ satisfying Inequalities (6) and (7). On each iteration k the method calculates the inverse Hessian matrix and gradient of the objective function, which gives a direction for the updated vector d_{k+1}. Evaluation of the objective function involves computation of mean task response time using the Pollaczek-Khinchine formula and computation of mean subtask dispersion, which in turn involves evaluating the expected value of the minimum and maximum heterogeneous order statistics by numerical integration (using the trapezoidal rule). The vector of optimal delays is deemed to be found as d_k when

$$\frac{\partial T(d_k, \lambda)}{\partial d_k} = 0.$$

4   Case Study

Consider a split-merge system with 3 parallel servers having heterogeneous service time density functions (as considered in [20]):

$f_1(t) = \text{Pareto}(\alpha = 3, b = 3.5)$ $(E[X_1] = 5.25, \text{Med}[X_1] = 4.41, \text{Var}[X_1] = 9.19)$
$f_2(t) = \text{Erlang}(n = 2, \lambda = 1)$ $(E[X_2] = 2, \text{Med}[X_2] = 1.68, \text{Var}[X_2] = 2)$
$f_3(t) = \text{Det}(5)$ $(E[X_3] = 5, \text{Med}[X_3] = 5, \text{Var}[X_3] = 0)$

Without adding any subtask delays, mean subtask dispersion is $E[D_{\mathbf{d}=\mathbf{0}}] = 3.576$ time units and maximum sustainable task throughput is $\lambda_{\max} = 0.182$ tasks per time unit.

For $\lambda = 0.01$, mean task response time is $E[R_{\mathbf{d}=\mathbf{0},0.01}] = 5.651$ time units, while for $\lambda = 0.15$, mean task response time is $E[R_{\mathbf{d}=\mathbf{0},0.15}] = 18.658$ time units.

Using our previously developed optimisation technique designed to minimise mean subtask dispersion without regard to impact on response time [19] we obtain the vector of optimal subtask delays:

$$\mathbf{d}_D = (0.608, 3.372, 0)$$

as shown in Fig. 2.

After applying these delays, maximum sustainable throughput drops to $\lambda_{\max} = 0.161$ and mean subtask dispersion improves to $E[D_{\mathbf{d}_D}] = 1.72354$. For $\lambda = 0.01$, mean task response time rises to $E[R_{\mathbf{d}_D,0.01}] = 6.434$, a 14% increase. For $\lambda = 0.15$ mean task response time dramatically increases to $E[R_{\mathbf{d}_D,0.15}] = 51.382$, an increase of 175%.

Now we optimise the same system except under the subtask dispersion–response time trade-off developed in the present paper. With $\lambda = 0.01$ we obtain the following vector of optimal subtask delays:

$$\mathbf{d}_T = (0.453, 3.002, 0)$$

as shown in Fig. 3. Mean subtask dispersion becomes $E[D_{\mathbf{d}_T}] = 1.755$ time units, which is only 1.8% higher than the dispersion obtained under delay vector $\mathbf{d}_D$. Mean task response time is $E[R_{\mathbf{d}_T,.001}] = 6.24$ (a 10% rise in comparison to a system without delays, but a 3% reduction compared to the response time under delay vector $\mathbf{d}_D$).

With $\lambda = 0.15$ the vector of optimal subtask delays drops to:

$$\mathbf{d}_T = (0.0928, 2.043, 0.0)$$

as shown in Fig. 4. Mean subtask dispersion is now $E[D_{\mathbf{d}_T}] = 2.079$, a 21% increase over the dispersion obtained under delay vector $\mathbf{d}_D$, but still a good improvement over the system without added delays (3.576). The corresponding distributions of subtask dispersion are shown in Fig. 5. It is apparent that the trade-off is able to maintain competitive dispersion with our previous methodology, especially for high percentiles. Mean task response time is $E[R_{\mathbf{d}_T,0.15}] = 22.875$, which is 23% worse than the system without delays but a 55% improvement on mean task response time under delay vector $\mathbf{d}_D$. The corresponding distributions of response time are shown in Fig. 6. It is apparent that the trade-off is able to maintain competitive response times as compared to the system with no delays, in stark contrast with our previous methodology.

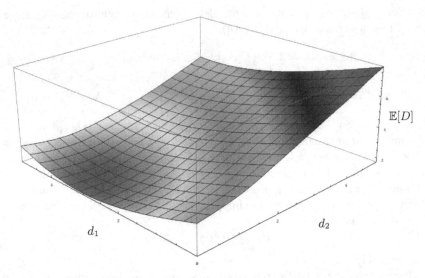

Fig. 2. Surface plot of mean subtask dispersion against subtask delays using our previous methodology from [19]

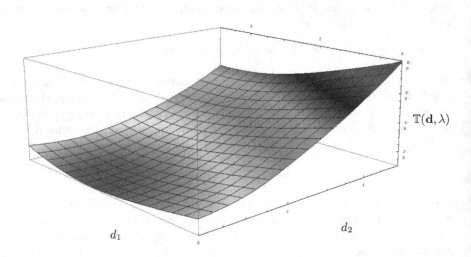

Fig. 3. Surface plot of subtask dispersion–response time trade-off objective function against subtask delays for $\lambda = 0.01$

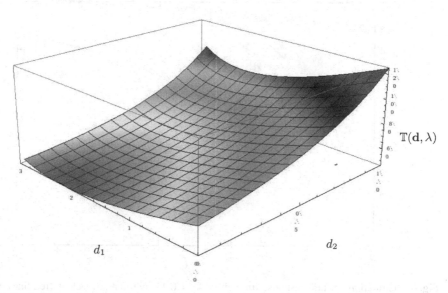

Fig. 4. Surface plot of subtask dispersion–response time trade-off objective function against subtask delays for $\lambda = 0.15$

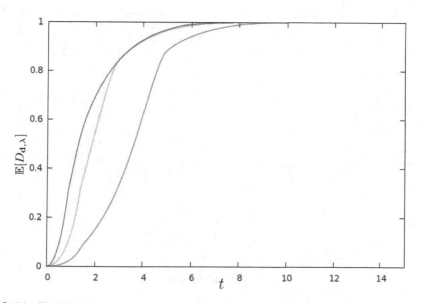

Fig. 5. Distributions of subtask dispersion with $\lambda = 0.15$ without any delays (red line) with delays optimised for $\mathbb{E}[T_{\mathbf{d},0.15}]$ (green line) and for $\mathbb{E}[D_{\mathbf{d}}]$ (blue line)

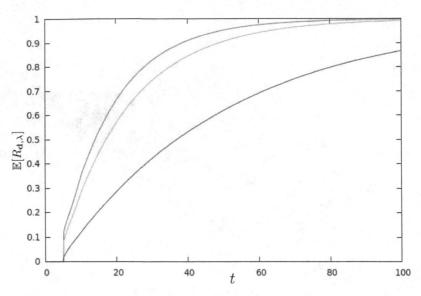

Fig. 6. Distributions of task response time given $\lambda = 0.15$, without any delays (red line), with delays optimised for $\mathbb{E}[T_{\mathbf{d},0.15}]$ (green line) and delays optimised for $\mathbb{E}[D]$ (blue line)

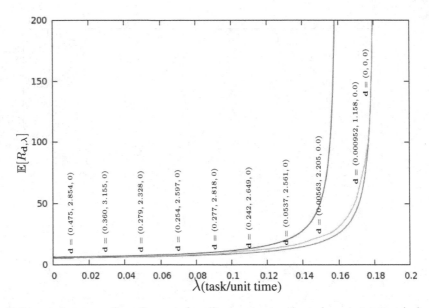

Fig. 7. Expected response time of case study split-merge system for various customer arrival rates without any delays (red line), with delays optimised for subtask dispersion–response time trade-off (green line), and with delays optimised for mean subtask dispersion alone (blue line). Subtask delay vectors are also shown for the subtask dispersion–response time trade-off.

It is interesting to note that under our trade-off as λ converges to the maximum sustainable throughput of the split-merge system without delays (cf. [12]), the vector of optimal subtask delays tends to a vector of zeros. Indeed, optimising the case study system for $\lambda = 0.18$ leads to the vector of optimal subtask delays:

$$\mathbf{d} = (0.0, 0.0, 0.0)$$

Fig. 7 shows how the vector of optimal delays changes with λ, and how it converges to the zero-vector as λ approaches $\lambda_{max} = 0.182$. We note that the maximum sustainable throughput of the system optimised under our previous methodology is rather less than that of the system without delays, whereas the maximum sustainable throughput of the system without delays is maintained using the present methodology.

5 Conclusion

In this paper we have described a framework for delaying the dispatch of subtasks to parallel servers in split-merge systems in order to manage the trade-off between response time and subtask dispersion. At the core of our technique is an objective function computed as the product of expected subtask dispersion and expected response time. Previous research has concentrated on the optimisation of each one of these metrics in isolation, frequently resulting in signficant deterioration in the other. By contrast, the present approach is able to achieve an excellent compromise between the two metrics, without adversely affecting the maximum sustainable throughput of the system.

One possible direction for future work is to develop similar techniques for the optimisation of fork-join systems, which are less synchronised – and much less analytically tractable – relatives of split-merge systems. Although this is expected to be a very challenging exercise, it would open up the application of our technique to an even broader range of real-world systems.

References

1. Armijo, L.: Minimization of functions having Lipschitz continuous first partial derivatives. Pacific Journal of Mathematics 16(1), 1–3 (1966)
2. Baccelli, F., Makowski, A.M., Shwartz, A.: The fork-join queue and related systems with synchronization constraints: Stochastic ordering and computable bounds. Advances in Applied Probability 21(3), 629–660 (1989)
3. Baccelli, F., Massey, W.A., Towsley, D.: Acyclic fork-join queuing networks. J. ACM 36(3), 615–642 (1989)
4. David, H.A., Nagaraja, H.N.: The non-IID case. In: Order Statistics, 3rd edn., ch. 5, pp. 95–120. J. Wiley & Sons, Inc. (2003)
5. David, H.A., Nagaraja, H.N.: Order Statistics, 3rd edn. Wiley Series in Probability and Mathematical Statistics. John Wiley (2003)
6. Flatto, L., Hahn, S.: Two parallel queues created by arrivals with two demands I. SIAM Journal on Applied Mathematics 44(5), 1041–1053 (1984)
7. Flatto, L.: Two parallel queues created by arrivals with two demands ii. SIAM Journal on Applied Mathematics 45(5), 861–878 (1985)

8. Gandhi, A., Gupta, V., Harchol-Balter, M., Kozuch, M.A.: Optimality analysis of energy-performance trade-off for server farm management. Performance Evaluation 67(11), 1155–1171 (2010)
9. Harrison, P.G., Zertal, S.: Queueing models of RAID systems with maxima of waiting times. Perf. Evaluation 64(7-8), 664–689 (2007)
10. Heidelberger, P., Trivedi, K.S.: Analytic queueing models for programs with internal concurrency. IEEE Transactions on Computers C-32(1), 73–82 (1983)
11. Kim, C., Agrawala, A.K.: Analysis of the fork-join queue. IEEE Transactions on Computers 38(2), 250–255 (1989)
12. Knottenbelt, W.J., Tsimashenka, I., Harrison, P.G.: Reducing subtask dispersion in parallel systems. In: Trends in Parallel, Distributed, Grid and Cloud Computing for Engineering. Saxe-Coburg Publications (2013)
13. Lebrecht, A., Knottenbelt, W.J.: Response Time Approximations in Fork-Join Queues. In: 23rd Annual UK Performance Engineering Workshop, UKPEW (July 2007)
14. Lui, J.C.S., Muntz, R.R., Towsley, D.: Computing performance bounds of fork-join parallel programs under a multiprocessing environment. IEEE Transactions on Parallel and Distributed Systems 9(3), 295–311 (1998)
15. Timothy Marler, R., Arora, J.S.: Survey of multi-objective optimization methods for engineering. Structural and Multidisciplinary Optimization 26(6), 369–395 (2004)
16. Nelson, R., Tantawi, A.N.: Approximate analysis of fork/join synchronization in parallel queues. IEEE Transactions on Computers 37(6), 739–743 (1988)
17. Nocedal, J., Wright, S.J.: Numerical optimization. Springer (1999)
18. Towsley, D., Rommel, C.G., Stankovic, J.A.: Analysis of fork-join program response times on multiprocessors. IEEE Transactions on Parallel and Distributed Systems 1(3), 286–303 (1990)
19. Tsimashenka, I., Knottenbelt, W.J.: Reduction of Variability in Split-Merge Systems. In: Imperial College Computing Student Workshop (ICCSW 2011), pp. 101–107 (2011)
20. Tsimashenka, I., Knottenbelt, W.J., Harrison, P.: Controlling variability in split-merge systems. In: Al-Begain, K., Fiems, D., Vincent, J.-M. (eds.) ASMTA 2012. LNCS, vol. 7314, pp. 165–177. Springer, Heidelberg (2012)
21. Varki, E.: Response time analysis of parallel computer and storage systems. IEEE Transactions on Parallel and Distributed Systems 12(11), 1146–1161 (2001)
22. Varma, S., Makowski, A.M.: Interpolation approximations for symmetric fork-join queues. Performance Evaluation 20(13), 245–265 (1994)
23. Wolfe, P.: Convergence conditions for ascent methods. SIAM Review 11(2), 226–235 (1969)
24. Wolfe, P.: Convergence conditions for ascent methods. II: Some corrections. SIAM Review 13(2), 185–188 (1971)

The Discrete-Time Queue with Geometrically Distributed Service Capacities Revisited

Joris Walraevens*, Herwig Bruneel, Dieter Claeys, and Sabine Wittevrongel

Department of Telecommunications and Information Processing
Ghent University - UGent
Sint-Pietersnieuwstraat 41, B-9000 Gent, Belgium
{jw,hb,dclaeys,sw}@telin.UGent.be

Abstract. We analyze a discrete-time queue with variable service capacity, such that the total amount of work that can be performed during each time slot is a stochastic variable that is geometrically distributed. We study the buffer occupancy by constructing an analogous model with fixed service capacity. In contrast with classical discrete-time queueing models, however, the service times in the fixed-capacity model can take the value zero with positive probability (service times are non-negative). We study the late arrival models with immediate and delayed access, the first model being the most natural model for a system with fixed capacity and non-negative service times and the second model the more practically relevant model for the variable-capacity model.

1 Introduction

In this manuscript, we are interested in the study of a discrete-time queue with variable service capacity. In classical discrete-time queueing systems, the service capacity is fixed, equal to 1 in single-server models (e.g. [1, 2]), and larger than 1 in multiserver models (e.g. [3, 4]). Furthermore, service capacity is expressed in number of customers that can be served simultaneously. Therefore, the *service time*, the number of time slots a customer is effectively being served, is equal to the *service requirement*, the total amount of work units his service requires. In the current model, however, a variable number of *work* units can be executed by the server in one time unit, and this in a purely FCFS (First Come First Served) manner, resulting in a service time that can be different from the service requirement of a customer. We assume the service capacity to be geometrically distributed, independent from slot to slot.

This paper is a follow-up paper to [5], where the unfinished work and the customer delay were analyzed for the same model. The buffer occupancy, however, was only obtained in case of geometric service requirements, as difficulties in analyzing the buffer occupancy were observed when service requirements have a general distribution. We here solve this latter problem, by looking at the problem from a different angle. We first construct an analogy with a fixed

---

* Corresponding author.

A. Dudin and K. De Turck (Eds.): ASMTA 2013, LNCS 7984, pp. 443–456, 2013.

service-capacity model with non-negative service times. We further show that, for the latter model, the same formulas can be found for the probability generating functions (pgfs) of buffer occupancy and customer delay as in a model with strictly positive service times, a model that is commonly adopted (see e.g. [2]). Therefore, a remarkable side result of our analysis is that the adoption of non-zero service times seems to be an unnecessary assumption in discrete-time queueing analyses (although analyses are usually heavily based on it).

The model we study is relevant in many application areas. We illustrate this through discussion of two important example domains. First, it is closely related to the 'effective bandwidth' or 'effective capacity' concepts in telecommunication networks, to model the time-varying capacity of stations in wireless networks/LANs [6–8]. A wireless station can indeed be regarded as a server with varying capacity, due to rate fluctuations of the physical channel or at the MAC layer. A second potential application domain is the modeling of varying production capacity of a production system with a single product line [9–11], in order to estimate the influence of this variability on the holding times.

The remainder of this paper is composed as follows. In section 2, the model is introduced in detail. The translation to a fixed service-capacity model is tackled in section 3. Next, we calculate the pgf of the buffer occupancy in section 4, and analyze moments and asymptotic probabilities in section 5. In section 6, we shortly comment on the customer delay, which although retrieved in full in [5], can also be found more elegantly with our approach. Some conclusions are drawn in section 7.

2   Model

We review the model of [5] in this section. We consider a discrete-time queueing system with infinite waiting room and a server which can deliver a variable amount of service as time goes by, called the service capacity. We assume that all events (arrivals and departures) occur at the end of slots. The arrival process of new customers in the system is characterized by means of a sequence of i.i.d. (independent and identically distributed) non-negative discrete random variables with common probability mass function (pmf) $a(n)$ and common probability generating function (pgf) $A(z)$ respectively. More specifically,

$$a(n) \triangleq \Pr[n \text{ customer arrivals in one slot}], \quad n \geq 0,$$

$$A(z) \triangleq \sum_{n=0}^{\infty} a(n)z^n.$$

The mean number of customer arrivals per slot is given by

$$\lambda \triangleq A'(1).$$

The service process of the customers is described in two steps. First, we characterize the *demand* that customers place upon the resources of the system, by attaching to each customer a corresponding *service requirement* (or service demand), which indicates the number of work units required to give complete service to the customer at hand. The service requirements of consecutive customers arriving at the system are modeled as a sequence of i.i.d. positive discrete random variables with common pmf $s(n)$ and common pgf $S(z)$ respectively. More specifically,

$$s(n) \triangleq \Pr[\text{service demand equals } n \text{ work units}], \quad n \geq 1,$$

$$S(z) \triangleq \sum_{n=1}^{\infty} s(n) z^n.$$

The mean service demand of the customers is given by

$$\frac{1}{\sigma} \triangleq S'(1).$$

Next, we describe the (variable) *resources* of the server, by attaching to each time slot a corresponding *service capacity*, which indicates the number of work units that the server is capable of delivering in this slot. We assume that service capacities are nonnegative random variables, independent from slot to slot and geometrically distributed, with common pmf $r(n)$ and common pgf $R(z)$ respectively. More specifically,

$$r(n) \triangleq \Pr[\text{service capacity equals } n \text{ work units}]$$

$$= \frac{1}{1+\mu} \left(\frac{\mu}{1+\mu} \right)^n, \quad n \geq 0, \tag{1}$$

$$R(z) \triangleq \sum_{n=0}^{\infty} r(n) z^n$$

$$= \frac{1}{1+\mu-\mu z}. \tag{2}$$

The mean service capacity of the system (per slot) is given by

$$\mu = R'(1).$$

3   Translation to a Fixed-Capacity Model

Up to now, we have not yet defined the concept *service time of a customer*, the number of slots a customer resides in the server. We have not even elaborated on the concept 'the server'. We note that this can be done in different ways. For instance, we could assume that all customers that are (partly) served during a slot are residing in the server during that slot; this slot would then count as a slot of the service times of all these customers and these service times would hence be overlapping. We opt here for a different approach. We assume that customers are served one by one and that the service time of a customer is the number of slots it resides in the server. Service times of consecutively served customers are therefore non-overlapping, as in a regular single-server model. Note that we assumed that all events occur at the end of the slots. Thus, in order to be able to model the fact that more than 1 customer can end service in the same slot, we have to assume that service times can be equal to zero slots. We clarify this with an example. Say that three customers are in the system in a given slot and that the remaining service requirement of the first customer equals 3 work units, while the service requirements of the other two equal 5 and 4 work units respectively. The slot then counts as a slot of the service time of the first customer only, regardless of the service capacity in that slot. If the service capacity in the slot is higher than 7, both the first and second customer leave the system at the end of the same slot, and the service time of the second customer is therefore assumed to be zero.

As said, service times of different customers are non-overlapping. Due to the i.i.d. service capacities and the memoryless nature of the geometric distribution, service times of consecutively served customers are also independent. Indeed, when a customer departs at the end of slot k (say), the remaining service capacity in slot k is spent on serving the following customer (and possibly even other customers). The remaining part of a random variable with a geometric distribution is independent of the elapsed part, and has the same distribution as the random variable itself. Furthermore, service capacities in later slots are independent of the service capacity in slot k, and therefore the actual service time of the second customer is independent of that of the first customer.

We now calculate the pgf of the service time v of a random customer, i.e., the required number of slots to serve a customer completely. Assume a random customer. Since the service capacity is geometrically distributed with parameter $\mu/(1 + \mu)$ (as in (1)), the service time of this customer can be written as

$$v = \sum_{i=1}^{s} v_i, \tag{3}$$

with s the service requirement of the tagged customer and v_i the number of slots required to process the i-th work unit of this customer. Due to the geometric distribution of the service capacity (and its memoryless property), the service capacity in a slot has expired with probability $1/(1 + \mu)$, or, capacity of at least one work unit remains with probability $\mu/(1 + \mu)$, irrespective of the elapsed part of the service capacity. Therefore, the v_i's are independent and are all geometrically distributed with parameter $1/(1 + \mu)$:

$$\Pr[v_i = n] = \frac{\mu}{1 + \mu} \left(\frac{1}{1 + \mu} \right)^n, \quad n \geq 0,$$

$$V_i(z) \triangleq \mathrm{E}[z^{v_i}]$$

$$= \frac{\mu}{1 + \mu - z}.$$

Since the v_i's are independent of s as well, (3) leads to

$$V(z) = S\left(\frac{\mu}{1 + \mu - z}\right). \tag{4}$$

We may conclude that our queueing system is equivalent to a discrete-time single-server system with a fixed service capacity of one work unit per slot, with the (uncommon) characteristic that service times can be zero, as the pgf of the service times is given by (4); specifically, the probability that the service time is zero equals

$$V(0) = S\left(\frac{\mu}{1 + \mu}\right).$$

For further use, we also define the mean service time:

$$\nu = V'(1) = \frac{1}{\sigma\mu}. \tag{5}$$

The analysis in the remainder makes extensive use of this analogy, in contrast with the analysis in [5]. In [5], first the unfinished work at random slot boundaries was analyzed, and from this unfinished work, buffer occupancy and customer delay were investigated. This is a primarily complex-analytic analysis and the analysis of the buffer occupancy is restricted to *geometric* service requirements (in addition to geometric service capacities). Here, we resort to a more intuitive, stochastic and direct analysis of the buffer occupancy and customer delay, for a *general* service-requirement distribution.

4   Buffer Occupancy

From the previous section, we conclude that we can resort to the analysis of a regular discrete-time single-server model with non-negative service times, the latter variable's distribution being characterized by a general pgf $V(z)$. Substitution of (4) in the resulting expressions will then finally deliver results for the variable service-capacity model.

The fact that service times might be zero is a complication. For instance in [2], the positiveness of the service times is explicitly used in the analysis, which makes those results invalid for our model. However, we will retrieve the same results for non-negative service times, through a different analysis.

We first analyze the buffer occupancy in the late arrival model with immediate access [12, 13], which is from a theoretical point of view the most natural model for a fixed service-capacity queueing system with non-negative service times (we will comment on this later). In a second model, we then adopt the model with delayed access, as in the variable service-capacity model of [5], which is a more practically-oriented model. We will use results of the first model in the analysis of the latter.

4.1 Immediate Access

We assume the following order of events at the end of each slot: first (i) customers arrive, then (ii) customers leave and finally (iii) we observe the buffer occupancy. Customers that have just arrived might depart directly, thus leading to zero customer delays. We first analyze the buffer occupancy as seen by a departing customer. We then relate this to the buffer occupancy as seen by an arriving customer and conclude with the analysis of the buffer occupancy at a random slot mark.

Denote the buffer occupancy seen by the k-th departing customer as $\tilde{b}_k$, $k \geq 1$. Note that, if customer $k + 1$ departs at the same time as customer k, we count customer $k + 1$ as part of $\tilde{b}_k$. The service time of customer k equals v_k and the number of customers arriving in the same batch as customer k is denoted by $\tilde{a}_k$. The number of customer arrivals during the i-th slot of the service time of customer k is given by $\tilde{a}_{k,i}$, $k \geq 1$. We then obtain following system equation:

$$\tilde{b}_{k+1} = \begin{cases} \tilde{b}_k - 1 + \sum_{i=1}^{v_{k+1}} \tilde{a}_{k+1,i} & \text{if } \tilde{b}_k > 0 \\ \tilde{a}_{k+1} - 1 + \sum_{i=1}^{v_{k+1}} \tilde{a}_{k+1,i} & \text{if } \tilde{b}_k = 0 \end{cases}, \tag{6}$$

$k \geq 1$. This is identical to equation (1.41) in [13], and thus the pgf of the steady-state buffer occupancy seen by a random departure equals (1.43) in the same book:

$$\tilde{B}(z) = \frac{1 - \lambda \nu}{\lambda} \cdot \frac{(A(z) - 1)V(A(z))}{z - V(A(z))}. \tag{7}$$

In a second step, we relate the pgfs of the buffer occupancies as seen by departing and arriving customers. As for the departures, we will assume that, if more than one customer arrives at the same time, there is a precise order in the arrival sequence. So, if the k-th and $(k + 1)$-st arriving customers arrive at the same epoch, the $(k + 1)$-st customer sees the k-th customer as part of the buffer occupancy upon arrival, while the opposite is not true. Since we therefore look at the buffer occupancy as left by departing customers and as seen by arriving customers (rather than at arrival and departure epochs), we can regard the system as a single-arrival, single-departure system (see [14] for a similar observation). For this type of systems, it holds that buffer occupancies as seen by departing and arriving customers are identically distributed (see [15]), and thus

$$\hat{B}(z) = \tilde{B}(z), \tag{8}$$

with $\hat{B}(z)$ the pgf of the buffer occupancy as seen by an arriving customer.

Finally, we connect the pgfs of the buffer occupancies as seen by an arriving customer and at random slot boundaries. We can write

$$\hat{b} = b + f, \tag{9}$$

with $\hat{b}$ the buffer occupancy as seen by a random arriving customer, b the buffer occupancy at the preceding slot boundary and f the number of customers arriving at the same epoch as the randomly selected customer and buffered 'before' him. It is clear that b and f are independent random variables. Since the pgf $F(z)$ of f is well-known (see e.g. [1]),

$$F(z) = \frac{A(z) - 1}{\lambda(z - 1)}, \tag{10}$$

we find

$$\begin{aligned}
B(z) &= \frac{\hat{B}(z)}{F(z)} \\
&= \frac{\lambda(z - 1)\hat{B}(z)}{A(z) - 1} \\
&= \frac{\lambda(z - 1)\tilde{B}(z)}{A(z) - 1} \\
&= (1 - \lambda\nu)\frac{(z - 1)V(A(z))}{z - V(A(z))}, \tag{11}
\end{aligned}$$

where we consecutively used (9), (10), (8) and (7). Substitution of (4) in (11) yields the pgf of the steady-state buffer occupancy for the varying service-capacity model with immediate access:

$$B(z) = (1 - \lambda\nu)\frac{(z - 1)S\left(\dfrac{\mu}{1 + \mu - A(z)}\right)}{z - S\left(\dfrac{\mu}{1 + \mu - A(z)}\right)}.$$

We remark that the final expression (11) is the same expression as the one of the pgf of the buffer occupancy in a discrete-time $\text{Geo}^X/G/1$ queue with pgf $A(z)$ for the number of arrivals at a slot boundary, $V(z)$ the pgf of *strictly positive* service times $(V(0) = 0)$ and *delayed access*, see e.g. [2]. Therefore, in our opinion, the latter case can be regarded as special case of the system analyzed in this subsection. In fact, we regard the term 'late arrival with delayed access' wrongly chosen in fixed service-capacity systems, as the late arriving customer can in fact enter the service unit immediately to start service, only his service time is at least one. Therefore, it would be better to use the terms 'systems with positive service times' or (the more general) 'systems with non-negative service times' instead.

4.2   Delayed Access

Here, we return to the model of [5]. In the model with delayed access, customers cannot leave the system immediately upon arrival. Therefore, we change the order of arrivals and departures (departures now occur just before new customers arrive, and then the system is observed), so that new arrivals have to wait at least one slot for the next service opportunity. In this model, the delay of customers is at least one slot, which is a natural assumption in practice (in contrast with the model with immediate access). This does not mean that service times of customers are at least one slot as well. Multiple customers can still leave at a particular slot boundary and the service times of all but the first are then equal to zero. For this reason, the delayed-access model is not as natural as the immediate-access model, from a single-sever fixed-capacity point of view. It also complicates the analysis slightly, but the system can still be regarded as a discrete-time fixed service-capacity system, only it is a model with delayed access.

We take the same approach as in previous subsection; we analyze the buffer occupancy respectively as seen by departing customers, as seen by arriving customers and at slot boundaries. In fact, the same system equations (6) for and pgf (7) of the buffer occupancy as seen by departing customers, and the same relation (8) between the distributions of the buffer occupancy as seen by departing and by arriving customers hold. The only difference is the relation (9) between the buffer occupancy as seen by arriving customers and the buffer occupancy at the preceding slot boundary. For the current model, we find

$$\hat{b} = b - c + f,$$

with $\hat{b}$ the buffer occupancy as seen by a random arrival, b the buffer occupancy at the previous slot boundary, f the number of arrivals at the arrival epoch of the customer and admitted before the tagged arrival, and c the total number of departures directly preceding these arrivals. The term $-c$ is thus the difference with the previous model (see (9)). Note that c and b are not independent which complicates analysis. We have

$$\hat{B}(z) = \mathrm{E}\left[z^{b-c}\right] F(z). \tag{12}$$

We calculate $\mathrm{E}\left[z^{b-c}\right]$. We can write

$$b^* = b - c + a, \tag{13}$$

with b the buffer occupancy at a random slot boundary, and c, a and b^* the number of departures, arrivals and customers in the system at the following slot boundary. Since b^* is also distributed as the buffer occupancy at a random slot boundary, we may write, from (13),

$$B(z) = \mathrm{E}\left[z^{b-c}\right] A(z). \tag{14}$$

Using (14), (12), (8), (7) and (10), we obtain the pgf $B(z)$ of the buffer occupancy at random slot boundaries as

$$B(z) = (1 - \lambda\nu)\,\frac{A(z)(z - 1)V(A(z))}{z - V(A(z))}. \tag{15}$$

Substitution of (4) in (15) yields the pgf of the steady-state buffer occupancy for the varying service-capacity model with delayed access:

$$B(z) = (1 - \lambda\nu) \frac{A(z)(z - 1)S\left(\dfrac{\mu}{1 + \mu - A(z)}\right)}{z - S\left(\dfrac{\mu}{1 + \mu - A(z)}\right)}. \tag{16}$$

For the special case of geometric service requirements, we obtain the same result as in [5].

5 Performance Measures and Discussion

In this section, we calculate and discuss some performance measures. We concentrate on the model with delayed access. In the discussion, we will focus on the influence of the mean service capacity μ, while keeping the mean service time ν constant (i.e., by scaling the mean service demand $1/\sigma$ accordingly, see (5)). If feasible, we also assume the normalized moments (such as the coefficient of variation and skewness) of mean number of per-slot arrivals and service demand to be constant.

5.1 Moments

First, the mean buffer occupancy is found by taking the first derivative of (16) in $z = 1$:

$$E[b] = B'(1) = \lambda + \frac{\lambda\nu\left[1 + \lambda\left(C_A^2 + 1\right) + \lambda\nu\left(C_S^2 - 1\right)\right]}{2(1 - \lambda\nu)} + \frac{\lambda^2\nu}{2(1 - \lambda\nu)\mu}, \tag{17}$$

where we introduced the coefficients of variation C_A and C_S of the number of per-slot arrivals and the service demands respectively. From this formula, we can conclude that if the service capacity μ goes to infinity (while scaling the mean and variance of the service demand in order to keep the mean service time ν per customer and the coefficient of variation C_S of the service demand constant), the mean buffer occupancy tends to a constant. This constant depends on the mean service time, the arrival rate and the coefficients of variations of the numbers of arrivals per slot and the service demands. If the mean service capacity goes to 0, the mean buffer occupancy tends to infinity, even though the service demand is scaled accordingly, and this according to a $1/\mu$-rule with a prefactor depending on both the arrival rate and the mean service time.

Higher moments of the buffer occupancy can be calculated by taking higher derivatives of $B(z)$ in 1. The expression for the variance, for instance, contains a constant term, a term in $1/\mu$ and a term in $1/\mu^2$. The constant term depends on the mean value, the coefficient of variation and the skewness of both the number of arrivals per slot and the service demand. The term in $1/\mu$ only depends on the mean values and coefficients of variations, while the term in $1/\mu^2$ only depends on the mean values. For reference, we give the latter as it is the dominant term for the mean service capacity μ going to 0:

$$\text{Var}[b] \sim \frac{\lambda^3 \nu(8 - 5\lambda\nu)}{12(1 - \lambda\nu)^2 \mu^2},$$

for $\mu \to 0$.

5.2 Asymptotic Probabilities

Asymptotics of probabilities $\Pr[b = n]$, $n \to \infty$ can be calculated by investigation of the dominant singularity of the pgf $B(z)$ (i.e. the singularity of lowest norm) and the behavior of $B(z)$ in the neighbourhood of this dominant singularity [16, 17]. This dominant zero is real and bigger than or equal to 1. The position of the dominant singularity is dependent on complete distributions, i.e., of $A(z)$ and $S(z)$. We will therefore investigate a special case instead of analyzing it in full generality. We consider the case that the functions $A(z)$ and $S(z)$ are meromorphic.

The dominant singularity z_0 of $B(z)$ is a single zero of the denominator, i.e.,

$$z_0 = S\left(\frac{\mu}{1 + \mu - A(z_0)}\right), \tag{18}$$

and the corresponding probabilities decay geometrically

$$\Pr[b = n] \sim (1 - \lambda\nu) \frac{A(z_0)(z_0 - 1)}{S'\left(\dfrac{\mu}{1 + \mu - A(z_0)}\right) \dfrac{\mu A'(z_0)}{(1 + \mu - A(z_0))^2} - 1} z_0^{-n}.$$

We now investigate the decay rate z_0 in terms of μ. We vary μ, while scaling σ in order to keep a constant mean service time ν. Since z_0 has to be calculated numerically, in general, cf. (18), we look at the following illustrative example. Assume Bernoulli arrivals, i.e., in each slot one customer arrives with probability λ and no customers arrive with complementary probability $1-\lambda$. The distribution of the service demand is chosen such that we can model deterministic as well as bursty demands, namely we assume that the service demand is either 1 slot or s slots, respectively with probability p and $1 - p$. The arrival rate λ is equal to 0.8, while the mean service time ν is equal to 0.5. Then μ varies linearly with s according to following relation:

$$\mu = \frac{p + (1 - p)s}{\nu}$$

In table 1, we list the values of z_0 for several combinations of p and s.

For $s = 1$, deterministic service demands of 1 slots each are required. In this case the probabilities decay with factor 2.5. This can be taken as reference case, as it is also the decay rate of a more general geometric distribution of the service demands. From [5], it is known that z_0 is indeed independent of the actual

Table 1. Asymptotic decay rate z_0 of probabilities $\Pr[b = n]$

s	1	2	3	4	5
$p = 0$	2.5	3.2087	3.6117	3.8705	4.0505
$p = 0.3$	2.5	2.9542	3.1771	3.3093	3.3970
$p = 0.6$	2.5	2.7155	2.7621	2.7655	2.7580
$p = 0.9$	2.5	2.5254	2.4387	2.3203	2.2103

value of μ, as long as $\nu = 1/\mu\sigma$ is fixed, in case of geometric service demands. Therefore, for increasing s, we can conclude on the impact of 'burstiness' (or lack thereof) on the decay rate of the mass function of the buffer occupancy, by comparing with the constant rate 2.5 of this geometric distribution. We see that for small p, the decay rate increases with s meaning that the probabilities decay faster for higher s. For $p = 0$, the service demand is not bursty at all, as it is deterministically equal to s. Higher service demands and service requirements are in this case advantageous. For higher p, one can see the impact of burstiness. For $p = 0.6$, the decay rate is almost constant, as is the case for the geometric distribution. For even higher p (and high s) the decay rate decreases for increasing s, which means that the probability tail decreases less fast. Therefore, lower service demands and service requirements are advantageous in this case. Another way to look at it is the following: if we assume that geometric capacity is perfectly fitted to geometric demands, we can conclude that geometric capacity is not fitted well to either smoother (too much capacity at times) or burstier (too less capacity at times) service demands.

6   Customer Delay

Let us now briefly turn to the customer delay. We assume a FCFS (First Come First Served) scheduling discipline. Customer delay is defined as the number of slots a customer sojourns in the system. The pgf of the customer delay was calculated in [5] for the model with delayed access, by means of an elaborate algebraic analysis (first the unfinished work was analyzed and customer delay was then related to the unfinished work). We here present a stochastic analysis by making use of the analogy with a fixed service-capacity model (see section 3). We first discuss the model with immediate access and then come back to the model with delayed access.

6.1   Immediate Access

Denote the customer delay of the k-th arriving customer as d_k, $k \geq 1$. Then the following Lindley-type equation is easily constructed:

$$d_{k+1} = [d_k - t_k]^+ + v_{k+1}, \tag{19}$$

with t_k the number of slots between the arrivals of customers k and $k+1$, v_{k+1} the service time of the $(k+1)$-st customer and $[.]^+$ shorthand for $\max(.,0)$. This is the same (Lindley-type) system equation as in the common discrete-time single-server queue with non-zero service times and thus the pgf $D(z)$ of the delay of a random customer is also the same [1], in this case,

$$D(z) = \frac{1-\lambda\nu}{\lambda} \cdot \frac{(z-1)V(z)\left[A\left(V(z)\right)-1\right]}{\left[z-A\left(V(z)\right)\right]\left[V(z)-1\right]}. \tag{20}$$

Substitution of (4) results in the pgf of the customer delay in the variable service-capacity system:

$$D(z) = \frac{1-\lambda\nu}{\lambda} \cdot \frac{(z-1)S\left(\dfrac{\mu}{1+\mu-z}\right)\left[A\left(S\left(\dfrac{\mu}{1+\mu-z}\right)\right)-1\right]}{\left[z-A\left(S\left(\dfrac{\mu}{1+\mu-z}\right)\right)\right]\left[S\left(\dfrac{\mu}{1+\mu-z}\right)-1\right]}.$$

6.2 Delayed Access

Now, we assume that arrivals occur after departures at a slot boundary, leading to a minimal delay of 1 slot for each customer. The system equation for the customer delay then reads

$$\tilde{d}_{k+1} = \begin{cases} \tilde{d}_k - t_k + v_{k+1} & \text{if } \tilde{d}_k > t_k \\ 1 + v_{k+1} & \text{if } \tilde{d}_k \le t_k \end{cases}, \tag{21}$$

with $\tilde{d}_k$ the delay of customer k in this model with delayed access, and the other variables as defined in the former subsection. The difference with the Lindley-equation (19) is the term '1' on the second line. We transform the system equation (21) as follows

$$\tilde{d}_{k+1} - 1 = \begin{cases} (\tilde{d}_k - 1) - t_k + v_{k+1} & \text{if } \tilde{d}_k > t_k \\ v_{k+1} & \text{if } \tilde{d}_k \le t_k \end{cases},$$

which can be rewritten as

$$(\tilde{d}_{k+1} - 1) = [(\tilde{d}_k - 1) - t_k]^+ + v_{k+1}.$$

From this equation and (19), it is easily seen that $\{d_k\}_{k\ge 1}$ and $\{\tilde{d}_k - 1\}_{k\ge 1}$ obey the same system equations and thus

$$D(z) = \frac{\tilde{D}(z)}{z},$$

with $\tilde{D}(z)$ the pgf of the customer delay in the model with delayed access. Using (20), we finally retrieve

$$\tilde{D}(z) = \frac{1-\lambda\nu}{\lambda} \cdot \frac{z(z-1)V(z)\left[A\left(V(z)\right)-1\right]}{\left[z-A\left(V(z)\right)\right]\left[V(z)-1\right]}.$$

Substitution of (4) results in the pgf of the customer delay in the variable service-capacity system:

$$\tilde{D}(z) = \frac{1 - \lambda \nu}{\lambda} \cdot \frac{z(z-1)S\left(\frac{\mu}{1+\mu-z}\right)\left[A\left(S\left(\frac{\mu}{1+\mu-z}\right)\right) - 1\right]}{\left[z - A\left(S\left(\frac{\mu}{1+\mu-z}\right)\right)\right]\left[S\left(\frac{\mu}{1+\mu-z}\right) - 1\right]},$$

which yields the expression obtained in [5].

7   Conclusions

In this paper, we studied a discrete-time queue with variable (geometric) service capacity. The analysis is heavily based on inventive stochastic arguments rather than heavy algebraic calculations. Indeed, we showed how this system is identical to a fixed service-capacity system with non-negative service times. The buffer occupancy was analyzed by connecting the buffer occupancy as seen by departing customers, the buffer occupancy as seen by arriving customers and the buffer occupancy at slot boundaries. The natural single-server model with non-negative service times is a late arrival system with immediate access, and therefore we studied this first. However, service requirements in the original model are strictly positive and a late arrival system with delayed access is more natural in that context. Therefore, we extended the fixed service-capacity analysis to this system. In future work, we plan on tackling non-geometric service capacities. However, the service times in the analog single-server system are then correlated, which is expected to greatly complicate the analysis.

Acknowledgments. This research has been funded by the Interuniversity Attraction Poles Programme initiated by the Belgian Science Policy Office.

References

1. Bruneel, H., Kim, B.: Discrete-time models for communication systems including ATM. Kluwer Academic Publisher, Boston (1993)
2. Bruneel, H.: Performance of discrete-time queueing systems. Computers and Operations Research 20, 303–320 (1993)
3. Gao, P., Wittevrongel, S., Bruneel, H.: Discrete-time multiserver queues with geometric service times. Computers and Operations Research 31, 81–99 (2004)
4. Janssen, A., van Leeuwaarden, J.: Analytic computation schemes for the discrete-time bulk service queue. Queueing Systems 50, 141–163 (2005)
5. Bruneel, H., Walraevens, J., Claeys, D., Wittevrongel, S.: Analysis of a discrete-time queue with geometrically distributed service capacities. In: Al-Begain, K., Fiems, D., Vincent, J.-M. (eds.) ASMTA 2012. LNCS, vol. 7314, pp. 121–135. Springer, Heidelberg (2012)
6. Kafetzakis, E., Kontovasilis, K., Stavrakakis, I.: Effective-capacity-based stochastic delay guarantees for systems with time-varying servers, with an application to IEEE 802.11 WLANs. Performance Evaluation 68, 614–628 (2011)

7. Chang, C., Thomas, J.: Effective bandwidth in high-speed digital networks. IEEE Journal on Selected Areas in Communications 13, 1091–1100 (1995)
8. Jin, X., Min, G., Velentzas, S.: An analytical queuing model for long range dependent arrivals and variable service capacity. In: Proceedings of IEEE International Conference on Communications (ICC 2008), Beijing, pp. 230–234 (2008)
9. Glock, C.: Batch sizing with controllable production rates. International Journal of Production Research 48, 5925–5942 (2010)
10. Balkhi, Z.: On the global optimal solution to an integrated inventory system with general time varying demand, production and deterioration rates. European Journal of Operations Research 114, 29–37 (1999)
11. Yang, H.L.: A partial backlogging production-inventory lot-size model for deteriorating items with time-varying production and demand rate over a finite time horizon. International Journal of Systems Science 42, 1397–1407 (2011)
12. Hunter, J.: Mathematical Techniques of Applied Probability. Discrete Time Models: Techniques and Applications, vol. 2. Academic Press, New York (1983)
13. Takagi, H.: Queueing analysis: a foundation of performance evaluation. Discrete-time systems, vol. 3. North-Holland (1991)
14. Fiems, D., Steyaert, B., Bruneel, H.: Discrete-time queues with generally distributed service times and renewal-type server interruptions. Performance Evaluation 55, 277–298 (2004)
15. Takagi, H.: Queueing analysis: a foundation of performance evaluation. Vacation and priority systems, vol. 1, part 1. North-Holland (1991)
16. Bruneel, H., Steyaert, B., Desmet, E., Petit, G.: An analytical technique for the derivation of the delay performance of ATM switches with multiserver output queues. International Journal of Digital and Analog Communication Systems 5, 193–201 (1992)
17. Flajolet, P., Sedgewick, R.: Analytic Combinatorics. Cambridge University Press (2008)

Resource Allocation
in a Multiple-Priority Buffered Link

Piotr Żuraniewski[1,4,*], Michel Mandjes[2],
Hans van den Berg[1], and Richa Malhotra[3]

[1] TNO, The Netherlands
[2] University of Amsterdam, The Netherlands
[3] SURFnet, The Netherlands
[4] AGH University, Poland

Abstract. In this paper we consider a multiple-priority buffered link, with a focus on resource allocation issues; our study was motivated by such issues in Carrier Ethernet, but the applicability of the results is by no means limited to this technology. In our model, the resource is shared by two priority classes. As the high-priority queue — intended for traffic generated by delay-sensitive applications — typically has a small buffer, the low-priority queue can be modelled as a queue with a time-varying service rate; this service rate behaves independently of the input of the low-priority queue. The analysis of the high-priority queue is standard, and we therefore provide an in-depth study of the performance of the low-priority queue. Assuming that all traffic offered to the system is subjected to a leaky-bucket type of policer, one of the approaches that we present borrows elements from the setup of [6], in which the notion of effective bandwidth plays a crucial role.

Keywords: resource allocation, priority queues, effective bandwidth, large deviations, Carrier Ethernet.

1 Introduction

In many networking technologies, in order to make efficient use of the network resources, traffic is split into multiple categories. In the most basic form, one distinguishes between just two classes: delay-sensitive and delay-tolerant traffic, which are dealt with differently: the delay-sensitive traffic enjoys service priority over the delay-tolerant traffic. The rationale behind this service differentiation is that, if one would have just one class, all traffic should be offered strict performance guarantees; the introduction of multiple classes thus enables a more efficient use of resources (in terms of the achievable utilization). The idea of traffic differentiation, as described above, has been proposed for Carrier Ethernet, but has been exploited several times before. For instance, in the 1990s this type of differentiation was pursued for ATM [15], later similar ideas were developed for IP (see for instance [18] for a recent account of the state of affairs). More detailed background on Carrier Ethernet is found in e.g. [7].

---

* This project was financially supported by SURFnet. Part of this work was done while PZ was with the University of Amsterdam.

A. Dudin and K. De Turck (Eds.): ASMTA 2013, LNCS 7984, pp. 457–471, 2013.

In an abstract form under the paradigm described above a network element can be modeled as a *two-level queueing system*. In the first place there is a high-priority queue, particularly intended to take care of the delay-sensitive traffic, getting strict priority, and being equipped with a small buffer (basically just to take care of packet-level queueing effects). Then there is a low-priority queue, meant to serve delay-tolerant traffic, which uses the residual service capacity, and which may be equipped with a sizeable buffer. Noticing that the high-priority queue essentially does not 'see' the low-priority traffic, the performance of this queue reduces to that of a single queue in isolation, which has been studied extensively in the literature. The performance of the low-priority queue is considerably more involved, though, as it is a queue that is served at a randomly fluctuating rate. It is the performance of this queue that we address in this paper.

In the literature several studies have been devoted to performance issues of priority queues. In this respect we mention [5], which considers the system assuming that the input streams behave as Markov fluid sources, and [10], which focuses on a fairly general light-tailed framework. These studies do not take into account, though, that the traffic streams are *policed*: incoming traffic is checked for compliance with a traffic contract, where 'excess traffic' is marked. In view of this, attention has been paid to identifying the 'worst case traffic' that complies with a given traffic contract; here 'worst case' refers to yielding the worst possible performance, according to a given metric [4,9,16].

In [6] it was argued that the worst case profile in case of a leaky bucket regulator (defined through a peak rate, a sustainable rate, and a maximum burst size) is a deterministic on-off stream. This stream alternates between transmitting at the peak rate and being silent, with the on- and off times chosen such that the burst size is exactly the maximum burst size, and the average rate equals the sustainable rate. When multiplexing several of these 'worst case sources', the only randomness in the model is the 'phasing' of the individual sources. For a single queue, [6] develops a technique to (conservatively) estimate the loss probability (for a given mix of sources, each with their specific leaky bucket triplet), which also leads to a computationally attractive call admission procedure.

This paper focuses on an operations-research based analysis of the two-level queueing system described above. We study mechanisms to guarantee given performance levels through adequate resource allocation; the techniques developed can in principle also be used for admission control purposes. It is emphasized that, despite the topic could be considered as classical, several open issues remained. In more detail, the contributions are the following:

· As mentioned above, the high-priority queue is not affected by low-priority traffic, and as a result adequate evaluation techniques are available; the low-priority queue, on the contrary, receives the bandwidth not used by the high-priority queue, and is therefore considerably harder to analyze. In Section 2 we consider a versatile model for the service rate consumed by the high-priority queue, based on a multi-rate model [8]. We give for it expressions for the loss-probability in the low-priority queue. We do so, both for the case of bufferless multiplexing in the low-priority queue (known as rate envelope multiplexing, or shortly REM, [17]), as well as the case of a moderately sized or large buffer fed by worst-case traffic (as was discussed above).

· These expressions being computationally demanding, we consider in Section 3 their asymptotics. After a rescaling of the resources (buffer and link rate), and focusing on the so-called many-flows scaling, manageable expressions are derived, in terms of exact asymptotics for REM, and in terms of logarithmic asymptotics for a (moderate or large buffer fed by worst case traffic. The latter expressions generalize the setup of [6] to our setting with two service classes.

· Then we perform in Section 4 numerical experimentation, underscoring the accuracy of our approximations. They also indicate that the effective bandwidth concept developed in [6] tends to be rather conservative, in line with earlier findings reported in [2] for the single-priority case.

2   Model

We consider a model with two classes of traffic: a high-priority (hp) class and a low-priority (lp) class. The hp traffic is fed into a buffered resource that is emptied at rate C. As this traffic is likely to be delay-sensitive, the buffer of this class is typically relatively small. For ease we assume the buffer has size 0, but in practice it is a small positive size (sufficiently large to absorb packet-level queueing). The remaining service capacity is used by the lp traffic. If temporarily the input rate of this lp traffic exceeds the available service rate, traffic can be stored in a buffer of size B.

The hp queue does not 'see' the lp traffic, and therefore the hp queue can be treated as a queue in isolation. These single-queues have been thoroughly studied in the literature, and accurate performance models are available. The analysis of the lp queue, though, is more involved; it can be regarded as a queue with time-varying service rate, where the stochastic characteristics of this service rate behave independently of the input of the lp queue. To model the lp queue, we first propose two models that describe the amount of bandwidth (that is, service capacity) taken away by the hp class.

2.1   Bandwidth Used by Hp Queue

In this subsection, we introduce a versatile model that describes the amount of bandwidth used by hp traffic. We chose this model because of its flexibility: it could be the case the hp class does not exactly obey the characteristics of the model we propose, but by tuning the models' parameters an accurate fit can be realized.

Multi-rate model. Let there be I types of hp flows. Flows of type i arrive according to a Poisson process of rate λ_i, and their mean duration is $\mathbb{E}D_i$; define the 'load' imposed by type i by $\nu_i := \lambda_i \mathbb{E}D_i$ (in terms of number of users). Let α_i be the traffic rate of a single type-i flow — we assume this rate to be more or less fixed during its duration (but it can also be thought of as some sort of effective bandwidth).

With $k \in \mathbb{N}$ denoting the (I-dimensional) vector representing the numbers of jobs of all types, such a vector is admissible if $\sum_i \alpha_i k_i \leq C$, where we recall that C denotes the link rate; we call this admissible set $\mathscr{S}(C)$. It is well-known from the theory of loss networks [8] that, with X_i denoting the number of type-i calls in steady state, for $k \in \mathscr{S}(C)$

$$\mathbb{P}(\boldsymbol{X} = \boldsymbol{k}) = \frac{1}{\mathbb{B}^{\mathscr{M}}(C)} \prod_{i=1}^{I} \frac{\nu_i^{k_i}}{k_i!},$$

where $1/\mathbb{B}^{\mathscr{M}}(C)$ serves as normalizing constant:

$$\mathbb{B}^{\mathscr{M}}(C) := \sum_{\boldsymbol{k} \in \mathscr{S}(C)} \prod_{i=1}^{I} \frac{\nu_i^{k_i}}{k_i!}.$$

We assume that $\sum_i \alpha_i \nu_i \leq C$; this is a highly natural assumption, as it entails that the 'point of operation' is admissible. The rate used by the hp class is $C_{\mathrm{hp}}^{\mathscr{M}} = \sum_i \alpha_i X_i$. Observe that $\boldsymbol{X}$ has a truncated Poisson distribution: with $\bar{X}_i$ having a Poisson distribution with mean ν_i, we have $\mathbb{P}(\boldsymbol{X} = \boldsymbol{k}) = \mathbb{P}(\bar{\boldsymbol{X}} = \boldsymbol{k} \mid \bar{\boldsymbol{X}} \in \mathscr{S})$. In this multi-rate model based approximation, one could try to choose the $\boldsymbol{\nu}$ and A such that models matches the measurement data; in case $I = 1$ this gives two parameters to tune (and in general $2I$).

2.2 Performance of the Lp Queue

Suppose now that the lp customers can use the remaining capacity, that is, $C - C_{\mathrm{hp}}$. Let there be J types of lp flows, and let ε be maximally allowable loss probability. We now consider two ways of dealing with the lp queue. The first approach does not exploit the buffering capability of the lp queue (traffic is assumed to be lost as soon as the input rate of the lp queue exceeds the available bandwidth); traffic is modeled as on-off (where no assumptions are imposed on the distributions of the on- and off times). The other approach explicitly takes into account the buffering capability, and in addition the lp traffic is assumed to be compliant with a leaky-bucket regulator. Due to the fact that the latter approach relies on the ideas of [6], on the computational level the procedures are very similar.

A more detailed description of these two options is as follows:
▷ In the first approach we abstract from the lp buffer B, and we do as if packets are lost as soon as the input rate of lp traffic exceeds $C - C_{\mathrm{hp}}$.
Suppose a flow of type j transmits traffic at rate β_j during random times $T_{\mathrm{on},j}$, and is silent during random times $T_{\mathrm{off},j}$; let the sequence of on- and off-times be i.i.d. random variables, and let both sequences be mutually independent. We define the probability of a source of type j transmitting traffic by $\pi_j := T_{\mathrm{on},j}/(T_{\mathrm{on},j} + T_{\mathrm{off},j})$. Supposing there are ℓ_j jobs of type j, then the aggregate rate required by type j is $\beta_j Y_j$, where Y_j has a binomial distribution with parameters ℓ_j and π_j. It is concluded that, in the multi-rate setting, a vector $\boldsymbol{\ell}$ can be admitted if

$$q_1^{\mathscr{M}} := \sum_{\boldsymbol{k} \in \mathscr{S}(C)} \mathbb{P}(\boldsymbol{X} = \boldsymbol{k}) \mathbb{P}\left(\sum_{j=1}^{J} \beta_j Y_j > C - \sum_{i=1}^{I} \alpha_i k_i \right) \leq \varepsilon.$$

This admission policy for the lp class is often referred to as *rate envelope multiplexing* (REM) [17]; the non-standard feature here is that the service rate is *random*.

▷ There are computationally attractive ways to take into account the buffering capability of the low-priority queue, though. One of these possibilities relies on the framework presented in [6] and will be referred to as EMW. In this framework the focus is on users transmitting traffic in the 'worst-case-manner' that still fits in the traffic profile characterized by a regulator. For customers of type j, this regulator is characterized by a sustainable rate r_j (or 'token rate'), a peak rate p_j (larger than or equal to r_j), and a token buffer size τ_j. In [6] it is argued that the worst-case profile corresponding to this regulator is a deterministic on-off pattern: it is alternatingly transmitting at rate p_j during $\tau_j/(p_j - r_j)$ time units, and then silent during τ_j/r_j, thus generating an average traffic rate r_j.

Then [6] shows, for a model with just one priority class, that it can be guaranteed that there is *zero loss* if one uses the following algorithm. Let (r_j, p_j, τ_j) be the regulator profile of a source of type j. Let K be the amount of bandwidth available, and B the amount of buffering capacity, and define

$$c_j \equiv c_j(K) := \max\left\{\frac{p_j}{1 + B(p_j - r_j)/(\tau_j K)}, r_j\right\}. \tag{1}$$

Let there be ℓ_j users of type j. Then the vector ℓ can be admitted with zero loss if $\sum_j c_j \ell_j \leq K$. This algorithm can be adapted to allow a fraction $\varepsilon > 0$ loss. The key observation is that the effective bandwidth c_j is needed just a fraction r_j/c_j of the time, and 0 in the remaining fraction. This means that we can use an algorithm similar to the one defined above: for the multi-rate model, we can admit a vector ℓ if

$$q_2^{\mathscr{M}} := \sum_{k \in \mathscr{S}(C)} \mathbb{P}(\boldsymbol{X} = \boldsymbol{k}) \mathbb{P}\left(\sum_{j=1}^{J} c_j Y_j > C - \sum_{i=1}^{I} \alpha_i k_i\right) \leq \varepsilon,$$

where $Y_j \equiv Y_j(\ell_j, \pi_j)$ has a binomial distribution with parameters ℓ_j and $\pi_j := r_j/c_j$. One needs to be careful with the precise definition of the effective bandwidth: instead of the K used in the case of a constant capacity, one now needs to take the available service capacity:

$$c_j \equiv c_j\left(C - \sum_{i=1}^{I} \alpha_i k_i\right).$$

For proper resource allocation, there is a need for fast an accurate techniques to determine the probabilities $q_1^{\mathscr{M}}$ and $q_2^{\mathscr{M}}$. These are developed in the next section.

3   Asymptotic Analysis

In the previous section we have distinguished essentially two cases: the lp class performing either rate envelope multiplexing (REM) or following the 'lossy' EMW algorithm. In both cases one still needs to evaluate the loss probability experienced by the lp queue, which boils down to evaluating sums over the state space $\mathscr{S}(C)$. We now develop accurate approximations of the loss probability. The idea is to introduce scaling by n: we replace $\nu_i \mapsto n\nu_i$, $C \mapsto nC$, $B \mapsto nB$, and $\ell_j \mapsto n\ell_j$; under this scaling the decay rate of the above probabilities can be determined, as we will show in this section.

Multi-rate and REM Model. We first focus on the setup of rate envelope multiplexing. X has now a truncated Poisson distribution with parameters $n\nu_1$ up to $n\nu_I$, and we therefore write $X^{(n)}$. The state space is now $\mathscr{S}(nC)$. The random variables Y_j have now binomial distributions with parameters $n\ell_j$ and π_j, and to express the dependence on n we write $Y_j^{(n)}$. The goal is to find the exact asymptotics of

$$q_1^{\mathscr{M}}(n) := \sum_{k \in \mathscr{S}(nC)} \mathbb{P}(X^{(n)} = k) \mathbb{P}\left(\sum_{j=1}^{J} \beta_j Y_j^{(n)} > nC - \sum_{i=1}^{I} \alpha_i k_i\right),$$

that is, an explicit function $\phi(\cdot)$ such that $\phi(n)q_1^{\mathscr{M}}(n) \to 1$ as $n \to \infty$. From the interpretation of the distribution of X being truncated Poisson, we can alternatively write, with $\bar{X}_i^{(n)}$ having a Poisson distribution with mean $n\nu_i$,

$$q_1^{\mathscr{M}}(n) = \mathbb{P}\left(\sum_{i=1}^{I} \alpha_i \bar{X}_i^{(n)} + \sum_{j=1}^{J} \beta_j Y_j^{(n)} > nC \,\middle|\, \sum_{i=1}^{I} \alpha_i \bar{X}_i^{(n)} \le nC\right).$$

With V_m, $m = 1, \ldots, n$, i.i.d. random variables distributed as $\sum_{i=1}^{I} \alpha_i \bar{X}_i^{(1)}$, and W_m, $m = 1, \ldots, n$, i.i.d. random variables distributed as $\sum_{j=1}^{J} \beta_j Y_j^{(1)}$, we have

$$q_1^{\mathscr{M}}(n) = \mathbb{P}\left(\frac{1}{n}\sum_{m=1}^{n}(V_m + W_m) > C \,\middle|\, \frac{1}{n}\sum_{m=1}^{n} V_m \le C\right).$$

Now consider the following lemma.

Lemma 1. *Let V_m, $m = 1, \ldots, n$, and W_m, $m = 1, \ldots, n$, be mutually independent sequences of i.i.d. random variables. Let $\mathbb{E}V_i < C$. Then*

$$\lim_{n \to \infty} \frac{1}{n} \log \mathbb{P}\left(\frac{1}{n}\sum_{m=1}^{n}(V_m + W_m) > C \,\middle|\, \frac{1}{n}\sum_{m=1}^{n} V_m \le C\right) \tag{2}$$

$$= \lim_{n \to \infty} \frac{1}{n} \log \mathbb{P}\left(\frac{1}{n}\sum_{m=1}^{n}(V_m + W_m) > C\right).$$

Proof: It is immediate that the decay rate (2) equals the difference of two decay rates:

$$\lim_{n \to \infty} \frac{1}{n} \log \mathbb{P}\left(\frac{1}{n}\sum_{m=1}^{n}(V_m + W_m) > C, \frac{1}{n}\sum_{m=1}^{n} V_m \le C\right) - \lim_{n \to \infty} \frac{1}{n} \log \mathbb{P}\left(\frac{1}{n}\sum_{m=1}^{n} V_m \le C\right),$$

The latter decay rate equals 0 due to the law of the large numbers. Noticing that

$$\mathbb{P}\left(\frac{1}{n}\sum_{m=1}^{n}(V_m + W_m) > C\right) \ge \mathbb{P}\left(\frac{1}{n}\sum_{m=1}^{n}(V_m + W_m) > C, \frac{1}{n}\sum_{m=1}^{n} V_m \le C\right)$$

$$= \mathbb{P}\left(\frac{1}{n}\sum_{m=1}^{n}(V_m + W_m) > C\right) - \mathbb{P}\left(\frac{1}{n}\sum_{m=1}^{n}(V_m + W_m) > C, \frac{1}{n}\sum_{m=1}^{n} V_m > C\right),$$

in conjunction with

$$\lim_{n\to\infty} \frac{1}{n} \log \mathbb{P} \left(\frac{1}{n} \sum_{m=1}^{n} (V_m + W_m) > C, \frac{1}{n} \sum_{m=1}^{n} V_m > C \right)$$

$$\leq \lim_{n\to\infty} \frac{1}{n} \log \mathbb{P} \left(\frac{1}{n} \sum_{m=1}^{n} V_m > C \right) < \lim_{n\to\infty} \frac{1}{n} \log \mathbb{P} \left(\frac{1}{n} \sum_{m=1}^{n} (V_m + W_m) > C \right),$$

the stated follows.   □

As we assumed that $\sum_i \alpha_i \nu_i < C$, Lemma 1 implies that the exact asymptotics of $q_1^{\mathscr{M}}(n)$ and those of

$$\bar{q}_1^{\mathscr{M}}(n) := \mathbb{P} \left(\frac{1}{n} \sum_{m=1}^{n} (V_m + W_m) > C \right)$$

coincide. Now we are in a position to apply the celebrated Bahadur-Rao result to establish the exact asymptotics of $q_1^{\mathscr{M}}(n)$. Let, with $Z_m := V_m + W_m$,

$$\Lambda(s) := \log \mathbb{E} e^{sZ_1} = \sum_{i=1}^{I} \nu_i (e^{s\alpha_i} - 1) + \sum_{j=1}^{J} \ell_j \log(1 - \pi_j + \pi_j e^{s\beta_j})$$

denote the log-moment generating function of Z_1, and

$$\Phi(C) := \sup_{s>0} (sC - \Lambda(s))$$

the corresponding Legendre transform. In addition, let $s^\star$ be the optimizing s in the definition of $\Phi(C)$. Let $f(n) \sim g(n)$ denote $f(n)/g(n) \to 1$ as $n \to \infty$.

Theorem 1. *Distinguish between Z_m being lattice and non-lattice.*

1. *Assume Z_1 is a lattice random variable, i.e., for certain numbers z_0 and d the random variable $(Z_1 - z_0)/d$ is an integer number, with d being the largest one with this property. Then, as $n \to \infty$*

$$q_1^{\mathscr{M}}(n) \sim \mathbb{P} \left(\frac{1}{n} \sum_{m=1}^{n} Z_m > C \right) \sim \frac{s^\star d}{1 - e^{-s^\star d}} \frac{e^{-n\Phi(C)}}{s^\star \sqrt{2\pi n \Lambda''(s^\star)}}.$$

2. *If Z_1 is non-lattice, then, as $n \to \infty$,*

$$q_1^{\mathscr{M}}(n) \sim \mathbb{P} \left(\frac{1}{n} \sum_{m=1}^{n} Z_m > C \right) \sim \frac{e^{-n\Phi(C)}}{s^\star \sqrt{2\pi n \Lambda''(s^\star)}}.$$

There are alternative ways to compute the exponent $\Phi(C)$ in the expansion of $q_1^{\mathscr{M}}(n)$. To this end, let $\mathscr{T}(C) := \{x \in \mathbb{R}^I : \alpha_i x_i \leq C\}$, and define

$$\xi(x) := \sup_{s \geq 0} \left(sx - \sum_{j=1}^{J} \ell_j \log(1 - \pi_j + \pi_j e^{s\beta_j}) \right);$$

$$\omega(x) := \sum_{i=1}^{I} \left(x_i \log \frac{\nu_i}{x_i} + x_i - \nu_i \right) - \xi \left(C - \sum_{i=1}^{I} \alpha_i x_i \right).$$

This following lemma is proven in the appendix.

Lemma 2. $\Phi(C) = \sup_{x \in \mathscr{T}(C)} \omega(\boldsymbol{x})$.

Example 1. Let us consider the most simple example: one traffic type within both classes, and $\alpha = \beta = 1$. The stability condition is then $\nu + \ell\pi < C$. It is easily verified that

$$\xi(x) = (C - x) \log \frac{C - x}{\pi} + (\ell - C + x) \log \frac{\ell - C + x}{1 - \pi} - \ell \log \ell$$

so that $\Phi(C)$ equals

$$\max_{x \leq C} \left(x \log \frac{\nu}{x} + x - \nu - (C - x) \log \frac{C - x}{\pi} - (\ell - C + x) \log \frac{\ell - C + x}{1 - \pi} + \ell \log \ell \right).$$

After tedious calculations, this eventually leads to the following optimizing x:

$$x^\star = -\frac{1}{2}(\ell - C) - \frac{1 - \pi}{2\pi}\nu + \sqrt{\left(\frac{\ell - C}{2}\right)^2 + \frac{1}{2}\frac{1 - \pi}{\pi}\nu(\ell + C) + \left(\frac{(1 - \pi)\nu}{2\pi}\right)^2}.$$

It is easy to show that $x^\star \leq C$ with equality being possible only if $\ell = 0$ or, trivially, $C = 0$.

If we wish to find the exact asymptotics, we have to work with the setup of Thm. 1. It can be calculated that

$$s^\star = \log\left(\frac{1}{2}\frac{z - \ell}{\nu} - \frac{1}{2}\frac{1 - \pi}{\pi} + \sqrt{\left(\frac{\ell - z}{2\nu} + \frac{1 - \pi}{2\pi}\right)^2 + \frac{(1 - \pi)z}{\pi\nu}}\right);$$

$$\Lambda''(s) = e^s \left(\nu + \frac{\ell\pi(1 - \pi)}{(1 - \pi + \pi e^s)^2}\right).$$

This enables the computation of the exact asymptotics.   ◇

Multi-rate and EMW Model. We now continue with the evaluation of $q_2^{\mathscr{M}}(n)$, i.e., the setting in which the lp traffic streams are allocated their effective bandwidth. As before, we use the scaling $\nu_i \mapsto n\nu_i$, $C \mapsto nC$, $B \mapsto nB$, and $\ell_j \mapsto n\ell_j$. The probability $q_2^{\mathscr{M}}(n)$ can alternatively be written as

$$\mathbb{P}\left(\sum_{i=1}^I \alpha_i \bar{X}_i^{(n)} + \sum_{j=1}^J c_j \left(nC - \sum_{i=1}^I \alpha_i \bar{X}_i^{(n)}\right) \cdot Y_j^{(n)} > nC \,\middle|\, \sum_{i=1}^I \alpha_i \bar{X}_i^{(n)} \leq nC\right).$$

As before, it can be argued that, when looking at asymptotics, the impact of the condition is negligible; we therefore leave it out in the sequel. Let $V_{m,i}$, $m = 1, \ldots, n$, be i.i.d. samples from a Poisson distribution with mean ν_i, and $W_{m,j}$ i.i.d. samples from a binomial distribution with parameters ℓ_j and π_j. We thus arrive at

$$\mathbb{P}\left(\sum_{i=1}^I \alpha_i \left(\frac{1}{n}\sum_{m=1}^n V_{m,i}\right) + \sum_{j=1}^J \gamma_j \cdot \left(\frac{1}{n}\sum_{m=1}^n W_{m,j}\right) > C\right),$$

where

$$\gamma_j := \max \left\{ \frac{p_j}{1 + nB(p_j - r_j)/(\tau_j(nC - \sum_{i=1}^{I} \alpha_i \sum_{m=1}^{n} V_{m,i}))}, r_j \right\}$$

$$= \max \left\{ \frac{p_j}{1 + B(p_j - r_j)/(\tau_j(C - \sum_{i=1}^{I} \alpha_i \cdot (1/n) \sum_{m=1}^{n} V_{m,i}))}, r_j \right\}.$$

Note, however that $(\alpha_i/n) \sum_{m=1}^{n} V_{m,i}$ and $(\gamma_j/n) \sum_{m=1}^{n} W_{m,j}$ are *not* independent: the value of $n^{-1} \sum_{m=1}^{n} V_{m,i}$ affects γ_j. As a result, we cannot write

$$\sum_{i=1}^{I} \alpha_i \left(\frac{1}{n} \sum_{m=1}^{n} V_{m,i} \right) + \sum_{j=1}^{J} \gamma_j \cdot \left(\frac{1}{n} \sum_{m=1}^{n} W_{m,j} \right) \tag{3}$$

as the sum of n i.i.d. random variables. This also means that Bahadur-Rao cannot be applied here, and therefore we cannot identify the exact asymptotics of $q_2^{\mathscr{M}}(n)$ in this way. Logarithmic asymptotics can be derived, though, as follows.

The idea is to condition on $n^{-1} \sum_{m=1}^{n} V_{m,i} \approx x_i$, for $i = 1, \ldots, I$, and then to find the most likely value of x corresponding to the event under consideration. To this end, first recall that

$$\lim_{n \to \infty} \frac{1}{n} \log \mathbb{P} \left(\frac{1}{n} \sum_{m=1}^{n} V_{m,i} \approx x_i \right) = -H_i(x_i), \quad \text{where } H_i(x) := x \log \frac{x}{\nu_i} + x - \nu_i.$$

Also, conditioning on $n^{-1} \sum_{m=1}^{n} V_{m,i} \approx x_i$, we have that, informally,

$$\lim_{n \to \infty} \frac{1}{n} \log \mathbb{P} \left(\sum_{j=1}^{J} \gamma_j \cdot \left(\frac{1}{n} \sum_{m=1}^{n} W_{m,j} \right) > C \right)$$

$$= - \sup_{s \geq 0} \left(sC - \sum_{j=1}^{J} \ell_j \log \left(1 - \pi_j(x) + \pi_j(x) e^{s\gamma_j(x)} \right) \right),$$

where, with $C(x) := C - \sum_{i=1}^{I} \alpha_i x_i$,

$$\gamma_j(x) := \max \left\{ \frac{p_j}{1 + B(p_j - r_j)/(\tau_j C(x))}, r_j \right\}, \quad \pi_j(x) := \frac{r_j}{\gamma_j(x)}.$$

Upon combining these decay rates, we find

$$\lim_{n \to \infty} \frac{1}{n} \log q_2^{\mathscr{M}}(n)$$

$$= -\inf_{x} \left(\sum_{i=1}^{I} H_i(x_i) + \sup_{s \geq 0} \left(sC(x) - \sum_{j=1}^{J} \ell_j \log \left(1 + \pi_j(x) \left(e^{s\gamma_j(x)} - 1 \right) \right) \right) \right).$$

Here the optimization over x essentially finds the most likely value of the sample mean $n^{-1} \sum_{m=1}^{n} V_{m,i}$, for $i = 1, \ldots, I$ such that (3) exceeds C. This result can be rigorously proven applying the same line of arguments as in the proof of Lemma 2. This proof is tedious, though, and does not add any new insights. It is stressed that the optimization over s can be performed explicitly, and therefore the computation of this decay rate just requires the evaluation of a I-dimensional minimization in order to find the optimal x; the objective function is rather 'ill-behaved' though, particularly due to the non-smooth behavior of the $\gamma_j(x)$.

4 Numerical Results

We now consider a number of experiments that assess the quality of the approximations made in our analysis. To shed light on resource allocation issues, our goal is to assess the impact of the various parameters on the performance. For the buffer parameters we based our choice on the values given by leading equipment vendors in their 10GE line cards specifications.

Regarding the scaling parameter n, the following remark is in place. The reason for introducing this parameter is that under the proportional scaling of resources as well as input, we could prove results on the various decay rates. In the present section, our objective is to perform numerical experiments with representative parameter values. From the formulas it is readily verified that we can then pick $n = 1$, such that we can take for B the 'real' (that is, non-scaled) buffer size, for C the real link rate, for the ν the real load generated by the hp traffic, and for the ℓ the real number of lp sources. It is left to the reader to verify that these approximations do not change when taking $n = K$ and buffer B/K, link rate C/K, load ν/K and number of lp sources ℓ/K; in other words, to evaluate the approximations the value of the scaling parameter is irrelevant; the only thing that matters is the *product* of the scaling parameters and the buffer size, link rate, load, and number of lp sources.

In our experiments we consider a non-buffered resource and rate envelope multiplexing, where the rate consumed by hp traffic is described by a multi-rate model. In the left panel of Fig. 1 we take $I = J = 1$ along with $\alpha = \beta = 1$, so that we are in the situation of Example 1. Picking as indicated above without loss of generality $n = 1$, the link speed of 1 Gbps results in $C = 1000$ (measured in Mbps). We choose $\nu = 420$, $\pi = 0.6$. Then we vary ℓ to see how the loss probability $q_1^{\mathscr{M}}(1)$ is affected. The stability region is $420 + 0.6\ell \le 1000$.

The graph shows the ^{10}log of the loss probability $q_1^{\mathscr{M}}(1)$ as a function of ℓ. There are three lines: (i) a curve based on the exact asymptotics of Thm. 1, relying on the Bahadur-Rao result, (ii) a curve based on the logarithmic approximation (only containing the exponential part of the Bahadur-Rao approximation), and (iii) a curve that estimates the loss probability $q_1^{\mathscr{M}}(1)$ based on simulations. Regarding the simulated values, it is mentioned that per data punt we used 10^6 independent samples.

The graphs show that the exact asymptotics match excellently with the simulated values. The fit is less good in the region close to saturation: there in reality the loss probabilities are still well below 1, but the exact asymptotics predict that $q_1^{\mathscr{M}}(1) \approx 1$. Notice, however, that this region close to saturation is practically irrelevant (as the target values of the loss probability are considerably lower than 10^{-1}). It is further observed that the curve that is based on just the exponent (that is, the logarithmic asymptotics) is too

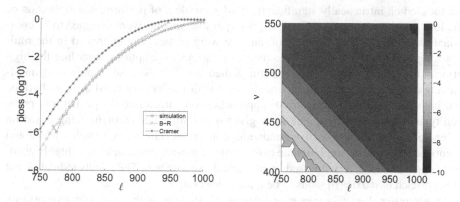

Fig. 1. Left: approximations of the loss probability; right: admissible region based on simulation

pessimistic: in reality $q_1^{\mathcal{M}}(n)$ is roughly one order of magnitude smaller; however, the curve captures the right trend excellently, as it is nearly parallel to the simulated curve.

The other three panels show admissible regions; this is done for different values of $^{10}\log q_1^{\mathcal{M}}(1)$. Fig. 1 (right panel) is based on simulation (with 10^6 runs per data point), Fig. 2 (left panel) is based on Thm. 1, and Fig. 2 (right panel) on just the decay rate (logarithmic asymptotics). Again we observe that the results based on Thm. 1 match the simulated regions excellently. The regions based on the logarithmic asymtotics are conservative. Also observe that the boundaries of the admissible regions (for various values of the loss probability) are (nearly) straight lines.

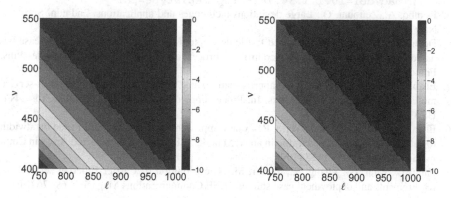

Fig. 2. Left: admissible region based on Thm. 1; right: admissible region based on just the decay rate

5   Concluding Remarks

This paper dealt with a classical problem in operations research: specific resource allocation issues for network links shared by two traffic classes through (strict) priority queueing at the network nodes. The analysis of the high-priority queue being standard, we focused on intrinsically significantly harder problem of performance evaluation of the low-priority traffic class. The low priority traffic streams were assumed to be compliant with a leaky-bucket regulator, and we were particularly interested in the multiplexing gain for these traffic streams under various assumptions regarding the high priority traffic. In our analysis we distinguished between the case of bufferless multiplexing at the low priority queue and the case with moderately sized or large buffers. For both cases we proposed various approaches to characterize the packet loss probability at the low priority queue for a given traffic mix, one of them being based on a generalization of the effective bandwidth concept developed by Elwalid, Mitra and Wentworth [6] for the single-queue case. Some of these approaches led to highly accurate approximations, as witnessed by numerical experiments. The results indicated that the approach based on [6] tends to be quite conservative.

Future research in this area may relate to (i) systems with more than two priority classes, for instance by introducing a third class, in between the low-priority and high-priority classes considered here, where only a certain fraction of the traffic is treated in the best effort manner, (ii) the impact of the non-homogeneity of hp traffic streams (for example voice and video mixed together) (iii) a more extensive measurement-based backing of the procedures proposed.

References

1. Botvich, D., Duffield, N.: Large deviations, economies of scale, and the shape of the loss curve in large multiplexers. Queueing Systems 20, 293–320 (1995)
2. Brichet, F., Mandjes, M., Sánchez-Cañabate, M.: COST 257, Mid-term seminar interim report on admission control (1999), http://citeseerx.ist.psu.edu/viewdoc/download?doi=10.1.1.34.7415&rep=rep1&type=pdf
3. Dembo, A., Zeitouni, O.: Large deviations techniques and applications, 2nd edn. Springer, New York (1998)
4. Doshi, B.: Deterministic rule based traffic descriptors for broadband ISDN: worst case behavior and connection acceptance control. In: Proceedings of ITC 14, Antibes Juan-les-Pins, France, pp. 591–600 (1997)
5. Elwalid, A., Mitra, D.: Analysis, approximations and admission control of a multi-service multiplexing system with priorities. In: Proceedings of IEEE INFOCOM, New York, NY, USA, pp. 463–472 (1995)
6. Elwalid, A., Mitra, D., Wentworth, R.: A new approach for allocating buffers and bandwidth to heterogeneous, regulated traffic in an ATM node. IEEE Journal on Selected Areas in Communications 13, 1115–1127 (1995)
7. Fang, L., Bitar, N., Zhang, R., Taylor, M.: The evolution of Carrier Ethernet services — requirements and deployment case studies. IEEE Communications Magazine, 69–76 (March 2008)
8. Kelly, F.: Loss networks. Annals of Applied Probability 1, 319–378 (1991)
9. Kesidis, G., Konstantopoulos, T.: Worst-case performance of a buffer with independent shaped arrival processes. IEEE Communications Letters 4, 26–28 (2000)
10. Kulkarni, V., Gautam, N.: Admission control of multi-class traffic with service priorities in high-speed networks. Queueing Systems 27, 79–97 (1997)
11. Mandjes, M.: Large Deviations of Gaussian Queues. Wiley, Chichester (2007)
12. Mandjes, M., Mannersalo, P., Norros, I.: Priority queues with Gaussian input: a path-space approach to loss and delay asymptotics. In: Proceedings of ITC 19, Beijing, China, pp. 1135–1144 (2005)

13. Mandjes, M., Ridder, A.: Optimal trajectory to overflow in a queue fed by a large number of sources. Queueing Systems 31, 137–170 (1999)
14. Mannersalo, P., Norros, I.: A most probable path approach to queueing systems with general Gaussian input. Computer Networks 40, 399–412 (2002)
15. Onvural, R.: Asynchronous Transfer Mode Networks: Performance Issues, 2nd edn. Artech House, Norwood (1995)
16. Reisslein, M., Ross, K.W., Rajagopal, S.: Packet multiplexers with adversarial regulated traffic. In: Proceedings of IEEE INFOCOM, San Francisco, CA, pp. 347–355 (1998)
17. Roberts, J., Mocci, U., Virtamo, J.: Broadband Network Teletraffic: Final Report of Action COST 242. Springer, Berlin (1996)
18. Sung, Y.-W.E., Lund, C., Lyn, M., Rao, S., Sen, S.: Modeling and understanding end-to-end class of service policies in operational networks. In: Proceedings of ACM Special Interest Group on Data Communications (SIGCOMM), Barcelona, Spain (2009)

A   Proof of Lemma 2

UPPER BOUND. It is evident that the number of elements $\#\{\mathscr{S}(nC)\}$ of $\mathscr{S}(nC)$ is $O(n^I)$. It follows that $q_1^{\mathscr{M}}(n)$ is bounded from above by

$$\#\{\mathscr{S}(nC)\} \times \max_{k \in \mathscr{S}(nC)} \mathbb{P}(X(n) = k)\mathbb{P}\left(\sum_{j=1}^{J} \beta_j Y_j(n) > nC - \sum_{i=1}^{I} \alpha_i k_i\right).$$

As $n^{-1}\log\#\{\mathscr{S}(nC)\} \to 0$ as $n \to \infty$, we are left with determining an upper bound to

$$\limsup_{n\to\infty} \frac{1}{n} \log\left(\max_{k\in\mathscr{S}(nC)} \mathbb{P}(X(n) = k)\mathbb{P}\left(\sum_{j=1}^{J} \beta_j Y_j(n) > nC - \sum_{i=1}^{I} \alpha_i k_i\right)\right). \quad (4)$$

First consider the normalizing constant; to this end observe that for all $\eta > 0$ and n large enough, with $Z_i(n)$ denoting a Poisson random variable with mean $n\nu_i$,

$$\sum_{k\in\mathscr{S}(nC)} \prod_{i=1}^{I} \frac{(n\nu_i)^{k_i}}{k_i!} = \left(e^{-n\sum_{i=1}^{I}\nu_i}\right) \times \left(\sum_{k\in\mathscr{S}(nC)} \prod_{i=1}^{I} \frac{(n\nu_i)^{k_i}}{k_i!} e^{-n\nu_i}\right)$$

$$= e^{-n\sum_{i=1}^{I}\nu_i}\mathbb{P}(Z(n) \in \mathscr{S}(nC)) \geq e^{-n\sum_{i=1}^{I}\nu_i}(1-\eta),$$

using that $Z_i(n)$ is distributed as the sum of n independent Poisson(ν_i) random variables, and applying the law of large numbers (recalling that $\nu \in \mathscr{S}(C)$). Using the Stirling-type bound (uniformly in k) $\log(k!) \leq k\log k - k + \kappa(k)$, with $\kappa(k) := \frac{1}{2}\log(2\pi k) + 1/(12k)$, we thus arrive at

$$\mathbb{P}(X(n) = k) \leq \prod_{i=1}^{I} \left(\frac{n\nu_i}{k_i}\right)^{k_i} e^{k_i - n\nu_i}e^{\kappa(k_i)}(1-\eta),$$

for all $k \in \mathscr{S}(nC)$. Due to the Chernoff bound, we also have

$$
\log \mathbb{P}\left(\sum_{j=1}^{J} \beta_j Y_j(n) > nC - \sum_{i=1}^{I} \alpha_i k_i\right) \leq -\xi_n\left(nC - \sum_{i=1}^{I} \alpha_i k_i\right),
$$

where $\xi(\cdot)$ is the associated Legendre transform:

$$
\xi(x) := \sup_{s \geq 0}\left(sx - n\sum_{j=1}^{J} \ell_j \log(1 - \pi_j + \pi_j e^{s\beta_j})\right).
$$

Now realize that the maximum over $\mathscr{S}(nC)$ is smaller than the maximum over the (bigger) set $\mathscr{T}(nC)$. It follows that the decay rate (4) is bounded from above by

$$
\limsup_{n\to\infty} \frac{1}{n} \log \max_{k\in\mathscr{T}(nC)} \prod_{i=1}^{I}\left(\frac{n\nu_i}{k_i}\right)^{k_i} e^{k_i - n\nu_i} e^{\kappa(k_i)}(1-\eta) \exp\left(-\xi\left(nC - \sum_{i=1}^{I} \alpha_i k_i\right)\right),
$$

which equals (note that the factor $(1-\eta)$ trivially cancels)

$$
\limsup_{n\to\infty} \frac{1}{n} \log \max_{x\in\mathscr{T}(C)} \prod_{i=1}^{I}\left(\frac{\nu_i}{x_i}\right)^{nx_i} e^{n(x_i - \nu_i)} e^{\kappa(nx_i)} \exp\left(-n\xi\left(C - \sum_{i=1}^{I} \alpha_i x_i\right)\right).
$$

This in turn can be majorized by

$$
\max_{x\in\mathscr{T}(C)} \omega(x) + \limsup_{n\to\infty} \frac{1}{n} \log \max_{x\in\mathscr{T}(C)} \prod_{i=1}^{I} e^{\kappa(nx_i)}.
$$

Now noting that $\kappa(nx) \leq \frac{1}{2}\log(2\pi nx) + 1$, and noting that there is a finite M such that $x_i \leq M$ for all $x \in \mathscr{T}(C)$ and all $i = 1, \ldots, I$, we conclude that

$$
\limsup_{n\to\infty} \frac{1}{n} \log q_1^{\mathcal{M}}(n) \leq \max_{x\in\mathscr{T}(C)} \omega(x). \tag{5}
$$

LOWER BOUND. Call the optimizer in the right-hand side of (5) $x^\star$. It is evident that for all $k \in \mathscr{S}(n)$,

$$
q_1^{\mathcal{M}}(n) \geq \mathbb{P}(X(n) = k)\mathbb{P}\left(\sum_{j=1}^{J} \beta_j Y_j(n) > nC - \sum_{i=1}^{I} \alpha_i k_i\right).
$$

Now take $k = \lfloor nx^\star \rfloor \in \mathscr{S}(nC)$. First observe that, again using $\nu \in \mathscr{S}(C)$ and 'Stirling'

$$
\liminf_{n\to\infty} \frac{1}{n} \log \mathbb{P}(X(n) = \lfloor nx^\star \rfloor) = \liminf_{n\to\infty} \frac{1}{n} \log \prod_{i=1}^{I} \frac{(n\nu_i)^{\lfloor nx_i^\star \rfloor}}{\lfloor nx_i^\star \rfloor!}
$$

$$
= \sum_{i=1}^{I}\left(x_i^\star \log \frac{\nu_i}{x_i^\star} + x_i^\star - \nu_i\right)
$$

The other decay rate we need to study is

$$\liminf_{n\to\infty} \frac{1}{n} \log \mathbb{P}\left(\frac{1}{n}\sum_{j=1}^{J}\beta_j Y_j(n) > C - \sum_{i=1}^{I}\alpha_i \frac{\lfloor nx_i^\star\rfloor}{n}\right).$$

Now choose n sufficiently large that $\lfloor nx_i^\star\rfloor/n > (1-\delta)x_i^\star$ for all $i = 1, \ldots, I$. Then, due to 'Cramér', the above decay rate majorizes

$$-\xi\left(C - \sum_{i=1}^{I}\alpha_i x_i^\star(1-\delta)\right).$$

Using the continuity of the Legendre transform $\xi(\cdot)$, it follows by letting $\delta \downarrow 0$ that

$$\limsup_{n\to\infty} \frac{1}{n} \log q_1^{\mathcal{M}}(n) \geq \omega(x^\star).$$

Combining the above bounds, we have proven the stated. $\qquad\qquad\square$

Author Index